NOETHER SYMMETRIES IN THEORIES OF GRAVITY

This volume summarizes the many alternatives and extensions to Einstein's General Theory of Relativity and shows how symmetry principles can be applied to identify physically viable models. The first part of the book establishes the foundations of classical field theory, providing an introduction to symmetry groups and the Noether theorems. A quick overview of general relativity is provided, including discussion of its successes and shortcomings; then several theories of gravity are presented, and their main features are summarized. In the second part, the "Noether Symmetry Approach" is applied to theories of gravity to identify those which contain symmetries. In the third part of the book, these selected models are tested through comparison with the latest experiments and observations. This constrains the free parameters in the selected models to fit the current data, demonstrating a useful approach that will allow researchers to construct and constrain modified gravity models for further applications.

FRANCESCO BAJARDI is a researcher at Scuola Superiore Meridionale di Napoli. His primary areas of research are general relativity, theoretical cosmology, and theories of gravity. He graduated at the University of Palermo and obtained his PhD at the Università di Napoli "Federico II."

SALVATORE CAPOZZIELLO is Full Professor in Astronomy and Astrophysics at the Università di Napoli "Federico II," and former President of the Italian Society for General Relativity and Gravitation. His primary areas of research are general relativity, cosmology, and relativistic astrophysics. He has authored more than 600 scientific publications with more than 34,000 citations.

CAMBRIDGE MONOGRAPHS ON MATHEMATICAL PHYSICS

S. J. Aarseth *Gravitational N-Body Simulations: Tools and Algorithms*†
D. Ahluwalia *Mass Dimension One Fermions*
J. Ambjørn, B. Durhuus and T. Jonsson *Quantum Geometry: A Statistical Field Theory Approach*†
A. M. Anile *Relativistic Fluids and Magneto-fluids: With Applications in Astrophysics and Plasma Physics*
J. A. de Azcárraga and J. M. Izquierdo *Lie Groups, Lie Algebras, Cohomology and Some Applications in Physics*†
O. Babelon, D. Bernard and M. Talon *Introduction to Classical Integrable Systems*†
F. Bajardi and S. Capozziello *Noether Symmetries in Theories of Gravity*
F. Bastianelli and P. van Nieuwenhuizen *Path Integrals and Anomalies in Curved Space*†
D. Baumann and L. McAllister *Inflation and String Theory*†
V. Belinski and M. Henneaux *The Cosmological Singularity*†
V. Belinski and E. Verdaguer *Gravitational Solitons*†
J. Bernstein *Kinetic Theory in the Expanding Universe*†
G. F. Bertsch and R. A. Broglia *Oscillations in Finite Quantum Systems*†
N. D. Birrell and P. C. W. Davies *Quantum Fields in Curved Space*†
K. Bolejko, A. Krasiński, C. Hellaby and M-N. Célérier *Structures in the Universe by Exact Methods: Formation, Evolution, Interactions*
D. M. Brink *Semi-Classical Methods for Nucleus-Nucleus Scattering*†
M. Burgess *Classical Covariant Fields*†
E. A. Calzetta and B.-L. B. Hu *Nonequilibrium Quantum Field Theory*
S. Carlip *Quantum Gravity in 2+1 Dimensions*†
P. Cartier and C. DeWitt-Morette *Functional Integration: Action and Symmetries*†
J. C. Collins *Renormalization: An Introduction to Renormalization, the Renormalization Group and the Operator-Product Expansion*†
P. D. B. Collins *An Introduction to Regge Theory and High Energy Physics*†
M. Creutz *Quarks, Gluons and Lattices*†
P. D. D'Eath *Supersymmetric Quantum Cosmology*†
J. Dereziński and C. Gérard *Mathematics of Quantization and Quantum Fields*
F. de Felice and D. Bini *Classical Measurements in Curved Space-Times*
F. de Felice and C. J. S Clarke *Relativity on Curved Manifolds*†
B. DeWitt *Supermanifolds, 2nd edition*†
P. G. O. Freund *Introduction to Supersymmetry*†
F. G. Friedlander *The Wave Equation on a Curved Space-Time*†
J. L. Friedman and N. Stergioulas *Rotating Relativistic Stars*
Y. Frishman and J. Sonnenschein *Non-Perturbative Field Theory: From Two Dimensional Conformal Field Theory to QCD in Four Dimensions*
J. A. Fuchs *Affine Lie Algebras and Quantum Groups: An Introduction, with Applications in Conformal Field Theory*†
J. Fuchs and C. Schweigert *Symmetries, Lie Algebras and Representations: A Graduate Course for Physicists*†
Y. Fujii and K. Maeda *The Scalar-Tensor Theory of Gravitation*†
J. A. H. Futterman, F. A. Handler, and R. A. Matzner *Scattering from Black Holes*†
A. S. Galperin, E. A. Ivanov, V. I. Ogievetsky and E. S. Sokatchev *Harmonic Superspace*†
R. Gambini and J. Pullin *Loops, Knots, Gauge Theories and Quantum Gravity*†
T. Gannon *Moonshine beyond the Monster: The Bridge Connecting Algebra, Modular Forms and Physics*†
A. García-Díaz *Exact Solutions in Three-Dimensional Gravity*
M. Göckeler and T. Schücker *Differential Geometry, Gauge Theories, and Gravity*†
C. Gómez, M. Ruiz-Altaba and G. Sierra *Quantum Groups in Two-Dimensional Physics*†
M. B. Green, J. H. Schwarz and E. Witten *Superstring Theory Volume 1: Introduction*
M. B. Green, J. H. Schwarz and E. Witten *Superstring Theory Volume 2: Loop Amplitudes, Anomalies and Phenomenology*
V. N. Gribov *The Theory of Complex Angular Momenta: Gribov Lectures on Theoretical Physics*†
J. B. Griffiths and J. Podolský *Exact Space-Times in Einstein's General Relativity*†
T. Harko and F. Lobo *Extensions of f(R) Gravity: Curvature-Matter Couplings and Hybrid Metric-Palatini Gravity*
S. W. Hawking and G. F. R. Ellis *The Large Scale Structure of Space-Time*†
J. Harnad and F. Balogh *Tau Functions and Their Applications*
B. B. Hu and E. Verdaguer *Semiclassical and Stochastic Gravity*
F. Iachello and A. Arima *The Interacting Boson Model*†
F. Iachello and P. van Isacker *The Interacting Boson-Fermion Model*†
C. Itzykson and J. M. Drouffe *Statistical Field Theory Volume 1: From Brownian Motion to Renormalization and Lattice Gauge Theory*†

C. Itzykson and J. M. Drouffe *Statistical Field Theory Volume 2: Strong Coupling, Monte Carlo Methods, Conformal Field Theory and Random Systems*†
G. Jaroszkiewicz *Principles of Discrete Time Mechanics*
G. Jaroszkiewicz *Quantized Detector Networks*
C. V. Johnson *D-Branes*†
P. S. Joshi *Gravitational Collapse and Spacetime Singularities*†
J. I. Kapusta and C. Gale *Finite-Temperature Field Theory: Principles and Applications, 2nd edition*†
V. E. Korepin, N. M. Bogoliubov and A. G. Izergin *Quantum Inverse Scattering Method and Correlation Functions*†
K. Krasnov *Formulations of General Relativity*
J. Kroon *Conformal Methods in General Relativity*
M. Le Bellac *Thermal Field Theory*†
L. Lusanna *Non-Inertial Frames and Dirac Observables in Relativity*
Y. Makeenko *Methods of Contemporary Gauge Theory*†
S. Mallik and S. Sarkar *Hadrons at Finite Temperature*
A. Malyarenko and M. Ostoja-Starzewski *Tensor-Valued Random Fields for Continuum Physics*
N. Manton and P. Sutcliffe *Topological Solitons*†
N. H. March *Liquid Metals: Concepts and Theory*†
I. Montvay and G. Münster *Quantum Fields on a Lattice*†
P. Nath *Supersymmetry, Supergravity, and Unification*
L. O'Raifeartaigh *Group Structure of Gauge Theories*†
T. Ortín *Gravity and Strings, 2nd edition*
A. M. Ozorio de Almeida *Hamiltonian Systems: Chaos and Quantization*†
M. Paranjape *The Theory and Applications of Instanton Calculations*
L. Parker and D. Toms *Quantum Field Theory in Curved Spacetime: Quantized Fields and Gravity*
R. Penrose and W. Rindler *Spinors and Space-Time Volume 1: Two-Spinor Calculus and Relativistic Fields*†
R. Penrose and W. Rindler *Spinors and Space-Time Volume 2: Spinor and Twistor Methods in Space-Time Geometry*†
S. Pokorski *Gauge Field Theories, 2nd edition*†
J. Polchinski *String Theory Volume 1: An Introduction to the Bosonic String*†
J. Polchinski *String Theory Volume 2: Superstring Theory and Beyond*†
J. C. Polkinghorne *Models of High Energy Processes*†
V. N. Popov *Functional Integrals and Collective Excitations*†
L. V. Prokhorov and S. V. Shabanov *Hamiltonian Mechanics of Gauge Systems*
S. Raychaudhuri and K. Sridhar *Particle Physics of Brane Worlds and Extra Dimensions*
A. Recknagel and V. Schiomerus *Boundary Conformal Field Theory and the Worldsheet Approach to D-Branes*
M. Reuter and F. Saueressig *Quantum Gravity and the Functional Renormalization Group*
R. J. Rivers *Path Integral Methods in Quantum Field Theory*†
R. G. Roberts *The Structure of the Proton: Deep Inelastic Scattering*†
P. Romatschke and U. Romatschke *Relativistic Fluid Dynamics In and Out of Equilibrium: And Applications to Relativistic Nuclear Collisions*
C. Rovelli *Quantum Gravity*†
W. C. Saslaw *Gravitational Physics of Stellar and Galactic Systems*†
R. N. Sen *Causality, Measurement Theory and the Differentiable Structure of Space-Time*
M. Shifman and A. Yung *Supersymmetric Solitons*
Y. M. Shnir *Topological and Non-Topological Solitons in Scalar Field Theories*
H. Stephani, D. Kramer, M. MacCallum, C. Hoenselaers and E. Herlt *Exact Solutions of Einstein's Field Equations, 2nd edition*†
J. Stewart *Advanced General Relativity*†
J. C. Taylor *Gauge Theories of Weak Interactions*†
T. Thiemann *Modern Canonical Quantum General Relativity*†
D. J. Toms *The Schwinger Action Principle and Effective Action*†
A. Vilenkin and E. P. S. Shellard *Cosmic Strings and Other Topological Defects*†
R. S. Ward and R. O. Wells, Jr *Twistor Geometry and Field Theory*†
E. J. Weinberg *Classical Solutions in Quantum Field Theory: Solitons and Instantons in High Energy Physics*†
J. R. Wilson and G. J. Mathews *Relativistic Numerical Hydrodynamics*†

† Available in paperback

Noether Symmetries in Theories of Gravity

With Applications to Astrophysics and Cosmology

FRANCESCO BAJARDI
Scuola Superiore Meridionale di Napoli

SALVATORE CAPOZZIELLO
Università degli Studi di Napoli "Federico II"

CAMBRIDGE
UNIVERSITY PRESS

CAMBRIDGE
UNIVERSITY PRESS

Shaftesbury Road, Cambridge CB2 8EA, United Kingdom

One Liberty Plaza, 20th Floor, New York, NY 10006, USA

477 Williamstown Road, Port Melbourne, VIC 3207, Australia

314–321, 3rd Floor, Plot 3, Splendor Forum, Jasola District Centre, New Delhi – 110025, India

103 Penang Road, #05–06/07, Visioncrest Commercial, Singapore 238467

Cambridge University Press is part of Cambridge University Press & Assessment, a department of the University of Cambridge.

We share the University's mission to contribute to society through the pursuit of education, learning and research at the highest international levels of excellence.

www.cambridge.org
Information on this title: www.cambridge.org/9781009208741
DOI: 10.1017/9781009208727

First published 2023

A catalogue record for this publication is available from the British Library.

A Cataloging-in-Publication data record for this book is available from the Library of Congress.

ISBN 978-1-009-20874-1 Hardback

Dedicated to all women who,
like Emmy Noether,
are fighting for Science and Knowledge.

Contents

Preface

This book aims to provide evidence of the importance of symmetries in Gravitation and Cosmology. In particular, we pay attention to the search for symmetries in the context of theories of gravity, outlining how their existence contributes to solve dynamics, to select physically viable models, and to achieve, eventually, exact solutions. We will take into account large classes of extensions and alternatives to Einstein's General Relativity (GR) considering, inside the gravitational action, functions of the Ricci scalar and other curvature invariants, higher-order geometric terms, minimal and non-minimal couplings to scalar fields, torsion, and non-metricity invariants. As often pointed out throughout the text, in particular in Parts I and II of the book, there are several physically motivated reasons to consider alternatives and extensions to GR. Indeed, although GR obtained a lot of confirmations and successes in astrophysics and cosmology, it suffers some shortcomings from the theoretical and experimental points of view that do not allow one to obtain a comprehensive and self-consistent picture of gravitational interaction, and then of cosmology, ranging from ultraviolet (UV) to infrared (IR) scales. Alternative and extended theories of gravity aim to fix such problems by relaxing some of the hypotheses lying behind Einstein's theory. Specifically, testing theories of gravity through observational cosmology is a good starting point in order to find consistent models and discard others. However, this approach cannot be satisfactory from a fundamental physics point of view. In fact, since the final form of the gravitational interaction is substantially unknown, we need general selection criteria, based on first principles, to constrain the starting action. The existence of Noether symmetries can be a useful approach to this issue for several reasons. For example, extending GR leads to hardly solvable field equations, so that, several times, the given theory ends up losing any predictive power. In this perspective, in order to reduce dynamics, it is possible to introduce the so-called *Noether Symmetry Approach*, by means of which theories of gravity containing symmetries can be selected. The corresponding conserved quantities can be used to reduce dynamics.

It is interesting to point out that, if symmetries exist, the related conserved quantities often have a physical meaning, as we will discuss throughout the book. To probe this statement, in Part II of the book, we apply the Noether Symmetry Approach to several theories of gravity, focusing on cosmological backgrounds and deriving exact solutions. Furthermore, Noether symmetries can be searched for in spherical and cylindrical symmetry, providing black hole solutions with further features besides the Schwarzschild radius.

It is worth noticing that the Noether approach can also be used in Hamiltonian formalism in view of Quantum Cosmology applications. In this perspective, it is possible to demonstrate that the existence of Noether symmetries allows one to select "observable" universes, according to the Hartle criterion, which means that a *Wave Function of the Universe*, giving rise to physically viable initial conditions, can be obtained.

The book is organized in three parts. Part I is devoted to preliminary concepts. We introduce the notion of symmetry group, conserved quantity, and the Noether Theorem. Then we discuss some applications of the latter, as well as the free-particle Lagrangian, the harmonic oscillator, the Dirac and electromagnetic fields, as well as the invariance under translations and rotations. In the two final chapters of the Part I, we outline the main aspects of some modified theories of gravity, pointing out the differences with respect to GR. In Chapter 5 we outline the *Canonical Quantization* of gravity, focusing on the application to a homogeneous and isotropic universe and constructing the formalism of *Quantum Cosmology*. We also analyze the link between Noether symmetry and canonical quantization, as well as the capability of the former to select physically viable models.

In Part II, we show that the First Noether Theorem can be formalized as a method to select models with symmetries, even without knowing a priori the form of the symmetry generator. Then, we use the Noether Symmetry Approach to select exact solutions for cosmological models. Specifically, in Chapter 7, we consider geometric extensions of GR, Teleparallel Equivalent of General Relativity, and Symmetric Teleparallel Equivalent of General Relativity, showing, in other words, that the methods work for any representation of gravity. In Chapters 8 and 9 second-order curvature invariants and higher-order terms in the gravitational action are respectively treated. In Chapter 10, non-minimal couplings between geometry and scalar fields are considered.

In Chapter 11, we deal with nonlocal terms in curvature and torsion in gravitational action. In fact, it is worth noticing that nonlocality can play a major role in selecting effective theories useful for both Quantum Gravity and Cosmology, as we will discuss. As reported, Noether symmetries are a useful criterion to fix such nonlocal terms. In Chapter 12 the Noether Symmetry Approach is considered within the context of Bianchi Universes. Therefore, starting from a scalar-tensor action (where the scalar field is minimally coupled with the geometry), several symmetries are provided. These results are useful to show that more than one Noether symmetry can be achieved for the same point-like Lagrangian. In Chapter 13, we apply the Noether approach to spherically symmetric Lagrangians in order to find out black hole solutions. In particular, we focus on $f(R)$, $f(T)$, and

$f(\mathcal{G})$ theories respectively in Sections 13.1, 13.2 and 13.3, and show how it is possible to generate the axial symmetry from the spherical one.

In the Part III, we conclude by considering applications of theories of gravity to classical and quantum cosmology, and to astrophysics through the Noether symmetries. Specifically, we show that the range of parameters from Solar System tests, stellar mass-radius diagram, and galaxy rotation curves can be fixed by the presence of Noether symmetries. It is worth noticing that not all models selected by the Noether approach are necessarily viable because some of them could suffer theoretical shortcomings or could be ruled out by experiments. However, the criterion of searching for symmetries allows one to select classes of reliable models that can eventually be constrained by experimental observations. The conserved quantities allow one, indeed, to reduce the complexity of the field equations and to find solutions whose physical meaning can be, in principle, investigated.

This book is not meant to be only a summary of solutions of theories of gravity with symmetries, but it is aimed at stressing the importance of symmetries, at comparing cosmological and astrophysical solutions of different theories, and at developing a coherent treatment that emphasizes the importance of symmetries in Physics of Gravitation. All theories discussed here take into account *scalar fields* of different origin. These scalar fields are often geometric invariants like curvature R, torsion T, or the Gauss–Bonnet term $\mathcal{G}$. The intent is to show that different dynamical (and geometrical) quantities can be related to each other, providing equivalent descriptions for the gravitational interaction. Therefore, comparing their symmetries and solutions in different backgrounds can constitute a sort of paradigm to address shortcomings of gravitational interaction at any scale of energy. This could contribute to constructing a comprehensive cosmic history from UV to IR epochs. Clearly, the discussion is sometimes not complete. We apologize for this and refer to the literature reported in the bibliography for detailed information.

Acknowledgments

We want to acknowledge many collaborators and colleagues who helped us to develop the topics presented in this book with useful exchanges of ideas, discussions, and suggestions.

One of us (SC) wants to dedicate this book to the beloved memory of Ruggiero de Ritis who, with the patience and dedication of a mentor and a friend, introduced him to the amazing world of symmetries.

Furthermore, we acknowledge J. Alcaniz, L. Amendola, K. Atazadeh, S. Bahamonde, K. Bamba, J. D. Barrow, F. Bascone, S. Basilakos, M. Benetti, T. Bernal, D. Borka, V. Borka Jovanović, A. Borowiec, W. G. Boskoff, M. Bouhmadi-López, U. Camci, V. F. Cardone, M. Capriolo, S. Carloni, R. Cianci, G. Cristofano, R. D'Agostino, F. Darabi, V. De Falco, A. De Felice, M. De Laurentis, C. Deliduman, I. De Martino, M. Demianski, K. F. Dialektopoulos, P. K. S. Dunsby, E. Elizalde, H. Farajollahi, V. Faraoni, L. Fatibene, M. Francaviglia, N. Frusciante, R. Garattini, G. Gionti, T. Harko, M. Jamil, J. C. Hidalgo, P. Jovanović, T. S. Koivisto, G. Lambiase, J. Levi Said, L. Lobato-Graef, F. S. N. Lobo, O. Luongo, C. A. Mantica, V. I. Man'ko, A. A. Marino, G. Marmo, P. Martin-Moruno, S. Mendoza, J. P. Mimoso, L. G. Molinari, H. Motavali, R. Myrzakulov, G. L. Nashed, S. Nesseris, S. Nojiri, A. N. Nurbaki, F. Occhionero, S. D. Odintsov, V. Oikonomou, G. J. Olmo, A. Paliathanasis, M. Paolella, L. Perivolaropoulos, E. Piedipalumbo, A. Ravanpak, M. Roshan, C. Rubano, V. Salzano, E. N. Saridakis, P. Scudellaro, A. Stabile, C. Stornaiolo, S. V. Sushkov, A. Troisi, M. Tsamparlis, S. Tsujikawa, Z. Urban, D. Vernieri, S. Vignolo, and A. M. Wojnar.

We wish to thank particularly the Editorial Director of Cambridge University Press, Simon Capelin, who proposed us to write the book, and the Commissioning Editors, Nicholas Gibbons and Vince Higgs, for their continuous support and suggestions that allowed us to realize this book.

Amalie Emmy Noether: A Life for Mathematics

Amalie Emmy Noether in 1905

Amalie Emmy Noether can be considered as one of the most prominent mathematicians of all time. With the theory of algebraic invariants and her theorems on symmetries and conservation laws, she revolutionized Mathematics and Physics.

She was born on March 23, 1882 in the German city of Erlangen. After completing secondary school, she could not immediately access the University, which was then closed to women. She managed to do so only in 1904 when the University of Erlangen definitively opened the doors to female enrollment. She quickly recovered the lost time, graduating in just three years with a thesis on the theory of algebraic invariants. The difficulties did not end there. After moving to Göttingen, she lived for more than ten years with various jobs, semiofficial and poorly paid lessons, and scientific assistance to some established professors. Despite the difficulties, she collaborated with the top-level mathematicians of the epoch, such as Felix Klein, David Hilbert, and Hermann Weyl, and, finally, she discovered some of the mathematical bases of Einstein's GR.

Emmy Noether was already distinguished by addressing several open questions of Mathematics, but it was the publication of two original theorems in the field of the calculus of variations, which we call *Noether's First and Second Theorems*, that brought her to the attention of the scientific community in 1918. By these results, Emmy Noether was capable of demonstrating the deep link between symmetries and conservation laws. In other words, Noether's theorems state that conserved quantities correspond to symmetries of Lagrangians defining dynamics of physical systems ranging from Classical to Quantum Mechanics up to GR and Quantum Field Theory. The theorems not only have important theoretical implications: they also offer a practical tool to derive the conserved quantities starting from the symmetries observed in a physical system. If a new theory is proposed, the Noether Theorems guarantee that, if symmetries exist, conserved quantities also must exist. In some sense, these theorems represent also a sort of protocol for experiments: the theory is proved true, if conserved quantities are observed.

The impact of the Noether Theorems in GR is enormous. After the Einstein and Hilbert formulation, many problems remained open. In particular, the lack of local energy conservation challenged the whole theory. Hilbert pointed out "the failure of the energy theorem" in GR, not being able to demonstrate the local energy conservation. The problem was so serious that it directly questioned the foundations of the theory. Emmy Noether, in a short time, provided a brilliant solution that had wide-reaching consequences not only for GR but also for the whole evolution of modern Physics, as Hilbert himself acknowledged in 1924.

Despite the relevance of her research, the doors of the academy did not open for Emmy Noether. Her main work on variational invariants was presented in 1918, at a meeting of the Royal Society of Sciences in Göttingen by her tutor Felix Klein. She was not only not admitted as a member of the Society but, furthermore, being a woman, she was not allowed to be present in the audience.

In 1919, she applied again for a teaching position at Göttingen. She had been refused years before, despite support by Hilbert. However, thanks to her outstanding results, she obtained the position. She slowly managed to get out of her "gender precariousness." Within a few years, she began to get paid for her lessons, which attracted so many students from all over Germany, and which she had long been forced to offer for free and on behalf of her mentor, David Hilbert, who concretely helped her despite the prejudices. It is worth noticing that, until 1923, she never had a proper salary.

Besides her famous theorems, Noether's research works concerned areas of Mathematics other than the calculus of variations, and most of them were developed during these fruitful years. In the meantime, the "Göttingen School," became attractive for the most prestigious physicists and mathematicians coming from all over Europe. Noether's contributions were outstanding, in particular, in abstract algebra: from the discovery of rings, today called Noetherian Rings, to the noncommutative algebra, passing through the theory of ideals and the algebraic theory of numbers. In some way, most of the language and results of modern Algebra can be attributed to Emmy Noether or traced from her fundamental contributions. Her students, the "Noether Boys," gave her the affectionate nickname of "Mother Algebra."

Despite the successes, her career was suddenly broken by the advent of Nazism. She was of Jewish origin and, in 1933, she was forced to flee to the United States, where she got a position at Bryn Mawr College in Pennsylvania. But once again, being a woman, she was prevented from accessing the most prestigious research centers, such as Princeton University, where Einstein, Weyl, and Gödel were welcomed at the same time.

She died on April 14, 1935 from the consequences of an infection contracted during a surgical operation for the removal of a carcinoma. Einstein wrote an obituary for her in *The New York Times*: "Noether was the most significant creative mathematical genius thus far produced since the higher education of women began." Weyl defined her as the most important woman in the history of Mathematics, equivalent to what Marie Curie had been for Physics. Starting from the 1950s, Noether's theorems were recognized as a milestone for Physics, and "Fräulein Emmy" is considered an absolute genius of Mathematics for all time.

Notation

We will set $\hbar = c = 8\pi G = 1$ unless otherwise indicated and we will use the following notation:

1. For the indexes:
   - Greek indexes $\{\alpha, \beta, \gamma... = 0,1,2,3\} \rightarrow$ label the four-dimensional curved space-time coordinates,
   - Latin indexes $\{a, b, c... = 0,1,2,3\} \rightarrow$ label the four-dimensional flat space-time coordinates,
   - Middle indexes $\{i, j, k... = 1,2,3\} \rightarrow$ label the spatial coordinates in curved space-time,
   - Symmetrization over the indexes will be indicated by the curly brackets, and anti-symmetrization by the square brackets.
2. Let A_μ be a generic four-vector, we adopt the following notation:
   - $D_\nu A_\mu = A_{\mu;\nu}$ is the covariant derivative in terms of the Levi–Civita connection,
   - $\partial_\nu A_\mu = A_{\mu,\nu}$ is the partial derivative,
   - Christoffel connections will be indicated equivalently by $\Gamma^\alpha_{\beta\gamma}$ or $g^{\alpha\sigma}\{\sigma, \beta\gamma\}$,
   - $\nabla_\nu A_\mu$ is the covariant derivative in terms of any connection except for the Levi–Civita connection.
3. We use the symbol $\mathscr{L}$ for Lagrangian density, while the Lagrangian will be denoted by $\mathcal{L}$.
4. For the Einstein tensor, we use the notation $G_{\mu\nu} = R_{\mu\nu} - \frac{1}{2}g_{\mu\nu}R$, with $R_{\mu\nu}$ being the Ricci tensor and $R = R^\mu{}_\mu$ the Ricci scalar.
5. The derivative with respect to a given variable will be indicated by the subscript variable or sometimes by the subscript variable in the partial derivative.
6. Derivatives of a given function f with respect to temporal and spatial coordinates will be denoted by $\dot{f}$ and f', respectively, unless otherwise indicated.
7. D labels the number of space-time dimensions, while spatial dimensions are denoted by d.

8. $\Box$ stands for the d'Alembert operator $\Box = g_{\mu\nu}\nabla^{\mu}\nabla^{\nu}$.
9. X represents the generator of a certain symmetry, while $X = X + \dot{\eta}^i \partial_{\dot{q}^i}$ is the Noether vector.

The adopted metric signature is $(+,-,-,-)$, unless otherwise indicated.

We will introduce other particular symbols during the discussion.

Acronyms

- ADM: *Arnowitt–Deser–Misner*
- AdS: *Anti de Sitter*
- CFT: *Conformal Field Theory*
- CMB: *Cosmic Microwave Background*
- DEC: *Dominant Energy Condition*
- EH: *Einstein–Hilbert*
- EL: *Euler–Lagrange*
- EoS: *Equation of State*
- FLRW: *Friedman–Lemaître–Robertson–Walker*
- GR: *General Relativity*
- GWs: *Gravitational Waves*
- IDGs: *Infinite Derivative Theories of Gravity*
- IKGs: *Integral Kernel Theories of Gravity*
- IR: *Infrared*
- LG: *Lorentz Group*
- LT: *Lorentz Transformation*
- NEC: *Null Energy Condition*
- PN: *Post-Newtonian*
- PPN: *Parametrized Post-Newtonian*
- QFT: *Quantum Field Theory*
- RLG: *Restricted Lorentz Group*
- SEC: *Strong Energy Condition*
- STEGR: *Symmetric Teleparallel Equivalent of General Relativity*
- SUSY: *Supersymmetry*
- TEGR: *Teleparallel Equivalent of General Relativity*
- TOV: *Tolman–Oppenheimer–Volkoff*
- UV: *Ultraviolct*
- WDW: *Wheeler–DeWitt*
- WEC: *Weak Energy Condition*

Part I
PRELIMINARIES

1
The Concept of Symmetry

> Symmetry, as wide or as narrow as you may define its meaning, is one idea by which man through the ages has tried to comprehend and create order, beauty and perfection.
>
> — *Hermann Weyl*
> (*Symmetry* [1952])

Since birth, all of us have grown up with the concept of symmetry rooted in our minds; broadly, symmetry can be defined as the repeating property of phenomena in time and space with regularity. Regardless of the field considered, this definition branches out into several different interpretations: for example in the arts, it assumes the meaning of "sense of harmonious and beautiful proportions." Specifically, symmetries can be used to obtain ideal representations of figures, respecting proportions. On the other hand, in Mathematics and Physics, they represent invariances under some transformations.

Certainly, many aspects of our world and life are ruled by symmetry relations from the smallest to the largest scale. It comes natural, indeed, wondering whether the symmetry of structures is a fundamental and essential property, or whether it is just a mere tool, only aimed at schematizing the laws of Nature and the corresponding dynamics. Is it actually surprising that the Universe enjoys certain symmetries, once we created the concept of symmetry by looking at the Universe itself? In other words: Are symmetries constructions of our mind or fundamental structures that reality is endowed with?

Reality, in which we exist, continuously manifests many expressions of symmetry, even in everyday life, such as the indistinguishability between right and left, or up and down. One may think that the latter could be distinguished by the natural propensity of objects to fall down from the top to the bottom. However, this is only a consequence of gravitational interaction, which when far away from massive bodies (as well as planets, stars, or galaxies) is not manifesting. Although at small scales the Universe seems to be inhomogeneous and anisotropic, such asymmetry disappears at properly large scales (cosmological scales), where no preferred direction occurs. The assumption of homogeneity and isotropy was proposed in the first half of the twentieth century and it is known as the "Cosmological Principle"; we will largely use such a principle throughout the second part of the book, where we will assume the space-time metric to be isotropic and homogeneous. Just as the large-scale Universe is dominated by symmetry properties, so also molecules and atoms assume certain preferred symmetry configurations, such as the invariance under translations or rotations. The latter concept is particularly useful in Chemistry since it reduces the complexity of the whole system made of many-bodies interactions.

Can the idea of symmetry be considered as the lens through which we watch the world, or is it a concept deep-rooted in our world, regardless of human perception and conception? In other words: Might it be just something created by the human mind with no real confirmation in Nature? In order to better explain this point, let us consider the symmetry of geometric figures. A "regular polygon" can be thought of as a figure having equal angles and equal sides. Equivalently, a regular polygon with n sides can be thought of as a geometric structure that is invariant under rotations of $\frac{2k\pi}{n}$ radians around its center, with k a natural number. In this way, the square is *at least* symmetric under rotation of $\frac{\pi}{2}$ radians, the pentagon under rotation of $\frac{2}{5}\pi$ radians, and so (see Fig. 1.1). In the limit $n \to \infty$ one gets a geometric figure that is invariant under any rotation of any arbitrary angle: the circle.

However, it has to be clear that this is just a mental abstraction, since in nature neither perfect circles nor regular polygons exist. Any geometric figure we find in everyday life is just similar to the hypothetical definition that we keep in our mind, since no figure having exactly the same sides or the same angles can be found in nature. Equivalently, we may argue that no natural symmetry can be completely perfect, so that any time a symmetry property is identified in nature, it is automatically addressed to something that seems to respect the idea of symmetry we created in our mind. For example, from this point of view, the Cosmological Principle would represent only an approximation of the perfect symmetry in the Universe; molecules would not be invariant under rotations of given angles, but their rotated configuration would just look similar to the initial one, instead of exactly identical. According to this vision, symmetry might be only a human creation not describing the real world, which we recognize whenever our observations look similar to what we cataloged as "symmetric" in our thought. The same argument can also be applied to Mathematics and other

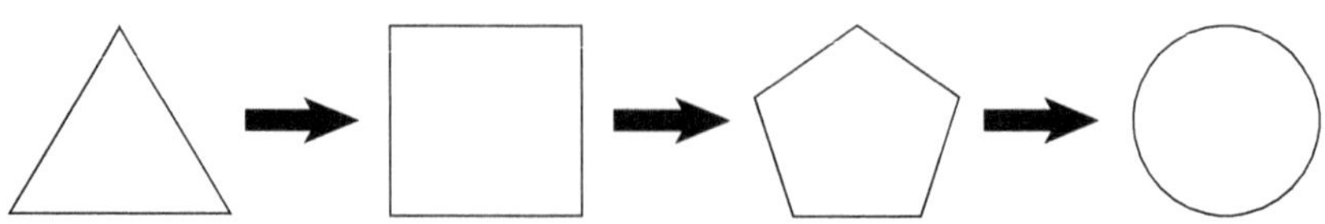

Figure 1.1 Polygons invariant under rotations of $\frac{2\pi}{3}$ rad, $\frac{\pi}{2}$ rad, $\frac{2\pi}{5}$, and any angle, respectively.

fields, where the utility of symmetries is undoubtedly independent of this issue. Just as it is true that a human being is too much influenced by the Universe in which she/he lives to answer these questions, so is it true that, although imperfect or ideal, symmetries play a crucial role in the comprehension of Physics; without them, it would be impossible to get several results that constitute the basis of the whole of Science. Besides simplifying dynamics of systems, continuous symmetries are always linked to quantities that remain constants in time or space. These quantities allow one to introduce cyclic variables, which reduces the complexity of the equations of motion. In Biology, they can be used to schematize animal and plant species, which can be classified with respect to their radial, bilateral, or spherical symmetry. In Chemistry, the stereochemistry studies the arrangements in space of atoms or atomic groups making up the molecules (molecular configuration and conformation) and the relationships between these structures and chemical properties. From this definition, it is easy to understand the importance of symmetries. Those just mentioned are only some examples of the importance of symmetry in Science.

In the next section, we discuss what is meant by "symmetry" in Physics, analyzing then some symmetry groups. For an exhaustive treatment of philosophical and physical aspects of symmetries, we suggest Refs. [1, 2], while mathematical aspects are considered in Ref. [3], in the context of differential equation solutions. Specifically, the latter reference uses mathematical techniques – based on symmetries – to find out solutions of nonlinear differential equations.

1.1 Symmetries in Physics

The concept of symmetry in Physics plays such an important role that it is hard to think that symmetric properties of the systems are the result of mere coincidence. Over the years, the concept has been so exaggerated as to begin to think that the asymmetries that we observe in the current Universe may be attributed to spontaneous symmetry breakings in early epochs. The best known theory that pursues such a prescription is called *supersymmetry* (SUSY), according to which for each particle there corresponds a symmetric partner with the same mass but different spin. This supersymmetric situation held in the early Universe. Subsequently, after a symmetry breaking, the mass of supersymmetric particles changed, unlike other quantum properties, which remained unchanged. Such supersymmetric particles are the best candidates to solve many problems of String Theory, Grand Unification of physical interactions, and Standard Model of Particles, but so far they have never been observed. The formal elegance of the theory and its promising results are the real reason for its reputation, but as long as SUSY is not supported by experimental observations it cannot be considered fully valid. Currently we have no evidence of such spontaneously broken symmetry in the early Universe. For this reason, the detection of new particles that

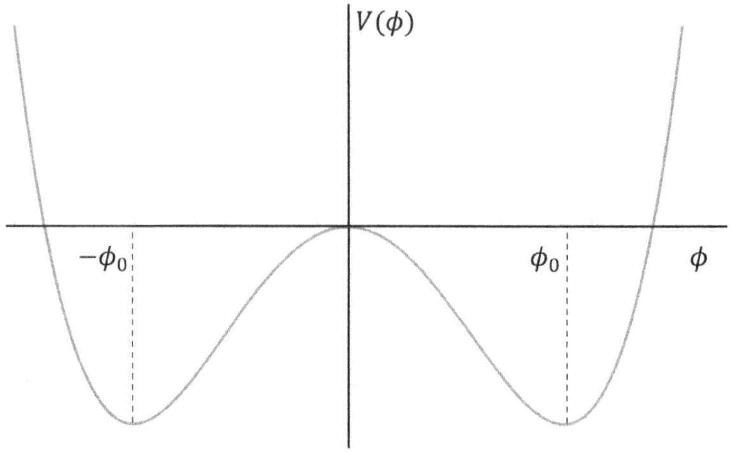

Figure 1.2 $\mu^2\phi^2 - \lambda\phi^4$ potential with $\mu^2 < 0$

can potentially confirm the validity of the theory is a very demanding research area.

Another example of the importance of symmetry is represented by the Brout–Englert–Higgs Mechanism (better known as *Higgs mechanism*). It is mostly used to explain the currently observed mass of elementary particles, which is attributed to a spontaneous symmetry breaking in the early epoch of the Universe. The fundamental mechanism states that when the symmetry of a certain system is spontaneously broken, a massive (Higgs boson) and a massless (Goldstone boson) particle arise. A standard example is given by the potential $V(\phi) = \mu^2\phi^2 - \lambda\phi^4$, whose graphical representation is shown in Fig. 1.2.

The symmetry can be broken after shifting the minimum by means of a reparameterization of the scalar field, as pointed out in Section 3.4. The Higgs mechanism is a clear example of how the presence of symmetry (and its breaking) allows one to fix dynamics and to give an overall picture of very complicated problems like the generation of particle masses.

Another important example of how symmetry is related to physical properties comes from the study of functions in Mathematics. One of the most basic distinctions in the framework of real functions depending on real variables is provided by the symmetry of such functions with respect to the ordinate axis or the origin point. In the former case, a function $f(x)$ respecting the symmetry condition $f(x) = f(-x)$ is said to be *even*; in the latter case, it is said to be *odd*. In the analysis of the function trend, such properties may largely simplify the treatment, allowing one to focus the procedure on particular intervals of the real axis rather than studying the trend for any real value of x. In this way, the integral of an even function over the real axis can be restricted to the positive axis and vice versa. By means of this manipulation, it is possible to compute, for example, the integral of the Gaussian function within the interval $[0, \infty[$. More precisely, the integral can be extended to the entire real domain and calculations can be computed in the complex plane. In the same context, many difficult integrals or differential equations can be easily calculated by considering the symmetry of the integrating function. This finds wide application in Quantum Mechanics, where the integral over the three-dimensional space of the spherical harmonics vanishes everywhere due to parity properties.

In Classical Mechanics, symmetries with respect to space or time translations allow one to define constants of motion that can be used to find the final configuration of a system, once the initial conditions are known. Without making use of invariance properties, it is still possible to solve differential equations of dynamics. Though this procedure has the great advantage of providing dynamics at any time, it turns out to be complicated because, sometimes, it leads to analytically unsolvable equations of motion. The invariance under time translation yields a time-independent Hamiltonian, while the invariance under space translations leads to the conservation of conjugate momenta. As shown in Section 3.2.1, these quantities can be thought as the on-diagonal components of the energy-momentum tensor.

Moreover, in Section 3.2.2, we will also show that, from the invariance of a system under rotations, another constant of motion arises: the *Angular Momentum*. As in Classical Mechanics, the Angular Momentum conservation can easily explain several phenomena, thereby avoiding a large amount of computations. In Quantum Mechanics, it plays a fundamental role in the development of theoretical atomic structure, including the theory of spin.

In the nineteenth century it was not clear whether conservation laws were valid even at microscopic scales, or if they were only the macroscopic result of microscopic averages. According to the currently accepted view, they hold even at the microscopic level and represent an essential tool in all fields of Science.

Another famous result based on symmetry considerations is the Gauss theorem, by means of which it is possible to get the electric field of given configurations without calculating the contribution of each charge. When a one-dimensional distribution is considered, such as a charged line, the surrounding electric field only depends on the radial distance from the line itself. It can be calculated either by integrating the contribution provided by any charge or by using the Gauss theorem. To compare the two procedures, let us consider a line with homogeneous linear density of charge λ and length h. The electric field is provided by an integration over the infinite set of charges at the distance r, namely:

$$\begin{cases} E_z = \dfrac{1}{4\pi\epsilon_0}\displaystyle\int \cos\theta \frac{dq}{R^2} = \frac{1}{4\pi\epsilon_0}\int \frac{z\,dq}{R^3} = \frac{\lambda}{4\pi\epsilon_0}\int_{-\frac{h}{2}}^{\frac{h}{2}} \frac{zdz}{\sqrt{(x^2+z^2)}^3} \\ E_x = \dfrac{1}{4\pi\epsilon_0}\displaystyle\int \sin\theta \frac{dq}{R^2} = \frac{1}{4\pi\epsilon_0}\int \frac{x\,dq}{R^3} = \frac{\lambda}{4\pi\epsilon_0}\int_{-\frac{h}{2}}^{\frac{h}{2}} \frac{xdz}{\sqrt{(x^2+z^2)}^3}. \end{cases} \tag{1.1}$$

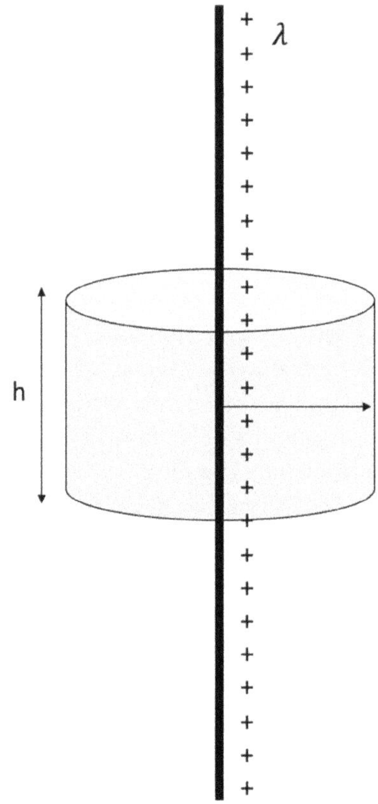

Figure 1.3 Closed cylindrical surface enclosing the charged line

After solving the integrals, it turns out that in the limit $h \to \infty$, the total electric field can be written as

$$\mathbf{E} = \frac{\lambda}{2\pi\epsilon_0|r|}\mathbf{e}_r, \tag{1.2}$$

with $\mathbf{e}_r$ being the unitary vector, which labels the field direction. The same result can be achieved by taking into account the Gauss theorem. Considering a cylindrical surface that encloses the line (as shown in Fig. 1.3), we notice that, in the case of infinite length, the total flux of the electric field over the base areas vanishes due to the same amount of incoming and outgoing contributions. Therefore only lateral areas contribute to the total flux, and the Gauss theorem can be written as

$$\int_{Cylinder} \mathbf{E} \cdot \mathbf{n}\, dS = \frac{\lambda h}{\epsilon_0}. \tag{1.3}$$

Thus the integral can be easily recast as a product, since the electric field is constant over the surface chosen. It finally reads

$$|E| = \frac{\lambda}{2\pi\epsilon_0|r|}. \tag{1.4}$$

The role of symmetries in this treatment lies behind the choice of the cylindrical surface. Though the theorem holds for every closed surface, most of them make the procedure even more complicated. The closed surface that best fits the standard of this example must satisfy the symmetry of the system. When the line is supposed to be infinitely long, the system acquires a cylindrical symmetry, so that a cylindrical surface allows one to neglect the flux over the base areas.

This concept can be further stressed by considering the two-dimensional case. The electric field generated by a continuous surface with charge surface density σ can be straightforwardly obtained by means of the Gauss theorem as follows:

$$\Phi_E = \frac{q}{\epsilon_0} \to 2|E|S = \frac{\sigma S}{\epsilon_0} \to |E| = \frac{\sigma}{2\epsilon_0}. \tag{1.5}$$

Nevertheless, it is worth noticing that a careful choice of the closed surface is always important. It must respect the symmetry of the distribution considered; otherwise no simplification can be provided by the theorem.

So far we have dealt with continuous symmetries generated by continuous transformations. Another physically relevant example is given by a set of three *discrete* transformations, very useful in Particle Physics.

The first one is the so-called *Charge Conjugation* or *C-Parity*, whose action of the corresponding operator $\hat{C}$ transforms a particle into the related antiparticle. A physical system that is symmetric with respect to such transformation is said to be *C-symmetric*. The only interaction violating the C-symmetry is the Weak Interaction. It goes without proving that once the C-Parity acts twice to a given state $|n>$, the final state must be the same as the initial one. Therefore the charge operator must have only two eigenvalues: ± 1.

Another discrete transformation is the so-called *Parity Transformation*, which reverses the spatial coordinates of the system. For the same reason as before, the eigenvalues of the corresponding operator can assume only the values ± 1. This transformation is very important since it allows one to distinguish scalar from pseudoscalar or vector from pseudovector. Scalar quantities that change sign under parity transformations are called *pseudoscalars*, while vector fields that do not change sign under parity transformations are called *pseudovectors*. Even in this case, the only fundamental force that violates parity symmetry is the Weak Interaction.

The last discrete transformation is the *Time Reversal*. It is a symmetry transformation of time $t \to -t$ whose violation can be attributed to the second principle of thermodynamics violation. Since the symmetry under time reversal (T-symmetry) implies the conservation of entropy, the macroscopic Universe is not symmetric under such transformation. However, at very small scales, some systems exhibit symmetric properties under time reversal.

In the twentieth century, it was supposed that the laws of physics were invariant under CP transformations, which can be obtained by combining charge conjugation and parity transformation. However, in 1964 the *CP* symmetry violation was discovered in the decays of neutral kaons. Nowadays CP symmetry is still deeply studied due to its quantum and cosmological implications and because it might potentially explain the dominance of matter with respect to antimatter. Even though the attempt to find a discrete symmetry for all physical systems failed in 1964, soon after it turned out that this feature could be attributed to CPT transformation. It represents the only discrete symmetry in nature, whose violation would imply the Lorentz invariance violation, as shown by Greenberg in [4]. Therefore it automatically follows that a system violating the CP symmetry must also violate T-symmetry. Basically, according to CPT symmetry, a hypothetical universe where matter is replaced with antimatter, where particles and fields have opposite positions, and where time flows in the opposite direction would evolve exactly like our observed Universe.

In light of these examples, given the large and different amounts of symmetries in Physics, it is worth distinguishing all possible cases that can characterize a given system. Continuous symmetries, first, are symmetries that can be described by infinitesimal transformations generated by continuous parameters. On the other hand, discrete symmetries describe noncontinuous changes in the system as well as parity or charge conjugation. Moreover, continuous symmetries can be further split into *local* symmetries, which depend on the space-time point of the given manifold, and *global* symmetries, which are independent of the local position. Invariance under translations or Lorentz invariance belongs to the latter category, and gauge invariance to the former. This last distinction is fundamental since, as pointed out in Chapter 4, it is responsible for the impossibility of treating gravity under the formalism of the Yang–Mills theory. Furthermore, an internal transformation acting on the field $\phi(x^a)$ does not involve the coordinates x^a, so that its variation reads as $\phi(x^a) \rightarrow \tilde{\phi}(x^a)$. On the contrary, an external transformation changes also the coordinates, so that the given scalar field turns out to transform as $\phi(x^a) \rightarrow \tilde{\phi}(\tilde{x}^a)$.

The need of treating gravity as a gauge theory comes from the fact that gauge invariance seems to be a crucial property toward the construction of a self-consistent theory of Quantum Gravity. Indeed, it turns out that theories that are not invariant under gauge transformations do not have any predictive power at UV scales. Nevertheless they may be perfectly acceptable theories at lower scales. The main issue lies behind the diffeomorphism invariance enjoyed by the gravitational action, which is an external symmetry. On the other hand, the nongravitational fundamental interactions are symmetric under local gauge transformations and this provides a non-Abelian structure that gravity in four dimensions cannot show. Reversing the argument, the gravitational action of GR cannot be written as a local gauge-invariant action (Yang–Mills theory) for the diffeomorphism group. Nevertheless, it is still possible to locally link the curved space-time to a flat tangent space-time point by point. This procedure makes standard GR invariant under the local Lorentz group (LG), where tetrad fields and spin connections are used to label the geometry. For a detailed discussion see

Section 5.2, of Chap. 5, where we show how to recast the gravitational interaction as a local gauge theory.

To conclude this short overview of symmetries in Physics, if the action is invariant under some transformation that does not act on the Euler–Lagrange (EL) equations, the system is said to enjoy an *on-shell* symmetry; otherwise, if the transformation changes the *EL* equations, it is said to be *off-shell*. To better clarify this point, let us consider a field transformation $\phi \to \tilde{\phi}$, whose Lagrangian variation yields a set of *EL* equations plus a total derivative *TD*: $\delta\mathcal{L} \to EL + TD$. On-shell symmetry occurs if $EL = TD = 0$; otherwise when $EL = -TD$ we have an off-shell symmetry. In the former case, the total derivative can be intended as a conserved quantity $TD = \partial_a j^a$, called Noether current (see the following chapters for details).

In what follows, we shall overview the main features of the symmetry groups that play a crucial role in the applications of the Noether Theorem. We mainly analyze the unitary $U(n)$ group, the LG – isomorphic to $O(1,3)$ – the translation and rotation group, and the Poincaré group.

1.1.1 The Unitary Group

The n-parameters unitary group $U(n)$ and the special unitary group $SU(n)$ represent an essential step in the framework of symmetries in fundamental Physics. Formally, a $n \times n$ matrix M is said to be *unitary* if $M^\dagger M = \mathbb{1}$, where $M^\dagger$ is the conjugate transpose of M. These matrices, whose determinant may only assume the two values ± 1, form the so-called $U(n)$ group. The set of unitary matrices with positive determinant form in turn the *special* subgroup $SU(n)$. It is the $n^2 - 1$ dimensional subgroup of the n^2 dimensional $U(n)$, which is, in turn, a subgroup of the general $GL(n, \mathbb{C})$. Of particular interest are the $U(1)$ transformations, which form the only Abelian subgroup of $U(n)$ with vanishing structure constants of the corresponding Lie algebra. $U(1)$ transformations are fundamental in all physical interactions since all Lagrangians depending on scalar combinations of the variables are $U(1)$-invariant. From the definition of $SU(n)$, it follows that the $U(n)$ group can be written as the semi-direct product between $SU(n)$ and $U(1)$. Other important subgroups of $U(n)$ having several physical implications are the $SU(3)$ group (whose algebra is defined by the Gell-Man matrix) for Quantum Chromodynamics and the $SU(2)$ group for the Electroweak Interaction. Regarding the latter, it can be straightforwardly shown that the Pauli Matrices satisfy all the constraints imposed by the $SU(2)$ Lie algebra and therefore can be chosen as generators of $SU(2)$ transformations. Being represented by all the 2×2 matrices with unitary determinant, the generator G of $SU(2)$ must satisfy the relations

$$G^\dagger G^{-1} = \mathbb{1}, \qquad Tr[G] = 0. \tag{1.6}$$

The second condition can be easily demonstrated by considering an element g of $SU(2)$ that weakly differs from the unitary matrix:

$$g = \mathbb{1} + iG \to det(g) = 1 + iTr[G] \to Tr[G] = 0. \tag{1.7}$$

Since the Pauli Matrices, σ^a, are traceless and Hermitian, they can be arbitrarily chosen as a basis for $SU(2)$. In this way, the generator G is represented by any linear combination of σ^a. Moreover, by means of the definition $G^\alpha = \sigma^\alpha/2$ the Lie algebra of $SU(2)$ turns out to be

$$[G^a, G^b] = i\epsilon^{abc} G_c. \tag{1.8}$$

Notice that, as better pointed out in Section 1.1.3, the above algebra is formally equivalent to that of the $SO(3)$ group.

1.1.2 The Translation Group

Let us now consider another set of transformations that act on the positions of all points. It is the Abelian subgroup of plane isometries. In order to find the transformation generator, we take into account the infinitesimal translation

$$x^a \to x^a + \delta x^a, \tag{1.9}$$

with respect to which a scalar field $\phi(x^a)$ changes as

$$\tilde{\phi}(x^a) = \phi(x^a + \delta x^b) \sim \phi(x^a) + \delta x^b \partial_b \phi(x^a). \tag{1.10}$$

The transformation, therefore, can be written as

$$T\phi = \phi + \partial_b \phi \delta x^b \to T = \mathbb{1} + \delta x^b \partial_b, \tag{1.11}$$

and coming back to the corresponding finite quantity, the preceding translation takes the form

$$T = e^{-i\epsilon^a \mathcal{T}_a}, \tag{1.12}$$

where $\epsilon^a \equiv \delta x^a$ and $\mathcal{T}_a$ the group generator is defined as $\mathcal{T}_a \equiv i\partial_a$.

1.1.3 The Rotation Group

Other transformations change the angles of vectors but preserve the distances. The Lie group that corresponds to such transformations is the so-called orthogonal group $O(n)$, where n is the number of dimensions. It admits as a subgroup the *special orthogonal* group $SO(n)$, called *Rotation Group*, whose determinant of the corresponding matrices is equal to $+1$. In addition, it can be connected to the preceding unitary group according to the relation $O(n) \supset U(n) \supset SU(n)$. The 3-D rotation group $SO(3)$ consists of 3×3 symmetric matrices whose elements represent the given rotation. In general, the elements of $O(n)$ are combinations of rotations.

To introduce the main features of the $SO(n)$ group, let $\mathcal{R}$ be an infinitesimal rotation transforming the vector r into another vector, $\tilde{r}$:

$$\tilde{r} = \mathcal{R}(\theta)\, r. \tag{1.13}$$

A finite rotation around the direction j can be written as $\mathcal{R}_j(\theta) = e^{-i\hat{L}_j\theta}$, where $\hat{L}_j$ is the generator of the transformation. By means of the preceding relation, it is possible to write the corresponding infinitesimal transformation as

$$\mathcal{R}_j(\delta\theta) = 1 - i\,\delta\theta\,\hat{L}_j. \tag{1.14}$$

The rotation transforms the vector $r(x_i, x_k, x_j)$ to $\tilde{r}(x_i, x_k, x_j) = r(x_i - x_k d\theta, x_k + x_i d\theta, x_j)$, which, up to the first order, can be written as follows[1]:

$$\begin{aligned} \tilde{r}(x_i, x_k, x_j) &= r(x_i, x_k, x_j) + d\theta\,(x_i\partial_k - x_k\partial_i)\, r(x_i, x_k, x_j), \\ \delta r &= d\theta\,(x_i\partial_k - x_k\partial_i)\, r(x_i, x_k, x_j). \end{aligned} \tag{1.15}$$

Equation (1.15) shows that the infinitesimal generator of the rotation is $\hat{L}_j = (x_i\partial_k - x_k\partial_i)$, with corresponding Lie algebra given by

$$[\hat{L}_i, \hat{L}_j] = i\epsilon_{ijk}\hat{L}_k. \tag{1.16}$$

As is well known, $\hat{L}_j$ represents the j-component of the orbital angular momentum operator. Note that the Lie algebra is formally equivalent to that of $SU(2)$ in Eq. (1.8). This means that the two groups are locally equivalent. Indeed, the $SU(2)$ group can be understood as the universal cover of $SO(3)$.

1.1.4 The Lorentz Group

Once the three-dimensional rotations have been defined and the main features of the $O(n)$ group have been set up, we can introduce the Lorentz transformations (LTs), which form the so-called *Lorentz group*, isomorphic to $O(1,3)$. Considering the properties of LTs (hereafter denoted by the symbol Λ), one can straightforwardly prove that they respect all the axioms needed to form a group. Specifically, they are *closed* since two consecutive LTs provide another LT, namely

$$\Lambda_1, \Lambda_2 \in LG \rightarrow \Lambda_1\Lambda_2 \in LG. \tag{1.17}$$

The *associativity* and the existence of the *identity* transformation can be similarly proven. Moreover, from the possibility of going back to the starting reference frame, it follows that LTs are also *invertible*. We mainly focus on the only connected subgroup of LG, namely the *Restricted Lorentz Group* (RLG), isomorphic

[1] In this paragraph and in 3.2.2, we only consider the three-dimensional case; thus the Einstein convention on indexes is not used.

to $SO(1,3)$ and characterized by the conditions $|\Lambda| = +1$ and $\Lambda_{00} \geq 1$. With the aim of obtaining the generator of the RLG, let us consider a LT weakly differing from the identity:

$$\Lambda^c_d = \delta^c_d + \lambda^c_d. \tag{1.18}$$

From the definition of LT, it follows that $\eta_{cd}\Lambda^c_a\Lambda^d_b = \eta_{ab}$, where η_{cd} is the Minkowski tensor. Substituting the preceding relation into Eq. (1.18), we obtain

$$\eta_{cd}(\delta^c_a + \lambda^c_a)(\delta^d_b + \lambda^d_b) = \eta_{ab}, \tag{1.19}$$

so that, neglecting higher-than-first-order terms, Eq. (1.19) yields:

$$\eta_{cd}(\delta^c_a\delta^d_b + \delta^c_a\lambda^d_b + \lambda^c_a\delta^d_b) = \eta_{ab} \to \lambda_{ab} = -\lambda_{ba}. \tag{1.20}$$

Finally, by means of the definition

$$\left(\mathcal{J}^{ab}\right)^c_d = \frac{i}{2}\left(\eta^{ac}\delta^b_d - \eta^{bc}\delta^a_d\right), \tag{1.21}$$

the infinitesimal LT reads

$$\Lambda^c_d = \delta^c_d - i\lambda_{ab}\left(\mathcal{J}^{ab}\right)^c_d. \tag{1.22}$$

The corresponding finite transformation therefore can be written as $\Lambda = e^{-i\lambda_{ab}\mathcal{J}^{ab}}$, with $\mathcal{J}$ being the generator of the RLG, satisfying the following Lie algebra:

$$\left[\mathcal{J}^{ab}, \mathcal{J}^{cd}\right] = \frac{i}{2}\left(\eta^{ad}\mathcal{J}^{bc} + \eta^{bc}\mathcal{J}^{ad} - \eta^{ac}\mathcal{J}^{bd} - \eta^{bd}\mathcal{J}^{ac}\right). \tag{1.23}$$

It is possible to split the contributions of boosts from those of rotations, by means of the definitions

$$\begin{cases} \mathcal{R}^i = \epsilon^{ij\ell}\mathcal{J}_{j\ell} \\ \mathcal{B}^i = \mathcal{J}^{0i}. \end{cases} \tag{1.24}$$

In this way, the general LT takes the form

$$\Lambda = \exp\left\{-i\left(\lambda_{0i}\mathcal{B}^i - \epsilon_{ij\ell}\lambda^{j\ell}\mathcal{R}^i\right)\right\}. \tag{1.25}$$

It is worth noticing that LTs can be intended as the combination of three-dimensional rotation and boosts. The LG plays a role of special interest in fundamental physics because LTs can be used to define the rank of tensor fields. In particular, depending on how tensor fields transform under LTs, the following distinctions can be made:

- The spin-n *tensor field* under LTs transforms as an n-rank tensor, namely

$$\phi^{b_1 b_2 b_3 \ldots b_n}(x^a) \xrightarrow{LT} \tilde{\phi}^{b_1 b_2 b_3 \ldots b_n}(\tilde{x}^a) = \Lambda^{b_1}_{c_1}\Lambda^{b_2}_{c_2}\Lambda^{b_3}_{c_3}\ldots\ldots\Lambda^{b_n}_{c_n}\phi^{c_1 c_2 c_3 \ldots c_n}(x^a). \tag{1.26}$$

Two main subcases follow:

1. The spin-0 *scalar field* is defined as the invariant under LTs, that is,

$$\phi(x^a) \xrightarrow{LT} \tilde{\phi}(\tilde{x}^a) = \phi(x^a). \tag{1.27}$$

2. The spin-1 *vector field* transforms as

$$\phi^b(x^a) \xrightarrow{LT} \tilde{\phi}^b(\tilde{x}^a) = \Lambda^b_c \phi^c(x^a). \tag{1.28}$$

- The spin-$\frac{1}{2}$ *spinor field*, whose transformation is defined through the Dirac matrices, is

$$\Phi(x^a) \xrightarrow{LT} \tilde{\Phi}(\tilde{x}^a) = e^{-\frac{i}{4}\lambda_{ab}\sigma^{ab}} \Phi(x^a), \tag{1.29}$$

where Φ is the spinor field, $\sigma_{ab} = \frac{i}{2}[\gamma_a, \gamma_b]$, and γ are the Dirac matrices.

1.1.5 The Poincaré Group

The Poincaré group is an extension of the preceding LG. It consists of the semi-direct product $T \otimes O(1,3)$ and forms a 10-parameter Lie group. Four parameters belong to the translations group, and the remaining six to the LG. When the Poincaré transformation acts on a vector field ϕ^c, the latter transforms as

$$\tilde{\phi}^c = \Lambda^c_d \phi^d + t^c. \tag{1.30}$$

The Poincaré group finds the most application in Quantum Mechanics, where the relativistic single-particle states are described by an irreducible unitary representation of the covering group of the Poincaré group. The Poincaré transformation can be obtained by merging the generic translation in Eq. (1.12), with an LT (1.22):

$$PT = \exp\left\{-i\left(\lambda_{ab}\mathcal{J}^{ab} + \epsilon_c \mathcal{T}^c\right)\right\}. \tag{1.31}$$

The corresponding Lie algebra is ruled by the relations

$$\begin{cases} [\mathcal{T}^a, \mathcal{J}^{cd}] = -\dfrac{i}{2}\eta^{a[c}\mathcal{T}^{d]} \\ [\mathcal{J}^{ab}, \mathcal{J}^{cd}] = i\left(\eta^{ca}\mathcal{J}^{db} - \eta^{cb}\mathcal{J}^{da} + \eta^{db}\mathcal{J}^{ca} - \eta^{da}\mathcal{J}^{cb}\right) \\ [\mathcal{T}^a, \mathcal{T}^b] = 0. \end{cases} \tag{1.32}$$

The summary proposed here is intended to point out the most important symmetry groups. Their main features are worth noticing in view of applications considered in Chapter 3, where the Noether Theorem will be applied to Lagrangian densities to find out the corresponding conserved quantities.

2
The Two Noether Theorems

My methods are really methods of working and thinking;
this is why they have crept in everywhere anonymously.

— *Emmy Noether*
(*"Letter to Helmut Hasse"* [1931])

This chapter is mainly devoted to one of the most important theorems of Mathematics, which finds a wide number of applications in Physics: Noether's First Theorem. Since its formulation in 1918, this theorem has represented a fundamental tool for theoretical and applied Physics, providing several results in many fields and leading to a better comprehension and formalization of theories. Now it represents an unmissable argument for the foundation of Quantum Field Theory (QFT) [5–8]. The reason for its glory lies behind the link provided between symmetries and conservation laws. Though in the previous chapter we mentioned the relations occurring between symmetry transformations and conservation laws (such as energy or angular momentum), Noether's First Theorem generalizes the concept and provides a method capable of finding integrals of motion with the corresponding generator of transformations.

The principle is so general that many physically relevant quantities are usually defined through their symmetry properties. For example, the energy can be defined as the integral of motion of a system that is invariant under time translations. The angular momentum is the invariant quantity under rotations and so on.

At the beginning of the twentieth century, many conserved quantities were associated to the symmetries of the corresponding actions. Due to this fact, it became natural to wonder whether it was possible to find a general criterion aimed at connecting symmetries and conservation laws. The answer arrived in 1918, thanks to the works by Emmy Noether. In [9], for the first time and in the most general way, she linked the invariance of Lagrangians under given transformations with integrals of motion, showing that any symmetry leads to a continuity equation of the form $\partial_\mu j^\mu = 0$. Soon after, Noether introduced a second theorem, according to which the symmetries of action functionals can be related to systems of differential equations. For a comprehensive discussion, see [10,11].

In this chapter we introduce both theorems by Noether, with the aim of using Noether's First Theorem to investigate theories of gravity. As discussed in Chapter 6, it is possible to develop a method capable of selecting physically motivated models and then solving the related dynamics starting from the statement of this theorem.

2.1 Noether's First Theorem

Let us enunciate and demonstrate Noether's First Theorem. It formally states that *to every differentiable symmetry generated by local actions, there corresponds a conserved current*, also called *Noether current*. Even though, before the twentieth century, it was clear that symmetries were related to conserved quantities, the theorem represents the first universal link between symmetries and integrals of motion. In order to demonstrate the theorem, we will take into account two different approaches, developed in Sections 2.1.1 and 2.1.2, respectively.

2.1.1 First Demonstration

Let us consider the following infinitesimal transformation, involving coordinates and fields, which acts on the Lagrangian density $\mathscr{L}(\phi^i, \partial_a\phi^i, x^a)$:

$$\begin{cases} \mathscr{L}(\phi^i, \partial_a\phi^i, x^a) \to \mathscr{L}(\tilde{\phi}^i, \partial_a\tilde{\phi}^i, \tilde{x}^a) \\ \tilde{x}^a = x^a + \delta x^a \\ \tilde{\phi}^i(\tilde{x}^a) = \phi^i(x^a) + \tilde{\delta}\phi^i(x^a). \end{cases} \tag{2.1}$$

We denote by $\tilde{\delta}$ a transformation that includes both coordinates and fields, and by δ a transformation involving the fields only, namely

$$\begin{cases} \tilde{\delta}\phi^i(x^a) = \tilde{\phi}^i(\tilde{x}^a) - \phi^i(x^a) \\ \delta\phi^i(x^a) = \tilde{\phi}^i(x^a) - \phi^i(x^a). \end{cases} \tag{2.2}$$

Substituting the third equation of (2.1) into the first of (2.2), up to first order, we get

$$\tilde{\delta}\phi^i(x^a) = \delta\phi^i(x^a) - \partial_b\phi^i(x^a)\delta x^b \;\to\; \tilde{\delta} = \delta - \delta x^a\partial_a. \tag{2.3}$$

Assuming that the transformation (2.1) leaves the **equations of motion** unchanged, the variational principle, in four dimensions, can be recast as

$$\tilde{\delta}S = 0 = \int \tilde{\delta}\mathscr{L}\, d^4x + \int \mathscr{L}\tilde{\delta}d^4x = \int \tilde{\delta}\mathscr{L}\, d^4x - \int \mathscr{L}\delta d^4x. \tag{2.4}$$

It is worth noticing that the invariance of EL equations is not directly addressed to the Lagrangian density, so that $\tilde{\delta}\mathscr{L} = 0$ turns out to be a subcase of that considered in Eq. (2.4). The second integral of the RHS can be computed by means of the relation

$$\delta d^4x = d^4\tilde{x} - d^4x = (|\partial_b \tilde{x}^a| - 1)\, d^4x = (|\delta^a_b + \partial_b \delta x^a| - 1)\, d^4x, \tag{2.5}$$

where the vertical straight lines denote the determinant. The latter can be made explicit by using two matrix properties. Given an n-dimensional matrix M with eigenvalues λ_i, the following relations hold:

$$\mathrm{Tr}\, M = \sum_i \lambda_i, \tag{2.6}$$

$$|M| = \prod_i \lambda_i. \tag{2.7}$$

By means of Eq. (2.6), it turns out that the condition $\mathrm{Tr}\, M \geq n|M|^{\frac{1}{n}}$ holds when n is strictly positive. In the first-order approximation, the preceding relation yields the equality $\mathrm{Tr}\, M = |M|$. On the other hand, it is worth noticing that the condition $\mathrm{Tr}\, M = |M|$ is also satisfied by any diagonal matrix with unitary elements. From these considerations, it follows that Eq. (2.5) can be recast as

$$\delta d^4x = \partial_a \delta x^a\, d^4x. \tag{2.8}$$

Using the relation (2.3), the Lagrangian variation occurring in the first integral of Eq. (2.4) takes the following form:

$$\begin{aligned}
\tilde{\delta}\mathscr{L} &= \frac{\partial \mathscr{L}}{\partial \phi^i}\tilde{\delta}\phi^i + \frac{\partial \mathscr{L}}{\partial(\partial_a \phi^i)}\partial_a \tilde{\delta}\phi^i + \partial_a \mathscr{L}\, \delta x^a \\
&= \left(\frac{\partial \mathscr{L}}{\partial \phi^i}\delta\phi^i + \frac{\partial \mathscr{L}}{\partial(\partial_a \phi^i)}\partial_a \delta\phi^i\right) - \left(\frac{\partial \mathscr{L}}{\partial \phi^i}\partial_a \phi^i\, \delta x^a + \frac{\partial \mathscr{L}}{\partial(\partial_a \phi^i)}\partial_b \partial_a \phi^i\, \delta x^b\right) \\
&\quad + \partial_a \mathscr{L} \delta x^a = \left(\partial_a \frac{\partial \mathscr{L}}{\partial(\partial_a \phi^i)}\delta\phi^i + \frac{\partial \mathscr{L}}{\partial(\partial_a \phi^i)}\partial_a \delta\phi^i\right) \\
&\quad - \left(\partial_a \frac{\partial \mathscr{L}}{\partial(\partial_a \phi^i)}\partial_b \phi + \frac{\partial \mathscr{L}}{\partial(\partial_a \phi^i)}\partial_a \partial_b \phi^i\right)\delta x^b + \partial_a \mathscr{L}\, \delta x^a \\
&= \partial_a \left(\frac{\partial \mathscr{L}}{\partial(\partial_a \phi^i)}\delta\phi^i\right) - \partial_a \left(\frac{\partial \mathscr{L}}{\partial(\partial_a \phi^i)}\partial_b \phi^i\right)\delta x^b + \partial_a \mathscr{L}\, \delta x^a.
\end{aligned} \tag{2.9}$$

By merging Eq. (2.9) and Eq. (2.8), one gets

$$\begin{aligned}
&\partial_a \left(\frac{\partial \mathscr{L}}{\partial(\partial_a \phi^i)}\delta\phi^i\right) - \partial_a \left(\frac{\partial \mathscr{L}}{\partial(\partial_a \phi^i)}\partial_b \phi^i\right)\delta x^b + \partial_a \mathscr{L}\, \delta x^a + \mathscr{L}\partial_a \delta x^a \\
&= \partial_a \left(\frac{\partial \mathscr{L}}{\partial(\partial_a \phi^i)}\delta\phi^l - \frac{\partial \mathscr{L}}{\partial(\partial_a \phi^i)}\partial_b \phi^i\, \delta x^b + \mathscr{L}\delta x^a\right) = \partial_a j^a = 0,
\end{aligned} \tag{2.10}$$

where j^a is the *Noether current*, defined as

$$j^a = \frac{\partial \mathscr{L}}{\partial(\partial_a \phi^i)}\delta\phi^i - \frac{\partial \mathscr{L}}{\partial(\partial_a \phi^i)}\partial_b \phi^i\, \delta x^b + \mathscr{L}\delta x^a - g^a, \tag{2.11}$$

with g^a being a vector field with vanishing divergence. This result shows that if the transformation (2.1) is a symmetry of the action, then there exists a set of conserved quantities, so that the initial statement is finally proven. Substituting Eq. (2.3) into Eq. (2.11), the Noether current can be equivalently written as

$$j^a = \frac{\partial \mathscr{L}}{\partial(\partial_a \phi^i)} \tilde{\delta}\phi^i + \mathscr{L}\delta x^a - g^a. \tag{2.12}$$

Equations (2.11) and (2.12) are completely equivalent; thus, the choice of which to use only relies on a matter of convenience.

2.1.2 Second Demonstration

Here we provide another proof for Noether's First Theorem. To this purpose, let us consider a transformation of the form

$$\begin{cases} \tilde{x}^a = x^a - \delta x^a \\ \tilde{\phi}^i = \phi^i - \delta\phi^i , \end{cases} \tag{2.13}$$

following which the Lagrangian density changes as $\mathscr{L}(x^a, \phi^i, \partial_a \phi^i) \to \mathscr{L}(\tilde{x}^a, \tilde{\phi}^i, \partial_a \tilde{\phi}^i)$. In order to find the generator of the transformation (2.13), we consider the transformation law for the prime derivatives of the fields ϕ^i:

$$\begin{aligned} \frac{d\tilde{\phi}^i}{d\tilde{x}^a} &= \frac{d\phi^i - \delta\phi^i}{dx^a - \delta x^a} = \left(\frac{d\phi^i}{dx^a} - \frac{d\delta\phi^i}{dx^a}\right)\left(1 - \frac{d\delta x^b}{dx^b}\right)^{-1} \\ &\sim \left(\frac{d\phi^i}{dx^a} - \frac{d\delta\phi^i}{dx^a}\right)\left(1 + \frac{d\delta x^b}{dx^b}\right), \end{aligned} \tag{2.14}$$

where the last equivalence holds up to the first order. Neglecting the higher-order terms, we get

$$\frac{d\tilde{\phi}^i}{d\tilde{x}^a} = \frac{d\phi^i}{dx^a} - \left(\frac{d\delta\phi^i}{dx^a} - \frac{d\phi^i}{dx^a}\frac{d\delta x^b}{dx^b}\right) = \partial_a\phi^i - \left(\partial_a\delta\phi^i - \partial_a\phi^i\partial_b\delta x^b\right). \tag{2.15}$$

From the preceding result, it follows that the generator of the transformation (2.13) involving coordinates, fields, and first derivatives has the form

$$X^{[1]} = \delta x^a \partial_a + \delta\phi^i \frac{\partial}{\partial\phi^i} + \left(\partial_a\delta\phi^i - \partial_a\phi^i\partial_b\delta x^b\right)\frac{\partial}{\partial(\partial_a\phi^i)}. \tag{2.16}$$

The symbol "[1]" denotes the *first prolongation* of the Noether vector, involving only first-order derivatives of fields. When canonical Lagrangians are considered, no further prolongation of the Noether vector is needed to find symmetries. For general systems, indeed, the procedure can be generalized by including the transformation rules of higher derivatives. In particular, for a generic Lagrangian of the form

$$\mathscr{L} \equiv \mathscr{L}(x^a, \phi^i, \partial_{a_1}\phi^i, \partial_{a_1 a_2}\phi^i \ldots \partial_{a_1 a_2 \ldots a_n}\phi^i), \tag{2.17}$$

the vector (2.16) can be further extended to

$$X^{[n]} = \delta x^a \partial_a + \delta\phi^i \frac{\partial}{\partial\phi^i} + \delta\phi^{i\,[1]}_{a_1} \frac{\partial}{\partial(\partial_{a_1}\phi^i)} + \ldots + \delta\phi^{i\,[n]}_{a_1\,a_2\ldots a_n} \frac{\partial}{\partial(\partial_{a_1\,a_2\ldots a_n}\phi^i)}, \tag{2.18}$$

with

$$\phi^{i\,[n]}_{a_1\,a_2\ldots a_n} = \partial_{[a_1}\delta\phi^{i\,[n-1]}_{a_2\,a_3\ldots a_{n-1}]} - \partial_b\delta x^b \partial_{a_1\,a_2\ldots a_n}\phi^i, \qquad \partial_a\delta\phi^{i\,[0]} = \partial\delta\phi^i. \tag{2.19}$$

In view of the applications of the Noether Theorem to canonical Lagrangians, there is no reason to further investigate higher prolongations; thus hereafter we mainly focus on the vector in Eq. (2.16).

If the transformations (2.13) and (2.15) leave the equations of motion invariant, then there exists a function $g^a = g^a(x^a, \phi^i)$ satisfying the condition

$$\tilde{\mathscr{L}} = \frac{d\tilde{x}^a}{dx^a}\mathscr{L} + \partial_a g^a, \tag{2.20}$$

which means that the starting Lagrangian and the transformed one differ only by a four-divergence. Using the complete set of transformations, the transformed Lagrangian density $\tilde{\mathscr{L}}$, up to the first order, can be written as

$$\tilde{\mathscr{L}}(\tilde{x}^a, \tilde{\phi}^i, \partial_a\tilde{\phi}^i) = \mathscr{L}(x^a, \phi^i, \partial_a\phi^i) + \partial_a\mathscr{L}\,\delta x^a + \frac{\partial\mathscr{L}}{\partial\phi^i}\delta\phi^i + \frac{\partial\mathscr{L}}{\partial(\partial_a\phi^i)}\delta(\partial_a\phi^i). \tag{2.21}$$

By means of the relation (2.15), namely $\delta(\partial_a\phi) = -\partial_a\delta\phi^i + \partial_a\phi^i\partial_b\delta x^b$, the preceding equation becomes

$$X^{[1]} = \delta x^a \partial_a + \delta\phi^i \frac{\partial}{\partial\phi^i} + (\partial_a\delta\phi^i - \partial_a\phi^i\partial_b\delta x^b)\frac{\partial}{\partial(\partial_a\phi^i)}. \tag{2.22}$$

In addition, substituting the identity $\dfrac{d\tilde{x}^a}{dx^a} = 1 - \partial_a\delta x^a$ in Eq. (2.22), the condition of invariance of EL equations finally reads

$$\left[\delta x^a \partial_a + \delta\phi^i \frac{\partial}{\partial\phi^i} + (\partial_a\delta\phi^i - \partial_a\phi^i\partial_b\delta x^b)\frac{\partial}{\partial(\partial_a\phi^i)}\right]\mathscr{L} + \partial_a\delta x^a\mathscr{L} = \partial_a g^a, \tag{2.23}$$

which, by means of Eq. (2.16), can be recast as

$$X^{[1]}\mathscr{L} + \partial_a\delta x^a\mathscr{L} = \partial_a g^a. \tag{2.24}$$

Summing up, we showed that if the set of transformations (2.13) is a symmetry for the equations of motion, then the condition (2.24) must be satisfied. In order to find the Noether current associated to the symmetry transformation, notice that the partial derivative of the Lagrangian density $(\partial_a \mathscr{L})$ can be written in terms of its total derivative $\left(\frac{d\mathscr{L}}{dx^a}\right)$ as

$$\delta x^a \partial_a \mathscr{L} = \frac{d\mathscr{L}}{dx^a}\delta x^a - \frac{\partial \mathscr{L}}{\partial \phi^i}\partial_a \phi^i \,\delta x^a - \frac{\partial \mathscr{L}}{\partial(\partial_b \phi^i)}\partial_b \partial_b \phi^i \,\delta x^a. \tag{2.25}$$

Moreover, with the help of the two identities

$$\begin{cases} \partial_a \delta\phi^i \dfrac{\partial \mathscr{L}}{\partial(\partial_a \phi^i)} = \partial_a \left(\dfrac{\partial \mathscr{L}}{\partial(\partial_a \phi^i)}\delta\phi^i\right) - \delta\phi^i \partial_a \left(\dfrac{\partial \mathscr{L}}{\partial(\partial_a \phi^i)}\right) \\ \\ \partial_a \delta x^a \dfrac{\partial \mathscr{L}}{\partial(\partial_b \phi^i)}\partial_b \phi^i = \partial_a \left(\dfrac{\partial \mathscr{L}}{\partial(\partial_b \phi^i)}\partial_b \phi^i \delta x^a\right) - \partial_a \left(\dfrac{\partial \mathscr{L}}{\partial(\partial_b \phi^i)}\partial_b \phi^i\right)\delta x^a, \end{cases} \tag{2.26}$$

we can plug Eqs. (2.25) and (2.26) into Eq. (2.24) to finally obtain

$$\partial_a \left[\frac{\partial \mathscr{L}}{\partial(\partial_a \phi^i)}\delta\phi^i - \frac{\partial \mathscr{L}}{\partial(\partial_a \phi^i)}\partial_b \phi^i \,\delta x^b + \mathscr{L}\delta x^a - g^a\right] = 0. \tag{2.27}$$

Consequently, the term in the square bracket is a conserved quantity. Specifically, by defining

$$j^a = \frac{\partial \mathscr{L}}{\partial(\partial_a \phi^i)}\delta\phi^i - \frac{\partial \mathscr{L}}{\partial(\partial_a \phi^i)}\partial_b \phi^i \,\delta x^b + \mathscr{L}\delta x^a - g^a, \tag{2.28}$$

it turns out that Eq. (2.28) is the same current previously obtained in Eq. (2.12). Thus the Noether Theorem can be equivalently recast in terms of the first prolongation of the Noether vector as follows:

If the condition $X^{[1]}\mathscr{L}(a, \phi^i, \partial_a \phi^i) + \delta x^a \mathscr{L}(a, \phi^i, \partial_a \phi^i) - \partial_a g^a(x^a, \phi^i) = 0$ *holds, then the quantity* $j^a = \dfrac{\partial \mathscr{L}}{\partial(\partial_a \phi^i)}\delta\phi^i - \dfrac{\partial \mathscr{L}}{\partial(\partial_a \phi^i)}\partial_b \phi^i \,\delta x^b + \mathscr{L}\delta x^a - g^a$ *is a first integral of the equations of motion.*

Finally, integrating $\partial_a j^a$ over the four-volume Ω and using the divergence theorem, it is possible to introduce the so-called *Noether charge* Q:

$$\int_\Omega \partial_a j^a \, d\Omega = \int_\Sigma j^a \, d\Sigma_a \equiv Q, \tag{2.29}$$

with Σ_a being the hypersurface. The Noether charge, defined as the flow of j^a, is a scalar quantity under symmetry transformations. For instance, when $U(1)$ invariance of the electromagnetic field is considered, Q represents the electric charge, as pointed out in Section 3.3.2.

2.1.3 Internal Symmetries

Let us investigate now a particular subcase of the Noether Theorem. We assume that the symmetry transformation is an on-shell internal symmetry, which does not involve the variation of the space-time coordinates. Thus the transformations (2.13) simply reduce to

$$\begin{cases} \delta x^a = 0 \\ \delta\phi^a = \tilde{\phi}^i - \phi^i \\ \tilde{\mathscr{L}}(\tilde{x}^a, \tilde{\phi}^i, \partial_a\tilde{\phi}^i) = \mathscr{L}(x^a, \phi^i, \partial_a\phi^i), \end{cases} \tag{2.30}$$

so that the Lagrangian variation takes the form

$$\delta\mathscr{L} = \frac{\partial\mathscr{L}}{\partial\phi^i}\delta\phi^i + \frac{\partial\mathscr{L}}{\partial(\partial_a\phi^i)}\delta\partial_a\phi^i = \partial_a\left(\frac{\partial\mathscr{L}}{\partial(\partial_a\phi^i)}\delta\phi^i\right) = 0. \tag{2.31}$$

Under this assumption, the Noether current and the corresponding Noether vector can be simplified as

$$\begin{aligned} j^a &= \frac{\partial\mathscr{L}}{\partial(\partial_a\phi^i)}\delta\phi^i, \\ X &= \frac{\partial}{\partial\phi^i}\delta\phi^i + \frac{\partial}{\partial(\partial_a\phi^i)}\delta\partial_a\phi^i. \end{aligned} \tag{2.32}$$

Though this is just a simplification of the general theorem, it yields a suitable relation for methodically introducing a cyclic variable in the Lagrangian system. The purpose is to find a new variable ψ^1 such that its conjugate momentum is a constant of motion, namely

$$\frac{\partial\mathscr{L}}{\partial(\partial_a\psi^1)} = \pi^a_{\psi^1} = \text{const.} \tag{2.33}$$

The first equation of (2.32) suggests that such a constant can be addressed to the Noether current only if the infinitesimal generator of ψ^1 is of the order of unity.

Therefore, let us consider a general change of variables $\phi^i \to \psi^i(\phi^j)$, such that ψ^1 is cyclic. The Noether vector X can be written in terms of the new set of variables ψ^i by means of the inner derivative i_X. Specifically, the infinitesimal generators of the old variables ϕ^i can be linked to those of the new variables ψ^i as

$$\delta\phi^i\frac{\partial}{\partial\phi^i} = \delta\phi^i\frac{\partial\psi^j}{\partial\phi^i}\frac{\partial}{\partial\psi^j} = i_X d\psi^j\frac{\partial}{\partial\psi^j}. \tag{2.34}$$

The preceding equation allows one to easily write the new Noether vector X' as

$$X' = \delta\phi^i\frac{\partial}{\partial\phi^i} + \partial_a\delta\phi^i\frac{\partial}{\partial(\partial_a\phi^i)} = (i_X d\psi^k)\frac{\partial}{\partial\psi^k} + \partial_a(i_X d\psi^k)\frac{\partial}{\partial(\partial_a\psi^k)}. \tag{2.35}$$

As a consequence, the conserved quantity j^a can be recast in terms of the fields ψ^i:

$$j^a = \delta\psi^i\frac{\partial\mathscr{L}}{\partial(\partial_a\psi^i)} = i_X d\psi^i\frac{\partial\mathscr{L}}{\partial(\partial_a\psi^i)}. \tag{2.36}$$

Requiring the conserved quantity to be equal to the conjugate momentum of ψ^1, the following conditions must be satisfied:

$$i_X d\psi^1 = \delta\phi^j \frac{\partial \psi^1}{\partial \phi^j} = 1, \qquad i_X d\psi^i = \delta\phi^j \frac{\partial \psi^i}{\partial \phi^j} = 0, \qquad i \neq 1, \tag{2.37}$$

so that, from Eq. (2.36), one gets

$$j^a = \delta\phi^i \frac{\partial \mathscr{L}}{\partial(\partial_a \phi^i)} = \frac{\partial \mathscr{L}}{\partial(\partial_a \psi^1)} \quad \to \quad \pi^a_{\psi^1} = \textbf{constant}. \tag{2.38}$$

Therefore the conjugate momentum $\pi^a_{\psi^1}$ of the new cyclic variable ψ^1 turns exactly into the Noether current, as expected by construction.

Notice that the condition $X'\mathscr{L}' = X\mathscr{L} = 0$ holds regardless of the variables considered, so that Noether symmetry is preserved under the change of variables. Nonetheless, it is worth remarking that the change of variables in Eq. (2.37) is not unique, but infinite possible field transformations can occur. Therefore, in order to reduce the dynamics, new variables must be chosen carefully.

2.2 Noether's Second Theorem

Despite its formal elegance, Noether's Second Theorem has never caught the full attention of physicists and mathematicians, probably due to the lack of a considerable amount of applications. In this section, for the sake of completeness, we want to briefly enunciate and demonstrate such a theorem, although it will not be used in the rest of the book. It formally relates symmetries of a given action with symmetries of the equations of motion and can be formulated as follows:

Let S be an action having an infinite-dimensional Lie algebra of infinitesimal symmetries linearly parameterized by m arbitrary functions $f^\kappa(x^a)$ $(\kappa = 0, 1, 2, ...)$ and let $\Gamma^m_\kappa \equiv \Gamma^m_\kappa(x^a, \phi, \partial_a\phi)$, $\zeta^{m\,a}_\kappa \equiv \zeta^{m\,a}_\kappa(x^a, \phi, \partial_a\phi)$ be two generic functions depending on the considered transformations. Then, there exists a set of functions Φ_m satisfying the following relations:

$$\Phi_m \Gamma^m_\kappa = \partial_a \left\{ \Phi_m \zeta^{m\,a}_\kappa \right\}. \tag{2.39}$$

The proof of the theorem is straightforward: since Γ^m_κ and $\zeta^{m\,a}_\kappa$ depend on the field ϕ, the decomposition of the latter in terms of the functions f^κ can be considered. In this way, and up to the first order, the field variation reads

$$\delta\phi^m = \Gamma^m_\kappa \delta f^\kappa + \zeta^{m\,a}_\kappa \partial_a(\delta f^\kappa). \tag{2.40}$$

Furthermore, recasting the quantity $\Phi_m \zeta^{m\,a}_\kappa \partial_a(\delta f^\kappa)$ as

$$\Phi_m \zeta^{m\,a}_\kappa \partial_a(\delta f^\kappa) = \partial_a \left(\Phi_m \zeta^{m\,a}_\kappa \delta f^\kappa \right) - \partial_a \left(\Phi_m \zeta^{m\,a}_\kappa \right) \delta f^\kappa, \tag{2.41}$$

and substituting Eq. (2.41) into Eq. (2.40), we obtain

$$\Phi_m \delta\phi^m = \Phi_m \Gamma^m_\kappa \delta f^\kappa + \partial_a \left(\Phi_m \zeta^{m\,a}_\kappa \delta f^\kappa \right) - \partial_a \left(\Phi_m \zeta^{m\,a}_\kappa \right) \delta f^\kappa. \tag{2.42}$$

Therefore, the action $S[\Phi(\phi), \mathscr{L}(x^a, \phi, \partial_a\phi]$ can be varied with respect to the field ϕ, providing

$$\delta S = \int \left(\Phi_m \delta\phi^n + \delta\mathcal{L}\right) d^4x. \tag{2.43}$$

As suggested by Noether's First Theorem, the variation of the Lagrangian for off-shell symmetries turns into a boundary term; thus the only remaining quantity in Eq. (2.43) can be written as a pure divergence:

$$\Phi_m \delta\phi^m = \partial_a g^a. \tag{2.44}$$

Substituting Eq. (2.44) into Eq. (2.42), the variational principle yields

$$\int \left[\Phi_m \Gamma^m_\kappa - \partial_a \left(\Phi_m \zeta^{m\,a}_{\kappa}\right)\right] \delta f^\kappa \, d^4x = \int \partial_a \left(g^a - \Phi_m \zeta^{m\,a}_{\kappa} \delta f^\kappa\right) d^4x. \tag{2.45}$$

Note that the quantity in the RHS is a surface term, which vanishes at the boundary. Moreover, the arbitrary functions f^κ can be chosen to satisfy the following condition:

$$\Phi_m \Gamma^m_\kappa - \partial_a \left(\Phi_m \zeta^{m\,a}_{\kappa}\right) = 0. \tag{2.46}$$

The relation (2.46) is exactly the same as that in Eq. (2.39) and finally proves Noether's Second Theorem. In order to discuss the physical applications, it is possible to find a relation between the two theorems. To this purpose, let us assume that the functions that parameterize the action are independent of the space-time coordinates. In such a case, the relation (2.40) takes the simpler form,

$$\delta\phi^m = \Gamma^m_\kappa \delta f^\kappa, \tag{2.47}$$

and, once substituted into Eq. (2.44), yields

$$\Phi_m \Gamma^m_\kappa \delta f^\kappa = \partial_a g^a. \tag{2.48}$$

Finally, taking into account Noether's Second Theorem (2.39) and considering that the functions f_κ are independent of x^a, we get

$$\partial_a \left\{\Phi_m \zeta^{m\,a}_{\kappa} \delta f^\kappa - g^a\right\} = 0. \tag{2.49}$$

By identifying Φ_m as a set of m EL equations, Eq. (2.49) basically means that the term in curly brackets is a conserved quantity. Notice that the latter is obtained without assuming the EL equations to be valid a priori. It means that the existence of a set of EL equations automatically guarantees the existence of a conserved quantity, without any further assumption. From this point of view, as the first theorem provides the explicit form of the Noether current, so the second theorem states that such a current is related to the existence of a set of equations of motion. Furthermore it is also worth noticing that the second theorem only relies on general improper transformations, which do not require the existence of a set of field equations to hold. A straightforward example of improper conserved quantity is given by the energy-momentum tensor, whose conservation is satisfied regardless of whether the Klein–Gordon equations of additional scalar fields are satisfied or not.

2.3 The Lie Derivative: Applications and Beyond

In this section we overview the main properties of the Lie derivative, with the aim to find a link between it and the Noether Theorem. In this regard, we show that imposing $\delta\mathscr{L} = 0$ implicitly relies on the application of the Lie derivative along the flux of the Noether vector X. More precisely, the Noether Theorem can be recast in the condition $L_X\mathscr{L} = 0$. This imposition yields a set of conserved quantities that close on the same algebra as the generators, but which undergo different commutation relations.

Let us consider a general linear Lie group $GL(n, \mathbb{R})$, whose representation $\mathcal{R}$ consists of a set of matrices $D_{\mathcal{R}}(\theta)$ depending on elements θ_j where θ_0 is the element that gives the identity. The following properties hold:

$$\begin{cases} D_{\mathcal{R}}(\theta_0) = 1 \\ D_{\mathcal{R}}(\theta_1)D_{\mathcal{R}}(\theta_2) = D_{\mathcal{R}}(\theta_1\theta_2). \end{cases} \tag{2.50}$$

By defining $-i\left.\dfrac{\partial D_{\mathcal{R}}}{\partial \theta_a}\right|_{\theta=0}\theta_a = X^a_{\mathcal{R}}$, the matrices $D_{\mathcal{R}}(\theta)$ can be written as

$$D_{\mathcal{R}}(\theta) = e^{i\theta_a X^a_{\mathcal{R}}}. \tag{2.51}$$

By means of the second equation of (2.50), we obtain

$$e^{i\beta_a X^a_{\mathcal{R}}}e^{i\gamma_a X^a_{\mathcal{R}}} = e^{i\delta_a X^a_{\mathcal{R}}}, \tag{2.52}$$

where, in general, $\delta_a \neq \beta_a + \gamma_a$, as the equality only holds for Abelian groups. Equation (2.52) can be expanded up to the second order, providing

$$i\delta_a X^a_{\mathcal{R}} \sim \ln\left\{\left[1 + i\beta_a X^a_{\mathcal{R}} + \frac{1}{2}\left(i\beta_a X^a_{\mathcal{R}}\right)^2\right]\left[1 + i\gamma_a X^a_{\mathcal{R}} + \frac{1}{2}\left(i\gamma_a X^a_{\mathcal{R}}\right)^2\right]\right\}, \tag{2.53}$$

from which we finally obtain

$$i\delta_a X^a_{\mathcal{R}} = \beta_b\gamma_c\left[X^b_{\mathcal{R}}, X^c_{\mathcal{R}}\right] \quad \rightarrow \quad \left[X^b_{\mathcal{R}}, X^c_{\mathcal{R}}\right] = if^{bc}_a X^a_{\mathcal{R}}, \tag{2.54}$$

with $\beta_b\gamma_c f^{bc}_a \equiv \delta_a$. If X represents the generator of a certain symmetry, then the conserved quantities close on the same algebra as the generators, namely

$$\begin{cases} \left[X^i_{\mathcal{N}}, X^j_{\mathcal{N}}\right] = if^{ij}_k X^k_{\mathcal{N}} \\ \left\{\Sigma^i_{\mathcal{N}}, \Sigma^j_{\mathcal{N}}\right\} = if^{ij}_k \Sigma^k_{\mathcal{N}}, \end{cases} \tag{2.55}$$

where the curly brackets are Poisson's brackets. In Section 2.1.3 we showed that internal symmetries of the starting Lagrangian density can be selected by the condition

$$\frac{\partial \mathscr{L}}{\partial \phi^i}\delta\phi^i + \frac{\partial \mathscr{L}}{\partial(\partial_a\phi^i)}\delta\partial_a\phi^i, \tag{2.56}$$

so that the quantity

$$j^a = \frac{\partial \mathscr{L}}{\partial(\partial_a\phi^i)}\delta\phi^i \tag{2.57}$$

is a constant of motion. Equations (2.56) and (2.57) occur after considering a transformation that leaves the Lagrangian invariant ($\delta\mathscr{L} = 0$). The internal symmetry condition $X\mathscr{L} = 0$ can be written in terms of the vanishing Lie derivative of the given Lagrangian along the flux of X, that is,

$$L_X\mathscr{L} = [X, \mathscr{L}] = X\mathscr{L} = 0 \;\; \rightarrow \;\; \Sigma = \text{Noether Charge}. \tag{2.58}$$

To outline the properties of the Lie derivative, let us introduce two functions f, g, such that $f, g \in C^\infty(\mathcal{M})$, three vector fields X, Y, Z, and two differential forms ω, u. In light of these definitions, the Lie derivative must respect the following six conditions:

- $L_X(fg) = (L_Xf)\, g + f(L_Xg)$
- $L_Xf = Xf$
- $L_X[Y, Z] = [L_XY, Z] + [Y, L_XZ]$
- $L_XY = adX(Y) = [X, Y]$
- $L_X(\omega \wedge u) = (L_X\omega) \wedge u + \omega \wedge (L_Xu)$
- $L_Xd\omega = dL_X\omega$

with d being the exterior derivative and ad the self-adjoint operator. Summing up, the preceding conditions suggest that the identity $L_X\mathscr{L} = 0$ implies that the phase flux along the vector field X is conserved and internal symmetries arise. This condition can also be extended to the first prolongation of the Noether vector, occurring when external symmetries are considered. To this purpose, the application of the Noether identity (2.24) to a given Lagrangian density can be recast in terms of the Lie derivative as

$$\begin{aligned}
&X^{[1]}\mathscr{L} + \partial_a\delta x^a\mathscr{L} = \partial_a g^a \\
&\rightarrow \left\{\delta x^a\partial_a + \delta\phi^i\frac{\partial}{\partial\phi^i} + (\partial_a\delta\phi^i - \partial_a\phi^i\partial_b\delta x^b)\frac{\partial}{\partial(\partial_a\phi^i)} + \partial_a\delta x^a\right\}\mathscr{L} = \partial_a g^a \\
&\rightarrow \left\{L_X - \partial_b\delta x^b\left(\partial_a\phi^i\frac{\partial}{\partial(\partial_a\phi^i)} - 1\right)\right\}\mathscr{L} = \partial_a g^a \\
&\rightarrow L_X\mathscr{L} - \partial_b\delta x^b\mathscr{T} = \partial_a g^a,
\end{aligned} \tag{2.59}$$

with $\mathscr{T}$ being the trace of the density of the energy-momentum tensor. (See the definition in Eq. (3.18).) The importance of this relation will be clear when the application to canonical Lagrangians is considered. In fact, for time-dependent scalar fields, Eq. (2.59) takes the following form:

$$L_X\mathcal{L} - \dot{\xi}\mathcal{H} = \dot{g}, \tag{2.60}$$

where $\mathcal{H}$ is the Hamiltonian and where we set $\xi = \delta t$. Note that the function g can always be chosen to be constant by means of an appropriate gauge choice. Furthermore, as soon as $\dot{\xi} = \xi_0$, Eq. (2.60) can be written as $L_X\mathcal{L} \sim$ const, and internal symmetries are recovered when const $= 0$.

The Lie derivative can also be used to find the symmetries of the metric tensor. Indeed, it is possible to show that the application of L_X to the metric tensor, over the flux of any conformal Killing vector X, yields the following relation:

$$L_X g_{ab} = 2g_{ab}\phi(x^a). \tag{2.61}$$

Conformal Killing vectors can be split into three main categories depending on the properties of the field ϕ:

- *Proper* Killing Vector $\to \phi(x^a) \neq 0 \to L_X g_{ab} = 2g_{ab}\phi(x^a)$
- *Special* Killing Vector $\to \phi(x^a) = 0 \to L_X g_{ab} = 0$
- *Homotetic* Killing Vector $\to \partial_a\phi(x^a) = 0 \to L_X g_{ab} \sim g_{ab}$.

For instance, when special Killing vectors are considered, Eq. (2.61) can be written as

$$L_X g_{ab} = [X, g_{ab}] = 0 \quad \to \quad D_b X_a + D_a X_b = 0. \tag{2.62}$$

As we will discuss, this property defines isometries on a given manifold.

To conclude, we show how the Lie derivative can be used to find symmetries of a given Lagrangian. Specifically, according to the result provided in Eq. (2.59), it is possible to classify the Noether symmetries into three main groups:

- $L_X\mathscr{L} = \psi(x^a, \phi^i, \partial_a\phi^i) = \partial_b\delta x^b\mathscr{T} + \partial_a g^a \to$ General Noether Symmetry
- $L_X\mathscr{L} =$ const $\to$ Noether Symmetry for canonical Lagrangians
- $L_X\mathscr{L} = 0 \to$ Internal Noether Symmetry

The second condition generally occurs for canonical Lagrangians. This is due to the fact that, when the Lagrangian is canonical, the infinitesimal generator of time translation does not depend on the configuration space variables. Moreover, though not always, the infinitesimal generator of symmetry transformations is often linearly dependent on the time.

In this section, we showed that Killing vectors and conserved quantities obey the same rules imposed by the Lie derivative. Although in the rest of the book we will mainly focus on the conserved quantities, the application of the Lie derivative to the metric tensor can be used as a selection criterion to find symmetries, as shown, for example, in [12,13].

2.4 The Noether–Bessel-Hagen Theorem: Symmetries of Equations of Motion

The main purpose of this section is to analyze Noether-type symmetries of dynamical (e.g. mechanical) systems with *external forces*, which leave the corresponding equations of motion *invariant*. These kinds of symmetries are known as the *Noether–Bessel-Hagen symmetries* [14–16]. We develop a method aimed at finding Noether–Bessel-Hagen symmetries, which can also be applied to those systems without any Lagrangian description, called *not-variational*, where the standard Noether's First Theorem cannot work. The application to general mechanical systems is motivated by the fact that they are usually not-variational.

The main idea is to solve the *Noether–Bessel-Hagen equation* of Killing type

$$L_X \varepsilon = 0, \tag{2.63}$$

with respect to the unknown vector field X, with ε being the source form, whose components are

$$\varepsilon_i = E_i(\mathcal{L}) - \varphi_i.$$

Here φ stands for the external force, and $E_i(\mathcal{L})$ are the EL expressions of the mechanical Lagrange function $\mathcal{L} = \mathcal{T} - \mathcal{U}$, with $\mathcal{U}$ being the potential.

This prescription allows one to search for mechanical system potentials arising from symmetry requirements as well as to identify the symmetry once given the equations of motion.

In what follows, we summarize the basic definitions of invariant differential forms on jet prolongations of a fibered manifold over a one-dimensional base (*fibered mechanics*). We restrict the analysis to second-order equations, coming from first-order Lagrangians, but the case of higher-order field equations can be found, for example, in [15, 17]. The main reference paper we follow is [18], whose geometric approach and Noether invariant variational structure are in turn based on [17,19–24]. Closely related works that use an alternative approach are [25,26], which study symmetries of dissipative systems by means of variational multipliers. This is called the *Neutral Action Method.*

Let us focus now on symmetries of mechanical systems with external forces, introduced as a 1-form on the tangent bundle TM [27]. In order to work with a global chart, the configuration space is considered to be the second jet prolongation J^2Y of a product fibered manifold $Y = \mathbb{R} \times M$, where M is an open subset of Euclidean space $\mathbb{R}^m$. Jet spaces are then canonically identified with products, namely $J^1Y \cong \mathbb{R} \times TM$, $J^2Y \cong \mathbb{R} \times T^2M$, where TM is the tangent space over M, endowed with global coordinates $(q^i, \dot{q}^i)$, and T^2M is the bundle of velocities of order 2 over M, endowed with global coordinates $(q^i, \dot{q}^i, \ddot{q}^i)$. More details of variational geometric structures can be found in [15, 28–30].

Let $\pi;\ Y \to X$ be a fibered manifold with $\dim X = 1$, so that J^1Y and J^2Y represent the first and the second jet prolongations, respectively. If (V, ψ), $\psi = (t, q^i)$ is a fibered chart on Y, the associated chart on J^1Y (respectively, J^2Y) reads (V^1, ψ^1), $\psi^1 = (t, q^i, \dot{q}^i)$ (respectively, (V^2, ψ^2), $\psi^2 = (t, q^i, \dot{q}^i, \ddot{q}^i)$).

Let

$$\lambda = \mathcal{L}\,dt \tag{2.64}$$

be the 1-form Lagrangian of order 1 for Y defined as a π^1-horizontal 1-form on an open subset $W^1 \subset J^1Y$. The associated EL form E_λ, which is a 1-contact 2-form on $W^2 \subset J^2Y$, is locally expressed as

$$E_\lambda = E_i(\mathcal{L})\omega^i \wedge dt, \tag{2.65}$$

where its coefficients $E_i(\mathcal{L})$ are the EL expressions

$$E_i(\mathcal{L}) = \frac{\partial \mathcal{L}}{\partial q^i} - \frac{d}{dt}\frac{\partial \mathcal{L}}{\partial \dot{q}^i} = \frac{\partial \mathcal{L}}{\partial q^i} - \frac{\partial^2 \mathcal{L}}{\partial t \partial \dot{q}^i} - \frac{\partial^2 \mathcal{L}}{\partial q^j \partial \dot{q}^i}\dot{q}^j - \frac{\partial^2 \mathcal{L}}{\partial \dot{q}^j \partial \dot{q}^i}\ddot{q}^j \tag{2.66}$$

and ω^i are contact 1-forms $\omega^i = dq^i - \dot{q}^i dt$. In Eq. (2.65), the symbol $\wedge$ denotes the wedge product (see App. B for details). It is worth remarking that the EL form (2.65) accounts for a particular source form ε, which is by definition a 1-contact, $\pi^{2,0}$-horizontal 2-form on $W^2 \subset J^2Y$.

For a π-projectable vector field X on $W \subset Y$, and for any section γ of $\pi : Y \to X$ with values in W, we have the *infinitesimal first variation formula*

$$J^1\gamma^* L_{J^1X}\lambda = J^2\gamma^* i_{J^2X}E_\lambda + d\left(J^1\gamma^* i_{J^1X}\Theta_\lambda\right), \tag{2.67}$$

where Θ represents the Cartan form of the first-order Lagrangian λ expressed by

$$\Theta_\lambda = \mathcal{L}\,dt + \frac{\partial \mathcal{L}}{\partial \dot{q}^i}\omega^i. \tag{2.68}$$

A diffeomorphism $\alpha : W \to Y$ is called an *invariance transformation* of λ with respect to ε, if

$$(J^r\alpha)^*\lambda = \lambda, \quad \text{resp.} \quad (J^r\alpha)^*\varepsilon = \varepsilon, \tag{2.69}$$

with $J^r\alpha : W^r \to J^rY$ being the r-jet prolongation of α. Thus, a π-projectable vector field X on Y is called a *generator of invariance transformations* of λ with respect to ε, if its local one-parameter group α_t^X consists of invariance transformations of λ with respect to ε.

The generators of invariance transformations of λ with respect to ε form a subalgebra of the algebra of vector fields on $W \subset Y$. Once the basic structures in this framework are obtained, Noether's First Theorem can be expressed as follows:

Let $\lambda \in \Omega^1_{1,X}W$ be a Lagrangian of order 1 *for Y, and let γ be an extremal for λ. Then for every generator X of invariance transformations of λ*

$$d\left(J^1\gamma^* i_{J^1X}\Theta_\lambda\right) = 0, \tag{2.70}$$

where Θ_λ *is the Cartan form (2.68) of* λ.

It also implies that if X is a generator of invariant transformations of λ, and a section γ of Y satisfies the conservation law equation, then the EL expressions of λ are linearly dependent along γ.

Now, if ε coincides with the EL form E_λ for some Lagrangian $\lambda \in \Omega^r_{1,X}W$, for any diffeomorphism $\alpha : W \to Y$, the EL form E_λ obeys the formula

$$J^{2r}\alpha^* E_\lambda = E_{J^r\alpha^*\lambda} \tag{2.71}$$

and for any vector field X on M

$$L_{J^2X}E_\lambda = E_{L_{J^1X}\lambda}. \tag{2.72}$$

As a consequence, every invariance transformation of λ is an invariance transformation of E_λ. Noether's First Theorem can be extended by considering invariant transformations of the EL equations E_λ as:

Let $\lambda \in \Omega^1_{1,X}W$ *be a Lagrangian of order* 1 *for* Y, *let* γ *be an extremal for* λ, *and let* X *be a generator of invariance transformations of the EL form* E_λ. *Then there exists a function* f *on* $W \subset Y$ *such that*

$$d\left(J^1\gamma^*(i_{J^1X}\Theta_\lambda + f)\right) = 0. \tag{2.73}$$

Note that the preceding theorem only holds for first-order Lagrangians. For second and higher-order Lagrangians, Eq. (2.73) provides *local* conservation laws.

The theorem can be applied to any first-order Lagrangian, but for the sake of simplicity, here we focus on mechanical Lagrangians given by linear combinations of a kinetic and a potential term as

$$\mathcal{L} = \mathcal{T} - \mathcal{U}. \tag{2.74}$$

The kinetic energy $\mathcal{T}$ can be seen as a real-valued function defined on TM by the formula

$$\mathcal{T} = \frac{1}{2} g_{ij}\dot{q}^i\dot{q}^j, \tag{2.75}$$

where g_{ij} is a metric tensor on M. The potential energy $\mathcal{U}$ is a real-valued function defined on the configuration space M, that is, $\mathcal{U} = \mathcal{U}(q^i)$, and describes properties of the mechanical system. The EL expressions, associated to $\mathcal{L}$, are functions on T^2M, defined as

$$E_i(\mathcal{L}) = -\frac{\partial \mathcal{L}}{\partial q^i} + \frac{d}{dt}\frac{\partial \mathcal{L}}{\partial \dot{q}^i}. \tag{2.76}$$

Substituting (2.74) into (2.76), we get

$$\begin{aligned} E_i(\mathcal{L}) &= E_i(\mathcal{T}) + \frac{\partial \mathcal{U}}{\partial q^i} = -\frac{\partial \mathcal{T}}{\partial q^i} + \frac{d}{dt}\frac{\partial \mathcal{T}}{\partial \dot{q}^i} + \frac{\partial \mathcal{U}}{\partial q^i} \\ &= \frac{1}{2}\left(\frac{\partial g_{ij}}{\partial q^k} + \frac{\partial g_{ik}}{\partial q^j} - \frac{\partial g_{jk}}{\partial q^i}\right)\dot{q}^j\dot{q}^k + g_{ij}\ddot{q}^j + \frac{\partial \mathcal{U}}{\partial q^i} \\ &= g_{ij}\left(\ddot{q}^j + \Gamma^j_{ks}\dot{q}^k\dot{q}^s\right) + \frac{\partial \mathcal{U}}{\partial q^i}, \end{aligned}$$

where Γ^j_{ks} are the Christoffel Symbols associated to g_{ij}. Since g_{ij} is a non-singular matrix of the form

$$g_{ij} = \frac{\partial^2 \mathcal{L}}{\partial \dot{q}^i \partial \dot{q}^j},$$

the Lagrange function $\mathcal{L}$ is regular. Hence the conservation law of energy along extremals reads

$$\frac{1}{2} g_{ij}\dot{q}^i\dot{q}^j + \mathcal{U} = 0. \tag{2.77}$$

Consider now equations of motion $E_i(\mathcal{L})$ under the influence of external forces $\varphi_i = \varphi_i(q^j, \dot{q}^j)$ with $\varepsilon = \varepsilon_i \omega^i \wedge dt$ a dynamical form on $\mathbb{R} \times T^2M$, where

$$\varepsilon_i = E_i(\mathcal{L}) - \varphi_i. \tag{2.78}$$

We call ε the source form associated with the Lagrangian $\lambda = \mathcal{L}dt$.

This definition allows one to formulate the main theorem of this section, describing the symmetries of the source form ε. More precisely, we set out to find conditions on a vector field X on M, a potential energy function $\mathcal{U} = \mathcal{U}(q^i)$, and force $\varphi = (\varphi_i)$, such that the *Noether–Bessel-Hagen equation*,

$$L_{J^2X}\varepsilon = 0, \tag{2.79}$$

holds. The theorem states that, given the source form ε and the force φ previously defined, the following conditions are equivalent:

(a) A vector field X on M is a generator of invariant transformations of ε, that is, the Noether–Bessel-Hagen equation (2.79) is satisfied identically.

(b) X, $\mathcal{U}$, and (φ_i) satisfy the following condition:

$$L_X g = 0, \tag{2.80}$$

and, for every i, it is

$$\frac{\partial}{\partial q^i}(L_X\mathcal{U}) + (L_{J^1X}\varphi)_i = 0, \tag{2.81}$$

where

$$(L_{J^1X}\varphi)_i = \frac{\partial X^j}{\partial q^i}\varphi_j + \frac{\partial \varphi_i}{\partial q^j}X^j + \frac{\partial \varphi_i}{\partial \dot{q}^j}\frac{\partial X^j}{\partial q^k}\dot{q}^k \tag{2.82}$$

represents the Lie derivative $L_{J^1X}\varphi$ of external force 1-form $\varphi = (\varphi_i)$ w.r.t. J^1X, and $L_X\mathscr{U}$ is the Lie derivative of function $\mathscr{U}$ w.r.t. X.

We do not prove the theorem here, but for further readings see [18] and references therein. Reducing the form of external force, we obtain useful simplifications of the theorem. For instance, let us distinguish five main cases:

I. φ is identically zero, that is, the source ε coincides with the EL equations.

II. φ is conservative, that is, its components φ_i are of the form

$$\varphi_i = -\frac{\partial \mathscr{U}}{\partial q^i}. \tag{2.83}$$

III. φ is defined on M, that is, its components φ_i are functions of q^j variables only.

IV. φ is dissipative, that is, its components φ_i are of the form

$$\varphi_i = -\varphi_{ij}\dot{q}^j, \tag{2.84}$$

where $\varphi_{ij} : M \to \mathbb{R}$ is a symmetric matrix, and $\varphi_i = -\partial F/\partial \dot{q}^i$ for a *dissipative function*

$$F = \frac{1}{2}\varphi_{ij}\dot{q}^i\dot{q}^j, \tag{2.85}$$

which is a positive definite quadratic form in the dot variables.

V. φ is variational, that is, φ_i coincide with the EL expressions of a Lagrange function, hence

$$\varphi_i = \frac{\partial h}{\partial q^i} + \left(\frac{\partial \eta_i}{\partial q^j} - \frac{\partial \eta_j}{\partial q^i}\right)\dot{q}^j \tag{2.86}$$

for some functions $h = h(q^j)$, $\eta_i = \eta_i(q^j)$, on M. An alternative approach using variational multipliers has been applied in [25, 26].

Once the different cases are outlined, it follows that, given a vector field X on M, it is a generator of invariance transformations of ε if and only if X, $\mathscr{U}$, and $\varphi = (\varphi_i)$ satisfy the following conditions:

(Case I)

$$L_X g = 0, \quad L_X\mathscr{U} = \text{constant} \tag{2.87}$$

(Case II)

$$L_X g = 0 \tag{2.88}$$

(Case III)

$$L_X g = 0, \quad (L_X\varphi)_i + \frac{\partial}{\partial q^i}(L_X\mathscr{U}) = 0 \tag{2.89}$$

(Case IV)

$$L_X g = 0, \quad L_{J^1X}\varphi = 0, \quad L_X\mathscr{U} = \text{constant} \tag{2.90}$$

(Case V)

$$L_X g = 0, \quad L_{J^1X}\tilde{\varphi} = 0, \quad L_X(\mathscr{U} + h) = \text{constant}, \tag{2.91}$$

where $\tilde{\varphi}$ is a 1-form given by

$$\tilde{\varphi}_i = \varphi_i - \frac{\partial h}{\partial q^i} = \left(\frac{\partial \eta_i}{\partial q^j} - \frac{\partial \eta_j}{\partial q^i}\right)\dot{q}^j. \tag{2.92}$$

Also, the following conditions hold for every generator X of invariant transformations of λ and E_λ, respectively:

(i)

$$dJ^1\gamma^*\left(\left(g_{ij}\dot{q}^j + \eta_i\right)X^i\right) = 0, \tag{2.93}$$

that is, γ is a geodesic with respect to the Levi–Civita connection as soon as $\eta_i = 0$.

(ii)

$$dJ^1\gamma^*\left(\left(g_{ij}\dot{q}^j + \eta_i\right)X^i + \mathscr{U}_0 t - f\right) = 0, \tag{2.94}$$

where $\mathscr{U}_0 = L_X\mathscr{U} \in \mathbb{R}$ and f is a solution of the first-order partial differential equation

$$\frac{\partial \eta_j}{\partial q^i}X^i + \eta_i\frac{\partial X^i}{\partial q^j} = \frac{\partial f}{\partial q^j}, \tag{2.95}$$

which is integrable, and on a star-shaped domain a solution of (2.95) reads

$$f = q^j\int_0^1\left(\frac{\partial \eta_j}{\partial q^i}X^i + \eta_i\frac{\partial X^i}{\partial q^j}\right)_{(sq^k)} ds.$$

Although we will mainly focus on the standard Noether Theorem,[1] the Noether–Bessel-Hagen symmetries are also used in the context of variational systems, which lack a Lagrangian description. We will provide an example of the application of the theorem in Appendix C, where we will find symmetries for a scalar-tensor theory of gravity.

[1] Hereafter Noether's First Theorem often will be referred to only as Noether's Theorem.

3

Applications of Noether's First Theorem to Fields and Particles

Longum iter est per praecepta,
breve et efficax per exempla.

— *Lucius Annaeus Seneca*
(*Epistulae ad Lucilium,* VI-5)

In this chapter we apply Noether's First Theorem to some main cases of physical interest. Specifically, we use the first and the second demonstrations of the theorem, discussed in Section 2.1, to show that both provide the same final results. In the former case, we find the conserved quantity without considering the generator of the symmetry, while in the latter case, we assume that a certain symmetry exists and we find the corresponding integral of motion.

3.1 The Free Scalar Field: The $U(1)$ Gauge Invariance

The simplest application of Noether's Theorem is provided by the scalar field model, with a Lagrangian consisting only of a linear combination of scalar fields and their derivatives. It is invariant under infinitesimal $U(1)$ transformations of the form $\tilde{\phi} = e^{-i\alpha}\phi$. The latter equation suggests that the infinitesimal generator of the transformation is $\delta\phi = -i\alpha\phi$. If the transformation is a symmetry for the given Lagrangian, Eq. (2.24) can be used to find the shape of the Noether current, that is

$$j^a = \frac{\partial \mathscr{L}}{\partial(\partial_a \phi^i)}\delta\phi^i - \frac{\partial \mathscr{L}}{\partial(\partial_a \phi^i)}\partial_b \phi^i \, \delta x^b + \mathscr{L}\delta x^a - g^a = -i\alpha\phi^i \frac{\partial \mathscr{L}}{\partial(\partial_a \phi^i)} - g^a. \quad (3.1)$$

As an example, the theory of a complex scalar field with a general potential $V(\phi^*, \phi)$ and Lagrangian,

$$\mathscr{L} = \partial_a \phi^* \partial^a \phi - V(\phi^*, \phi), \quad (3.2)$$

is invariant under $U(1)$ transformations. The transformed complex field reads as

$$\begin{cases} \tilde{\phi} = \phi - i\alpha\phi \\ \tilde{\phi}^* = \phi^* + i\alpha\phi^*. \end{cases} \quad (3.3)$$

Replacing Eq. (3.3) and setting $g^a = 0$, the current in Eq. (3.1) takes the form

$$j^a = -i\alpha\phi^i \frac{\partial \mathscr{L}}{\partial(\partial_a \phi^i)} = -i\alpha\left(\phi^* \partial^a \phi - \phi \partial^a \phi^*\right). \tag{3.4}$$

The same result can be achieved even by using the first prolongation of Noether vector $X^{[1]}$, which, in this case, can be written as

$$X^{[1]} = \frac{\partial \mathscr{L}}{\partial \phi}\delta\phi + \frac{\partial \mathscr{L}}{\partial \phi^*}\delta\phi^* + \partial_a \delta\phi \frac{\partial \mathscr{L}}{\partial(\partial_a \phi)} + \partial_a \delta\phi^* \frac{\partial \mathscr{L}}{\partial(\partial_a \phi^*)}. \tag{3.5}$$

By applying the identity (2.24) with the Noether vector (3.5) to Lagrangian (3.2), we obtain

$$\frac{\partial V(\phi, \phi^*)}{\partial \phi}\delta\phi + \frac{\partial V(\phi, \phi^*)}{\partial \phi^*}\delta\phi^* - \partial_a \delta\phi\, \partial^a \phi^* - \partial_a \delta\phi^*\, \partial^a \phi = 0. \tag{3.6}$$

In order to solve the preceding equation, let us define two scalar fields ψ and χ as

$$\begin{cases} \phi = \psi + i\chi \\ \phi^* = \psi - i\chi. \end{cases} \tag{3.7}$$

Substituting the definitions (3.7) into Eq. (3.6), we find:

$$\begin{aligned} &(\delta\phi + i\delta\chi)\frac{\partial V(\psi, \chi)}{\partial \phi} + (\delta\phi - i\delta\chi)\frac{\partial V(\psi, \chi)}{\partial \phi^*} \\ &-\partial_a(\delta\psi + i\delta\chi)\partial^a(\psi - i\chi) - \partial_a(\delta\psi - i\delta\chi)\partial^a(\psi + i\chi) = 0. \end{aligned} \tag{3.8}$$

Note that as the first two terms depend on the fields, the last two depend on the fields' prime derivatives. Therefore, the whole equation vanishes only if both vanish separately:

$$\begin{cases} (\delta\phi + i\delta\chi)\dfrac{\partial V(\psi, \chi)}{\partial \phi} + (\delta\phi - i\delta\chi)\dfrac{\partial V(\psi, \chi)}{\partial \phi^*} = 0 \\ \partial_a(\delta\psi + i\delta\chi)\partial^a(\psi - i\chi) + \partial_a(\delta\psi - i\delta\chi)\partial^a(\psi + i\chi) = 0. \end{cases} \tag{3.9}$$

While the first equation does not provide any physical information, the second equation, after performing some computations, can be recast as

$$\partial_a \delta\psi\, \partial^a \psi + \partial_a \delta\chi\, \partial^a \chi = 0, \tag{3.10}$$

whose solution is $\delta\psi = \alpha(\chi + \beta)$ and $\delta\chi = -\alpha(\psi + \gamma)$ with $\partial_a \beta = \partial_a \gamma = \partial_a \alpha = 0$. Substituting the fields' variations into Eq. (3.7), we finally obtain the infinitesimal generators:

$$\begin{cases} \delta\phi = \delta\psi + i\delta\chi = \alpha(\chi - i\psi) + \alpha(d - \gamma) = -i\alpha\phi \\ \delta\phi^* = \delta\psi - i\delta\chi = \alpha(\psi - i\chi) + \alpha(d - \gamma) = i\alpha\phi^*, \end{cases} \tag{3.11}$$

where we set $\gamma = \beta = 0$. This result shows that, from Eq. (2.24), it is possible to find the symmetry generator and the conserved quantity even when the former is not known a priori. It is worth noticing that, in this case, the first prolongation of Noether vector $X^{[1]}$ trivially reduces to X since the considered transformation does not involve space-time coordinates.

3.2 Invariance under Translations and Rotations

In this section we find the Noether current associated to translation and rotation group. With respect to the previously discussed $U(1)$ transformation, here we do not consider a particular Lagrangian, rather only the two transformations and the corresponding constants of motion. The final results will be particularly useful in the following subsections, where we will treat particle Lagrangians, which are invariant under translations and rotations.

3.2.1 Space-Time Translations

Even though the invariance with respect to space-time translations leads to a general conservation law, it is worth discussing the conserved quantities arising from the invariance under time translation and under spatial translation separately. Le us begin by considering the former case. Given a time-dependent Lagrangian of the form $\mathscr{L} = \mathscr{L}(\phi^i(t), \dot{\phi}^i(t))$, if the transformation $\tilde{t} = t - \delta t$ is a symmetry for $\mathscr{L}$, then the infinitesimal transformation

$$\delta_t \tilde{\mathscr{L}} = \frac{d\mathscr{L}}{dt}\delta t = \left(\frac{\partial \mathscr{L}}{\partial \phi^i}\dot{\phi}^i + \frac{\partial \mathscr{L}}{\partial \dot{\phi}^i}\ddot{\phi}^i\right)\delta t = \frac{d}{dt}\left(\dot{\phi}^i \frac{\partial \mathscr{L}}{\partial \dot{\phi}^i}\right)\delta t \tag{3.12}$$

yields the condition

$$\frac{d}{dt}\left(\dot{\phi}^i \frac{\partial \mathscr{L}}{\partial \dot{\phi}^i} - \mathscr{L}\right) = 0, \tag{3.13}$$

from which it follows that $\mathscr{H} = \pi_i \dot{\phi}^i - \mathscr{L}$ is a conserved quantity. As mentioned in the previous chapter, $\mathscr{H}$ is the Hamiltonian density of the system, which is conserved whenever the latter is invariant under time translations. A similar procedure shows that the spatial translation $\tilde{r}^i = r^i - \delta r^i$ leads to the conserved quantity

$$\pi_i = \frac{\partial \mathscr{L}}{\partial \dot{r}^i}. \tag{3.14}$$

As we have mentioned, a more general result can be obtained, considering the invariance under general space-time translations. In this case, the Noether theorem can be used to get a conservation law that comprehends Eqs. (3.13) and (3.14) as particular subcases. To this purpose, let us consider a translation acting on the space-time coordinates as

$$\tilde{x}^a = x^a - \delta x^a. \tag{3.15}$$

With respect to the transformation (3.15), the scalar fields ϕ^i remain unchanged point by point, namely $\tilde{\phi}^i(\tilde{x}^j) = \phi^i(x^j)$. Notice that this relation does not automatically imply the condition $\delta\phi = 0$; rather, it implies $\tilde{\delta}\phi = 0$. Therefore, according to Eqs. (2.2) and (2.3), the infinitesimal generator $\delta\phi$ can be written as $\delta\phi^i = -\partial_a\phi^i\delta x^a$. The invariance of the system under transformation (3.15) leads to the following condition for the Lagrangian density:

$$\mathscr{L}((\phi^i(x^a - \delta x^a), \partial_b\phi^i(x^a - \delta x^a)) = \mathscr{L}(x^a, \phi^i(x^a), \partial_b\phi^i(x^a)). \tag{3.16}$$

The Noether current can be found by replacing the forms of $\delta\phi$ and δx^a in the general definition (2.11):

$$j^a = -\left(\frac{\partial\mathscr{L}}{\partial(\partial_a\phi^i)}\partial_b\phi^i - \mathscr{L}\delta^a_b\right)\delta x^b, \tag{3.17}$$

where we set $g^a = 0$. The term in the brackets is the so-called energy-momentum tensor, defined as

$$\mathcal{T}^a_b = \frac{\partial\mathscr{L}}{\partial(\partial_a\phi^i)}\partial_b\phi^i - \mathscr{L}\delta^a_b. \tag{3.18}$$

Moreover, knowing that the generator of the space-time translation is the operator ∂_a, the first prolongation of the Noether vector $X^{[1]}$ must have the form $X^{[1]} = \partial_a$. Hence, taking into account Eq. (2.16), the following equalities automatically hold:

$$\delta x^a = \mathbb{1}, \qquad \delta\phi^i = 0, \qquad (\partial_a\delta\phi^i - \partial_a\phi^i\partial_b\delta x^b) = 0, \tag{3.19}$$

which means that the Noether current conservation implies the energy-momentum tensor conservation when space-time translations are considered, namely

$$\partial_a\mathcal{T}^a_b = 0. \tag{3.20}$$

The conservation of energy can be recovered from Eq. (3.20) by considering time-dependent scalar fields, whose sole nonvanishing contribution of the energy-momentum tensor is

$$\mathcal{T}^0_0 = \dot{\phi}^i\frac{\partial\mathscr{L}}{\partial\dot{\phi}^i} - \mathscr{L} = \mathcal{H}. \tag{3.21}$$

It is also worth noticing that space-time translations are external transformations so that the first prolongation of the Noether vector turns out to be necessary in order to get nontrivial symmetries. The non-extended vector X, in fact, does not contain the infinitesimal generators δx, and the Noether theorem trivially provides $X = 0$ as a solution. As better pointed out in the following chapters, trivial results often occur also in the framework of cosmological point-like Lagrangians, where the set of symmetries provided by the vector $X^{[1]}$ is usually larger that that provided by X. In particular, being X contained in $X^{[1]}$, trivial solutions coming from the latter automatically implies the absence of physical symmetries, while the opposite is not true in general.

As a final remark, it is worth noticing that the energy-momentum tensor can be defined independently of the Noether current, starting from the variational principle. To this purpose, let us consider a Lagrangian density $\mathscr{L}$ that depends on a rank-n tensor field $A_{c_1,c_2,\ldots,c_n}$ and its first derivative $\partial_a A_{c_1,c_2,\ldots,c_n}$ and let us evaluate the quantity

$$\begin{aligned}\delta^a_b \partial_a \mathscr{L} = \partial_b \mathscr{L} &= \frac{\partial \mathscr{L}}{\partial A_{c_1,c_2,\ldots,c_n}} \partial_b A_{c_1,c_2,\ldots,c_n} + \frac{\partial \mathscr{L}}{\partial(\partial_a A_{c_1,c_2,\ldots,c_n})} \partial_b \partial_a A_{c_1,c_2,\ldots,c_n} \\ &= \partial_a \frac{\partial \mathscr{L}}{\partial(\partial_a A_{c_1,c_2,\ldots,c_n})} \partial_b A_{c_1,c_2,\ldots,c_n} + \frac{\partial \mathscr{L}}{\partial(\partial_a A_{c_1,c_2,\ldots,c_n})} \partial_b \partial_a A_{c_1,c_2,\ldots,c_n} \\ &= \partial_a \left[\frac{\partial \mathscr{L}}{\partial(\partial_a A_{c_1,c_2,\ldots,c_n})} \partial_b A_{c_1,c_2,\ldots,c_n} \right],\end{aligned} \tag{3.22}$$

where we used the EL equation

$$\frac{\partial \mathscr{L}}{\partial A_{c_1,c_2,\ldots,c_n}} - \partial_a \frac{\partial \mathscr{L}}{\partial(\partial_a A_{c_1,c_2,\ldots,c_n})} = 0. \tag{3.23}$$

Equation (3.22) yields

$$\partial_a \left[\frac{\partial \mathscr{L}}{\partial(\partial_a A_{c_1,c_2,\ldots,c_n})} \partial_b A_{c_1,c_2,\ldots,c_n} - \delta^a_b \mathscr{L} \right] = 0, \tag{3.24}$$

which is nothing else but the generalization of Eq. (3.20), with $\mathcal{T}^a_b$ defined as the term in the square brackets.

3.2.2 Rotations

Another standard example is provided by the invariance under rotations. It is well known that, for standard particles, the symmetry under rotations leads to the variation:

$$\delta \mathscr{L}(\dot{q}^i, q^i) = \frac{\partial \mathscr{L}}{\partial q^i} \delta q^i + \frac{\partial \mathscr{L}}{\partial \dot{q}^i} \delta \dot{q}^i = \frac{d}{dt} \left(\frac{\partial \mathscr{L}}{\partial \dot{q}^i} \delta q^i \right) = \frac{d}{dt} (\pi_i \delta q^i) = 0, \tag{3.25}$$

with q^i being the generalized coordinates. A rotation by an angle θ yields a variation in the position δq^i that can be written in terms of the unitary vector $\mathbf{n}$ as $\delta \mathbf{q} = \mathbf{n} \times \mathbf{q}$ or, equivalently, in terms of the Levi–Civita Tensor, as

$$\delta q^i = \varepsilon^{ijk} n_j q_k. \tag{3.26}$$

Substituting the above relation into Eq. (3.25), we get

$$\frac{d}{dt} \left(\pi_i \varepsilon^{ijk} n_j q_k \right) = 0. \tag{3.27}$$

Being independent of time and having a unitary norm, the unitary vector n_j does not provide any contribution. Therefore the quantity

$$L^i \equiv \varepsilon^{ijk} \pi_i q_k = -\varepsilon^{ikj} \pi_i q_k = \varepsilon^{ikj} q_i \pi_k \tag{3.28}$$

is conserved with respect to the affine parameter t. L^i is the angular momentum density and represents the Noether current coming from the invariance under rotations. A more general result can be obtained by considering the LG generator. As shown in Section 1.1.4, the latter can be split into two contributions, related to rotations and boosts, respectively. Moreover, in Section 1.1.3, we also found the explicit expression of the rotation generator, namely

$$\mathcal{R}^i = \epsilon^{ij\ell} \mathcal{J}_{j\ell}. \tag{3.29}$$

Therefore, neglecting the contribution of boosts, LTs reduce to spatial rotations of the form

$$\Lambda = e^{i\epsilon_{ij\ell}\lambda^{j\ell}\mathcal{R}^i}. \tag{3.30}$$

Thus, the matrix elements of Λ are

$$\begin{aligned}\Lambda^k_n &= \delta^k_n + i\epsilon_{ij\ell}\lambda^{j\ell}\epsilon^{ihm}\left(\mathcal{J}_{hm}\right)^k_n = \delta^k_n + i\lambda^{j\ell}\left(\delta^h_j\delta^m_\ell - \delta^h_\ell\delta^m_j\right)\left(\mathcal{J}_{hm}\right)^k_n \\ &= \delta^k_n + i\lambda^{j\ell}\left(\mathcal{J}_{j\ell} - \mathcal{J}_{\ell j}\right)^k_n,\end{aligned} \tag{3.31}$$

where we used the antisymmetric properties of the three-dimensional Levi–Civita tensor. Taking into account Eq. (1.21), the application of rotation (3.31) to the vector x^n yields

$$\Lambda^k_n x^n = x^k + i\lambda^{j\ell}\left[\left(\mathcal{J}_{j\ell}\right)^k_n - \left(\mathcal{J}_{\ell j}\right)^k_n\right] x^n = x^k - \lambda_{j\ell}\left(\eta^{jk}x^\ell - \eta^{\ell k}x^j\right). \tag{3.32}$$

Not affecting the scalar field but only the coordinates, the variation $\delta\phi$ vanishes identically[1] and the infinitesimal variation δx^k can be written as

$$\begin{cases}\delta\phi = 0\\ \delta x^k = -\lambda_{j\ell}\left(\eta^{jk}x^\ell - \eta^{\ell k}x^j\right).\end{cases} \tag{3.33}$$

From the preceding equations, it follows that the Noether current (2.11) can be written in terms of the energy-momentum tensor as

$$j^i = -\left(\frac{\partial\mathcal{L}}{\partial(\partial_i\phi)}\partial_k\phi - \mathcal{L}\right) = -\mathcal{T}^i_k\delta x^k = \lambda_{\ell j}\left(\mathcal{T}^{ij}x^\ell - \mathcal{T}^{i\ell}x^j\right). \tag{3.34}$$

Since $\lambda_{\ell j}$ is independent of the coordinates, the preceding result shows that the quantity $\mathcal{T}^{ij}x^\ell - \mathcal{T}^{i\ell}x^j$ is conserved under rotations. The Noether charge can be obtained by integrating Eq. (3.34) over the hypersurface, and it provides the covariant representation of the angular momentum.

Equivalently, it is possible to straightforwardly consider the rotation group $SO(3)$, whose main features are discussed in Section 1.1.3. We showed that the $SO(3)$ generator can be written in terms of the spatial coordinates x^i and their derivatives as $\hat{L}_j = (x_i\partial_k - x_k\partial_i)$. The Noether current turns out to be the same as that in Eq. (3.28), namely

$$j^i = \varepsilon^{ikj} q_i \pi_k. \tag{3.35}$$

[1] The variation $\tilde{\delta}\phi = \phi(\tilde{x}^i) - \phi(x^i)$ is indeed different from zero.

The angular momentum density, defined in Eq. (3.28), can be integrated over the hypersurface to provide the corresponding Noether charge. By choosing the surface $d\Sigma_a = (d^3x, 0, 0, 0)$, we have

$$\mathbf{L} = \int \boldsymbol{q} \times \boldsymbol{\pi}_q \, d^3x, \tag{3.36}$$

which shows that the Noether charge of the $SO(3)$ group is the orbital angular momentum. This treatment will be particularly useful in Section 3.5.2, where the Noether Theorem will be applied to the Harmonic Oscillator, showing that the system is invariant under rotations.

3.3 The Lorentz Invariance

The purpose of this section is to find the Noether current associated with the general LT

$$\Lambda^c_d = \delta^c_d - i\lambda_{ab} \left(\mathcal{J}^{ab}\right)^c_d, \qquad \text{with} \qquad \left(\mathcal{J}^{ab}\right)^c_d = \frac{i}{2}\left(\eta^{ac}\delta^b_d - \eta^{bc}\delta^a_d\right). \tag{3.37}$$

The preceding relation yields

$$\tilde{x}^c = \Lambda^c_d x^c = x^c + \frac{1}{2}\lambda_{ab}\left(\eta^{ac}x^b - \eta^{bc}x^a\right) = x^c - \lambda_{ab}\eta^{bc}x^a, \tag{3.38}$$

from which it is possible to find the transformation laws of coordinates and scalar fields, namely

$$\begin{cases} \delta x^c = -\lambda_{ab}\eta^{bc}x^a \\ \tilde{\delta}\phi^c = -i\lambda_{ab}\left(\mathcal{J}^{ab}\right)^c_d \phi^d. \end{cases} \tag{3.39}$$

Once the transformation laws have been obtained, Eq. (2.3) can be used to find the associated Noether current. The same analysis, as in the previous section for a set of scalar fields ϕ^c, yields

$$j^a = -\lambda_{bc}\left[-i\frac{\partial \mathcal{L}}{\partial_a \phi^e}\left(\mathcal{J}^{bc}\right)^e_d \phi^d - \left(\delta^c_d x^d \mathcal{T}^{ab} - \delta^b_d x^d \mathcal{T}^{ac}\right)\right]. \tag{3.40}$$

Since λ_{bc} is independent of the coordinates, it turns out that the quantity

$$J^{abc} = i\frac{\partial \mathcal{L}}{\partial_a \phi^e}\left(\mathcal{J}^{bc}\right)^e_d \phi^d + \left(x^c \mathcal{T}^{ab} - x^b \mathcal{T}^{ac}\right) \tag{3.41}$$

is conserved: $\partial_a J^{abc} = 0$. Also notice that the rank-3 tensor J^{abc} can be split into two contributions. One of them depends on the coordinates and corresponds to the orbital angular momentum, $L^{abc} \equiv x^c\mathcal{T}^{ab} - x^b\mathcal{T}^{ac}$. The other is an intrinsic contribution called *spin angular momentum:* $S^{abc} = i\dfrac{\partial \mathcal{L}}{\partial_a \phi^e}\left(\mathcal{J}^{bc}\right)^e_d \phi^d$. Therefore J^{abc} can be written as the sum $J^{abc} = L^{abc} + S^{abc}$ and represents the total angular momentum.

3.3.1 The Lorentz Invariance for the Electromagnetic Field

One of the most important examples of Lorentz-invariant Lagrangians is the electromagnetic free Lagrangian, that is

$$\mathscr{L} = -\frac{1}{4\pi} F^{ab} F_{ab}, \tag{3.42}$$

where F^{ab} is the coordinate representation of the electromagnetic 2-form $F = dA$, and d the exterior derivative. As previously shown, the Noether current can be thought of as a sum of the orbital angular momentum L^{abc} and the spin angular momentum S^{abc}, which, in the present case, can be written in terms of the potential A^c as

$$\begin{cases} L^{abc} = x^c \mathcal{T}^{ab} - x^b \mathcal{T}^{ac} \\ = -\frac{1}{4\pi}\left(x^c F^{ad} F^b_{\ d} - \frac{1}{4} x^c \eta^{ab} F^{de} F_{de} + x^b F^{ad} F^c_{\ d} - \frac{1}{4} x^b \eta^{ac} F^{de} F_{de} \right) \\ S^{abc} = -\frac{1}{4\pi} F^{ad} \left(\delta^b_d \delta^c_e - \delta^c_d \delta^a_e \right) A^e = -\frac{1}{4\pi} F^{a[b} A^{c]}, \end{cases} \tag{3.43}$$

where the energy-momentum tensor of the electromagnetic field is derived from Eq. (3.24) with Lagrangian (3.42) and reads as

$$\mathcal{T}^{ab} = -\frac{1}{4\pi}\left(F^{ad} F^b_{\ d} - \frac{1}{4} \eta^{ab} F^{de} F_{de} \right). \tag{3.44}$$

3.3.2 The Gauge Invariance for the Electromagnetic Field

After treating the invariance of the electromagnetic field under LTs, let us consider the $U(1)$ gauge invariance of the electromagnetic Lagrangian. Notice that the potential A can be recast as a closed form $\mathbf{A}$ so that the equations of motion are invariant under the gauge transformation $\tilde{A}_a \to A_a + \partial_a \chi$. Since the gauge symmetry is an internal symmetry, the Noether current associated with the gauge transformation can be easily computed by means of the non-extended Noether vector X, as pointed out in Section 2.1.3. The Noether current associated with the $U(1)$ invariance of the electromagnetic field is

$$j^a = \frac{\partial \mathscr{L}}{\partial(\partial_a A_b)} \delta A_b = \frac{\partial \mathscr{L}}{\partial(\partial_a A_b)} \partial_b \chi = F^{ab} \partial_b \chi. \tag{3.45}$$

The continuity equation therefore yields

$$\partial_b \chi \partial_a F^{ab} + F^{ab} \partial_a \partial_b \chi = 0. \tag{3.46}$$

Notice that Eq. (3.46) is identically trivial since the first term vanishes due to the Maxwell field equations, while the second is the contraction of the symmetric tensor $\partial_a \partial_b \chi$ with the antisymmetric tensor F^{ab}. The Noether charge associated with $U(1)$ transformations of the electromagnetic Lagrangian can be easily found by integrating the 0-component of Eq. (3.45) with respect to the surface $\sigma^a = (d^3x, 0, 0, 0)$, namely:

$$Q = \int j^0 \, d^3x = \int F^{0a} \partial_a \chi \, d^3x = \int_S E \cdot n \, dS. \tag{3.47}$$

Equation (3.47) suggests that the Noether charge can be addressed to the electric charge, and the Gauss theorem automatically follows. Moreover, the same result can also be obtained by means of the condition (2.24), namely by assuming the system to be invariant under some symmetry transformations. Then, the symmetry generator can be constructed and the conserved quantity can be found. The first prolongation of the Noether vector takes the following form:

$$\begin{aligned} X^{[1]} &= \partial_a \delta x^a + \delta x^a \partial_a + \frac{\partial}{\partial_a} + \delta A_a \frac{\partial}{\partial A_a} \\ &\quad + \partial_a \delta A_b \frac{\partial}{\partial(\partial_a A_b)} - \partial_a A_b \partial_c \delta x^c \frac{\partial}{\partial(\partial_a A_b)}. \end{aligned} \tag{3.48}$$

Since $\mathscr{L}$ is independent of the space-time coordinates, the application of $X^{[1]}$ to the Electromagnetic Lagrangian yields the identity

$$F^{ab} F_{ab} \partial_c \delta x^c + 2 \partial_a \delta A_b F^{ab} - 2 \partial_a A_b F^{ab} \partial_c \delta x^c = 0, \tag{3.49}$$

which leads to the following system:

$$\begin{cases} F^{ab} \partial_a \delta A_b = 0 \\ \partial_a \delta x^a = 0. \end{cases} \tag{3.50}$$

The second equation confirms that the $U(1)$ symmetry is an internal symmetry and the first prolongation of the Noether vector $X^{[1]}$ identically reduces to X. On the other hand, in order for the first equation to be satisfied, the rank-2 tensor $\partial_a \delta A_b$ must be symmetric. One possible choice is $\delta A_b = \partial_b \chi$, so that the first equation of the system becomes

$$F^{ab} \partial_a \partial_b \chi = 0, \tag{3.51}$$

which is equivalent to Eq. (3.46) and identically holds for any F^{ab}. Therefore, this application shows that the form of the symmetry generator can be unveiled by the Noether Theorem. From this point of view, the example suggests that the Noether theorem can be used as a criterion to select symmetry generators starting from general theories. This important feature will be better clarified in the second part of the book, where the application of the Noether Theorem to theories of gravity will be treated.

Let us finally sum up the results obtained in this chapter by means of Table 3.1.

Table 3.1 Symmetries and corresponding conserved quantities

Conserved Quantity	Symmetry
Energy	Time Translation
Momentum	Space Translation
Orbital Angular Momentum	Spatial Rotations $O(3)$
Total Angular Momentum	Spacetime Rotations $SO(1,3)$
Electric Charge	Unitary Transformations $U(1)$

3.4 The Spontaneous Symmetry Breaking

Let now use the results provided by the Noether theorem to briefly show the application to the spontaneous symmetry breaking mechanism. To this purpose, we adopt a quantum formalism, where the Hamiltonian and the Noether charge must be promoted to operators. The latter was introduced in Chapter 2.1 as the integral of motion, which arises in a system enjoying a certain symmetry, and this property is also preserved in quantum systems. As a consequence, in QFT the operator $\hat{Q}$ must commute with the Hamiltonian $\hat{H}$. Though it is possible to formally prove that the Noether charge can always be promoted to a quantum operator, Table 3.1 provides some straightforward examples of the best known Noether charges, as all the conserved quantities, listed in the table, are well-known operators in QFT. This correspondence holds as long as the classical symmetry is represented by a unitary transformation upon the Hilbert space, otherwise quantum states cannot be normalized and Q cannot be recast as a Hermitian operator. Summing up, in the quantum formalism, if the transformation is a symmetry for the Hamiltonian, then there exists a corresponding Noether charge Q such that the condition $[\hat{Q}, \hat{H}] = 0$ is satisfied.

Before considering the spontaneous symmetry breaking mechanism, we start by demonstrating the following:

If the Lagrangian of a given theory is symmetric under a certain transformation that changes the fundamental state, then there exists a massless boson called the Goldstone boson.

Reversing the argument, this means that if there exists a symmetry of the Lagrangian that is not a symmetry of the fundamental state, then a massless boson arises.

Let us consider a Hamiltonian operator $\hat{H}$ with ground state $|0>$ labeled by a null energy, namely $\hat{H}|0>=0$. The choice of null energy for the fundamental state is motivated by a desire for simplicity, but the demonstration holds regardless of such an assumption. The theorem just mentioned can be straightforwardly proved by considering a symmetry transformation with the Noether charge $\hat{Q}$ and the state $|q>$ defined as $|q>=\hat{Q}|0>$. Therefore, the energy of the newly defined state $|q>$ is

$$\hat{H}|q>=\hat{H}\hat{Q}|0>=\hat{Q}\hat{H}|0>=0. \tag{3.52}$$

Let us now define the state $|s>$ as $|s>=\int d^N x\, j^0(x,t)e^{ik\cdot x}|0>$, with N being the dimension, k the momentum, and j^0 the time-component of Noether current. Notice that as soon as the momentum k vanishes, we have

$$k=0\rightarrow |s>=\int d^N x\, j^0(x,t)|0>=\hat{Q}|0>=|q>. \tag{3.53}$$

Hence, when the momentum of $|s>$ vanishes, the states $|q>$ and $|s>$ are equivalent, which implies that the energy of $|k>$ also must be zero. This is possible only if no mass is associated with the state $|k>$. Therefore, if the fundamental state $|0>$ is not invariant under the transformation, which is a symmetry for the Lagrangian, a new massless particle (a Goldstone boson) arises. This result is a fundamental step toward comprehension of the symmetry breaking and the corresponding Higgs Mechanism.

In order to show that a massive particle arises under a spontaneous symmetry breaking procedure, let us consider the general Lagrangian

$$\begin{aligned}\mathcal{L}=&-\frac{1}{4}F^{\mu\nu}F_{\mu\nu}+\partial_\mu\phi^*\partial^\mu\phi-\mu^2\phi^*\phi-\lambda(\phi^*\phi)^2\\&-ie(\phi\partial^\mu\phi^*-\phi^*\partial^\mu\phi)A_\mu+e^2A_\mu A^\mu\phi^*\phi,\end{aligned} \tag{3.54}$$

invariant with respect to $U(1)$ transformations of the scalar field $\phi\rightarrow\phi'=e^{i\theta(x)}\phi$ and under gauge transformation of the electomagnetic potential $A^\mu(x)\rightarrow A'_\mu(x)=A_\mu+\frac{i}{e}\partial_\mu\theta(x)$. The standard approach to spontaneously breaking the symmetry is to redefine the scalar field ϕ through an appropriate transformation shifting one of the two ground states $\phi_0=\pm\sqrt{\frac{-\mu^2}{\lambda}}$. Assuming $\mu^2<0$ and selecting the positive ground state, the transformation of the scalar field ϕ can be performed by introducing two new scalar fields, ξ and η:

$$\phi=e^{i\frac{\xi}{\phi_0}}\left(\frac{\phi_0+\eta}{\sqrt{2}}\right), \tag{3.55}$$

which, up to the first order, provides

$$\phi=\frac{1}{\sqrt{2}}(\phi_0+\eta+i\xi), \qquad \partial_\mu\phi=\frac{1}{\sqrt{2}}(\partial_\mu\eta+i\partial\mu\xi), \tag{3.56}$$

$$\phi^*=\frac{1}{\sqrt{2}}(\phi_0+\eta-i\xi), \qquad \partial_\mu\phi^*=\frac{1}{\sqrt{2}}(\partial_\mu\eta-i\partial\mu\xi), \tag{3.57}$$

and the Lagrangian (3.54) becomes

$$\begin{aligned}\mathcal{L}=&-\frac{1}{4}F^{\mu\nu}F_{\mu\nu}+\frac{1}{2}\partial_\mu\eta\partial^\mu\eta+\frac{1}{2}\partial_\mu\xi\partial^\mu\xi\\&+\frac{1}{2}\phi_0^2A_\mu A^\mu-e\phi_0A_\mu\partial^\mu\xi+\mu^2\eta^2.\end{aligned} \tag{3.58}$$

Due to the mixing terms, the mass of the new scalar fields does not appear manifestly. However, since the Lagrangian invariant under gauge transformations is of the form

$$\begin{cases} \phi' = e^{-i\frac{\xi}{\phi_0}}\phi = e^{-i\frac{\xi}{\phi_0}}e^{i\frac{\xi}{\phi_0}}\left(\frac{\eta+\phi_0}{\sqrt{2}}\right) = \frac{\eta+\phi_0}{\sqrt{2}} \\ A'_\mu = A^\mu + \frac{i}{e}\partial_\mu\theta = A^\mu - \frac{i}{e\phi_0}\partial_\mu\xi, \end{cases} \tag{3.59}$$

the transformed Lagrangian $\mathcal{L}'$ must be equivalent to that of Eq. (3.58), namely

$$\mathcal{L}'(\xi', \partial_\mu\xi', \eta', \partial_\mu\eta') = \mathcal{L}(\xi, \partial_\mu\xi, \eta, \partial_\mu\eta). \tag{3.60}$$

The transformed Lagrangian can be written in terms of the new scalar fields as

$$\begin{aligned} \mathcal{L} = &-\frac{1}{4}F'^{\mu\nu}F'_{\mu\nu} + \frac{1}{2}\partial_\mu\eta\partial^\mu\eta + \frac{1}{2}e^2\phi_0^2A'_\mu A'^\mu + \frac{1}{2}e^2A'_\mu A'^\mu\eta(2\phi_0+\eta) \\ &-\frac{1}{2}\eta^2(3\lambda\phi_0^2+\mu^2) - \lambda\phi_0\eta^3 - \frac{1}{4}\lambda\eta^4. \end{aligned} \tag{3.61}$$

Note that, after using the gauge transformations (3.59) to transform the Lagrangian, no mixed quantity occurs and each term can be straightforwardly interpreted. Specifically, carrying zero mass, ξ is a Goldstone boson and does not provide any contribution to Eq. (3.61).

The mass of the fields η and A_μ is $m_\eta = \sqrt{(3\lambda v^2 + \mu^2)} = \sqrt{-2\mu^2}$ and $m_{A_\mu} = e\phi_0$, respectively. This result shows that the gauge-invariant Lagrangian (3.54) undergoes a Higgs Mechanism where both massless and massive bosons are generated by the symmetry breaking. Note that the presence of symmetries plays a fundamental role in this description: on one hand, the Noether charge permits one to demonstrate the Goldstone theorem in the general case; on the other hand, the starting gauge-invariant Lagrangian (3.54) allows one to use local gauge symmetries to disentangle the mixing terms. This can be considered as another feature of the Noether theorem thanks to which the demonstration of the Higgs Mechanism turns out to be straightforward. As a matter of fact, without using the symmetry properties of the Lagrangian, the Higgs Mechanism requires the computation of the propagators of ξ and A_μ and the poles of the S-matrix to be well interpreted.

3.5 The Noether Theorem for Particles

We now apply the Noether Theorem to the free particle Lagrangian and to the harmonic oscillator. The purpose is to use Eq. (2.24) to get the infinitesimal generators, following the same procedure provided in Sections 3.1 and 3.3.2. As pointed out in what follows, starting from symmetry generators, integrals of motion can be found and the dynamics can be reduced. More precisely, we want to show that this prescription can be pursued with the aim of recovering the invariance under translations for the free particle Lagrangian, and the invariance under rotations for the harmonic oscillator (see Section 3.2 for details).

3.5.1 Free Particle

Let us begin by considering a free-particle Lagrangian, depending on the generalized coordinates q^i and velocity $\dot{q}^i$. Even though the results provided in this paragraph can be easily extended for a many-particles system, for the sake of simplicity, we focus on the following single-particle Lagrangian:

$$\mathcal{L} \equiv \mathcal{L}(q^i(t), \dot{q}^i(t)) = \dot{q}^2. \tag{3.62}$$

We start from the assumption that the system contains a certain unknown symmetry with a Noether vector of the form

$$X = \delta q^i \frac{\partial}{\partial q^i} + \delta \dot{q}^i \frac{\partial}{\partial \dot{q}^i}. \tag{3.63}$$

By applying the condition (3.63) to Lagrangian (3.62), we obtain

$$2\dot{q}\,\delta\dot{q} = 0. \tag{3.64}$$

By means of the identity

$$\delta\dot{q} = \frac{d}{dt}\delta q = \frac{\partial \delta q}{\partial q}\frac{\partial q}{\partial t}, \tag{3.65}$$

the condition in Eq. (3.64) can be recast as

$$\frac{\partial \delta q}{\partial q} = 0 \quad \rightarrow \quad \delta q = \text{const} \equiv q_0. \tag{3.66}$$

Therefore the generator of the symmetry and the conserved quantity are

$$\mathcal{X} = q_0 \frac{\partial}{\partial q}, \quad j = q_0 \frac{\partial \mathcal{L}}{\partial \dot{q}} = q_0 \pi_q. \tag{3.67}$$

The preceding result shows that the Lagrangian is invariant under local translation so that the conserved quantity turns out to be the conjugate momentum of the particle q, as demonstrated in Section 3.2.

3.5.2 Harmonic Oscillator

Another simple application of the Noether theorem is given by the harmonic oscillator. We start from a two-particle Lagrangian of the form

$$\mathcal{L} = \dot{q}_1^2 + \dot{q}_2^2 - q_1^2 - q_2^2, \tag{3.68}$$

labeled by the harmonic potential:

$$V(q_1, q_2) = q_1^2 + q_2^2. \tag{3.69}$$

Assuming that the Lagrangian (3.68) is invariant under some symmetry transformation, the vector field X takes the following form:

$$X = \delta q_1 \frac{\partial}{\partial q_1} + \delta q_2 \frac{\partial}{\partial q_2} + \delta \dot{q}_1 \frac{\partial}{\partial \dot{q}_1} + \delta \dot{q}_2 \frac{\partial}{\partial \dot{q}_2}. \tag{3.70}$$

The condition $L_X \mathcal{L} = 0$ leads to a single equation containing mixed derivatives of the variables q_1 and q_2, which can be solved by setting to zero all the terms containing different time-derivatives. This procedure is very common in the framework of partial differential equations and will be widely used in the application of the Noether method to cosmological point-like Lagrangians. In the present case, we get a system of four differential equations:

$$\begin{cases} 2\,\delta q_1\, q_1 + 2\,\delta q_2\, q_2 = 0 \\ 2\partial_{q_1} \delta q_1 = 0 \\ 2\partial_{q_2} \delta q_2 = 0 \\ 2\partial_{q_2} \delta q_1 + 2\partial_{q_1} \delta q_2 = 0. \end{cases} \tag{3.71}$$

From the second and third conditions, we notice that the variation δq_1 (δq_2) is independent of q_2 (q_1), while the remaining two equations provide $\delta q_1 = q_2$ and $\delta q_2 = q_1$. The symmetry generator (3.70), therefore, reduces to

$$\mathcal{X} = q_2 \frac{\partial}{\partial q_1} - q_1 \frac{\partial}{\partial q_2} \tag{3.72}$$

and the conserved quantity takes the form

$$j = q_2 \frac{\partial \mathcal{L}}{\partial \dot{q}_1} - q_1 \frac{\partial \mathcal{L}}{\partial \dot{q}_2} = q_2 \pi_1 - q_1 \pi_2. \tag{3.73}$$

By comparing Eqs. (3.72) and (3.73) with Eqs. (3.36) and (3.35), we notice that the symmetry generator confirms the invariance of the system under rotations, and the conserved quantity turns out to be the angular momentum.

4
Theories of Gravity: An Overview

I believe that whatever we do or live for has its causality;
it is good, however, that we cannot see through to it.

— *Albert Einstein*
(*"Interview with Rabindranath Tagore,"*
published in *The Religion of Man* [1930]
by Rabindranath Tagore)

4.1 An Overview of General Relativity: Successes and Shortcomings

Before discussing theories of gravity from a general point of view, we will give a quick overview of GR, providing its main results and discussing successes and shortcomings. For a complete and dedicated treatment see, for example, [31–36].

Let us begin by premising that any relativistic theory of gravity must satisfy certain minimal requirements from the phenomenological point of view. First, it must explain astronomical observations, mapping the orbits of planets and the potential well of self-gravitating structures. This implies that Newtonian dynamics must be obtained in the weak-field, slow-motion limit. At the post-Newtonian (PN) level, the theory has to pass the classical Solar System tests [34], to reproduce correctly the observed galactic dynamics accounting for the known baryonic constituents and to recover the Newtonian potential. Moreover, the theory has to address the problem of generation and evolution of the large-scale structure of the universe and it has to reproduce the cosmological dynamics, namely predict in a self-consistent way the main cosmological parameters (like the Hubble parameter H_0, the deceleration parameter q_0, and the density parameters). The simplest theory satisfying the preceding requirements, at least at a certain degree of precision, is GR. It was formulated between 1909 and 1915 by Albert Einstein and still today provides the best description for the gravitational interaction, which is conceived as the geometry of space-time.

Specifically, GR relies on four basic assumptions:

1. The *Principle of Relativity*, which requires all observers to be equally valid for describing Physics. In particular, inertial frames (which do not exist globally) are not a priori preferred.
2. The *Equivalence Principle*, according to which the dynamics of massive bodies in non-inertial systems is the same as that in inertial systems in the presence of a gravitational field. In other words, any non-inertial frame can be locally reduced to an inertial frame.
3. The *Principle of General Covariance*, which states that the field equations must be "generally covariant" tensor equations, whose form is the same in all coordinate systems.
4. The *Principle of Causality*, which imposes that causality is preserved, namely that each space-time point should admit a notion of past, present, and future that is the same for all physical observers.

In light of the preceding postulates, when choosing a non-inertial system, the Minkowski metric $\eta_{\mu\nu}$ must be generalized to a general rank-2 tensor $g_{\mu\nu}$, whose elements are functions of space-time coordinates containing, eventually, off-diagonal terms. This tensor, symmetric with respect to the lowest indexes, is called a *metric tensor*, by means of which the invariant line element can be expressed as $ds^2 = g_{\mu\nu}\,dx^\mu dx^\nu$, with x^μ and x^ν being the space-time coordinates. In other words, this is the generalization of the Pythagoras Theorem on any curved space-time (see [35] for a detailed geometric discussion).

According to the Equivalence Principle, the latter relation must hold also in inertial systems in the presence of a gravitational field. Therefore, the space-time is influenced by the presence of massive objects, which can curve the surrounding space-time, yielding a metric tensor that is not Minowskian. As a consequence, in curvilinear coordinates, the derivative of a given vector A_μ, that is, $\partial_\nu A_\mu$, does not give rise to a rank-2 tensor. The quantity that transforms covariantly under a general coordinate transformation is the covariant derivative, defined by means of the Christoffel connection $\Gamma^\alpha_{\nu\mu}$ as $D_\nu A_\mu = \partial_\nu A_\mu - \Gamma^\alpha_{\nu\mu} A_\alpha$. Using the covariant derivative, it is possible to show that the parallel transport of the vector A_μ along a closed path provides another vector, rotated with respect to the initial one. The parallel displacement ΔA_μ is only due to the presence of the Christoffel connection in the covariant derivative, so that we have

$$\Delta A_\mu = \oint \delta A_\mu \, dx^\alpha = \oint \Gamma^\alpha_{\mu\nu} A_\alpha \, dx^\nu . \tag{4.1}$$

Using Stokes' theorem, the preceding relation can be recast as

$$\Delta A_\mu = \frac{1}{2} R^\alpha{}_{\mu\nu\lambda} A_\alpha \Delta S^{\nu\lambda}, \tag{4.2}$$

where $S^{\nu\lambda}$ is the surface tensor and $R^{\alpha}{}_{\mu\nu\lambda}$ is the *Riemann Tensor*, defined as

$$R^{\alpha}{}_{\mu\nu\lambda} = \partial_{\nu}\Gamma^{\alpha}{}_{\mu\lambda} - \partial_{\mu}\Gamma^{\alpha}{}_{\nu\lambda} + \Gamma^{\alpha}{}_{\beta\nu}\Gamma^{\beta}{}_{\mu\lambda} - \Gamma^{\alpha}{}_{\beta\lambda}\Gamma^{\beta}{}_{\mu\nu}. \tag{4.3}$$

It is straightforward to verify that the Riemann tensor satisfies the following relations:

$$\begin{aligned} R^{\alpha}{}_{\mu\nu\lambda} + R^{\alpha}{}_{\nu\lambda\mu} + R^{\alpha}{}_{\lambda\mu\nu} &= 0, \\ R^{\alpha}{}_{\mu\nu\lambda;\,\rho} + R^{\alpha}{}_{\mu\lambda\rho;\,\nu} + R^{\alpha}{}_{\mu\rho\nu;\,\lambda} &= 0, \end{aligned} \tag{4.4}$$

respectively called the "first" and "second" Bianchi identity. The first Bianchi identity can be equivalently written as

$$d\mathcal{R} + \omega \wedge \mathcal{R} + \mathcal{R} \wedge \omega = 0, \quad \text{with } \mathcal{R} = d\omega + \omega \wedge \omega, \tag{4.5}$$

with ω being the connection 1-form and R the curvature 2-form. By properly contracting the Riemann tensor, it is possible to define the *Ricci tensor* $R_{\mu\nu}$ and the *Ricci scalar* R:

$$R_{\mu\nu} \equiv R^{\alpha}{}_{\mu\alpha\nu}, \qquad R \equiv R^{\alpha}_{\alpha}. \tag{4.6}$$

Notice that for the space-time to be Minkowskian, the condition $R_{\mu\nu} = 0$ is not sufficient, while it is required that $R^{\alpha}{}_{\mu\nu\lambda}$ vanishes everywhere.

The gravitational action, called the Einstein–Hilbert (EH) action and first postulated by David Hilbert, is

$$S = \frac{1}{16\pi G}\int \sqrt{-g}R\, d^4x, \tag{4.7}$$

with g being the determinant of the metric tensor and G the Newton gravitational constant.[1] It is worth pointing out that Hilbert's and Einstein's choice was rather arbitrary, as it became clear a few years later, but it was certainly the simplest from both the mathematical and the physical points of view.

As clarified by Levi–Civita in 1919, curvature is not a purely metric notion, but it is also related to the linear connection defining parallel transport and covariant differentiation. But in Einstein's day only the metric concept was available to mathematicians and physicists alike and his formulation was the only viable one. It was later clarified that the three principles of relativity, equivalence, and covariance, together with causality, require that the space-time structure is determined by either one or both of the two fields, namely $g_{\mu\nu}$ and the connection $\Gamma^{\alpha}{}_{\mu\nu}$, assumed to be torsionless for simplicity. As better pointed out in the remainder of this chapter, the preceding physical constraints do not necessarily demand that the connection is the Levi–Civita one.

However, considering the action (4.7), its variation with respect to the metric tensor yields, in vacuum,

$$R_{\mu\nu} - \frac{1}{2}g_{\mu\nu}R = 0, \tag{4.8}$$

[1] In this section, we explicitly adopt the Newton coupling. In the rest of the book, we will set $1/(8\pi G)$ equal to 1, unless otherwise indicated.

called the *Einstein field equations*. See Appendix A for details on variational calculus. The LHS of these equations is called the *Einstein tensor*, denoted by the symbol $G_{\mu\nu}$. If matter fluids are included in the gravitational action through the Lagrangian density $\mathscr{L}_M(g_{\mu\nu}, \partial_\alpha g_{\mu\nu})$, the field equations can be recast by introducing the corresponding energy-momentum tensor. Equation (4.8) becomes

$$R_{\mu\nu} - \frac{1}{2} g_{\mu\nu} R = 8\pi G \mathcal{T}_{\mu\nu}, \tag{4.9}$$

where $\mathcal{T}_{\mu\nu}$ comes directly from the variation of matter Lagrangian with respect to the metric. It is

$$\begin{aligned} \delta S &= \int \delta(\sqrt{-g}\mathscr{L}_M)\, d^4x \\ &= \int \left[\frac{\partial(\sqrt{-g}\mathscr{L}_M)}{\partial g^{\mu\nu}} \delta g^{\mu\nu} + \frac{\partial(\sqrt{-g}\mathscr{L}_M)}{\partial(\partial_\alpha g^{\mu\nu})} \delta \partial_\alpha g^{\mu\nu} \right] d^4x \\ &= \int \left[\frac{\partial(\sqrt{-g}\mathscr{L}_M)}{\partial g^{\mu\nu}} - \partial_\alpha \frac{\partial(\sqrt{-g}\mathscr{L}_M)}{\partial(\partial_\alpha g^{\mu\nu})} \right] \delta g^{\mu\nu}\, d^4x \equiv \int \frac{1}{2}\sqrt{-g}\mathcal{T}_{\mu\nu}\, \delta g^{\mu\nu}\, d^4x, \end{aligned} \tag{4.10}$$

where boundary terms have been integrated out and it is possible to define

$$\mathcal{T}_{\mu\nu} = \frac{2}{\sqrt{-g}} \left[\frac{\partial(\sqrt{-g}\mathscr{L}_M)}{\partial g^{\mu\nu}} - \partial_\alpha \frac{\partial(\sqrt{-g}\mathscr{L}_M)}{\partial(\partial_\alpha g^{\mu\nu})} \right]. \tag{4.11}$$

Notice that the definition in Eq. (4.11) is formally different from that in Eq. (3.18). In fact, the energy-momentum tensor conservation, expressed by Eq. (3.24), is not valid in the presence of a gravitational field and must be generalized considering the covariant derivative, that is, $D_\mu \mathcal{T}^{\mu\nu} = 0$. Moreover, Eq. (4.11) provides directly a symmetric rank-2 tensor, and unlike the energy-momentum tensor in Eq. (3.18) has to be rendered symmetric. As pointed out in Ref. [37], the equivalence between the Noether canonical energy-momentum tensor and the gravitational energy-momentum tensor occurs when matter fields do not couple to the metric derivatives, with the consequence that both lead to the same conservation laws. On the other hand, if matter fields are coupled to the metric derivatives, the equivalence only holds in the Minkowskian limit. If the starting Lagrangian depends both on the metric and on a set of scalar fields ϕ^i, the same procedure as in Section 3.2.1 yields

$$\begin{aligned} \partial_\mu(\sqrt{-g}\mathscr{L}_M) = &\frac{\partial(\sqrt{-g}\mathscr{L}_M)}{\partial \phi^i} \partial_\mu \phi^i + \frac{\partial(\sqrt{-g}\mathscr{L}_M)}{\partial(\partial_\nu \phi^i)} \partial_\nu \partial_\mu \phi^i \\ &+ \frac{\partial(\sqrt{-g}\mathscr{L}_M)}{\partial g_{\alpha\beta}} \partial_\mu g_{\alpha\beta} + \frac{\partial(\sqrt{-g}\mathscr{L}_M)}{\partial(\partial_\nu g_{\alpha\beta})} \partial_\mu \partial_\nu g_{\alpha\beta}, \end{aligned} \tag{4.12}$$

from which

$$\partial_\mu \left[\sqrt{-g} \left(\mathcal{T}^{\mu\nu}_{Noether} - \mathcal{T}^{\mu\nu}_{GR} + \frac{\partial \mathscr{L}_M}{\partial(\partial_\nu g_{\alpha\beta})} g^{\sigma\mu} \partial_\sigma g_{\alpha\beta} \right) \right] = 0, \tag{4.13}$$

where we used the condition $D_\mu \mathcal{T}^{\mu\nu}_{GR} = 0$. Equation (4.13) finally shows that, in flat space-time and when the Lagrangian is independent of the metric derivatives, the two definitions of the energy-momentum tensor lead to the same conserved quantities. In the case of perfect fluids, the matter energy momentum tensor has the form

$$\mathcal{T}_{\mu\nu} = (\rho + p)u_\mu u_\nu - p g_{\mu\nu}\,, \tag{4.14}$$

where ρ is the energy density, p the pressure, and u_ν the 4-velocity.

Linearizing the field equations, Einstein himself showed that they reduce to a wave equation [34] giving rise to gravitational waves (GW) solutions propagating outward from their source at the speed of light. Specifically, considering a small perturbation $h_{\mu\nu}$ of the Minkowski flat metric tensor

$$g_{\mu\nu} = \eta_{\mu\nu} + h_{\mu\nu}, \qquad |h_{\mu\nu}| \ll 1\,, \tag{4.15}$$

in the so-called *de Donder gauge*, where $\partial_\nu(h^\nu_\mu - 1/2\delta^\nu_\mu h^\alpha_{\ \alpha}) = 0$, a d'Alembert equation of the form

$$\Box h^\mu_\nu = -16\pi G\left(\mathcal{T}^\mu_\nu - \frac{1}{2}\delta^\mu_\nu \mathcal{T}\right) \tag{4.16}$$

can be obtained from the field equations [32]. Here $\mathcal{T}$ is the trace of the energy-momentum tensor. The operator $\Box$ is the d'Alembert operator in the flat space-time, that is, $\Box \equiv \partial_\mu \partial^\mu$, whose generalization in a curved background is $\Box \equiv D_\mu D^\mu$. In vacuum, the preceding equation describes propagating waves at the speed of light. The general solution, in the so-called *TT* gauge condition, reads

$$h_{\mu\nu} = e^+_{\mu\nu} h_+ + e^\times_{\mu\nu} h_\times\,,$$

where

$$e^+_{\mu\nu} = \begin{pmatrix} 0 & 0 & 0 & 0 \\ 0 & 1 & 0 & 0 \\ 0 & 0 & -1 & 0 \\ 0 & 0 & 0 & 0 \end{pmatrix}, \qquad e^\times_{\mu\nu} = \begin{pmatrix} 0 & 0 & 0 & 0 \\ 0 & 0 & 1 & 0 \\ 0 & 1 & 0 & 0 \\ 0 & 0 & 0 & 0 \end{pmatrix}. \tag{4.17}$$

This result means that GWs are described by spin-2 massless particles, with spin orientated in the same direction of motion.

For many years, GWs represented only a theoretical solution of field equations. In 2015, the Laser Interferometer Gravitational-Wave Observatory (LIGO) revealed a GW event (GW150914) and opened a new window in astrophysics and cosmology [38]. The GW production occurred during the merging of two black holes with masses of 29 $M_\odot$ and 36 $M_\odot$. The merging process produced a black hole of 62 $M_\odot$. The remaining 3 $M_\odot$ mass-energy was released in the form of gravitational radiation. The observation gave a double result: it confirmed the existence of GWs and was a direct evidence for stellar-mass black holes. This observational fact is a cornerstone in stating the validity of GR in the strong field regime.

After this first detection, several other events have been observed thanks to the LIGO-VIRGO collaboration, and further detections are expected in forthcoming years.

Furthermore, at the astrophysical scale, GR obtained great successes after observations of light deflection, followed by the Radar Echo Delay, the exact estimation of the precession of Mercury's perihelion in its orbit around the Sun, and many other fine observational measurements [34, 39].

Specifically, most of the successes just mentioned come from the application of the theory to a spherically symmetric space-time of the form

$$ds^2 = P(r,t)^2 dt^2 - Q(r,t)^2 dr^2 - r^2 d\Omega^2, \tag{4.18}$$

with Ω being the 2-sphere defined as $d\Omega^2 \equiv d\theta^2 + \sin^2\theta d\phi^2$. Once the interval (4.18) is replaced in the Einstein field equations, it turns out that the only possible solution is the Schwarzschild solution:

$$ds^2 = \left(1 - \frac{r_S}{r}\right) dt^2 - \left(1 - \frac{r_S}{r}\right)^{-1} dr^2 - r^2(\sin^2\theta d\phi^2 + d\theta^2), \tag{4.19}$$

which, thanks to the Birkhoff theorem, is static and stationary [34]. This solution contains two intrinsic singularities. One of them is the divergence occurring at $r = 0$, due to the curvature generated by the compact object. The other singularity occurs when the radius is equal to the "Schwarzschild radius" r_S, defined as

$$r_S = \frac{2GM}{c^2}, \tag{4.20}$$

with M being the mass of the compact object.[2] It can be shown that the latter singularity is coordinate-dependent and can be eliminated by means of an appropriate transformation (Kruskal–Szekeres coordinates [34]). The plane $r = r_S$ is the event horizon and can be interpreted as the boundary beyond which events cannot affect an observer. The recent black hole image at the center of M87 galaxy, reported from the Event Horizon Telescope collaboration (EHT), showed that these theoretical predictions are consistent with observations [40].

The application of GR to homogeneous and isotropic space-times led to the formulation of consistent cosmological models. The aim of relativistic cosmology is to derive, in principle, a self-consistent cosmic history ranging from the Big Bang to the dark energy–dominated era. Considering a Friedman–Lemaître–Robertson–Walker (FLRW) metric of the form

$$ds^2 = dt^2 - a(t)^2 \left[\frac{dr^2}{1 - kr^2} + r^2 d\Omega^2\right], \tag{4.21}$$

[2] Here, we explicitly indicate the speed of light c for dimensional reasons. In the rest of the book, we set $c = 1$, unless otherwise indicated.

and a diagonal energy-momentum tensor of the form

$$\mathcal{T}^{\mu}_{\nu} = \mathrm{diag}(\rho, -p, -p, -p), \tag{4.22}$$

the only nonvanishing components of the Einstein field equations are

$$\left(\frac{\dot{a}}{a}\right)^2 + \frac{k}{a^2} = \frac{8\pi G}{3}\rho, \tag{4.23}$$

$$\frac{\ddot{a}}{a} = -\frac{4\pi G}{3}(\rho + 3p). \tag{4.24}$$

They are called *Friedman cosmological equations* and are capable of describing a homogeneous and isotropic expanding universe in terms of cosmic time. In the preceding equations, $a(t)$ is the *cosmological scale factor* of the universe. The spatial curvature constant k can only assume the values $k = -1$ (hyperbolic universe), $k = 0$ (flat universe), and $k = 1$ (closed universe). Equations (4.23) and (4.24) can be implemented by considering the energy-momentum tensor conservation $D_\mu \mathcal{T}^{\mu\nu} = 0$, which, for the metric (4.21) and the tensor (4.22), reads

$$\frac{\partial(a^3\rho)}{\partial t} + p\frac{\partial(a^3)}{\partial t} = 0. \tag{4.25}$$

Using a cosmological perfect fluid with Equation of State (EoS) $p = w\rho$, where w is a constant indicating the nature of the cosmic fluid, and setting $k = 0$, Eqs. (4.23), (4.24), and (4.25) admit the solutions

$$\frac{a}{a_0} = \left(\frac{t}{t_0}\right)^{\frac{2}{3(w+1)}}, \qquad \rho(t) = \left[6\pi G(1 + w^2)t^2\right]^{-1}, \tag{4.26}$$

with t_0 and a_0 being real constants. It is worth noticing that the spatial flatness ($k = 0$) is usually taken for granted in cosmology, as it is suggested with high precision by most recent observations [41]. Depending on the value of w, different fluids can be identified. For example:

- $w = \frac{1}{3} \rightarrow a(t) \sim t^{1/2}$: radiation
- $w = 1 \rightarrow a(t) \sim t^{1/3}$: stiff matter
- $w = 0 \rightarrow a(t) \sim t^{2/3}$: dust matter.

It is worth noticing that it has to be $0 \leq w \leq 1$ (Zeldovich interval), for realistic perfect fluids characterized by $w = (c_s/c)^2$, where c_s is the speed of sound, which cannot exceed the speed of light. This means that Eq. (4.24) can give only decelerated expansions for standard matter perfect fluids.

The first observational basis for the expansion of the universe was provided by Edwin Hubble between 1924 and 1929, who formulated a law that states that galaxies are moving away at a radial speed v proportional to their distance d, according to the formula $v = H_0 d$. H_0 is the Hubble constant $H = \dot{a}/a$ calculated at the time of observation t_0.

Notice that, according to this law, the speed of the farthest galaxies may be higher than c. This result is still in agreement with the principles of Special Relativity since the speed of light is the upper limit of any object moving in a fixed space-time and this fact cannot be applied to the space-time itself. As a consequence, it is possible to define the boundary between particles that are moving slower and faster than the speed of light relative to an observer at a given time, namely $r_H(t) = c/H(t)$, which is called the Hubble horizon.

Some important remarks are necessary at this point. The Hubble constant defines the rate of expansion of the universe. According to the estimated present value $H_0 \sim 70\,\mathrm{km/s/Mpc}$, without considering problems related with its tension [42], it is not enough to separate the components of local structures, still held together by the gravitational interaction. If, at a certain epoch, the value of H increased, the acceleration would become dominant with respect to gravity and all the cosmic structures would be destroyed, up to disintegrating into elementary particles. The only surviving objects would be black holes, which in turn would disappear by emitting Hawking radiation [43]. This represents a possible ultimate fate of the Universe, called *Big Rip* [44]. In combination with the latter, the expansion could also yield a universe that asymptotically approaches the absolute zero temperature, giving rise to a *Big Freeze* [45]. Another possibility, called *Big Crunch*, is that at a certain point in the future, the average density will be able to stop the expansion, so that the universe will start collapsing up to a dimensionless singularity. This case cannot be fully explained within the current understanding of gravity, as the quantum effect would become dominant at small scales. According to the *Big Bounce* model, the universe might continuously repeat the cycle of a Big Bang, followed up with a Big Crunch [46].

Apart from the preceding considerations, starting from 1998, several observations showed that the cosmic flow is expanding at an accelerating rate. This fact is in contradiction with Eq. (4.24) if only perfect fluids in the Zeldovich interval are considered. The issue can be solved assuming a positive *cosmological constant* Λ in the gravitational action (4.7). Λ was first introduced by Einstein himself in 1917 to stabilize his cosmological model, but neglected after the Hubble discovery.

Including Λ in the Einstein field equations, we have

$$R_{\mu\nu} - \frac{1}{2} g_{\mu\nu} R + \Lambda g_{\mu\nu} = 8\pi G \mathcal{T}_{\mu\nu}. \tag{4.27}$$

In a general FLRW background, Eq. (4.27) yields the following system of differential equations:

$$\left(\frac{\dot{a}}{a}\right)^2 + \frac{k}{a^2} = \frac{8\pi G\rho + \Lambda}{3}, \tag{4.28}$$

$$\frac{\ddot{a}}{a} = -\frac{4\pi G}{3}(\rho + 3p) + \frac{\Lambda}{3}. \tag{4.29}$$

When the cosmological constant is the dominant source, the cosmological equations give a scale factor of the form

$$a(t) = \begin{cases} a_0 \sinh\left(\sqrt{\frac{\Lambda}{3}}\, t\right) & k=-1 \\ a_0 e^{\sqrt{\frac{\Lambda}{3}}\, t} & k=0 \\ a_0 \cosh\left(\sqrt{\frac{\Lambda}{3}}\, t\right) & k=1, \end{cases} \tag{4.30}$$

denoting an exponentially accelerating universe. The first solution denotes the so-called *Anti-de Sitter* (AdS) space-time, which is the only case where the cosmological constant takes negative values. Notice that, in vacuum and when $k = 0$, it is possible to define effective energy density and pressure, provided by Λ, of the form

$$\begin{aligned} \rho_\Lambda &= \frac{\Lambda}{8\pi G}, \\ p_\Lambda &= -\frac{\Lambda}{8\pi G}, \end{aligned} \tag{4.31}$$

so that the EoS becomes $p = -\rho$, namely $w = -1$. The cosmological constant is a particular case of a more general problem because the cosmic accelerating expansion can be physically addressed by an unclustered fluid with negative pressure. It represents 68 percent of the bulk of the universe. It is dubbed *dark energy* and its fundamental nature is today unknown. Λ is a possibility for such a fluid and it is used as a paradigm for the so-called ΛCDM model, which agrees very well with observations [47].

Those mentioned in this section are only a few of the positive results gained by GR during more than one hundred years. Despite these successes, some results disagree with experiments and astrophysical observations. Specifically, there are good theoretical and experimental reasons to consider seriously the possibility that gravity is not fully described by Einstein's GR and alternative or extended theory has to be invoked.

For example, GR cannot predict the right correlation between mass and radius of some neutron stars [48–50]. Furthermore, galaxy rotation curves point out a missing matter problem, which cannot be explained without introducing exotic forms of matter [51, 52]. The latter relies on the speed of the farthest stars orbiting around the center of a given galaxy, which is experimentally higher than theoretically expected, considering the weak field limit of GR in agreement with Newton gravity. In order to theoretically fix this issue, the missing matter can be attributed to a zero pressure fluid, called *dark matter*, which does not interact electromagnetically.

Dark matter is supposed to represent 26.8 percent of the matter-energy content of the universe [53] (see Fig. 4.1), but, despite several particle candidates, it has

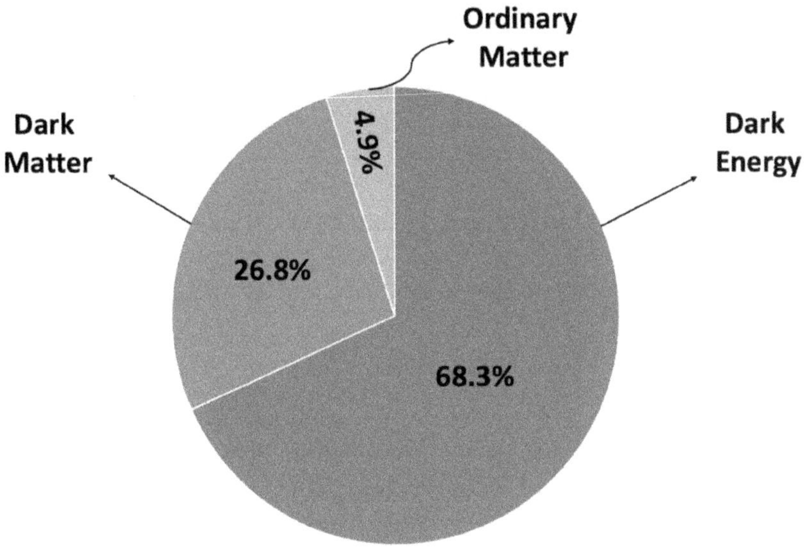

Figure 4.1 Energy-matter content in the universe. See Ref. [53]

never been directly detected, up to now, at a fundamental level. Moreover, gravity is still very poorly tested at the scales of galaxies and clusters, witnessing the fact that even Newtonian gravity is doubted at large scales. This fact led to the introduction of Modified Newtonian Dynamics (MOND) [54–56]. The approach consists of an alternative paradigm to dynamics, seeking to replace the standard Newtonian dynamics and taking into account the ubiquitous mass discrepancies without invoking any dark component. The main ingredient is that, at certain astrophysical scales, a new acceleration a_0 emerges and fixes the discrepancies.

On the other hand, cosmology cannot be advocated as a precise test of GR at large scales: in fact, almost all theories of gravity admit the FLRW line element as a solution of their field equations, with perfect fluids or other reasonable matter sources. Indeed, it is from cosmology that indications come that gravity at any scale may not be described exactly by GR.

As said before, the 1998 discovery that the present expansion of the universe, according to the luminosity distance versus redshift relation of type Ia supernovae, appears to be accelerating, has left cosmologists scrambling for an explanation. As the accelerating expansion cannot be predicted by GR without invoking the presence of dark energy, the galaxy rotation curve cannot be fitted without any dark matter or some improved dynamics [54]. Attempts to explain away dark energy using, for example, the backreaction of inhomogeneities on the background universe dynamics have been, so far, unconvincing [57].

However, even admitting that dark energy is actually described by the cosmological constant, it turns out that the theoretically predicted value of Λ by QFT is 120 orders of magnitude higher than the experimentally calculated one [58]. This disagreement arises by considering the vacuum energy density coming from the first Friedman equation (4.28), that is,

$$H^2 + \frac{k}{a^2} = \frac{\Lambda + 8\pi G\rho}{3}. \tag{4.32}$$

Therefore, the quantity $\frac{\Lambda}{8\pi G}$ can be intended as an effective vacuum energy density ρ_Λ.

Evaluating Eq. (4.32) at the present time and setting $\rho = 0$, we get

$$\rho_\Lambda = \frac{3H_0^2}{8\pi G}. \tag{4.33}$$

On the other hand, the vacuum energy density can also be derived by the zero-point energy fluctuations of a free quantum field with mass m and momentum P. This field, in turn, can be formally thought as a set of infinite harmonic oscillators in momentum space, whose zero point energy diverges. Nonetheless, as standard in QFT, this obstacle is overcome by considering a cutoff P_{Max} for the momentum, so that the zero-point energy density can be computed as

$$\rho'_\Lambda = \rho_{vac} = \int_0^{P_{Max}} \frac{1}{2}\frac{d^3P}{(2\pi)^3}\sqrt{P^2 + m^2} \sim \left.\frac{P^4}{16\pi^2}\right|_{P=P_{Max}}, \tag{4.34}$$

where we assumed $m \ll P$. Reasonably, the cutoff P_{Max} should be of the order of the Planck momentum $P_{Planck} = (\hbar/4\pi G)^{1/2}$ [58], so that Eq. (4.34) finally yields

$$\rho'_\Lambda = \frac{\hbar^2}{(16\pi^2 G)^2}, \tag{4.35}$$

and comparing Eq. (4.35) with Eq. (4.33) one gets

$$\frac{\rho'_\Lambda}{\rho_\Lambda} = \frac{\hbar^2}{96\pi^3 G H_0^2} \sim 10^{121}. \tag{4.36}$$

Other GR shortcomings are the space-time singularities. For example, the Kruskal–Szekeres maximal extension of the Schwarzschild solution exhibits a true space-time singularity, which cannot be removed by further coordinate transformations [34]. This yields several problems in the interpretation of black holes since, according to GR, all the matter content is supposed to be concentrated at the gravitational center. Such an interpretation is physically unrealistic, as the energy-matter density inside the black hole would increase up to infinity.

Several attempts toward the resolution of this problem have been recently proposed, such as the so-called "non-singular black hole models" [59], which avoid some theoretical shortcomings like information loss and singularity at the gravitational center [60]. The simplest viable description of a non-singular black hole is provided by the Hayward metric [61–63], which is regular, static, spherically symmetric, and asymptotically flat. Today some of the most viable candidates include the gravastars [64] and the dark energy stars [65]. However, the majority of non-singular models are ruled out by theoretical incompatibilities, such as instabilities in extremely fast rotation conditions.

Recently, another attempt to solve the singularity problem was proposed in Ref. [66], where the authors hypothesize the existence of a compact object without physical singularities, but with an event horizon, called a *black star*. It should be experimentally indistinguishable from a black hole, and its formation should be driven by a slow gravitational collapse.

In spite of these attempts, most likely, a reasonable solution can be addressed to the lack of a quantum theory of gravity, capable of settling GR issues even at the Planck scale. Loop Quantum Gravity, for instance, uses the canonical quantization prescription to provide suitable explanations for these issues. By means of quantum discreteness, this theory addresses the thermal properties of black holes close to equilibrium and the fate of information in black hole evaporation [67–70].

A brief overview of the problem of Quantum Gravity is given in Chapter 5. However, at the quantum level the "local" formulation of GR seems completely in disagreement with the intrinsic "nonlocality" of Quantum Mechanics.

Quantum Mechanics is the most revolutionary theory of the last century, which opened the doors to a completely new vision of Physics. The determinism of Classical Mechanics was replaced by a probabilistic interpretation of small-scale phenomena, which seemed to be the only way to fit experimental results coming from microphysics. As we gained a theory capable of describing almost all the evidences provided by the quantum world, we lost the capability to exactly predict the time evolution of systems. Soon after, QFT arose with the purpose of describing all the fundamental interactions under the standard of quantum fields. It was soon clear that this prescription could not be applied to the gravitational interaction. As a matter of fact, as Quantum Mechanics is probabilistic by nature, gravity is in turn described by Einstein's GR, where nonlocal interactions are not allowed. So far, a theory capable of describing both the large-scale structure and the UV scale results is still missing. Moreover, neither QFT nor GR hold at the Planck scale, where a new physics is needed.

More precisely, unlike the other fundamental interactions, GR cannot be treated under the Yang–Mills formalism, due to the lack of a Hilbert space and a probabilistic interpretation of the wave function. As a matter of fact, while the electroweak and the strong interactions are Yang–Mills gauge theories, GR is invariant under diffeomorphism transformations, which involve coordinates instead of fields. Moreover, according to the geometric description of gravity, in view of a possible quantum approach, the space-time metric represents both the dynamical field and the background; this is not the case with other interactions, whose treatment is simplified by the assumption that the space-time is given and supposed to be flat.

In 1988 Lasenby, Doran, and Gull proposed to deal with the flat tangent space of the Riemannian manifold, treating GR as a gauge theory with respect to the local LG [71]; in order to pass from the curved to the flat space-time, *tetrad fields* are necessary, which, in turn, become the fundamental dynamical fields. The formalism adopted is the so-called Einstein–Cartan formalism [72–75], where the connection is generally independent of the metric and the two-form curvature is found by the Cartan structure equations [72, 76]. This implies that curvature can be used along with torsion to simultaneously label the space-time so that the theory reduces to standard GR as soon as the antisymmetric part of connection vanishes. This approach is not aimed at solving all the problems occurring in GR at the small scales since even under an Einstein–Cartan formalism, several shortcomings are still suffered. As an example, neglecting the asymptotic safety scenario [77–79], the theory can be renormalized only up to the one-loop level.

Attempts to renormalize GR in the 1960s and 1970s showed clearly that counterterms must be introduced that alter the theory significantly and make its field equations of fourth order instead of second. From the physical point of view, this fact implies that extra degrees of freedom, in addition to the usual spin two graviton, need to be introduced. Unfortunately, the corrected theory is not ghost-free, which makes it non-unitary. Furthermore, once trying to handle gravity (and the other interactions) in the high-energy framework, Einstein's theory cannot be recovered in its weak field limit. It seems that first loop corrections or attempts to fully quantize gravity necessarily introduce significant deviations from GR and extra degrees of freedom.

Even though QFT in curved space-time addressed several results provided by the small-scale phenomenology (such as the Hawking Radiation, the Unruh effect, or cosmic inflation), it suffers several shortcomings. As we have just mentioned, one of them is given by the value of the cosmological constant. Another problem is the so-called *information paradox* [80], which relies on the loss of information occurring when an object approaches the event horizon and enters the black hole. More precisely, according to the no-hair theorem, black hole solutions are completely characterized by mass, electric charge, and angular momentum. Therefore, Hawking radiation should be completely independent of the material entering the black hole. However, at the quantum level, the transformation of such material into Hawking radiation would destroy information about the original state, violating the law of conservation of information.

For these reasons, a coherent and self-consistent theory of Quantum Gravity is today's one of the most studied topics of Theoretical Physics [81–86]. All the attempts pursued so far lack at least one of the necessary requirements for a Quantum Theory such as renormalizability, unitarity, or predictability. This state of the art points out the necessity to go beyond GR.

However, in the last few years, the quantum formalism has been adapted to cosmology, where the dynamics can be reduced by considering *minisuperspaces*. Quantum Cosmology is an approach that does not claim to be the complete theory of Quantum Gravity, but yields several interesting results to understand the early stages of the universe [87–89].

As better pointed out in the next section, modified theories of gravity arose with the purpose of formulating at least a semiclassical scheme in which GR and its successes could be replicated. One of the most fruitful approaches is the so-called Extended Theories of Gravity, which have, in some sense, become a paradigm in the study of gravitational interaction. They are based on corrections of the Einstein theory, which as to be recovered in any case due to its fundamental results. This is the case, for example, with of $f(R)$ gravity, which introduces into the action a function of the scalar curvature R. However, the EH Lagrangian can be extended in several ways, such as introducing the coupling between geometry and scalar fields, higher-than-fourth-order terms involving powers of the d'Alembert operator $\Box^n$ or higher-order curvature invariants (as well as $R^{\mu\nu}R_{\mu\nu}$ or $R^{\mu\nu\rho\sigma}R_{\mu\nu\rho\sigma}$). Indeed, every scheme unifying the fundamental interactions (such as Supergravity or String Theory) produces effective actions where non-minimal couplings to the geometry or higher-order terms in the curvature invariants are necessarily present. Such contributions are due to first or higher loop corrections in the high curvature regime approaching the full (and still speculative) Quantum Gravity regime. In this perspective, Einstein GR can be either a particular case among a wider class of theory or the theory to be recovered in some weak field regime (the so-called post-Einsteinian limit [90]).

From the conceptual point of view, there is no a priori reason to restrict the gravitational Lagrangian to being a linear function of the Ricci scalar. Exhaustive discussions on modified theories of gravity can be found in [91]. See also, for example, [92–96] for $f(R)$ gravity, [97–102] for teleparallel gravity, [103–106] for scalar-tensor gravity, [107–109] for actions depending on second-order curvature scalars. In the following sections, we will introduce the main classes of extended/modified theories of gravity. It is worth pointing out that claiming credit for a final and self-consistent theory of gravity at any scale is premature and unjustified at this moment. However, even if none of these alternative approaches ultimately proves to be the final one, they solve several problems of Einstein gravity and allow one to understand many ways in which gravity can be enlarged and modified.

4.2 Different Viewpoints in Theories of Gravity

The previous considerations illustrate the fact that one can envisage alternative gravitational theories starting from the basic assumptions used by Einstein and Hilbert in the construction of GR and, eventually, adding or relaxing some of them (see the beginning of Section 4.1). In these theories gravity is described by either a metric (purely metric theories) or a linear connection (purely affine theories) or by both fields (metric-affine theories). An open issue is identifying the fundamental variables (and the number of degrees of freedom) characterizing the gravitational field: they can be the metric tensor, the linear connection, and the tetrad fields in various combinations as we will discuss in the following sections [110].

A number of attempts to generalize GR and unify it with Electromagnetism along these lines were followed by Einstein and many others, including Eddington, Weyl, and Schrödinger, to mention a few. These attempts were eventually abandoned in the 1950s, though the search for a coherent quantum theory of gravitation, or the belief that gravity has to be considered as a sort of low-energy limit of some quantum theory, has recently revived the idea that it is not mandatory to adhere to the simple prescription of Einstein and Hilbert and to assume that classical gravity is governed by a Lagrangian linear in the curvature. Further curvature and torsion invariants or nonlinear functions of them can also be contemplated, especially in view of the fact that their inclusion is required in both the semiclassical expansion of a quantum Lagrangian and in the low-energy limit of stringy actions. Moreover, it is clear from recent astrophysical observations and from the current cosmological investigations that it is legitimate to doubt the paradigmatic role played by the Einstein equations at Solar System, galactic, extragalactic, and cosmological scales, unless one is willing to admit the presence of some types of exotic energy or matter.

The idea discussed in this section is, in principle, much simpler. Instead of changing the matter side of the Einstein equations and introducing the missing matter-energy content of the observed universe, while adding mysterious and odd-behaving states of the matter fields, one can take into account the fact that it could be simpler and more convenient to change the geometric/gravitational sector of these equations by inserting, into the gravitational Lagrangian, corrections coming from some effective fundamental theory. This procedure could be regarded as a mere matter of taste; however, there is no a priori reason to consider a gravitational Lagrangian linear in the Ricci scalar R and the possibility of considering more general and alternative approaches is intriguing and worthy of being explored. In principle, the action belongs to a vast family of permissible actions and, from the purely phenomenological point of view, this freedom allows it to be chosen on the basis of best-fits with the available observational data at IR scales (solar, galactic, extragalactic, and cosmological) as well as at UV scales (quantum gravitational effects). The downside of this approach is that too many models can fit observations and experiments because of the relatively large number of free functions and parameters that they contain. In this perspective predictive power may be lost. For this reason, using some general criteria, like the existence of Noether symmetries, can provide guidelines pointing to discriminate among large classes of models fitting data coming from observations and experiments.

GR can be modified in several ways since, as previously pointed out, it does not account for phenomenology of gravitational interaction at any scale. However the description of gravity through the space-time geometry is perhaps one of the greatest intuitions of the twentieth century, and the consequent experimental verifications marked a turning point in the physical comprehension of phenomena. In order to gain such a predictive power and to obtain consistent results, many hypotheses were adopted; in what follows we analyze two main assumptions lying behind GR, namely the Equivalence Principle and second-order field equations. As we will see, by relaxing these requirements, one can obtain different classes of modified theories of gravity, which will be analyzed in the rest of this chapter.

- Equivalence Principle and Symmetric Connection.

Let us first consider the assumption of symmetric connection, based on the requirement for the validity of the Equivalence Principle. The Weak Equivalence Principle states that there is no difference, locally, between gravitational field and accelerating systems, so that a free-falling reference frame is completely equivalent to a system with no gravitational field. In other words, according to the Weak Equivalence Principle, it is always possible to locally link the curved space-time to a flat tangent Minkowski space-time. The equivalence between the gravitational and the inertial mass is then automatically implied, and nowadays several experiments confirm such an equivalence with a precision of 1 part over 10^{14} [111]. Despite these results motivated by observations, which surely hold at macroscopic scales, it is still unclear whether they keep at quantum scales. In order to treat the gravitational interaction under the same standard as the other fundamental interactions and to construct a coherent theory of Quantum Gravity, such a precision of 10^{-14} makes a crucial difference, since it admits the possibility that, at higher energies or quantum scales, the ratio m_g/m_I might pull rapidly away from 1 (see, e.g., [112]).

In GR, the form of the affine connection can be found by imposing the "metricity condition" $D_\alpha g_{\mu\nu} = 0$, by means of which the identity

$$D_\alpha g_{\mu\nu} + D_\nu g_{\alpha\mu} - D_\mu g_{\nu\alpha} = 0 \tag{4.37}$$

holds. Considering the definition of the covariant derivative and assuming $\Gamma^\alpha_{[\mu\nu]} = 0$, it turns out that the only possible connection in GR is the Levi–Civita connection; that is:[3]

$$\hat{\Gamma}^\alpha{}_{\mu\nu} = \frac{1}{2} g^{\alpha\rho} \left(\partial_\mu g_{\rho\nu} + \partial_\nu g_{\mu\rho} - \partial_\rho g_{\mu\nu} \right). \tag{4.38}$$

However, once the metricity condition and the hypothesis of symmetric connection are relaxed, the same computation leads to a more general form of $\Gamma^\alpha_{\mu\nu}$, including other nontrivial terms. It reads as follows [113–115]:

[3] Hereafter in this chapter, the Levi–Civita connection will be denoted by a hat on the top.

$$\Gamma^{\alpha}{}_{\mu\nu} = \hat{\Gamma}^{\alpha}{}_{\mu\nu} + \frac{1}{2}g^{\alpha\lambda}(T_{\mu\lambda\nu} + T_{\nu\lambda\mu} + T_{\lambda\mu\nu}) + \frac{1}{2}g^{\alpha\lambda}(-Q_{\mu\nu\lambda} - Q_{\nu\mu\lambda} + Q_{\lambda\mu\nu}), \quad (4.39)$$

where we denote with $\hat{\Gamma}^{\alpha}{}_{\mu\nu}$ the Levi–Civita connection and by Q, T rank-3 tensors defined as

$$Q_{\beta\mu\nu} = \nabla_{\beta} g_{\mu\nu}, \qquad T^{\alpha}{}_{\mu\nu} = 2\Gamma^{\alpha}{}_{[\mu\nu]}. \quad (4.40)$$

The formalism in which the metric is disentangled from the connection, so that the latter is no longer "metric compatible," is known as *Einstein–Cartan–Sciama–Kibble* formalism. Standard GR is recovered when $\Gamma^{\alpha}{}_{\mu\nu} = \hat{\Gamma}^{\alpha}{}_{\mu\nu}$, namely when both $T^{\alpha}{}_{\mu\nu}$ and $Q_{\beta\mu\nu}$ vanish. In order to show that the Equivalence Principle implies symmetric connection as the only possible connection, let us consider a generic vector field A_{μ} and the rank-two tensor $\nabla_{\nu} A_{\mu}$, namely the covariant derivative of A_{μ} with respect to a general connection. Requiring that A_{μ} can be written as the gradient of a scalar field ϕ, namely $A_{\mu} = \partial_{\mu}\phi$, the quantity $\nabla_{[\nu} A_{\mu]}$ reads as

$$\nabla_{[\nu} A_{\mu]} = -\Gamma^{\alpha}{}_{[\mu\nu]} \partial_{\alpha}\phi. \quad (4.41)$$

If the Equivalence Principle holds, there exists a free fall reference frame where the covariant derivative turns into the standard partial derivative, so that the LHS must vanish. Being a rank-two tensor, the LHS must be zero in all reference frames as well as the antisymmetric part of the Christoffel connection $\Gamma^{\alpha}{}_{[\mu\nu]}$. As a consequence, $\Gamma^{\alpha}{}_{\mu\nu} = \hat{\Gamma}^{\alpha}{}_{\mu\nu}$ and, therefore, $\nabla_{\mu} = D_{\mu}$. In other words, if the Equivalence Principle holds, the geodesic structure of space-time coincides with its metric structure.

On the contrary, the antisymmetric contribution must be taken into account in the general form of the connection, when the Equivalence Principle is not assumed a priori. Nevertheless, field equations arising from general affine connections require other additional conditions to be solved. In this case, the metric is not enough to uniquely determine the dynamics of the system, which needs to be further constrained by other external impositions. Reversing the argument, by choosing a metric-compatible connection, the metric tensor turns out to be the only fundamental field needed to solve the field equations. The consequences of such an assumption affect the description of space-time. When the rank-3 tensor $T^{\alpha}{}_{\mu\nu}$ is included in the connection, torsion arises in the geometric description of space-time. $T^{\alpha}{}_{\mu\nu}$ is called the *Torsion Tensor*. According to GR, torsion is not admitted in space-time dynamics, which is fully described by curvature only.

Similarly, by relaxing the Metricity Principle, according to which the covariant derivative of the metric tensor must vanish, the rank-3 tensor $Q^{\alpha}{}_{\mu\nu}$ arises in the affine connection. As better pointed out in Section 4.7, this implies a space-time in which the norm of a vector changes while being parallel transported along a closed path. This means a violation of isometry of the given space-time manifold.

Therefore, if all the three terms in Eq. (4.39) are considered at the same level, it is possible to construct theories where the space-time is ruled by torsion or nonmetricity (or both). The former theory is usually called *Teleparallel Equivalent of General Relativity* (TEGR), while the latter is the *Symmetric Teleparallel Equivalent of General Relativity* (STEGR). Hereafter, up to Section 4.6, all scalar, vector, and tensor quantities are written in terms of the Levi–Civita connection.

- Second-Order Field Equations

Another assumption that is not based on strong theoretical foundations is that field equations have to be of second order. According to this request, the EH action is chosen linear in the scalar curvature. This choice is historically motivated by the requirement that gravitational equations of Classical Mechanics, like Poisson's and Laplace's, are of second order as well as those of Electromagnetism. According to this analogy, Einstein and Hilbert imposed the gravitational field equations to be of second order with respect to the metric. The only nontrivial four-dimensional action respecting the general covariance and leading to second-order field equations must be linear in the scalar curvature. It is the well-known EH action, namely[4]

$$S_{E-H} = \frac{1}{2}\int \sqrt{-g}R\, d^4x. \tag{4.42}$$

By relaxing this hypothesis, several extensions of GR can be developed, with actions also involving higher-order geometric terms, like, for example, R^2, $R_{\mu\nu}R^{\mu\nu}$, or $R^{\mu\nu\rho\sigma}R_{\mu\nu\rho\sigma}$. More precisely, functions of such curvature invariants can be generally considered as well as the coupling between geometry and scalar fields. For example, the action

$$S = \int \sqrt{-g}F(\phi, R, \Box R, \Box^2 R, \dots, \Box^z R, R^{\mu\nu}R_{\mu\nu}, R^{\mu\nu\rho\sigma}R_{\mu\nu\rho\sigma}), \qquad z \in \mathbb{Z}, \tag{4.43}$$

which depends on the scalar curvature R, on the quantity $\Box^z R$, on higher-order terms like $R_{\mu\nu}R^{\mu\nu}$, $R^{\mu\nu\rho\sigma}R_{\mu\nu\rho\sigma}$, and on couplings between geometry and scalar fields, leads generally to $(2z+4)$-th order field equations, with z being a relative number.[5] Moreover, Klein–Gordon-like equations can be obtained by varying the action with respect to the scalar field.

As pointed out in the previous section, though GR achieved several successes at large scales, it manifests some shortcomings that call its validity into question. Therefore, it seems natural to wonder whether GR could be the low-energy limit of some unified theory that holds from the Planck scale up to the cosmological scale. Modified theories of gravity might represent good candidates toward a final and comprehensive theory of gravitational field since most of them are capable of solving several issues at small and large scales. In what follows, we

[4] We set $G = 1/8\pi$, as in the rest of the book, unless otherwise indicated.
[5] As we will see in what follows, negative z give rise to nonlocal terms.

discuss the most-known GR modifications, sketching their main features and the most important results. Then, in the second part of the book, we apply the Noether Symmetry Approach to modified gravity actions depending on general unknown functions of some geometric term in order to select models containing symmetries.

4.3 Curvature Extensions

The simplest extension of GR is the so-called $f(R)$ gravity. It includes a function of the scalar curvature into the action

$$S = \int \sqrt{-g} f(R)\, d^4x. \tag{4.44}$$

By varying the action with respect to the metric, one gets:[6]

$$G_{\mu\nu} = \frac{1}{f_R(R)} \left\{ \frac{1}{2} g_{\mu\nu} \left[f(R) - R f_R(R) \right] + f_R(R)_{;\mu;\nu} - g_{\mu\nu} \Box f_R(R) \right\}, \tag{4.45}$$

being $G_{\mu\nu}$ the Einstein tensor $G_{\mu\nu} = R_{\mu\nu} - \frac{1}{2} g_{\mu\nu} R$ and $f_R(R)$ the first derivative of $f(R)$ with respect to R. Notice that when $f_R = 1$, Einstein field equations are recovered, that is $f(R) = R$. In the presence of matter, with energy momentum $\mathcal{T}_{\mu\nu}$, Eq. (4.45) becomes

$$f_R(R)\, G_{\mu\nu} = \frac{1}{2} g_{\mu\nu} \left[f(R) - R f_R(R) \right] + f_R(R)_{;\mu;\nu} - g_{\mu\nu} \Box f_R(R) + \mathcal{T}_{\mu\nu}, \tag{4.46}$$

so that the trace equation reads

$$R f_R(R) - 2 f(R) + 3 \Box f_R(R) = \mathcal{T}. \tag{4.47}$$

By comparing Eq. (4.45) with GR field equations in vacuum, it is possible to interpret the quantity

$$\overline{\mathcal{T}}^{Curv}_{\mu\nu} = \frac{1}{f_R(R)} \left\{ \frac{1}{2} g_{\mu\nu} \left[f(R) - R f_R(R) \right] + f_R(R)_{;\mu;\nu} - g_{\mu\nu} \Box f_R(R) \right\} \tag{4.48}$$

as an effective energy-momentum tensor of the gravitational field, so that the $f(R)$ field equations take the form $G_{\mu\nu} = \overline{\mathcal{T}}^{Curv}_{\mu\nu}$. In this way, the RHS can play the role of effective energy density and pressure, violating the energy conditions (see Chapter 16) and leading to an accelerating expansion at cosmological scales at late-time [116]. Therefore, as pointed out in the third part of this book, dark matter and dark energy can be mimicked by geometric contributions, and often no exotic matter is needed to fit some of the current astrometric data [93]. In [116], for instance, the dark energy behavior can be reproduced by the geometry provided by $f(R)$ while, in [117], it is shown that (4.48) is definitely a perfect fluid that can accelerate the cosmos.

[6] See Appendix A for the details of calculations.

However, solving simultaneously the dark matter and dark energy problems is a difficult task because data at galactic and cluster scales can imply different models with respect to the dark energy problem. For instance, $f(R)$ extensions of GR can reasonably fix some of these issues by considering specific shapes of the function into the starting action. Though these specific choices turn out to be valid at some scales, they are often ruled out by experiments at different scales, where other modifications result more appropriate. In this situation, achieving a comprehensive $f(R)$ action matching the whole cosmic history and phenomenology could be extremely difficult. However, valid attempts in this direction have been proposed. See, for example, [118, 119].

Although so far no modified theory of gravity can fit the whole set of cosmic observations at once, the hope is that the gravitational action may include different terms, with different coupling constants that become dominant at certain scales [120]. In other words, the action could be made of more contributions, which become subdominant at given large (or small) scales. This would explain why different modified theories are needed to fit observations at different scales.

In this perspective, as we have pointed out, from the conceptual point of view there is no a priori reason to restrict the gravitational Lagrangian. Furthermore, the idea that there are no "exact" laws of physics has been seriously considered: in such a case, the effective Lagrangians of physical interactions would be average quantities arising from stochastic behavior at a more microscopic level. This feature would mean that local gauge invariances and the related conservation laws are well approximated only in the low-energy limit, and the fundamental constants of physics can vary [121].

Among the $f(R)$ extensions, great success was gained by the *Starobinsky* model, first treated in [122] and whose action is

$$S = \frac{1}{2}\int \sqrt{-g}\,(R + \alpha R^2)\, d^4x. \tag{4.49}$$

The preceding action was considered in order to explain the evolution of the Universe in its early stage. A logarithmic-corrected extension of such an action is reported in Ref. [123], where the authors propose a model capable of unifying the constant-roll inflation with the dark energy era. Moreover, considering the action in Eq. (4.49), the field equation (4.46) can be recast as

$$G_{\mu\nu} + \alpha\left[2R\left(R_{\mu\nu} - \frac{1}{4}g_{\mu\nu}R\right) + 2\left(g_{\mu\nu}\Box R - D_\nu D_\mu R\right)\right] = \mathcal{T}_{\mu\nu}, \tag{4.50}$$

thus the trace equation reads

$$6\alpha\Box R - R = \mathcal{T}\,. \tag{4.51}$$

One can also define a frequency ω (equivalent to a mass m) so that

$$-\frac{1}{6\alpha} = \omega^2 = m^2. \tag{4.52}$$

After this definition, Eq. (4.51) becomes:

$$\Box R + m^2(R + \mathcal{T}) = 0\,, \tag{4.53}$$

which shows that α must be negative in $(+---)$ signature.[7] After the early phases of the universe, as the temperature decreases with the expansion, the term proportional to m^2 becomes dominant. Equation (4.53) can be seen as an effective Klein–Gordon equation for the effective scalar field R (called the *Starobinsky scalaron*).

While GR field equations are of the second order, $f(R)$ gravity, in metric formalism, leads to fourth-order field equations unless $f(R) = R$.

Higher than fourth-order field equations can be obtained by further generalizing the action (4.44), including higher derivatives of the scalar curvature, namely

$$S = \int \sqrt{-g}\, f(R, \Box R, \Box^2 R, \ldots \Box^k R)\; d^4x. \tag{4.54}$$

The variational principle (for $k > 0$) yields the following field equations:

$$\begin{aligned} G^{\mu\nu} = \frac{1}{\mathcal{M}} \Bigg\{ & \frac{1}{2} g^{\mu\nu}(F - \mathcal{M}R) + (g^{\mu\lambda}g^{\nu\sigma} - g^{\mu\nu}g^{\lambda\sigma})\mathcal{M}_{;\lambda;\sigma} \\ & + \frac{1}{2}\sum_{i=1}^{k}\sum_{j=1}^{i}(g^{\mu\nu}g^{\lambda\sigma} + g^{\mu\lambda}g^{\nu\sigma})(\Box^{j-1}R)_{;\sigma}\left(\Box^{i-j}\frac{\partial F}{\partial \Box^i R}\right)_{;\lambda} \\ & - g^{\mu\nu}g^{\lambda\sigma}\left[(\Box^{j-1}R)_{;\sigma}\left(\Box^{i-j}\frac{\partial F}{\partial \Box^i R}\right)\right]_{;\lambda}\Bigg\}, \end{aligned} \tag{4.55}$$

where $\mathcal{M}$ is defined as

$$\mathcal{M} \equiv \sum_{j=0}^{k}\left(\Box^j \frac{\partial F}{\partial \Box^j R}\right). \tag{4.56}$$

The differential equations (4.55), which reduce to those of $f(R)$ gravity when $k = 0$, are of the $(2k + 4)$-th order as we have already noticed. These theories can be considered in order to provide solutions for the renormalizability problem of gravity at IR scales and corrections to the Newtonian potential, which could have astrophysical interest [124]. The coupling constants of higher-order terms in the geometry have the dimensions of nonnegative mass, which make the corresponding theory power-counting renormalizable. As a matter of fact, when the gravitational action is expanded around a Gaussian fixed point, the coupling constant mass dimensions suggests whether the theory can be renormalized. In GR, the coupling constant turns out to have the dimensions of a negative mass, though the asymptotic safety argument endows a possible convergence at UV scales. Nonetheless, as mentioned at the beginning of this chapter, GR cannot be recast as a gauge invariant theory with respect to some unitary group,

[7] The opposite holds for the $(-+++)$ signature [36].

like other interactions. The action (4.54), under given limits, is said to be *non-local* [125, 126]. In this context, the parameter k is chosen to be negative, so that the operator $\Box^k$ can be recast as a propagator. A further discussion of this argument is reported in Chapter 11, where the form of the unknown nonlocal functions is selected by the Noether approach.

However, apart from introducing functions of the Ricci scalar (and its derivatives), extended actions can also contain other curvature invariants. GR assumes that the Riemann tensor must be contracted with the metric tensor such that the starting action is

$$S_{E-H} = \frac{1}{2}\int \sqrt{-g}\, g^{\mu\rho} g^{\nu\sigma} R_{\mu\nu\rho\sigma}\, d^4x. \tag{4.57}$$

In this way, several other curvature invariants are neglected, which in principle may contribute to the dynamics at the same level as the scalar curvature. For instance, with the definitions

$$P \equiv g^{\mu\rho} R^{\nu\sigma} R_{\mu\nu\rho\sigma}, \qquad \mathcal{Q} \equiv R^{\mu\nu\rho\sigma} R_{\mu\nu\rho\sigma} \tag{4.58}$$

by hand, the general action

$$S = \int \sqrt{-g} f(R, P, \mathcal{Q})\, d^4x \tag{4.59}$$

can be considered. The corresponding field equations read as

$$\begin{aligned} f_R(R, P, \mathcal{Q})\, G_{\mu\nu} = & \left[\frac{1}{2} g_{\mu\nu} f(R, P, \mathcal{Q}) - R f_R(R, P, \mathcal{Q})\right] \\ & - \left(g_{\mu\nu}\Box - D_\mu D_\nu\right) f_R(R, P, \mathcal{Q}) \\ & - 2\left[f_P(R, P, \mathcal{Q}) R^\alpha_\mu R_{\alpha\nu} + f_\mathcal{Q}(R, P, \mathcal{Q}) R_{\rho\sigma\alpha\mu} R^{\rho\sigma\alpha}{}_\nu\right] \\ & - g_{\mu\nu} D_\rho D_\sigma \left[f_P(R, P, \mathcal{Q}) R^{\rho\sigma}\right] - \Box\left[f_P(R, P, \mathcal{Q}) R_{\mu\nu}\right] \\ & + 2 D_\sigma D_\rho \left[f_P(R, P, \mathcal{Q}) R^\rho_{\{\mu}\, \delta^\sigma_{\nu\}} + 2 f_\mathcal{Q}(R, P, \mathcal{Q}) R^\rho{}_{\{\mu\nu\}}{}^\sigma\right], \end{aligned} \tag{4.60}$$

where $\{\}$ stands for the anti-commutator and where f_P and $f_\mathcal{Q}$ denote the derivative of f with respect to P and Q, respectively. When $f_P = f_\mathcal{Q} = 0$, $f(R)$ field Eqs. (4.45) are restored. As shown in [127], the action (4.59) yields new polarization spin-2 modes in the GWs. Among all the possible combinations of R, P, and $\mathcal{Q}$, the only action leading to ghost-free modes is

$$S = \int \sqrt{-g}\,(R^2 - 4P + \mathcal{Q})\, d^4x. \tag{4.61}$$

The quantity $\mathcal{G} \equiv R^2 - 4P + \mathcal{Q}$ is the *Gauss–Bonnet topological invariant* and it is of particular interest in the context of modified theories of gravity. In four dimensions, $\mathcal{G}$ is a topological surface term that, according to the Gauss–Bonnet theorem, integrated over the manifold provides a topological invariant. Specifically, the four-dimensional representation of the Gauss–Bonnet scalar is the Euler density, so that the identity

$$\int_{\mathcal{M}} \sqrt{-g}\,\mathcal{G}\, d^4x = \chi(\mathcal{M}) \tag{4.62}$$

holds, with $\chi(\mathcal{M})$ being the Euler characteristic. Therefore, a linear contribution of the Gauss–Bonnet topological surface term yields trivial dynamics in four dimensions.

Despite this fact, considering any function $f(\mathcal{G}) \neq \mathcal{G}$ is mathematically and physically relevant, also in 4D. In general, $f(R, \mathcal{G})$ gravity is taken into account, often assuming a starting action of the form

$$S = \int \sqrt{-g} \left(\frac{1}{2} R + f(\mathcal{G}) \right) d^4x, \tag{4.63}$$

so that GR is safely recovered when $f(\mathcal{G}) = 0$. On the other hand, actions that only contain a function of $\mathcal{G}$ also can be considered. In this case, GR can be restored in some cosmological contexts, where it turns out that P and $\mathcal{Q}$ are comparable with R^2. As a consequence, an action like $S \sim \int \sqrt{-g}\sqrt{\mathcal{G}}\, d^4x$ is dynamically equivalent to the EH one. This means that $\mathcal{G}$ and R^2 can be considered dynamically equivalent both in cosmology [107] and in spherical symmetry [128]. See Chapter 8 for a detailed discussion.

4.4 Scalar-Tensor Gravity

Let us consider now extensions of the EH action, including the non-minimal couplings between geometry and scalar fields. In this context, the interesting models were introduced by N. A. Chernikov and E. A. Tagirov in [129] and, subsequently, by C. G. Callan, Jr., S. R. Coleman, and R. Jackiw in [130]. Since the development of the inflationary models by A. Linde and A. Guth [131, 132], scalar-tensor theories have been taken into account by F. Lucchin, S. Matarrese, and M. D. Pollock [133] to address dynamics of the early universe, in the framework of inflationary paradigm. In fact, inflation is usually conceived as generated by a scalar field ϕ, called *inflaton*, which is supposed to be responsible for the accelerating expansion of the early universe. According to this picture, inflation should be led by some scalar field driving the cosmic acceleration between 10^{-34} and 10^{-35} s after the initial expansion, generating an isotropic and homogeneous universe. The theory also describes the production of particles after the early-time accelerating expansion (reheating) [134–136]. Moreover, in order to tackle the cosmological constant problem or the evolution of cosmological vacuum energy, new degrees of freedom for the gravitational field have to be considered.

Even though a lot of work has been done on the fact that scalar fields may couple with matter, the predominant opinion is that matter-fields do couple, with the field being "screened" at small scales. Three screening mechanisms are known: (i) the chameleon, (ii) the symmetron, and (iii) the Vainshtein mechanism [137]. The latter, specifically, "hides" the effects of nonlinear kinetic terms inside the so-called Vainshtein radius, allowing them to play an important role only at large IR scales, that is, at cosmological scales, as pointed out in [138].

A general starting action including the coupling $f(\phi)$, the kinetic term $u(\phi)$, and the potential $V(\phi)$ is

$$S = \int \sqrt{-g} \left\{ f(\phi)\, R + \frac{u(\phi)}{2} g^{\mu\nu} \phi_{,\mu} \phi_{,\nu} - V(\phi) \right\} d^4x. \tag{4.64}$$

As reported in Appendix A, by varying the preceding action with respect to the metric, one gets

$$f(\phi)\, G_{\mu\nu} = \frac{u(\phi)}{4} g_{\mu\nu}\, \phi^{;\alpha} \phi_{,\alpha} - \frac{u(\phi)}{2} \phi_{,\mu} \phi_{\nu} - \frac{1}{2} V(\phi) g_{\mu\nu} - g_{\mu\nu} \Box f(\phi) + f(\phi)_{,\mu,\nu}. \tag{4.65}$$

The limit in which the potential is null and the kinetic term reduces to $u(\phi) = \phi^{-1}$, and the coupling is $f(\phi) = \phi$, gives the *Brans–Dicke* gravity. The latter theory was first developed in [139] with the aim to make GR more coherent with Mach's Principle, which states that the inertia experienced by bodies in nonuniform motion is determined by the distribution of matter in the universe. According to this statement, Brans and Dicke proposed that the gravitational coupling could vary with respect to the space-time. The model was then considered in different contexts (see, e.g., [140–144]). It is worth noticing that even in this case, by means of the definition

$$\begin{aligned} \overline{\mathcal{T}}^{\phi}_{\mu\nu} = -\frac{1}{2f(\phi)} \Bigg[& -\frac{u(\phi)}{2} g_{\mu\nu}\, \phi^{;\alpha} \phi_{,\alpha} + u(\phi) \phi_{,\mu} \phi_{\nu} \\ & + V(\phi) g_{\mu\nu} + 2 g_{\mu\nu} \Box f(\phi) - 2 f(\phi)_{,\mu,\nu} \Bigg], \end{aligned} \tag{4.66}$$

the field equations can be recast as $G_{\mu\nu} = \overline{\mathcal{T}}^{\phi}_{\mu\nu}$, with $\overline{\mathcal{T}}^{\phi}_{\mu\nu}$ being the effective energy-momentum tensor provided by the dynamical scalar field and geometry. The variation of the action with respect to the scalar field ϕ provides the Klein–Gordon equation

$$2u(\phi)\Box\phi - f_\phi(\phi) R + V_\phi(\phi) + u_\phi(\phi) \partial^\mu \phi \partial_\mu \phi = 0. \tag{4.67}$$

Therefore, the scalar field can be interpreted as an effective Newton's coupling whose value depends on the considered space-time point. Coupling gravity to a scalar field can also describe, in principle, the evolution of the universe across different epochs, from the very early stage up to the late time. In other words, a scalar field (or more scalar fields) could trigger cosmological dynamics passing from the various accelerated/decelerated eras.

Moreover, it can be shown that fourth-order theories provide the same dynamics as second-order theories non-minimally coupled to a scalar field. In particular, the former can be addressed to the latter (and vice versa) by means of conformal transformations. In what follows we show how the $f(R)$ action can be obtained from Eq. (4.64) with a proper definition of the potential and the coupling functions. Conformal transformations are mostly used in modified theories of gravity in order to disentangle the further gravitational degrees of freedom coming from extended actions. They consist of the conformal rescaling of the metric through the relation $g_{\mu\nu} \to \tilde{g}_{\mu\nu} = e^{2\omega} g_{\mu\nu}$. The new set of conformally transformed variables $\{\tilde{g}_{\mu\nu}, \tilde{\phi}\}$ belongs to the so-called *Einstein conformal frame*, while the initial set $\{g_{\mu\nu}, \phi\}$ belongs the *Jordan frame*. Under the conformal rescaling, Christoffel connection, Ricci tensor, and Ricci scalar vary according to the relations,

$$
\begin{aligned}
\tilde{\hat{\Gamma}}^{\sigma}_{\lambda\mu} &= \hat{\Gamma}^{\sigma}_{\lambda\mu} + g^{\nu\sigma}\left(\partial_\lambda \omega\, g_{\mu\nu} + \partial_\mu \omega\, g_{\lambda\nu} - \partial_\nu \omega\, g_{\lambda\mu}\right), \\
\tilde{R}_{\alpha\beta} &= R_{\alpha\beta} - 2\omega_{;\alpha;\beta} + 2\omega_{;\alpha}\omega_{;\beta} - g_{\alpha\beta}\Box\omega - 2g_{\alpha\beta}\,\omega_{;\gamma}\omega^{;\gamma}, \\
\tilde{R} &= e^{-2\omega}\left(R - 6\Box\omega - 6\omega_{;\gamma}\omega^{;\gamma}\right),
\end{aligned} \tag{4.68}
$$

respectively [145].

Notice that the GR vacuum field equations, in the Jordan frame $R_{\alpha\beta} = 0$, are not equivalent to the corresponding equations $\tilde{R}_{\alpha\beta} = 0$ in the Einstein frame. This is due to the fact that gravity and matter are frame-dependent, so that the conformal factor in the Jordan frame plays the role of an effective form of matter. In order to show the equivalence between the two frames, let us consider the $f(R)$ extension of GR. Starting from Eq. (4.45) and performing a conformal transformation, the field equations in the Einstein frame read as

$$
\begin{aligned}
\tilde{G}_{\alpha\beta} = &\frac{1}{f_R(R)}\left\{\frac{1}{2}g_{\alpha\beta}[f(R) - Rf_R(R)] + f_R(R)_{;\alpha;\beta} - g_{\alpha\beta}\Box f_R(R)\right\} \\
&+2\left(\omega_{;\alpha;\beta} + g_{\alpha\beta}\Box\omega - \omega_{;\alpha}\omega_{;\beta} + \frac{1}{2}g_{\alpha\beta}\omega_{;\gamma}\omega^{;\gamma}\right).
\end{aligned} \tag{4.69}
$$

By means of the definition $\omega = \frac{1}{2}\ln|f_R(R)|$ and considering the conformal rescaling $\omega = \sqrt{\frac{1}{6}}\phi$, Eq. (4.69) can be recast as

$$
\tilde{G}_{\alpha\beta} = \tilde{\mathcal{T}}^{\phi}_{\alpha\beta}, \tag{4.70}
$$

with

$$
\tilde{\mathcal{T}}^{\phi}_{\alpha\beta} = \phi_{;\alpha}\phi_{;\beta} - \frac{1}{2}\tilde{g}_{\alpha\beta}\phi_{;\gamma}\phi^{;\gamma} + \tilde{g}_{\alpha\beta}V(\phi), \qquad V(\phi) = \frac{1}{2}\frac{f(R) - Rf_R}{[f_R]^2}. \tag{4.71}
$$

Clearly, Eq. (4.71) can be recast in the form of Eq. (4.48). Field equations therefore are equivalent to those written in Eq. (4.45), showing that $f(R)$ gravity can be recovered as a particular case of scalar-tensor gravity once a suitable scalar field is defined from the $f(R)$ function. This result points out the capability of the Starobinsky model to give rise to the early-time inflation of the universe. Specifically, the Starobinsky action can be rewritten in terms of a scalar tensor theory with a potential

$$
V(\phi) = V_0\left(1 - e^{-2\sqrt{\alpha}\phi}\right)^2, \tag{4.72}
$$

providing an inflationary behavior.

In light of these considerations, it is worth mentioning that the most general second-order scalar-tensor theory was proposed by Horndeski in [146] and the corresponding action reads as follows [147–149]:

$$S_{Horndeski} = \sum_{i=2}^{5} \int d^4x\sqrt{-g}\,\mathcal{L}_i\,, \tag{4.73}$$

with the definitions

$$\begin{aligned}
\mathcal{L}_2 &\equiv G_2(\phi, X)\,,\\
\mathcal{L}_3 &\equiv -G_3(\phi, X)\,\Box\phi\,,\\
\mathcal{L}_4 &\equiv G_4(\phi, X)R + G_{4X}\left[(\Box\phi)^2 - (D_\mu D_\nu\phi)^2\right],\\
\mathcal{L}_5 &\equiv G_5(\phi, X)\,G_{\mu\nu}D^\mu D^\nu\phi\\
&\quad -\frac{1}{6}\,G_{5X}\left[(\Box\phi)^3 - 3\,\Box\phi\,(D_\mu D_\nu\phi)^2 + 2\,(D_\mu D_\nu\phi)^3\right].
\end{aligned}$$

The related field equations can be found in [149]. The functions $G_i(\phi, X)$ are arbitrary functions of the scalar field and the kinetic term: $X \equiv \frac{1}{2}g^{\mu\nu}D_\mu\phi D_\nu\phi$, $G_{iX} \equiv \dfrac{\partial G_i}{\partial X}$.

As an example, by setting $G_2 = \frac{u(\phi)}{\phi}X$, $G_3 = 0$, $G_4 = \phi$, and $G_5 = 0$, the Brans–Dicke gravity is recovered, while by setting $G_2 = X - V(\phi)$, $G_3 = 0$, $G_4 = F(\phi)$, and $G_5 = 0$, Horndeski action turns into the standard scalar-tensor action with coupling function $F(\phi)$ and potential $V(\phi)$. Further details will be provided in Section 10.6, where the Noether Symmetry Approach will be applied to a subclass of the general Horndeski model.

4.5 The Palatini Formalism

Another important assumption of GR is that connection is derived from the metric thanks to the Levi–Civita connection (4.38). As discussed before, this choice is mandatory due to the Equivalence Principle, which implies that the geodesic structure and the metric (causal) structure of space-time have to coincide. By relaxing such a constraint, the metric can be chosen as independent of the connection, which consequently does not have to be assumed symmetric with respect to the lowest indexes.

This prescription, where metric and connection are treated as separated fields, is known as the *Palatini formalism*, first introduced by Einstein himself [150–152]. It was, however, called the Palatini approach as a consequence of a historical misunderstanding [153]. In 1919, the same approach was independently proposed by A. Palatini in [154].

The fundamental idea at the base of Palatini formalism is to consider the (usually torsion-less) connection Γ, entering the definition of the Ricci tensor, to be independent of the metric g. The Palatini formalism for the standard

Einstein–Hilbert torsionless theory results in being equivalent to the purely metric theory: this follows from the fact that the field equation for the connection fields states exactly that the same connection Γ, first considered to be independent, must be the Levi–Civita connection of the metric g. There is consequently no reason to impose the Palatini variational principle in the standard Einstein theory instead of the metric variational principle. The situation, however, completely changes when we consider the case of Extended Theories of Gravity, depending on analytic functions of curvature invariants as $f(R)$, or non-minimally coupled to a scalar field. In these cases, as we will show in what follows, the Palatini and the metric variational principle provide different field equations, and the theories thus derived differ [155].

From a physical viewpoint, considering the metric g and the connection Γ as independent fields is equivalent to decoupling the metric structure of space-time and its geodesic structure (i.e. the connection is not the Levi–Civita connection of g), governing, respectively, the chronological structure of space-time and the trajectories of particles moving in it. This decoupling enriches the geometrical structure and generalizes the purely metric formalism. This metric-affine structure is naturally translated, by means of the Palatini field equations, into a bimetric structure of space-time.

In $f(R)$ gravity, the Palatini variational principle yields the following set of field equations:[8]

$$\begin{cases} f_{\mathcal{R}}(\mathcal{R})\mathcal{R}_{\mu\nu} - \dfrac{1}{2}g_{\mu\nu}f(\mathcal{R}) = 0 \\ \nabla_\sigma[\sqrt{-g}f_{\mathcal{R}}(\mathcal{R})g^{\mu\nu}] = 0, \end{cases} \tag{4.74}$$

where $\mathcal{R}$ is the curvature scalar depending on the sole affine connection: $\mathcal{R} = \mathcal{R}(\Gamma)$. As the former equation comes from the variation of the action with respect to the metric, so the latter can be obtained from the application of the variational principle with respect to the connection. The $f(R)$ field equations, in the Palatini formalism, formally reduce to the GR ones when $f(R) = R$. Although a further degree of freedom is included into the action, the trace equation suggests that no further dynamical degree of freedom occurs in the field equations, with respect to $f(R)$ gravity. This leads to some puzzles at the astrophysical and cosmological levels [156–159]. Nevertheless, the Palatini approach is particularly relevant among GR extensions, due to the new vision exhibited by the formalism, according to which the connection is independent of the metric. Similarly to the metric $f(R)$ gravity, the Palatini $f(R)$ action can be recast in the Jordan frame by means of conformal transformations. To this purpose, it is convenient to rewrite the action as

$$S = \int \sqrt{-g}\,[f(\chi) - f_\chi(\chi)(\mathcal{R} - \chi)]\,d^4x, \tag{4.75}$$

[8] The variation of the action is performed step by step in Appendix A.

so that the variation with respect to the field χ provides the equation $\mathcal{R} = \chi$. Setting then $f_\chi(\chi) \equiv \phi$ and considering the conformal rescaling $\tilde{g}_{\mu\nu} = \phi g_{\mu\nu}$, the Palatini–Ricci scalar, in the Jordan frame, takes the form [160]

$$\mathcal{R} = R + \frac{3}{2\phi}\partial_\alpha\phi\partial_\alpha\phi - \frac{3}{2}\Box\phi. \tag{4.76}$$

Substituting Eq. (4.76) into Eq. (4.75) and neglecting the boundary terms, Eq. (4.75) can be written as

$$S = \int \sqrt{-g}\left[\phi R + \frac{3}{2\phi}\partial^\alpha\phi\partial_\alpha\phi - V(\phi)\right] d^4x, \tag{4.77}$$

with the definition $V(\phi) = \phi\chi(\phi) - f[\chi(\phi)]$. Though the potential is defined differently than in the previous section, the preceding action is formally equivalent to the Brans–Dicke one with a potential. Several studies (see, e.g., [161–165]) consider a starting action containing both the Palatini curvature $\mathcal{R}$ and the metric Ricci scalar R combined into the form

$$S = \sqrt{-g}\,[R + f(\mathcal{R})]\, d^4x. \tag{4.78}$$

In this way, standard GR is recovered when $f(\mathcal{R}) = 0$, and the extra function can play the role of an effective dark energy/matter term due to the affine connection [166]. The corresponding theory is called *Hybrid Gravity* (first considered in [167]) since it unifies the contributions of metric and metric-affine terms (see [168] for a detailed review).

4.6 Teleparallel Equivalent of General Relativity and the Extension to $f(T)$ Gravity

In the previous sections, we discussed some possible geometric modifications of GR, which extend the EH action by introducing functions of curvature invariants. All these theories share the common feature of having the same structure of GR, namely maintaining the general covariance and the invariance under diffeomorphism transformations with a metric or a metric-affine structure. They can be classified as *extensions* of GR, and the EH action can be recovered both in some limit or as a particular case of a more extended class of theories [91].

Here we introduce an approach to gravity that cannot be thought as a straightforward extension of GR but describes the gravitational interaction from a different point of view. At the beginning of this chapter we mentioned the possibility of considering affine connections more general than the Levi–Civita one and the possibility that nonmetric contributions can be taken into account; see (4.39). In principle, this gives rise to a theory where the space-time dynamics can be described by torsion instead of curvature: the so-called TEGR [102].

With a view to deal with such an approach, we can start from the fact that space-time is endowed with two fundamental properties: curvature and torsion.

In the building up of his theory, Einstein chose a particular space-time with curvature and no torsion. Such space-time, as we have discussed, is equipped with the specific Levi–Civita connection, derived from the metric, where gravitation is represented by curvature only.

On the other hand, it is perfectly conceivable to construct a space-time with vanishing curvature but nonvanishing torsion. Such space is known as Weitzenböck space, after the Austrian mathematician Roland Weitzenböck who, together with the French mathematician Elie Cartan, considered spaces with torsion (see [102] and references therein). Similarly to curvature, torsion is also capable of generating gravity but, in such a case, the geodesic structure can be replaced by a teleparallel structure called *teleparallelism*. The mathematical structure of a space equipped with torsion only had been known since the 1920s, and it was used by Einstein himself to unify gravitational and electromagnetic interactions without success [102].

Specifically, by defining the *Contorsion Tensor* $K^{\rho}{}_{\mu\nu}$ as

$$K^{\rho}{}_{\mu\nu} \equiv \frac{1}{2} g^{\rho\lambda} (T_{\mu\lambda\nu} + T_{\nu\lambda\mu} + T_{\lambda\mu\nu}), \quad \text{with} \quad T^{\alpha}_{\ \mu\nu} = 2\Gamma^{\alpha}_{[\mu\nu]} \tag{4.79}$$

and setting $\nabla_{\beta} g_{\mu\nu} = 0$, that is preserving isometries, the connection (4.39) turns out to be

$$\Gamma^{\alpha}{}_{\mu\nu} = \hat{\Gamma}^{\alpha}{}_{\mu\nu} + K^{\alpha}{}_{\mu\nu}. \tag{4.80}$$

Therefore, the Riemann curvature tensor, written in terms of the general connection (4.39), is

$$\begin{aligned} R^{\alpha}{}_{\mu\nu\rho} &= \partial_{\nu} \Gamma^{\alpha}{}_{\mu\rho} - \partial_{\rho} \Gamma^{\alpha}{}_{\mu\nu} + \Gamma^{\lambda}{}_{\mu\rho} \Gamma^{\alpha}{}_{\nu\lambda} - \Gamma^{\lambda}{}_{\mu\nu} \Gamma^{\alpha}{}_{\rho\lambda} \\ &= \partial_{\nu} \hat{\Gamma}^{\alpha}{}_{\mu\rho} - \partial_{\rho} \hat{\Gamma}^{\alpha}{}_{\mu\nu} + \hat{\Gamma}^{\lambda}{}_{\mu\rho} \hat{\Gamma}^{\alpha}{}_{\nu\lambda} - \hat{\Gamma}^{\lambda}{}_{\mu\nu} \hat{\Gamma}^{\alpha}{}_{\rho\lambda} \\ &\quad + \partial_{\nu} K^{\alpha}{}_{\mu\rho} - \partial_{\rho} K^{\alpha}{}_{\mu\nu} + K^{\lambda}{}_{\mu\rho} K^{\alpha}{}_{\nu\lambda} - K^{\lambda}{}_{\mu\nu} K^{\alpha}{}_{\rho\lambda}. \end{aligned} \tag{4.81}$$

Denoting by $\hat{R}^{\alpha}{}_{\mu\nu\rho}$ the second line of Eq. (4.81) (namely the Rieman curvature tensor written in terms of the Levi–Civita connection), in the so-called *Teleparallel Gauge* where $R^{\alpha}_{\mu\nu\rho} = 0$, we have

$$\hat{R}^{\alpha}{}_{\mu\nu\rho} = -\overset{\bullet}{R}{}^{\alpha}{}_{\mu\nu\rho}, \tag{4.82}$$

with $\overset{\bullet}{R}{}^{\alpha}{}_{\mu\nu\rho}$ defined as the third line of Eq. (4.81) (namely the Rieman curvature tensor written in terms of the teleparallel connection). In order to distinguish quantities written in terms of different connections, hereafter in this chapter, the teleparallel connection and all quantities related to it will be denoted by a bullet on top, to be distinguished from those with a hat, defined by the Levi–Civita connection. Finally, by defining a *Torsion Superpotential* as

$$\overset{\bullet}{S}{}^{\rho\mu\nu} \equiv \overset{\bullet}{K}{}^{\mu\nu\rho} - g^{\rho\nu} \overset{\bullet}{T}{}^{\sigma\mu}{}_{\sigma} + g^{\rho\mu} \overset{\bullet}{T}{}^{\sigma\nu}{}_{\sigma}, \tag{4.83}$$

and a *Torsion Scalar* as

$$\overset{\bullet}{T} \equiv \overset{\bullet}{T}{}^{\rho\mu\nu} \overset{\bullet}{S}{}_{\rho\mu\nu}, \tag{4.84}$$

it turns out that the actions

$$S_{GR} \equiv \frac{1}{2} \int d^4x \sqrt{-g}\, \hat{R}, \tag{4.85}$$

$$S_{TEGR} \equiv -\frac{1}{2} \int d^4x \sqrt{-g}\, \overset{\bullet}{T} \tag{4.86}$$

differ from each other only by a four-divergence.

Interestingly, the same result can be obtained by considering local translations in a flat space-time, which means that TEGR can be understood as a gauge theory of translation group in the locally flat tangent space-time. This is the reason for the term *teleparallelism.* The flat and the curved space-times can be linked by means of a mathematical tool called *tetrad fields.* They are rank-2 tensor with mixed indexes, belonging to the anholonomic frame (flat space-time) and to the holonomic frame (curved space-time), respectively. They can be easily introduced by considering a coordinate transformation $x^\mu \to \tilde{x}^\mu$, by means of which the metric tensor varies as

$$g_{\mu\nu} \to \tilde{g}_{\mu\nu} = \partial_\mu \tilde{x}^\alpha \partial_\nu \tilde{x}^\beta g_{\alpha\beta}. \tag{4.87}$$

Imposing the transformed reference frame to be the Minkowski space-time, namely $\tilde{g}_{\mu\nu} \to \eta_{ab}$, the preceding transformation takes the following form:

$$\eta_{ab} = \partial_a \tilde{x}^\alpha \partial_b \tilde{x}^\beta g_{\alpha\beta} \tag{4.88}$$

and, equivalently,

$$g_{\mu\nu} = \partial_\mu \tilde{x}^a \partial_\nu \tilde{x}^b \eta_{ab}. \tag{4.89}$$

Defining the tetrad fields as

$$e^a_\mu = \partial_\mu \tilde{x}^a \tag{4.90}$$

naturally yields the following relations:

$$e^a_\mu e^\mu_b = \delta^a_b\,; \qquad e^a_\mu e^\nu_a = \delta^\nu_\mu. \tag{4.91}$$

Thanks to tetrad fields, any quantity in the Riemannian manifold can be projected into the flat tangent space-time. Tetrad fields can be used to formally recast TEGR as a gauge theory of translation group. To this purpose, let us consider a translation of the form (1.12), namely

$$T = e^{-i\epsilon^a(x)\mathcal{T}_a}, \tag{4.92}$$

with infinitesimal generator of the form

$$\delta\phi = \epsilon^a \partial_a \phi, \qquad \text{with} \qquad \delta(\partial_a \phi) \neq \epsilon^a \partial_a \partial_b \phi, \tag{4.93}$$

where $\mathcal{T}_a$ is the generator of the transformation, ϕ is a generic scalar field, and where the latter relation is due to the dependence of ϵ^a on the coordinates. In order to restore the general covariance, let us introduce a new gauge field B^a_μ, whose variation is $\delta B^a_\mu = -\partial_\mu \epsilon^a$ so that the covariant derivative can be recast as

$$\nabla_\mu \phi = (\partial_\mu + B^a_\mu \partial_a)\phi. \tag{4.94}$$

By varying Eq. (4.94), one gets the relation $\delta(\nabla_\mu \phi) = \epsilon^a \partial_a \nabla_\mu \phi$. Therefore, as the standard derivative must be extended when the translation group is considered, nontrivial tetrads h^a_μ,

$$h^{a\,(Trasl)}_\mu = \nabla_\mu x^a = \partial_\mu x^a + B^a_\mu, \tag{4.95}$$

can be defined as a natural extension of Eq. (4.90).

So far, along with the translation group, we implicitly considered a class of frame with vanishing inertial effects. In general, they can be included by considering a LT acting on the gauge field B^a_μ as

$$\Lambda^a_b B^b_\mu \to B^a_\mu. \tag{4.96}$$

In this way, Eq. (4.95) can be generally extended by including a purely inertial Lorentz connection of the form

$$\overset{\bullet}{\omega}{}^a{}_{b\mu} = \Lambda^a_c \partial_\mu \lambda^c_b, \tag{4.97}$$

yielding

$$h^a_\mu = \partial_\mu x^a + B^a_\mu + \overset{\bullet}{\omega}{}^a{}_{b\mu} x^b \equiv \overset{\bullet}{\mathcal{D}}_\mu x^a + B^a_\mu, \tag{4.98}$$

which can be intended as the extension of Eq. (4.95) and where we have defined the Lorentz covariant derivative $\overset{\bullet}{\mathcal{D}}_\mu$ as

$$\overset{\bullet}{\mathcal{D}}_\mu x^a = \partial_\mu x^a + \overset{\bullet}{\omega}{}^a{}_{b\mu} x^b. \tag{4.99}$$

It is worth noticing that the Lorentz connection is not uniquely determined, but it can be arbitrarily chosen among infinite choices. This is due to the fact that the general covariance principle cannot fix uniquely the Lorentz connection, so that the latter can be characterized by different values of curvature and torsion. GR, for instance, assumes the contorsion tersor to vanish.

Using the tetrad postulate, according to which the total covariant derivative $\overset{\bullet}{\nabla}_\mu$ (including both spin and linear connections) of the tetrad field must vanish, the following relation holds [102, 169, 170] :

$$\overset{\bullet}{\nabla}_\mu h^a_\nu \equiv \overset{\bullet}{\mathcal{D}}_\mu h^a_\nu - \overset{\bullet}{\Gamma}{}^\alpha{}_{\mu\nu} h^a_\alpha = 0, \tag{4.100}$$

where we recast the covariant derivative as a difference between the Lorentz covariant derivative and a general Christoffel connection. The form of $\overset{\bullet}{\Gamma}{}^{\alpha}{}_{\mu\nu}$ is straightforwardly provided by Eq. (4.100),

$$\overset{\bullet}{\Gamma}{}^{\rho}{}_{\mu\nu} = h^{\rho}_{a}\overset{\bullet}{\mathcal{D}}_{\mu}h^{a}_{\nu}, \tag{4.101}$$

and is called the *Weitzenböck connection*. Neglecting the contribution of the local LG, a class of frame in which the spin connection vanishes is naturally selected, and Eq. (4.102) can be recast as

$$\overset{\bullet}{\Gamma}{}^{\rho}{}_{\mu\nu} = h^{\rho}_{a}\partial_{\mu}h^{a}_{\nu}. \tag{4.102}$$

However, this choice leads inevitably to the Lorentz invariance violation so that both tetrad fields and spin connection should be considered to preserve the frame-independence of TEGR [171].

In standard TEGR, this issue is usually neglected, as it does not affect directly the field equations. Nonetheless, it becomes nontrivial when extensions of TEGR are considered (such as $f(T)$ gravity, introduced in what follows). The assumption of vanishing spin connection is twofold: on the one hand, it is well motivated by the sake of simplifying the theory; on the other hand, it relies on the possibility to choose a proper class of frames in which the spin connection vanishes, due to the non-tensorial nature of the latter. However, this assumption introduces an unrealistic frame-dependence, as the field equations turn out to depend on the choice of the frame. This problem was first addressed in Ref. [171], where it is shown that including spin connections does not change any physically relevant field equation so that the Lorentz invariance can be formally preserved and dynamics results unchanged.

By means of Eq. (4.101), the torsion tensor, defined in Eq. (4.40), can be recast as a function of tetrad fields as

$$\overset{\bullet}{T}{}^{\rho}{}_{\mu\nu} = (\overset{\bullet}{\Gamma}{}^{\rho}{}_{\mu\nu} - \overset{\bullet}{\Gamma}{}^{\rho}{}_{\nu\mu}) = (h^{\rho}_{a}\overset{\bullet}{\mathcal{D}}_{\mu}h^{a}_{\nu} - h^{\rho}_{a}\overset{\bullet}{\mathcal{D}}_{\nu}h^{a}_{\mu}). \tag{4.103}$$

Therefore, the teleparallel action can be constructed by means of the definitions (4.79), (4.83), and (4.84). When TEGR is recast as a gauge theory of translation group, the determinant of tetrad fields h is included in the action, so that the latter reads as

$$S = \frac{1}{2}\int h\,\overset{\bullet}{T}\,d^4x. \tag{4.104}$$

Moreover, from Eq. (4.82), it is possible to show that the standard scalar curvature and the torsion scalar satisfy the relation

$$\hat{R} - \frac{2}{h}\partial_{\mu}(h\overset{\bullet}{T}{}^{\nu\mu}{}_{\nu}) = -\overset{\bullet}{T}, \tag{4.105}$$

namely they only differ for a boundary term. Therefore, the teleparallel action can be written in terms of the EH action S_{GR} as

$$S_{TEGR} = -S_{GR} + 2\int \partial_\mu (h \overset{\bullet}{T}^{\nu\mu}{}_{\nu})\, d^4x. \tag{4.106}$$

This implies that S_{TEGR} and S_{GR} formally yield the same equations of motion. In the first-order formalism, where the Lagrangian depends on the fields and on their first derivatives, the EL equations with respect to the tetrad fields, that is,

$$\frac{\partial \mathscr{L}_{TEGR}}{\partial h^a_\rho} - \partial_\sigma \frac{\partial \mathscr{L}_{TEGR}}{\partial(\partial_\sigma h^a_\rho)} = 0, \tag{4.107}$$

yield in vacuum[9]

$$\frac{4}{h}\partial_\mu \left(h \overset{\bullet}{S}_a{}^{\mu\beta}\right) - 4\overset{\bullet}{T}{}^\sigma{}_{\mu a}\overset{\bullet}{S}_\sigma{}^{\beta\mu} - \overset{\bullet}{T} h^\beta_a = 0. \tag{4.108}$$

When the Contorsion Tensor is included as a fundamental part of the Christoffel connection, the space-time turns out to be labeled by torsion. This means that when a vector is parallel transported around a closed path, its final position will be shifted with respect to the initial one. If curvature also is included, the final vector results are shifted and rotated.

The boundary term in Eq. (4.105) does not play any role in the dynamics provided by the standard teleparallel action. However, extensions of TEGR consider the boundary term as a fundamental degree of freedom, which allows one to exactly recover $f(\hat{R})$ gravity (see, e.g., [172–175] for detailed discussions). In this context, TEGR can be extended in several ways, for example, by including into the action a function of the torsion and of the boundary term $f(\overset{\bullet}{T}, \overset{\bullet}{B})$. Although TEGR is completely equivalent to GR at the level of equations, modifications of the former are generally different from modifications of the latter. For instance, $f(\overset{\bullet}{T})$ gravity is a second-order theory, while $f(\hat{R})$ leads to fourth-order field equations in metric formalism.

Recently, the $f(\overset{\bullet}{T})$ extension of the teleparallel gravity has been considered because its dynamics is substantially different with respect to $f(\hat{R})$ gravity. See, for example, [97, 101, 176–181]. Also, in the case of $f(\overset{\bullet}{T})$ gravity it is possible to achieve an effective energy-momentum tensor originated by torsion, which allows one to recast dark matter and dark energy as geometric effects [97].

More technically, by varying the action

$$S = \int h f(\overset{\bullet}{T})\, d^4x \tag{4.109}$$

with respect to tetrad fields, one gets the following field equations:

$$\begin{aligned}
&\frac{1}{h}\partial_\mu (h\, h^\rho_a \overset{\bullet}{S}_\rho{}^{\mu\nu}) f_T(\overset{\bullet}{T}) - h^\lambda_a \overset{\bullet}{T}{}^\rho{}_{\mu\lambda}\overset{\bullet}{S}_\rho{}^{\nu\mu} f_T(\overset{\bullet}{T}) \\
&+ h^\rho_a S_\rho{}^{\mu\nu}(\partial_\mu \overset{\bullet}{T}) f_{TT}(\overset{\bullet}{T}) + \frac{1}{4} h^\nu_a f(\overset{\bullet}{T}) = 0,
\end{aligned} \tag{4.110}$$

[9] See Appendix A for detailed calculations.

with $f_T(\overset{\bullet}{T})$ being the first derivative of $f(\overset{\bullet}{T})$ with respect to $\overset{\bullet}{T}$. Clearly, for $f(\overset{\bullet}{T}) = \overset{\bullet}{T}$ the standard TEGR is recovered. It is worth remarking an important difference between GR and TEGR that will be further stressed in the application of the Noether Symmetry Approach to teleparallel Lagrangians. While, in the Einstein theory, the starting line element is sufficient to uniquely determine the form of the metric, TEGR dynamics is described by tetrad fields, which cannot be uniquely settled only from the form of the interval. To overcome this issue, criteria for "good" and "bad" tetrads can be used, as shown, for example, in [100]. In applications to cosmology, we will use the simplest choice among all possible tetrads leading to a FLRW tetrads interval, namely $h^a_\mu = diag(1, a(t), a(t), a(t))$. Consequently, the cosmological equations coming from (4.110) take the simple form

$$\overset{\bullet}{T} f_T(\overset{\bullet}{T}) - \frac{1}{2} f(\overset{\bullet}{T}) = 0\,, \tag{4.111}$$

$$2\overset{\bullet}{T} f_{TT}(\overset{\bullet}{T}) + f_T(\overset{\bullet}{T}) = 0\,, \tag{4.112}$$

where the torsion scalar reads as

$$\overset{\bullet}{T} = -6\left(\frac{\dot{a}}{a}\right)^2 . \tag{4.113}$$

Though in this overview we only discussed the $f(\overset{\bullet}{T})$ extension of TEGR, several extended teleparallel theories containing nonlocal terms [182], non-minimally coupled scalar field [105], second-order torsion invariants [183], or higher-order terms [184] can be considered.

4.7 Symmetric Teleparallel Equivalent of General Relativity

So far, we have discussed modifications of Einstein GR including the antisymmetric part of the affine connection or extra geometric terms in the gravitational action. However, GR can be modified by considering also theories where the Metricity Principle, and then the isometry invariance, is violated. It is possible to show that a theory described only by non-metricity, without torsion and curvature, is formally equivalent to GR and TEGR at the level of field equations. This theory is the previously defined STEGR, in which the general connection (4.39) is reduced to

$$\Gamma^\alpha{}_{\mu\nu} = \hat{\Gamma}^\alpha{}_{\mu\nu} + \frac{1}{2} g^{\alpha\lambda}(-Q_{\mu\nu\lambda} - Q_{\nu\mu\lambda} + Q_{\lambda\mu\nu}) \equiv \hat{\Gamma}^\alpha{}_{\mu\nu} + L^\alpha{}_{\mu\nu}, \tag{4.114}$$

where $Q_{\beta\mu\nu}$ is defined as

$$Q_{\beta\mu\nu} = \nabla_\beta g_{\mu\nu} \tag{4.115}$$

and where the rank-3 tensor $L^{\alpha}{}_{\mu\nu}$ is called *Disformation Tensor*. By setting $\Gamma^{\alpha}{}_{\mu\nu} = 0$, the disformation tensor turns out to be equivalent to the Levi–Civita connection, up to a sign:

$$L^{\alpha}{}_{\mu\nu} = -\hat{\Gamma}^{\alpha}{}_{\mu\nu}. \tag{4.116}$$

This choice permits one to recast the nonmetric action in terms of the EH action up to a boundary term. More precisely, even when curvature and torsion vanish, there is still an inertial affine connection, but it can be made to vanish by a gauge choice. The gauge in which the connection is trivialized is called *coincident gauge*. Considering the coincident gauge and introducing the quantities

$$\begin{aligned}
Q &\equiv -\frac{1}{4} Q_{\alpha\mu\nu}\left[-2L^{\alpha\mu\nu} + g^{\mu\nu}\left(Q^{\alpha} - \tilde{Q}^{\alpha}\right) - \frac{1}{2}\left(g^{\alpha\mu}Q^{\nu} + g^{\alpha\nu}Q^{\mu}\right)\right],\\
Q_{\mu} &\equiv Q_{\mu}{}^{\lambda}{}_{\lambda},\\
\tilde{Q}_{\mu} &\equiv Q_{\alpha\mu}{}^{\alpha},
\end{aligned}$$

it turns out that the action

$$S_{STEGR} \equiv \frac{1}{2}\int d^4x \sqrt{-g}\, Q \tag{4.117}$$

leads to the same field equations as the GR and TEGR actions. This means that the torsion scalar T, the nonmetricity scalar Q, and the Ricci scalar R differ from each other for boundary terms. This can be easily seen from the relation occurring between the Levi–Civita connection and the disformation tensor in Eq. (4.116). In order to compare STEGR and GR, let us consider the connection

$$\Gamma^{\alpha}{}_{\mu\nu} = \hat{\Gamma}^{\alpha}{}_{\mu\nu} + L^{\alpha}{}_{\mu\nu}, \tag{4.118}$$

including the Levi–Civita connection and the disformation tensor. Writing the explicit form of the scalar curvature R, it turns out that the Ricci scalar $\hat{R}$ of GR and the nonmetricity scalar, in the coincident gauge, can be linked through the relation

$$\hat{R} = Q - \hat{\nabla}_{\alpha}\left(Q^{\alpha} + \tilde{Q}^{\alpha}\right). \tag{4.119}$$

Further details about nonmetric theories can be found, for example, in [185–187]. As $f(\overset{\bullet}{T})$ gravity and $f(\hat{R})$ gravity, the STEGR action (4.117) can be generalized considering a function of the nonmetricity scalar, namely

$$S = \int \sqrt{-g}\, f(Q)\, d^4x. \tag{4.120}$$

Considering the metric tensor as the only degree of freedom, the variation of the action yields [115, 188]

$$\frac{2}{\sqrt{-g}}\hat{\nabla}_\alpha\left(\sqrt{-g}\,g_{\beta\nu}f_Q\left[L^{\alpha\mu\beta}-\frac{1}{2}g^{\mu\beta}\left(Q^\alpha-\tilde{Q}^\alpha\right)+\frac{1}{4}\left(g^{\alpha\mu}Q^\beta+g^{\alpha\beta}Q^\mu\right)\right]\right)$$
$$+f_Q\left[L^{\mu\alpha\beta}-\frac{1}{2}g^{\alpha\beta}\left(Q^\mu-\tilde{Q}^\mu\right)+\frac{1}{4}\left(g^{\mu\alpha}Q^\beta+g^{\mu\beta}Q^\alpha\right)\right]Q_{\nu\alpha\beta}-\delta^\mu_\nu f=\mathcal{T}^\mu{}_\nu\,, \tag{4.121}$$

where, as previously, $\mathcal{T}^\mu{}_\nu$ is the matter stress-energy tensor. In a FLRW universe, the set of 16 equations (4.121) reduce to two equations, which are exactly the same as those of $f(\overset{\bullet}{T})$ cosmology in Eqs. (4.111) and (4.112). This is due to the expression of Q in a spatially flat cosmological background, which turns out to be the same as that of $\overset{\bullet}{T}$ (in the coincident gauge). Both, in fact, are

$$\overset{\bullet}{T}=Q=-6\left(\frac{\dot{a}}{a}\right)^2. \tag{4.122}$$

In this gauge, the diffeomorphism gauge symmetry is lost; thus time parameterization cannot arbitrarily be chosen as in cosmology derived from GR. Moreover, $\overset{\bullet}{T}$ and Q are completely equivalent and differ from $\hat{R}$ for a boundary term $\overset{\bullet}{B}$, so that the relations

$$\hat{R}(t)=-\overset{\bullet}{T}(t)+\overset{\bullet}{B}(t), \tag{4.123}$$
$$\hat{R}(t)=-Q(t)+B(t) \tag{4.124}$$

hold. Therefore, $B(t)$ and $\overset{\bullet}{B}(t)$ are totally equivalent in cosmology. Considering the cosmological expression of $\hat{R}(t)$ in a FLRW space-time

$$\hat{R}(t)=-6\left(\frac{\ddot{a}}{a}+\frac{\dot{a}^2}{a^2}\right), \tag{4.125}$$

the term $\overset{\bullet}{B}(t)$ can be easily calculated from its definition in Eq. (4.105) and reads as

$$\overset{\bullet}{B}=-6\left(\frac{\ddot{a}}{a}+2\frac{\dot{a}^2}{a^2}\right). \tag{4.126}$$

It can be written as a total-derivative when multiplied by the determinant of tetrad fields, that is,

$$h\,\overset{\bullet}{B}(t)=-6\frac{d}{dt}(a^2\dot{a}). \tag{4.127}$$

In this section and in Section 4.6 we introduced the basic foundations of TEGR and STEGR, respectively. Geometric implications, differences, and common points among the three equivalent theories are discussed in Section 4.8.

Notice that the same procedure as the previous section cannot be applied for STEGR, which cannot be thought of as a gauge theory in the locally flat tangent space-time. As better pointed out in Section 5.2, while the torsion and the curvature are respectively the gauge fields associated to local translation and to LTs, no gauge transformation can be associated to nonmetricity. This formally yields a fundamental difference among the three theories, although they lead to the same equations of motion.

4.8 The Geometric Trinity of Gravity

Taking into account the results achieved in previous sections, it is possible to introduce the so-called *Trinity of Gravity* [110]. We showed that TEGR and STEGR are totally equivalent to GR up to a boundary term in the starting Lagrangian. They can be obtained from the general affine connection

$$\Gamma^\rho{}_{\mu\nu} = \hat{\Gamma}^\rho{}_{\mu\nu} + \overset{\bullet}{K}{}^\rho{}_{\mu\nu} + L^\rho{}_{\mu\nu} \,. \tag{4.128}$$

Specifically, we have:

$$\begin{aligned} &\text{GR} \to L^\rho{}_{\mu\nu} = \overset{\bullet}{K}{}^\rho{}_{\mu\nu} = 0\,, \\ &\text{TEGR} \to \hat{\Gamma}^\rho{}_{\mu\nu},\ \overset{\bullet}{K}{}^\rho{}_{\mu\nu} \neq 0, \quad L^\rho{}_{\mu\nu} = 0\,, \\ &\text{STEGR} \to \hat{\Gamma}^\rho{}_{\mu\nu}, L^\rho{}_{\mu\nu} \neq 0, \quad \overset{\bullet}{K}{}^\rho{}_{\mu\nu} = 0\,. \end{aligned} \tag{4.129}$$

It follows that the most general form of the scalar curvature, written in terms of the general connection (4.128), is

$$\begin{aligned} R = \hat{R} &+ \left(\overset{\bullet}{K}{}^\rho{}_{\nu\mu} + L^\rho{}_{\nu\mu}\right)\left(\overset{\bullet}{K}{}^\beta{}_{\beta\rho} + L^\beta{}_{\beta\rho}\right) g^{\nu\mu} \\ &- \left(\overset{\bullet}{K}{}^\rho{}_{\beta\mu} + L^\rho{}_{\beta\mu}\right)\left(\overset{\bullet}{K}{}^\beta{}_{\nu\rho} + L^\beta{}_{\nu\rho}\right) g^{\nu\mu} \\ &+ \hat{\nabla}_\beta \left[\left(\overset{\bullet}{K}{}^\beta{}_{\nu\mu} + L^\beta{}_{\nu\mu}\right) g^{\nu\mu} - \left(\overset{\bullet}{K}{}^\nu{}_{\nu\mu} + L^\nu{}_{\nu\mu}\right) g^{\nu\mu}\right], \end{aligned} \tag{4.130}$$

with $\hat{R}$ being the standard Ricci scalar. When the space-time is labeled only by the Levi–Civita connection, the dynamics is ruled by the curvature, while torsion and nonmetricity vanish everywhere. In this case, the parallel transport of a given vector around a closed path yields a vector with the same norm but different orientation. An angular displacement therefore occurs due to the curvature of the manifold, as shown in Fig. **4.2**.

When the contorsion tensor occurs in the general connection, the parallel transport yields a shift of the vector after performing a closed path (see Fig. 4.3). The result is a radial displacement between the starting and the final point.

Finally, when both curvature and torsion vanish, nonmetricity describes a space-time in which a propagating vector continuously changes its modulus (see Fig. 4.4).

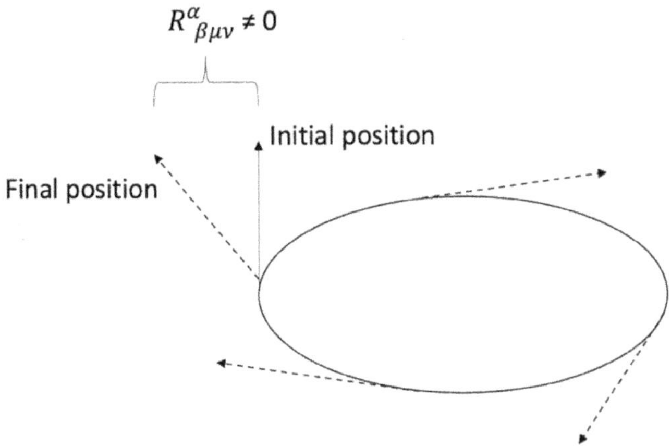

Figure 4.2 Parallel transport of a vector around a closed path in a curved space-time

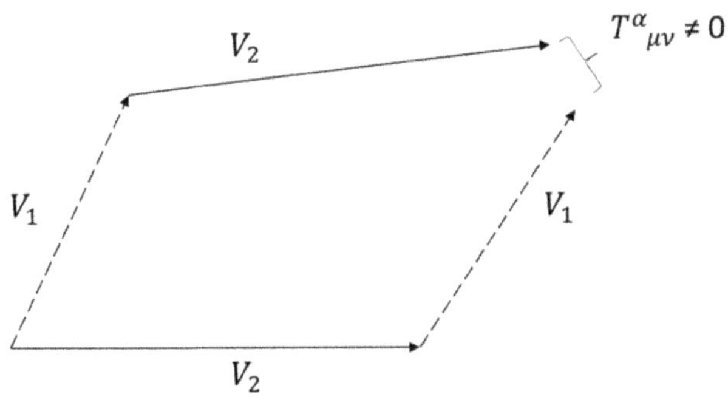

Figure 4.3 Parallel transport of the two vectors V_1 (dashed line) and V_2 (solid line) in a space-time with torsion

As pointed out in [189], nonmetricity, torsion, and curvature can be thought of as a geometric *Trinity* of gravity since all of them lead to totally equivalent theories by different formalism. Despite this, $f(\hat{R})$ gravity is quite different with respect to $f(\overset{\bullet}{T})$ and $f(Q)$ extensions, where the boundary terms B_Q and $\overset{\bullet}{B}_T$ play a nontrivial role in the dynamics. In Fig. **4.5**, a useful scheme of all possible theories with curvature, torsion, and nonmetricity is outlined.

To conclude this chapter, it is worth noticing that several extended and alternative theories of gravity can be formulated after GR. Most of them are capable of providing solutions for GR shortcomings at astrophysical and cosmological scales (see Part II and Part III of the book for a detailed discussion).

Furthermore, many other approaches have been developed to address gravity issues at fundamental scales, such as *String Theory* [190–194], *Kaluza–Klein Theory* [195–197], *Loop Quantum Gravity* [198–200], *Horava–Lifshitz Gravity*

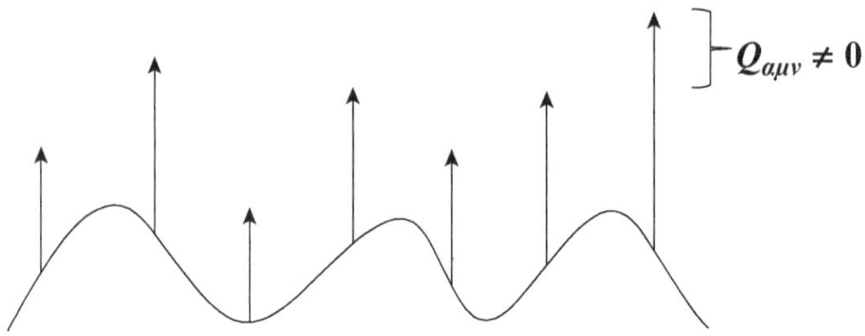

Figure 4.4 Space-time with nonmetricity: a propagating vector field maintains the same direction but changes its length

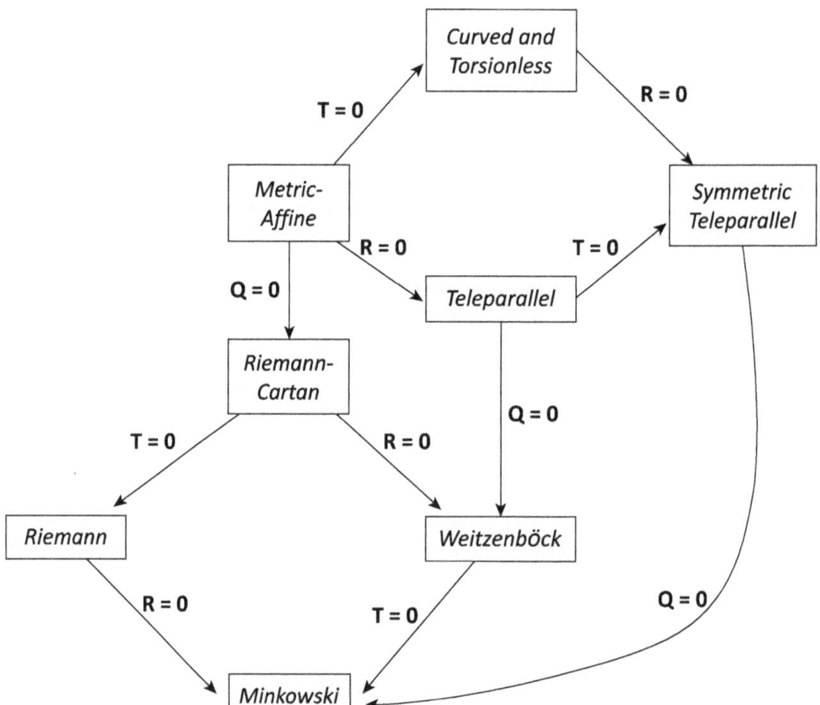

Figure 4.5 Classification of theories of gravity containing curvature, torsion, and nonmetricity

[201–204], *Nonlocal Gravity* [205, 206], *M-Theory* [207, 208], and so on. While GR turns out to be non-renormalizable from the two-loop level, the theories just mentioned can be both renormalizable and unitary.

Finally, we point out that the notation used in this chapter to denote scalar quantities, written in terms of different connections (i.e. the bullet and the hat on the top of the given scalar, tensor, vector), will not be considered in the rest of the book. Indeed, as in this chapter, the notation is needed to distinguish scalars with different connections; in the other chapters, the connections will be understood from context. This can be pursued because, hereafter, both curvature and torsion will not be dealt simultaneously, except for Section 7.2.2, where both R and T appear. However, the scalar curvature arises as a combination of torsion and boundary term B; thus the issue of different connection is easily addressed. Generally, in metric theories (such as $f(R)$ and $f(\mathcal{G})$) we will always consider the Levi–Civita connection, as in teleparallel theories (such as $f(T)$ and $f(T_{\mathcal{G}})$) we will consider the Weitzenböck connection. This allows one to use the cosmological expression of R (4.125) and the cosmological expression of T (4.113).

5

Toward Quantum Gravity

I'd rather attempt to do something great and fail
than to attempt to do nothing and succeed.
— *Robert Harold Schuller*
(*Tough Times Never Last, But Tough People Do!* [1984])

Modified theories of gravity, discussed in Chapter 4, are mostly considered to solve large-scale (IR) problems suffered by GR. In order to settle the UV scale issues, the first step is to introduce a *Canonical Quantization* procedure for gravity and then search for possible applications like Quantum Cosmology. The aim is to deal with gravity under the same standard as the other fundamental interactions.

Despite the formalism, the canonical approach is far from yielding a self-consistent theory of Quantum Gravity. However, the application to cosmology gained great interest because it can represent the semiclassical limit for a full, yet unknown, fundamental theory. In this perspective, Quantum Cosmology is a paradigm to be pursued for the problem of initial conditions of the universe, when the classical prescriptions to describe the cosmic evolution fail. This application will be deeply discussed in Section 5.1 and in the third part of the book, where Quantum Cosmology will be treated in the context of theories of gravity containing symmetries.

However, several issues also occur in the quantum description of the Universe. Some of them must be attributed to the interpretation of the wave function, which cannot be intended as a straightforward probability amplitude, whose squared modulus, integrated over the variable space, provides the probability to get a certain configuration. Possible solutions to this problem will be discussed in Section 5.1.

However, in order to merge the formalism of Quantum Mechanics with that of GR, a Hamiltonian description of the gravitational field must be developed. Performing a Legendre transformation to the EH Lagrangian is not straightforward since the 3+1 dimensional evolution of the gravitational field cannot be disentangled from the background. Reversing the argument, as in standard QFT, it is possible to consider the evolution of the fields in a fixed background. From this point of view, the canonical quantization of gravity must take into account that the fundamental field coincides with the background itself.

This shortcoming can be partially overcome using the *foliation* method, by means of which the evolution of the manifold $\mathcal{M}$ can be intended as the evolution of a class of hypersurfaces $\mathcal{M}_3$ evolving through a fixed timeline: $\mathcal{M} = \mathcal{M}_3 \times \mathbb{R}$. The Hamiltonian description is in turn based on the Arnowitt–Deser–Misner (ADM) formalism, thanks to which a Schrödinger-like equation for the gravitational field can be obtained, the so-called *Wheeler–DeWitt* (WDW) equation.

The ADM formalism was developed by R. Arnowitt, S. Deser, and C. W. Misner in 1962 [209], while the WDW equation was first obtained by J. A. Wheeler and B. DeWitt in 1967 [210–212]. The solution of the equation provides the so-called *Wave Function of the Universe*, which we will discuss in Section 5.1.

Before introducing the main aspects of canonical quantization of gravity, it is worth stressing the differences between the standard quantization procedure and the ADM formalism. In QFT, a fixed background allows one to quantize the theory by replacing Poisson's parentheses with commutators and by imposing the corresponding commutation relations. Then, divergences occurring at the UV scales can be deleted through the renormalization scheme, by means of which finite quantities can be introduced to describe dynamics. This procedure cannot be applied to the gravitational interaction and also in the ADM formalism several issues occur. For instance, infinite-dimensional *Superspace*, *Supermetric*, *Super Hamiltonian*, and *Supermomentum* arise in the canonical quantization procedure and cannot provide any prediction at the formal level. For this reason, the entire Superspace is usually reduced to a finite-dimensional *minisuperspace* of configurations, where solutions can be found and the WDW equation can be solved, becoming a partial differential equation instead of a functional one.

With the aim of introducing the basic foundations of the ADM formalism, let us now consider the most general form of EH action, that is,

$$S = \frac{1}{2}\int_V \sqrt{-g}[R-2\Lambda]\,d^4x + \int_{\partial V} \sqrt{h}K\,dx^3, \tag{5.1}$$

where h[1] is the determinant of the three-dimensional metric h_{ij}, Λ is the cosmological constant, and $K = h^{ij}K_{ij}$ is the trace of extrinsic curvature tensor K_{ij} of the three-dimensional hypersurface ∂V of the four-dimensional manifold V. The extrinsic curvature tensor arises because of the embedding of the three-dimensional surface in four dimensions. The Hamiltonian formulation can be obtained after performing a 3+1 decomposition of the metric $g_{\mu\nu}$ so that the spatial components become dynamical degrees of freedom. According to the

[1] This is the only section in which h denotes the determinant of the three-dimensional metric. In any other section, h is the determinant of nontrivial tetrad fields.

foliation method, a coordinate transformation $X^\alpha \to X'^\alpha$ can be seen as a transformation of hypersurfaces with local coordinates x^i. To any coordinate x^0, there corresponds a space-like hypersurface $x^0 = k$, and the variation of x^0 provides the required foliation. Moreover, to any point of the hypersurface, there corresponds a three-dimensional vectors basis X_i^α, tangent to the surface and perpendicular to the normal surface vector n^ν. In light of these considerations, the following relations hold:

$$g_{\mu\nu} X_i^\mu n^\nu = 0, \qquad g_{\mu\nu} n^\mu n^\nu = -1. \tag{5.2}$$

Let us now define the deformation tensor N^α, which links points with the same coordinates belonging to contiguous surfaces:

$$N^\alpha = \dot{X}^\alpha = \partial_0 X^\alpha(x^0, x^i). \tag{5.3}$$

By decomposing N^α with respect to the previously defined basis of normal and tangent vectors, namely

$$N^\alpha = N n^\alpha + N^i X_i^\alpha, \tag{5.4}$$

two new quantities can be introduced:

- $N \to$ Lapse function: labels the local displacements
- $N^i \to$ Shift vector: labels tangential displacements.

N and N^i describe the deformation relative to the foliation hypersurface and allow to pass from a given hypersurface to another as we will see below.

The coordinates X_i^μ, therefore, link the three-dimensional metric h_{ij} to the four-dimensional metric $g_{\mu\nu}$ through the only nontrivial projection, that is,

$$h_{ij} = g_{\mu\nu} X_i^\mu X_j^\nu. \tag{5.5}$$

In Fig. **5.1** two hypersurfaces are represented. Lapse function and the shift vector are graphically denoted by vectors that link the points P_1, P_2, and P_3.
In Fig. **5.1**, the distance between the points P_1 and P_4 needs both functions to be properly settled. Specifically, such a distance is represented by the interval ds^2, which can be obtained from the definition (5.2) and reads as

$$ds^2 = g_{\mu\nu} dX^\mu dX^\nu = (N^2 - N_i N^i) dt^2 - 2N^i dx_i dt - h_{ij} dx^i dx^j. \tag{5.6}$$

Therefore, the metric matrix and its inverse read, respectively,

$$g_{\mu\nu} = \begin{pmatrix} (N^2 - N_i N^i) & -N_i \\ -N_i & -h_{ij} \end{pmatrix}, \quad g^{\mu\nu} = \begin{pmatrix} 1/N^2 & -N_i/N^2 \\ -N_i/N^2 & N^i N^j/N^2 - h^{ij} \end{pmatrix}, \tag{5.7}$$

where $\beta^i \equiv h^{ij}\beta_j$ and $h^{ij}h_{jk} = \delta^i_k$, namely h^{ij} is the inverse of h_{ij}. Considering the above decomposition of the metric, it is convenient to split the three-dimensional spatial contributions from the four-dimensional ones. As a consequence, the three-dimensional Riemann tensor can be written as

$${}^{(3)}R^i_{jkl} = \partial_k \Gamma^i{}_{jl} - \partial_l \Gamma^i{}_{jk} + \Gamma^i{}_{km}\Gamma^m{}_{jl} - \Gamma^i{}_{lm}\Gamma^m{}_{jk}, \tag{5.8}$$

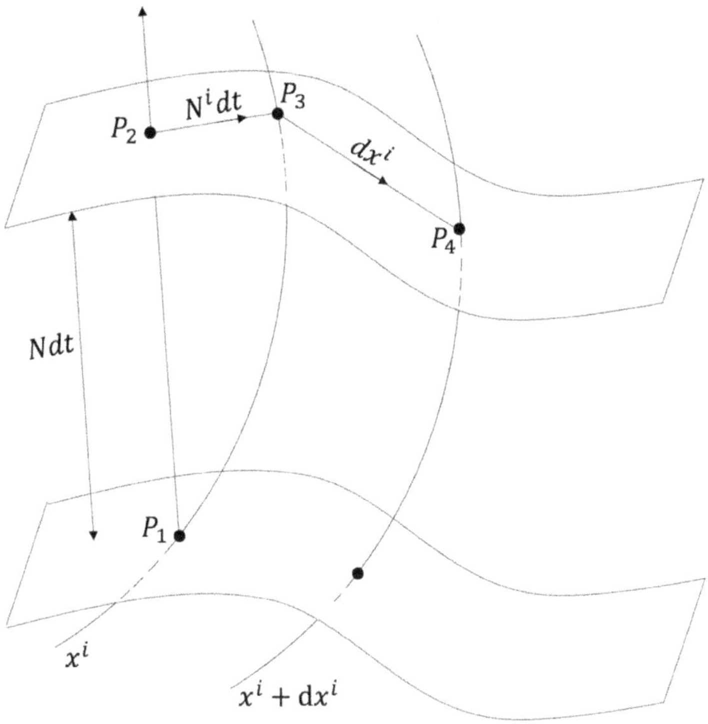

Figure 5.1 The figure shows the role of the lapse function and the shift vector in the ADM formalism. The former projects the point P_1 into P_2 and the latter P_2 into P_3.

with $\Gamma^i{}_{jk} = 1/2h^{il}(\partial_j h_{kl} + \partial_k h_{jl} - \partial_l h_{jk})$ being the connection related to the spatial coordinates, defined by means of the three-dimensional covariant derivative $\mathcal{D}^{(h)}$ as $\mathcal{D}^{(h)}_i V_j = \partial_i V_j - \Gamma^k_{ij} V_k$. By means of the Gauss–Codazzi theorem, the scalar curvature in Eq. (5.1) (up to total derivatives) can be decomposed as

$$R = K^{ij}K_{ij} - K^2 +^{(3)} R, \tag{5.9}$$

with ${}^{(3)}R = h^{jl}\,{}^{(3)}R^i_{jil}$ being the intrinsic three-dimensional curvature. Neglecting the cosmological constant Λ and the integral over the three-dimensional surface, the action becomes

$$S(N, N^i, h_{ij}) = \int N\mathcal{L}\, d^4x, \tag{5.10}$$

with

$$\mathcal{L} = \frac{1}{2}\sqrt{h}N(K^{ij}K_{ij} - K^2 +^{(3)} R) + s.t.\,, \tag{5.11}$$

where *s.t.* is the surface term. The extrinsic curvature tensor can be written in terms of the lapse function and shift vector as

$$K_{ij} = \frac{1}{2N}(\mathcal{D}^{(h)}_i N_j + \mathcal{D}^{(h)}_j N_i - \dot{h}_{ij}), \tag{5.12}$$

where the RHS can be obtained by considering the line element (5.6). The conjugate momenta with respect to 3-D metric, lapse function, and shift vector are read as follows:

$$\pi \equiv \frac{\delta \mathscr{L}}{\delta \dot{N}} = 0, \qquad \pi^i \equiv \frac{\delta \mathscr{L}}{\delta \dot{N}_i} = 0, \qquad \pi^{ij} \equiv \frac{\delta \mathscr{L}}{\delta \dot{h}_{ij}} = \frac{\sqrt{h}}{16\pi G}(Kh^{ij} - K^{ij}). \tag{5.13}$$

The first and the second relations in Eq. (5.13) constitute the so-called *primary constraints*. A straightforward Legendre transformation of the Lagrangian density (5.11) yields the Hamiltonian density

$$\mathscr{H} = \pi^{ij}\dot{h}_{ij} - \mathscr{L}. \tag{5.14}$$

The total Hamiltonian can be obtained by integrating Eq. (5.14) with respect to the three-dimensional surface d^3x:

$$\mathcal{H} = \int (\pi^{ij}\dot{h}_{ij} - \mathscr{L})\, d^3x. \tag{5.15}$$

With the aim to use the lapse function and the shift vector in the canonical quantization procedure, we notice that the Hamiltonian density can be decomposed as

$$\mathscr{H} = N\mathscr{H}_0 + N^i\mathscr{H}_i, \tag{5.16}$$

with

$$\begin{cases} \mathscr{H}_0 \equiv 2\, G_{ijkl}\pi^{ij}\pi^{kl} - 1/2\sqrt{h}\,{}^3R \\ \mathscr{H}_i = -\mathcal{D}_j\pi^j_i. \end{cases} \tag{5.17}$$

The former quantity is called *Super Hamiltonian*, while the latter is called *Supermomentum*. The rank-4 tensor G_{ijkl} is in turn the *Super-metric*, defined as

$$G_{ijkl} = \frac{1}{\sqrt{h}}(h_{ik}h_{jl} + h_{il}h_{jk} - h_{ij}h_{kl}). \tag{5.18}$$

By these definitions, the action can be written as

$$S = \int (\pi^{ij}\dot{h}_{ij} \quad \mathscr{H})\, d^4x = \int (\pi^{ij}\dot{h}_{ij} - N\mathscr{H}_0 - N^i\mathscr{H}_i)\, d^4x. \tag{5.19}$$

From the Hamilton equations, two important relations follow, namely $\mathcal{H}_0 = \mathcal{H}_i = 0$. They are called *secondary constraints*. They are useful in Quantum Cosmology for the construction of the WDW equation. These constraints are related to the condition of null energy and correspond to the field equations $G_{00} = 0$ and $G_{0i} = 0$, respectively.

The first step needed to quantize the theory is to promote the dynamical variables of the classic theory to operators, namely

$$\begin{cases} h_{ij} \to \hat{h}_{ij} \\ \pi \to \hat{\pi} = -i\dfrac{\delta}{\delta N} \\ \pi^i \to \hat{\pi}^i = -i\dfrac{\delta}{\delta N_i} \\ \pi^{ij} \to \hat{\pi}^{ij} = -i\dfrac{\delta}{\delta h_{ij}}, \end{cases} \tag{5.20}$$

and to impose the commutation relations:

$$\begin{cases} [\hat{h}_{ij}(x), \hat{\pi}^{kl}(x')] = i\,\delta_{ij}^{kl}\,\delta^3(x - x') \\ \delta_{ij}^{kl} = \dfrac{1}{2}(\delta_i^k \delta_j^l + \delta_i^l \delta_j^k) \\ [\hat{h}_{ij}, \hat{h}_{kl}] = 0 \\ [\hat{\pi}^{ij}, \hat{\pi}^{kl}] = 0. \end{cases} \tag{5.21}$$

From the above conditions, the quantum counterpart of secondary constraints $\mathcal{H}_0 = \mathcal{H}_i = 0$ automatically follows:

$$\begin{cases} \hat{\mathcal{H}}_0 |\psi >= 0 \\ \hat{\mathcal{H}}^i |\psi >= 0, \end{cases} \tag{5.22}$$

where the wave function ψ is a function of the three-dimensional metric $\psi \equiv \psi[h_{ij}]$. Using the quantum expressions of the Super Hamiltonian and the Super-momentum in terms of the operators in Eq. (5.20), one gets two differential equations of the form

$$\begin{cases} \left(2\,G_{ijkl}\dfrac{\delta}{\delta h_{ij}}\dfrac{\delta}{\delta h_{kl}} - \dfrac{1}{2}\sqrt{h}\,{}^3R\right)|\psi >= 0 \\ \left(2ih_{km}\mathcal{D}_l\dfrac{\delta}{\delta h_{kl}}\right)|\psi >= 0. \end{cases} \tag{5.23}$$

The former is the WDW equation, while the latter is the quantum version of the invariance under diffeomorphism transformations. Moreover, by defining

$$\nabla^2 \equiv G_{ijkl}\frac{\delta}{\delta h_{ij}}\frac{\delta}{\delta h_{kl}}, \tag{5.24}$$

the WDW equation can be suitably written as

$$\left(2\,\nabla^2 - \frac{1}{2}\sqrt{h}\,{}^3R\right)|\psi >= 0, \tag{5.25}$$

from which the analogy with the Schrödinger equation is straightforward. It is worth noticing that the matter field can be included in the treatment as a coordinate-dependent scalar field ϕ, with conjugate momentum

$$\pi_\phi = -i\frac{\delta}{\delta\phi}. \tag{5.26}$$

This is due to the fact that length scales are of the order of Planck scale so we can describe the matter content as given by scalar fields. The formalism is developed in an infinite-dimensional configuration Superspace, which is the space of all the possible 3-metrics. However, restrictions to finite-dimensional minisuperspaces are needed to handle the problem and obtain solutions for dynamics. This corresponds to the choice of suitable cosmological backgrounds (classes of universes) where the WDW equation can be solved, giving the Wave Function of the Universe. Solutions are related to the initial conditions from which, eventually, observable universes can originate.

Nevertheless, even reducing the dynamics, the Wave Function of the Universe cannot be interpreted in the standard way of a probability amplitude. In non-relativistic Quantum Mechanics, the dynamics of the system is described by the Schrödinger equation

$$[\nabla^2 + U(x)]\psi(x,t) = i\partial_t\psi, \tag{5.27}$$

which allows one to define a time-independent, positive-defined internal product of the form

$$<\psi|\psi> \equiv \int \psi^* \psi \, dx^3. \tag{5.28}$$

As a consequence, the solutions of the Schrödinger equation form an infinite-dimensional space in which the product $\psi^*\psi$ can be interpreted as the probability density of localizing a particle in the given configuration space. On the contrary, the WDW equation is more similar to the Klein–Gordon equation, where the scalar product (5.28) is not positive-defined everywhere, though a continuity equation can still be defined. This yields several problems in interpreting ψ as a probability amplitude, and any attempt to develop a second quantization scheme, in this formalism, failed. However, the solutions of the WDW equation give useful indications in Quantum Cosmology, as we shall discuss in what follows.

5.1 Quantum Cosmology: Everything from Nothing

Despite the mathematical difficulties of dealing with an infinite-dimensional Superspace and the absence of a complete and self-consistent theory of Quantum Gravity, it is still possible to use canonical quantization rules in a cosmological background, where the finite number of degrees of freedom allows one to find reliable solutions. The minisuperspace restriction turns out to be a useful tool in quantum cosmology, yielding several predictions in the early stages of the Universe. However, the latter cannot be treated as a standard system so that the same results as quantum mechanics cannot be expected. The main problem lies behind the interpretation of the wave function, which cannot be intended as a probability amplitude. Such an interpretation indeed requires many copies of the same system to make sense; otherwise the concept of probability itself stops being valid. This cannot be applied to gravity and cosmology since we lack a final theory of quantum gravity and a self-consistent interpretation of probability for the space-time. The cosmological point-like Lagrangian of GR, in a spatially closed FLRW universe, can be obtained by considering the general cosmological expression of the scalar curvature, namely

$$R(t) = -6\left(\frac{\ddot{a}}{a} + \frac{\dot{a}^2}{a^2} + \frac{k}{a^2}\right), \tag{5.29}$$

with $k = 1$. Thus, recasting the cosmological constant in terms of a_0, we have:

$$\mathcal{L} = 6\left[a\dot{a}^2 - a\left(1 - \frac{a^2}{a_0^2}\right)\right], \tag{5.30}$$

and a Legendre transformation yields the Hamiltonian

$$\mathcal{H} = \left[\frac{\pi_a^2}{24a} + 6a\left(1 - \frac{a^2}{a_0^2}\right)\right] = 0. \tag{5.31}$$

The application of the constraints (5.22) and (5.13) to the Hamiltonian (5.31) yields a WDW equation of the form

$$\mathcal{H}\psi(a) = 0 \rightarrow \left[\frac{1}{24}\frac{\partial^2}{\partial a^2} - 6a^2\left(1 - \frac{a^2}{a_0^2}\right)\right]\psi(a) = 0. \tag{5.32}$$

The Hamiltonian can be seen as the sum of the kinetic term $\frac{\partial^2}{\partial a^2}$ and the potential $V(a) \sim a_0^2\left(\frac{a^2}{a_0^2} - \frac{a^4}{a_0^4}\right)$. The latter can be qualitatively plotted as a function of the scale factor, as shown in Fig. **5.2**.

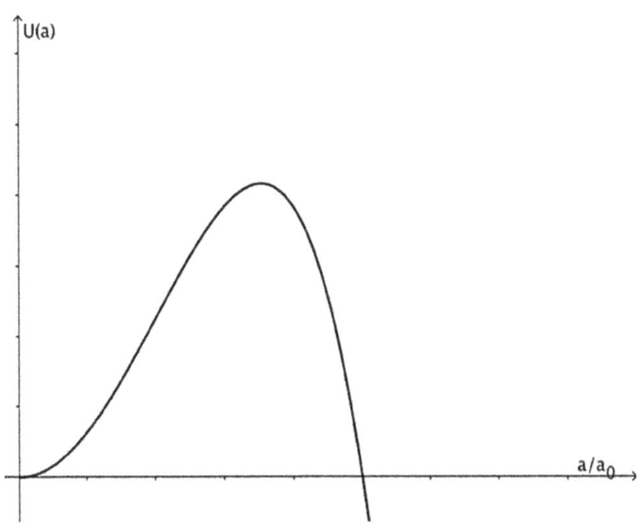

Figure 5.2 Potential barrier in the early Universe

The graph in Fig. **5.2** shows a potential barrier with length a_0, which can be crossed by the Universe through tunnel effect. The interval $a \in [0, a_0]$ is classically forbidden, while $]a_0, \infty[$ represents the classical region where the observable universe can evolve.

Although the real meaning of the wave function is still unclear, many interpretations have been given over the years. For instance, according to the so-called *Many World Interpretation*, the wave function comes from quantum measurements that are simultaneously realized in different universes, without showing any collapse of the wave function, as in standard quantum mechanics [213]. Another interpretation was provided by Hawking, according to whom the wave function is supposed to be related to the probability for the early Universe to develop toward our classical Universe [214,215]. In [216,217], the enucleation of the Universe from "nothing" is studied so that the Universe itself is described under a quantum point of view. This vision allows one to partially solve the problem of setting out the boundary condition, according to which the latter must be the fundamental laws of physics and cannot be imposed from outside [218]. In this scheme of interpretation, J. B. Hartle proposed a criterion to gain information from the wave function based on its trend in the late-time. Specifically, according to the *Hartle criterion*, the wave function must have an oscillating behavior in the classically permitted area, namely where it describes our classical Universe [219]. The analogy with nonrelativistic quantum mechanics is straightforward: considering a particle in a potential barrier, the solution of the Schrödinger equation provides an oscillating wave function inside the barrier and a decreasing exponential behavior outside, in the quantum allowed region. In principle, thanks to the Hartle criterion and in the semiclassical limit, the Wentzel–Kramers–Brillouin (WKB) approximation, it is possible to write the wave function in terms of the classical action S as $\psi = e^{im_P^2 S}$, with m_P being the Planck mass. At the lowest order, the Hamilton–Jacobi equation,

$$\nabla S \cdot \nabla S + V = 0, \tag{5.33}$$

must hold. Therefore, if the wave function is strongly peaked, classical trajectories can be found; otherwise a classically forbidden universe can evolve. In the context of GR, previous calculations show that, up to a normalization factor, an oscillating wave function must occur in the interval $]a_0, \infty[$. From Eq. (5.32) indeed we get

$$\psi(a > a_0) \sim \frac{N}{\sqrt[4]{-V(a)}} \exp\{iS\} - \frac{N}{\sqrt[4]{-V(a)}} \exp\left\{-i \int_{a_0}^{a} \sqrt{-V(a)}\, da\right\}. \tag{5.34}$$

Substituting the expression of the potential $V(a)$ in Eq. (5.34), the wave function takes the following form:

$$\psi_T(a > a_0) = \frac{N}{\sqrt[4]{-V(a)}} e^{-i4\pi^2 a_0^2 \left(\frac{a^2}{a_0^2} - 1\right)^{3/2}}. \tag{5.35}$$

Regarding the quantum allowed region, the wave function can be decomposed in progressive and regressive components, with a general exponential decreasing trend:

$$\psi\left(0 < a < a_0\right) = \frac{1}{\sqrt[4]{-V(a)}}\left[Ae^{-4\pi^2 a_0^2\left(1-\frac{a^2}{a_0^2}\right)^{3/2}} + Be^{4\pi^2 a_0^2\left(1-\frac{a^2}{a_0^2}\right)^{3/2}}\right] \tag{5.36}$$

with A, B complex coefficients. Therefore, finally, the total wave function is

$$\psi(a) = \begin{cases} \dfrac{1}{\sqrt[4]{-V(a)}}\left[Ae^{-4\pi^2 a_0^2\left(1-\frac{a^2}{a_0^2}\right)^{3/2}} + Be^{4\pi^2 a_0^2\left(1-\frac{a^2}{a_0^2}\right)^{3/2}}\right] & 0 < a < a_0 \\ \dfrac{N}{\sqrt[4]{-V(a)}}e^{-i4\pi^2 a_0^2\left(\frac{a^2}{a_0^2}-1\right)^{3/2}} & a > a_0. \end{cases} \tag{5.37}$$

The constants A, B, and N can be found by imposing the normalization condition and by requiring ψ to be a Lagrange function. This treatment holds for a spatially closed universe with $k = 1$, and in GR cannot be generalized to the case $k = 0$. This is due to the fact that the scale factor is the only dynamical degree of freedom of the theory. As shown in Chapter 17, in modified theories of gravity the formalism of quantum cosmology can be extended to a spatially flat background, and the relation (5.33) can be used to recover the currently observed cosmic expansion from the analytic wave function. These results make quantum cosmology an important connection point between classical and quantum gravity; while we are waiting for a complete theory of quantum gravity, the application to cosmology represents a sort of interpretative model capable of reducing the infinite-dimensional superspace coming from the ADM formalism to minisuperspaces where the equations of motion can be interpreted and, eventually, integrated. However, quantum cosmology does not aim to provide UV and IR quantum corrections, since it cannot settle, for example, the renormalizability problem or the lack of the Yang–Mills description. Therefore, the canonical quantization of gravity is not a complete theory, but only aims to solve parts of the high-energy issues arising in standard GR.

To conclude this section, it is worth investigating the link among quantum cosmology, Hartle criterion, and Noether's theorem. The latter assures that a set of conserved quantities j_i is associated with symmetry transformations and permits one to introduce a cyclic variable into the theory by means of the relations (2.37). In this way, the conjugate momenta of the cyclic variables can be written as

$$j_i \equiv \pi^i_{Q^i} = \frac{\partial \mathcal{L}}{\partial \dot{Q}^i}. \tag{5.38}$$

In the canonical quantization scheme, Eq. (5.38) can be written in terms of the corresponding quantized quantities as

$$-i\partial_i|\psi>=j_i|\psi>. \tag{5.39}$$

By considering a theory that carries m Noether symmetries, with corresponding conserved quantities $j_1, j_2 \dots j_m$, the WDW equation and the constraint (5.39) automatically yield $m+1$ equations of the form

$$\begin{cases} \hat{\mathscr{H}}|\psi>=0 \\ -i\partial_1|\psi>=j_1|\psi> \\ -i\partial_2|\psi>=j_2|\psi> \\ \vdots \\ -i\partial_m|\psi>=j_m|\psi>. \end{cases} \tag{5.40}$$

The preceding set of equations can be immediately integrated and, being j_i real constants, we obtain oscillatory behaviors for $|\psi>$ in the directions of symmetries. More precisely, the latter m equations share the following common solution [220]:

$$|\psi>=e^{ij_kQ^k}|\chi(Q^\ell)>, \qquad m<\ell<n, \tag{5.41}$$

where m is the number of symmetries, ℓ are the directions where symmetries do not exist, and n is the total dimension of minisuperspace. It is worthwhile to note that the component $|\chi>$ of the wave function could also depend on j_i, but it is not possible to state "in general" if it is oscillating. Dynamics given by the first equation of the system (5.40) can be reduced if and only if it is possible to define constant conjugate momenta, that is, oscillatory behaviors of a subset of solutions $|\psi>$ exist only if Noether symmetry exists for dynamics. The m symmetries give first integrals of motion and then the possibility to select classical trajectories. In one- and two-dimensional minisuperspaces, the existence of a Noether symmetry allows the complete solution of the problem and to get the full semiclassical limit of minisuperspace quantum cosmology. By these arguments, the Halliwell request that an oscillatory wave function predict correlations between coordinates and canonical conjugate momenta [221, 222] is fully recovered.

Therefore, in the semiclassical limit of quantum cosmology and in the framework of minisuperspace approximation, the reduction procedure of dynamics, due to the existence of Noether symmetries, allows one to select a subset of the solutions of the WDW equation where oscillatory behaviors are found. As a consequence, correlations between coordinates and canonical conjugate momenta emerge so that classical cosmological solutions can be recovered. On the contrary, if a subset of the solutions of the WDW equation has an oscillatory behavior, conserved momenta have to exist and Noether symmetries are present. In other words, Noether symmetries select classical universes.

In summary, the Hartle criterion can be always related to a Noether symmetry and then to the classical trajectories.

Since the analytic solution of the wave function is often missing, the existence of Noether symmetries turns out to be a useful tool for selecting those theories admitting exact solutions. Reducing the dynamics of the system, the Noether Theorem implies the existence of a subset of analytic solutions of the WDW equation, where oscillating wave functions are selected. Consequently, by means of Hamilton–Jacobi equations, classical trajectories can be found even without solving the equations of motion.

For further readings about quantum cosmology and applications to modified theories of gravity see [36, 223, 224].

5.2 Gauge Theories of Gravity

A fundamental step toward the development of a quantum theory for the gravitational interaction is to merge the formalism of the Standard Model with the geometric description of the gravitational field. To this aim, gravity must be treated as a gauge theory. In the covariant formalism, GR can be considered as a diffeomorphism invariant gauge theory, but such an invariance can be recast as a Lorentz invariance in the locally flat tangent space-time. We showed that a similar procedure permits one to describe TEGR as a gauge invariant theory with respect to the local translation group. Unfortunately, neither translations nor Lorentz invariance are internal symmetry, and this does not allow one to address gravity in a Yang–Mills scheme. Moreover, a theory that well satisfies these requirements at the high energy must also fit the results provided at large scales by GR in the low energy limit. In this section we overview the main features of gauge theories of gravity, finding the corresponding gauge Lagrangians in the flat space-time and showing that no unitary renormalizable Yang–Mills theory of gravity occurs in four dimensions.

Let us begin by considering a local LT of the form $\Lambda = e^{\frac{i}{2}\omega^{ab}J_{ab}}$, with ω^{ab} being the one-form Lorentz connection. In this case, the covariant derivative of a generic n-form $\mathcal{P}$ provides the $n+1$ form

$$D\mathcal{P} = d\mathcal{P} - \frac{i}{2}\omega^{ab}J_{ab}\mathcal{P} \ , \tag{5.42}$$

so that

$$D \wedge D\mathcal{P} = -\frac{i}{2}R^{ab}J_{ab}\mathcal{P} \quad \text{with} \quad R^{ab} = d\omega^{ab} + \omega^{a}_{\ c} \wedge \omega^{cb} \ . \tag{5.43}$$

R^{ab} is called *Two-Form Curvature* and the second equation of (5.43) is the *first Cartan structure equation*. Similarly, thanks to the vielbein[2] postulation, the *Contorsion Tensor* can be defined as

[2] Generalization of tetrad fields in higher dimensions.

$$D_{[\mu} e^a_{\nu]} = \partial_{[\mu} e^a_{\nu]} + \omega^a_{[\mu\nu]} = K^a{}_{\mu\nu} \,. \tag{5.44}$$

Notice that the $K^a{}_{\mu\nu}$ just defined is a generalization of the Contorsion Tensor (4.79), including the spin connection. The definition of the *Two Form Torsion* by means of the vielbein postulation provides the *second Cartan Structure Equation*:

$$T^a = De^a = de^a + \omega^a_b \wedge e^b \,. \tag{5.45}$$

This formalism manifestly shows that as torsion is the gauge field associated to local translations, curvature is the gauge field related to LTs. In those reference frames with vanishing spin connection, the two-form curvature identically vanishes and the space-time structure is described by TEGR. On the other hand, by imposing $T^\alpha = 0$ a torsionless theory ruled by curvature arises.

In Chapter 4 we stressed the equivalence among GR, TEGR, and STEGR, showing that the corresponding Lagrangians are formally equivalent among each other in the description of gravity. More precisely they only differ for a boundary term, which does not change the equations of motion. However, as R^{ab} and T^a are the gauge fields linked to LTs and translation, respectively, no gauge field associated to STEGR can be defined.

Therefore, in the development of a gauge theory of gravity, the contribution of nonmetricity can be neglected. In the flat tangent space-time, the action must be constructed by means of curvature, torsion, Levi–Civita tensor, and Minkowski metric, which are all invariants with respect to $SO(1, D-1)$. The most general torsionless action containing all these terms and leading to second-order field equations is the so-called *Lovelock Action*, which reads as follows [225–227]:

$$S = \kappa \int_{\mathcal{M}} \sum_{i=0}^{\frac{D}{2}} \alpha_i \mathscr{L}^{(D,i)}, \tag{5.46}$$

with κ dimensionless constant and where $\mathscr{L}^{(D,i)}$ is the *Lovelock Lagrangian*, defined as

$$\mathscr{L}^{(D,i)} = \epsilon_{a_1, a_2 \dots a_D} R^{a_1 a_2} \wedge R^{a_3 a_4} \wedge \dots \wedge R^{a_{2i-1} a_{2i}} \wedge e^{a_{2i+1}} \wedge e^{a_{2i+2}} \wedge \dots \wedge e^{a_D} \,. \tag{5.47}$$

In four dimensions, the Lovelock action turns out to be

$$\begin{aligned} S_L &= \kappa \int_{\mathcal{M}} \sum_{i=0}^{2} \alpha_i \mathscr{L}^{(4,i)} \\ &= \kappa \int \epsilon_{abcd} \left(\alpha_0 e^a \wedge e^b \wedge e^c \wedge e^d + \alpha_1 R^{ab} \wedge e^c \wedge e^d + \alpha_2 R^{ab} \wedge R^{cd} \right). \end{aligned} \tag{5.48}$$

The action can be equivalently expressed in a covariant form as

$$S = \kappa \int \sqrt{-g} \left(\alpha_0 + \alpha_1 R + \alpha_2 \mathcal{G} \right) d^4x, \tag{5.49}$$

namely the EH action with cosmological constant plus the Gauss–Bonnet term. Since $\mathcal{G}$ is a topological surface term, in four dimensions the Lovelock Lagrangian

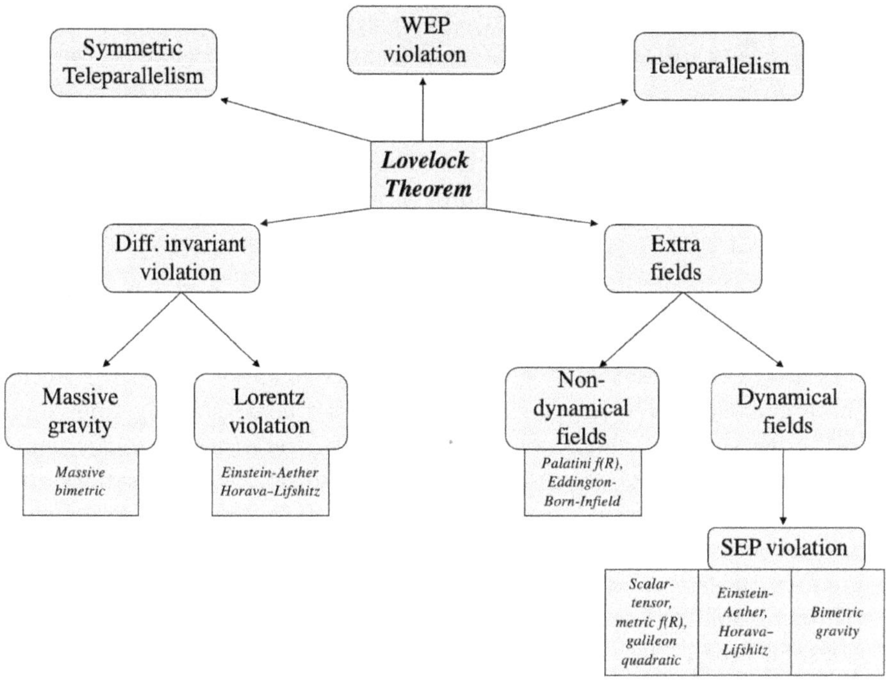

Figure 5.3 The figure shows modified/extended theories of gravity that can be derived from the Lovelock Theorem. According to the violation of some of the Lovelock assumptions, it is possible to classify theories of gravity. Specifically, each theory in the boxes, connected to the central one, arises from different violations of the Lovelock theorem.

yields the same field equations as GR. From these considerations, it follows the *Lovelock Theorem*, which states the following:

The only possible equations of motion coming from a local gravitational action that contains only second derivatives with respect to the metric and where the torsion does not explicitly appear are the Einstein field equations.

Most modified/extended theories of gravity can be derived from the Lovelock theory by relaxing some of the Lovelock hypotheses, as shown in Fig. **5.3** [228].

However, starting from Eq. (5.47), the free parameters α_i can be suitably set in order to select Lagrangians that are invariant under some gauge transformation. Given the general form of Eq. (5.47), several combinations that yield gauge invariant actions can be found. Nevertheless, this procedure lacks a formal structure and several combinations are missed when proceeding by trial and error.

A straightforward way to overcome this issue is to construct gauge Langrangians by topological invariants. Thus, let E be an n-form characteristic class which, by construction, is invariant under local gauge transformations, that is, $\delta E = 0$. If there exists a $D-1$ form $\mathscr{L}$ such that $d\mathscr{L} = E$, then $\mathscr{L}$ is invariant under gauge transformations up to a boundary term. The proof of this statement is straightforward: since E is invariant under gauge transformations, the relation

$$\delta E = \delta(d\mathscr{L}) = d(\delta\mathscr{L}) = 0 \tag{5.50}$$

holds. The preceding equation suggests that the quantity $\delta\mathscr{L}$ can be written as the exterior derivative of a $D-2$ form B, namely

$$d(\delta\mathscr{L}) = 0 \quad \rightarrow \quad \delta\mathscr{L} = dB. \tag{5.51}$$

This finally shows that $\mathscr{L}$ is invariant up to a total derivative. As $\mathscr{L}$ can be used as a Lagrangian of a topological field theory, E represents a topological surface term that can be found from the condition $E = d\mathcal{L}$. Reversing the argument, gauge Lagrangians can be constructed starting from topological surface terms, whose integration over the space-time provides nontrivial topological invariants. Consequently, the Lagrangian will be invariant up to a boundary term, namely *quasi-gauge invariant.* In general, D-dimensional topological surface terms permit one to construct $D-1$ dimensional Lagrangians; however, the lack of nontrivial topological surface terms in odd dimensions yields the impossibility of constructing nontrivial gauge Lagrangians in even dimensions. Such odd-dimensional Lagrangians are called *Chern–Simons Lagrangians* or *Chern–Simons forms.* Chern–Simons theories were first introduced by S. S. Chern and J. H. Simons in 1974 [229], with the aim of developing a topological field theory capable of describing all fundamental interactions as Yang–Mills theories. Basic foundations of Chern–Simons theory can be found, for example, in Refs. [226, 230–234] and applications in Refs. [235–238].

For example, the Chern–Simons 3-form coming from the Pontryagin density $R^{ab}R_{ab}$ is

$$CS^P_{(3)} = \omega^a_b d\omega^b_a + \frac{2}{3}\omega^a_b\omega^b_c\omega^c_a \,. \tag{5.52}$$

Another example is given by the previously mentioned Euler density, which, in four dimensions, is equivalent to the Gauss–Bonnet invariant and turns out to be the exterior derivative of the Chern–Simons Lagrangian invariant with respect to the AdS group

$$\mathscr{L}_3^{AdS} = \epsilon_{abc}\left(R^{ab} \wedge e^c + \frac{1}{3l^2} e^a \wedge e^b \wedge e^c\right). \tag{5.53}$$

The Euler density E_4 can be equivalently written in terms of the two-form torsion T^c, defined in Eq. (5.45) as

$$E_4 = \epsilon_{abc}\left(R^{ab} \wedge T^c + \frac{1}{3l^2} e^a \wedge e^b \wedge T^c\right), \tag{5.54}$$

where l is a constant with dimension of length. Notice that the presence of torsion in the Euler density does not contradict Lovelock's hypothesis of torsionless Lagrangians. Torsion disappears after applying the exterior derivative, so that the corresponding Chern–Simons form turns out to respect Lovelock assumptions. In general, the $2D-1$-dimensional AdS-invariant Lagrangian reads

$$\mathscr{L}^{AdS}_{2D-1} = \sum_{i=0}^{D-1} \tilde{\alpha}_i \mathscr{L}^{(2D-1,i)}, \qquad \tilde{\alpha}_i = \frac{(\pm 1)^{i+1}\ell^{2i-D}}{D-2i}\binom{D-1}{i}, \tag{5.55}$$

and the exterior derivative yields the $2D$-dimensional Euler density E_{2D}, that is,

$$E_{2D} = \epsilon_{a_1 a_2 \dots a_{2D-1} a_{2D}} R^{a_1 a_2} \wedge \dots \wedge R^{a_{2D-1} a_{2D}}, \tag{5.56}$$

providing the Ricci scalar in two dimensions. When integrated over the manifold, Eq. (5.56) provides a topological invariant, namely the $2D$ Euler characteristic.

In what follows, we focus mainly on the five-dimensional Lovelock Lagrangian, with particular interest in those coupling constants yielding the invariance under the local AdS group. The five-dimensional Chern–Simons form is

$$\begin{aligned}\mathscr{L}^{AdS}_5 = \frac{1}{\ell}\epsilon_{abcde} \Big(& R^{ab} \wedge R^{cd} \wedge e^e + \frac{2}{3\ell^2} R^{ab} \wedge e^c \wedge e^d \wedge e^e \\ & + \frac{1}{5\ell^4} e^a \wedge e^b \wedge e^c \wedge e^d \wedge e^e\Big),\end{aligned} \tag{5.57}$$

whose exterior derivative provides the 6-D Euler density [239]. Notice that the Lagrangian in Eq. (5.57) is contained in the general five-dimensional Lovelock Lagrangian

$$\begin{aligned}\mathscr{L}^{L}_5 = \epsilon_{abcde} \big(& \alpha_0\, e^a \wedge e^b \wedge e^c \wedge e^d \wedge e^e + \alpha_1 R^{ab} \wedge e^c \wedge e^d \wedge e^e \\ & + \alpha_2 R^{ab} \wedge R^{cd} \wedge e^e\big) ,\end{aligned} \tag{5.58}$$

with α_0, α_1, and α_2 defined according to the relation (5.55). It is worth remarking that, from Eq. (5.47), it is possible to consider several other gauge Lagrangians, belonging to different gauge groups. Most of them are mainly studied due to the results provided in Supergravity and String Theory [226, 232, 240–242].

Finally, the four-dimensional *Pontryagin density* $P_4 = dA\, dA$ yields the $U(1)$ invariant three-dimensional Lagrangian

$$\mathscr{L} = AdA, \tag{5.59}$$

where A is the one-form connection. Similarly, from the $SU(4)$ Pontryagin density, it is possible to construct the $SU(n)$ invariant 3D Lagrangian, namely

$$\mathscr{L} = \mathrm{Tr}\left(\mathbf{A} \wedge d\mathbf{A} + \frac{2}{3}\mathbf{A} \wedge \mathbf{A} \wedge \mathbf{A}\right). \tag{5.60}$$

Field equations and gauge invariance of Eqs. (5.59) and (5.60) are discussed in Appendix A. The models considered in this section are of particular interest for gravity theories since they are capable of fixing some UV and IR shortcomings suffered by GR.

Conclusion of Part I

In this preliminary part we overviewed the role played by symmetries in theoretical physics introducing the foundations of Noether theorems and their main applications. In order to find conserved quantities coming from the most important Lie groups, we started from well-defined theories and applied given symmetry transformations to relate the latter with integrals of motion. This procedure cannot be straightforwardly pursued in modified theories of gravity, which extend or modify the EH action to some generalization including GR as a limit. Indeed, alternatives to GR often exhibit unknown general functions to be constrained by physical criteria as well as experimental observations. Our aim, in this book, is to select the functional forms of gravitational Lagrangian functions by symmetry considerations, using Noether's first theorem. With this objective in mind, in Chapter 4, we outlined the main features of GR modifications, briefly explaining the reasons why they can be used to cure GR shortcomings, mostly in the low-energy limit. In this context, as we will show in the second part of the book, the application of Noether's first theorem is of main interest, since it provides a starting point to reduce higher-order field equations or to constrain coupling and potentials. Although GR gained great approval as the best accepted theory of gravity, its predictions in extreme regimes or very large scales often lack experimental confirmations. For example, GR predicts, with extreme precision, phenomena at the Solar System level, but provides disagreements with observations at larger scales, which led to the introduction of concepts like dark energy and dark matter. Furthermore, GR is not compatible with Quantum Mechanics by construction. However, as discussed in Chapter 5, gauge theories of gravity aim to relate the formalism of GR with QFT. While waiting for a self-consistent theory of Quantum Gravity, it is worth investigating its low-energy limit, that is, Quantum Cosmology. The latter represents a paradigm from which information regarding the evolution of the very early Universe can be achieved. Also in this framework, Noether symmetries gain great interest since they can lead to suitable cosmological Hamiltonian and analytic forms for the Wave Functions of the Universe. Moreover, they are strictly connected to the Hartle criterion, according to which oscillating wave functions describe observable universes. This treatment is further simplified by the capability of the Noether theorem to reduce the minisuperspace dimensions, introducing cyclic variables into the system. In the next part of the book, we shall apply the previously developed theoretical structure to modified Lagrangians in cosmological and spherically symmetric backgrounds.

Part II

THE NOETHER SYMMETRY APPROACH

6

From the Noether Theorem to the Noether Symmetry Approach

The desire for symmetry, for balance,
for rhythm in form as well as in sound,
is one of the most inveterate of human instincts.

– *Edith Wharton*
(*The Decoration of Houses,* [2015])

In this chapter we discuss the central issue of this book, namely the possibility of selecting theories containing symmetries, given a large class of possible Lagrangians. In the previous sections, we showed how to find the conserved quantity and the symmetry generator, once the starting Lagrangian is known. Specifically, in the standard approach, the symmetry of the system can be inferred by the shape of the given Lagrangian, and this allows one to get the transformation that leaves the equations of motion invariant. Noether current and Noether charge, ruled by Eqs. (2.11) and (2.29), automatically follow. On the other hand, we also selected the form of symmetry generator, starting from general Lagrangians and performing a general transformation. It turns out that the condition (2.24) can yield the infinitesimal generators and the corresponding integrals of motion. In all cases, we started from coordinate and field transformations, with the aim of using the Noether vector to reveal the form of the generator of the given transformation.

The purpose of this chapter is to show that the theorem can also be used as a *method* for choosing physically consistent models that, eventually, can be exactly integrated. It is worth noticing that, in the previous sections, we considered well-defined Lagrangians with given symmetries. Here we start from large classes of Lagrangians where the specific functional forms are selected by the existence of symmetries. Specifically, we apply the condition (2.24) in order to reduce the dynamics of the system, selecting only those Lagrangians with respect to the symmetry condition. Not knowing the exact form of the initial transformation a priori, we assume that there exists *at least* one transformation that leaves the EL equations unchanged. However, it is worth remarking that only some classes of theories can be selected by Noether's method. In particular, the amount of unknown functions contained in the Lagrangian must be such that the system is not overdetermined. In this way, the method provides the dynamics that can be exactly solved, and this is a sort of selection rule for theories. Unfortunately, there are no general indications to follow in order to establish whether the starting Lagrangian suits the requirements of the Noether Approach but, in most cases, the final result is physically consistent.

In the search for symmetries we will use both the Noether vector X and its first prolongation $X^{[1]}$, stressing the differences between the two, case by case. The infinitesimal generators δx, $\delta\phi$, and $\delta\partial_a\phi$ cannot be set from the beginning, but they will naturally arise from the Noether system resolution. This means that the infinitesimal generators are functions of the variables contained in the space of configuration, which must be determined by Noether's theorem. To conform to the usual notation we define

$$\begin{gathered}\delta x^a \equiv \xi^a(x^a, \phi^i, \partial_a\phi^i), \qquad \delta\phi^i = \eta^i(x^a, \phi^i, \partial_a\phi^i),\\ \delta\partial_a\phi^i = \partial_a\eta^i(x^a, \phi^i, \partial_a\phi^i),\end{gathered} \tag{6.1}$$

so that the first prolongation of Noether vector reads as

$$X^{[1]} = \xi^a\partial_a + \eta^i\frac{\partial}{\partial\phi^i} + \left(\partial_a\eta^i - \partial_a\phi^i\partial_b\xi^b\right)\frac{\partial}{\partial(\partial_a\phi^i)}. \tag{6.2}$$

The standard vector X can be easily obtained by setting $\xi^a = 0$. Furthermore, since the transformation related to the vector X does not involve coordinates, the non-extended Noether vector can be written as

$$X = \eta^i(\phi^i, \partial_a\phi^i)\frac{\partial}{\partial\phi^i} + \partial_a\eta^i(\phi^i, \partial_a\phi^i)\frac{\partial}{\partial(\partial_a\phi^i)}. \tag{6.3}$$

Conditions to find external and internal symmetries are, respectively,

$$\begin{cases}\left(X^{[1]} + \partial_a\xi^a\right)\mathcal{L} = \partial_a g^a\\ L_X\mathcal{L} = X\mathcal{L} = 0,\end{cases} \tag{6.4}$$

and the corresponding conserved quantities become

$$\begin{cases} j^{a\,[1]} = \dfrac{\partial\mathscr{L}}{\partial(\partial_a\phi^i)}\eta^i - \dfrac{\partial\mathscr{L}}{\partial(\partial_a\phi^i)}\partial_b\phi^i\xi^b + \mathscr{L}\xi^a - g^a\\ j^a = \dfrac{\partial\mathscr{L}}{\partial(\partial_a\phi^i)}\eta^i.\end{cases} \tag{6.5}$$

Both the preceding conditions give a homogeneous polynomial of second degree in the velocities plus an inhomogeneous term in the ϕ^i. Such a polynomial has to be identically zero, and then each coefficient must be independently null. Such a symmetry is overdetermined and, if a solution exists, it is expressed in terms of integration constants instead of boundary conditions.

6.1 The First Noether Theorem for Canonical Lagrangians

In what follows, we sketch the main steps leading to the general Noether identity (2.24) and to the vanishing Lie derivative condition $L_X\mathcal{L} = 0$, in the case of Lagrangians dependent on time. Then we obtain the rules that permit one to methodically find a suitable change of variables. Though Noether's theorem has been already demonstrated in Section 2.1, the subcases considered in this section represent useful and pedagogical schemes that allow one to better comprehend the applications to cosmological Lagrangians.

Let us then consider a Lagrangian $\mathcal{L}(q^i, \dot{q}^i)$, which depends on the coordinates and the related velocities. Let us assume the EL equations to be invariant with respect to the following change of coordinates:

$$\begin{cases} \tilde{t} = t + \epsilon\xi(t, q^i) + O(\epsilon^2) \\ \tilde{q}^i = q^i + \epsilon\eta^i(t, q^i) + O(\epsilon^2)\,. \end{cases} \tag{6.6}$$

As pointed out in Section 2.1, in order to find the generator of transformations, we need to find the transformation law of the first derivative since, due to the time involved in the transformation, the quantity $\dot{\tilde{q}}^i$ does not trivially correspond to $\dfrac{d\tilde{q}^i}{dt}$. For the first derivative, we have

$$\begin{aligned} \dot{\tilde{q}}^i &= \frac{d\tilde{q}^i}{d\tilde{t}} = \frac{dq^i + \epsilon\, d\eta^i(t, q^i)}{dt + \epsilon\, d\xi(t, q^i)} = \frac{\frac{dq^i}{dt} + \epsilon\frac{d\eta^i(t,q^i)}{dt}}{1 + \epsilon\frac{d\xi(t,q^i)}{dt}} \\ &= \left(\frac{dq^i}{dt} + \epsilon\frac{d\eta^i(t, q^i)}{dt}\right)\left(1 + \epsilon\frac{d\xi(t, q^i)}{dt}\right)^{-1}, \end{aligned} \tag{6.7}$$

which, up to the first order, takes the form

$$\begin{aligned} \dot{\tilde{q}}^i &= \left(\frac{dq^i}{dt} + \epsilon\frac{d\eta^i(t, q^i)}{dt}\right)\left(1 - \epsilon\frac{d\xi(t, q^i)}{dt}\right) + O(\epsilon^2) \\ &= \frac{dq^i}{dt} + \epsilon\left[\frac{d\eta^i(t, q^i)}{dt} - \frac{dq^i}{dt}\frac{d\xi(t, q^i)}{dt}\right] + O(\epsilon^2) \sim \dot{q}^i + \epsilon\left[\dot{\eta}^i - \dot{q}^i\dot{\xi}\right]. \end{aligned} \tag{6.8}$$

Defining $\eta^{i\,[1]} = \dot{\eta}^i - \dot{q}^i\dot{\xi}$, the system (6.6) can be implemented with Eq. (6.8), providing

$$\begin{cases} \tilde{t} = t + \epsilon\xi(t, q^i) + O(\epsilon^2) \\ \tilde{q}^i = q^i + \epsilon\eta^i(t, q^i) + O(\epsilon^2) \\ \dot{\tilde{q}}^i = \dot{q}^i + \epsilon\eta^{i\,[1]}, \end{cases} \tag{6.9}$$

and finally the generator of transformation takes the form

$$X^{[1]} = \xi\frac{\partial}{\partial t} + \eta^i\frac{\partial}{\partial q^i} + \eta^{i\,[1]}\frac{\partial}{\partial \dot{q}^i} = \xi\frac{\partial}{\partial t} + \eta^i\frac{\partial}{\partial q^i} + (\dot{\eta}^i - \dot{q}^i\dot{\xi})\frac{\partial}{\partial \dot{q}^i}\,. \tag{6.10}$$

Notice that we assumed our Lagrangian to be independent of higher-order derivatives, and hence it is not necessary to calculate the transformation laws related to $\ddot{q}^i$. Let us now show that if the coordinates transformation (6.9) leaves the equations of motion invariant, then the system satisfies the Noether identity

$$X^{[1]}\mathcal{L} + \dot{\xi}\mathcal{L} = \dot{g}(t, q^i), \tag{6.11}$$

where g is a generic function depending on coordinates and time. In order to prove the condition (6.11), we recall that the EL equations are invariant if the following condition holds:

$$\frac{d\tilde{t}}{dt}\tilde{\mathcal{L}} = \mathcal{L} + \epsilon\dot{g}\,. \tag{6.12}$$

Deriving Eq. (6.12) with respect to ϵ and then setting $\epsilon = 0$, we obtain

$$\frac{d\tilde{t}}{dt}\frac{\partial\tilde{\mathcal{L}}}{\partial\epsilon} + \tilde{\mathcal{L}}\frac{\partial}{\partial\epsilon}\left(\frac{d\tilde{t}}{dt}\right) = \dot{g}\,. \tag{6.13}$$

We can observe that $\frac{d\tilde{t}}{dt} = \frac{\partial\tilde{t}}{\partial t} + \frac{\partial\tilde{t}}{\partial q^i}\dot{q}^i = 1 + \epsilon\frac{\partial\xi}{\partial q^i}\dot{q}^i$, which, for $\epsilon = 0$, is equal to 1. Furthermore, assuming that it is possible to swap the order of derivatives, we have $\frac{\partial}{\partial\epsilon}\left(\frac{d\tilde{t}}{dt}\right) = \frac{d}{dt}\frac{\partial\tilde{t}}{\partial\epsilon} = \dot{\xi}$. Substituting these results into Eq. (6.13) one gets

$$\xi\frac{\partial\tilde{\mathcal{L}}}{\partial t} + \eta^i\frac{\partial\tilde{\mathcal{L}}}{\partial\tilde{q}^i} + \eta^{[1]\,i}\frac{\partial\tilde{\mathcal{L}}}{\partial\dot{\tilde{q}}^i} + \dot{\xi}\tilde{\mathcal{L}} = \dot{g}\,, \tag{6.14}$$

which is nothing else but Eq. (6.11). By the same procedure as Section 2.1, it follows that systems satisfying the condition (6.11) lead to the conserved quantity

$$I(t, q^i, \dot{q}^i) = \xi\left(\dot{q}^i\frac{\partial\mathcal{L}}{\partial\dot{q}^i} - \mathcal{L}\right) - \eta^i\frac{\partial\mathcal{L}}{\partial\dot{q}^i} + g(t, q^i)\,, \tag{6.15}$$

which is a first integral of motion.

6.1.1 Internal Symmetries

Let us assume the Lagrangian $\mathcal{L}$ to be non-dissipative and nondegenerate, that is,

$$\frac{\partial\mathcal{L}}{\partial t} = 0, \quad \left|\frac{\partial^2\mathcal{L}}{\partial\dot{q}^i\partial\dot{q}^j}\right| = 0. \tag{6.16}$$

Under these conditions, as shown in Section 2.1, the general theorem (6.11) reduces to the vanishing Lie derivative condition. Although the latter can be obtained as a subcase of Eq. (6.11), here, for pedagogical purposes, we start directly from the EL equations related to the Lagrangian $\mathcal{L}$ satisfying the assumptions in Eq. (6.16). Multiplying the EL equations by the function $\eta^i(q)$, we get

$$\eta^i\left(\frac{d}{dt}\frac{\partial\mathcal{L}}{\partial\dot{q}^i} - \frac{\partial\mathcal{L}}{\partial q^i}\right) = 0. \tag{6.17}$$

Since

$$\eta^i \frac{d}{dt}\frac{\partial \mathcal{L}}{\partial \dot{q}^i} = \frac{d}{dt}\left(\eta^i \frac{\partial \mathcal{L}}{\partial \dot{q}^i}\right) - \dot{\eta}^i \frac{\partial \mathcal{L}}{\partial \dot{q}^i}, \tag{6.18}$$

from (6.17) we have

$$\frac{d}{dt}\left(\eta^i \frac{\partial \mathcal{L}}{\partial \dot{q}^i}\right) = \eta^i \frac{\partial \mathcal{L}}{\partial q^i} + \dot{\eta}^i \frac{\partial \mathcal{L}}{\partial \dot{q}^i}. \tag{6.19}$$

Notice that the RHS of Eq. (6.19) is the Noether vector associated with the transformation, so that we obtain

$$\frac{d}{dt}\left(\eta^i \frac{\partial \mathcal{L}}{\partial \dot{q}^i}\right) = X\mathcal{L} = L_X\mathcal{L} \tag{6.20}$$

and the Noether theorem can be recast as follows:

If the condition $L_X\mathcal{L} = 0$ holds, then the quantity $j_0 \equiv \eta^i \frac{\partial \mathcal{L}}{\partial \dot{q}^i}$ is a first integral of motion.

Moreover, under the point transformation $q^j \to Q^j(q^i)$, the vector field X becomes

$$X' = (i_x dQ^j)\frac{\partial}{\partial Q^j} + \frac{d(i_x dQ^j)}{dt}\frac{\partial}{\partial \dot{Q}^j}, \tag{6.21}$$

with $i_X dQ^j$ being the inner derivative. Therefore, for the same reasons provided in Section 2.1.3, if X is the generator of a certain symmetry and we choose a point transformation in such a way that

$$i_x dQ^1, \;\; i_X dQ^j = 0, \;\; j \neq 1, \tag{6.22}$$

we get

$$X' = \frac{\partial \mathcal{L}}{\partial Q^1}, \quad \frac{\partial \mathcal{L}}{\partial Q^1} = 0; \tag{6.23}$$

thus Q^1 is a cyclic coordinate and the dynamics can be reduced. It is worth pointing out that, in principle, more than one X can be found (say, for instance, X_1,X_2). If they commute, that is, $[X_1, X_2] = 0$, then it is possible to obtain two cyclic coordinates by solving the system

$$i_{X_1} dQ^1 = 1, \;\; i_{X_2} dQ^2 = 1, \;\; i_{X_1} dQ^j = 0, \;\; i_{X_2} dQ^j = 0, \;\; j \neq 1, 2. \tag{6.24}$$

On the other hand, if the two Noether vectors do not commute, this procedure is clearly not applicable since commutation relations are preserved by diffeomorphisms. However, in such a case, notice that $X_3 = [X_1, X_2]$ is also a symmetry; indeed $L_{X_3}\mathcal{L} = L_{X_1}L_{X_2}\mathcal{L} - L_{X_2}L_{X_1}\mathcal{L} = 0$. Therefore, if X_3 is independent of X_1, X_2, we can go on until the vector fields close a Lie algebra. The standard way to deal with this situation is to make a Legendre transformation, going to the Hamiltonian formalism and to a Lie algebra of Poisson brackets.

If we want to look for the reduction with cyclic coordinates, the procedure is possible in the following way:

- We arbitrarily choose one of the symmetries, or a linear combination of them, and get new coordinates as above. After the reduction, we get a new Lagrangian $\mathcal{L}'(Q^j)$.
- We search again for symmetries in this new space, make a new reduction, and so on until it is possible.
- If the search fails, we try again with another of the existing symmetries.

The preceding procedure shows that, given a dynamical system, it is always possible to search for Noether symmetries. If they exist, the dynamics can be reduced by a change of variables since one (or more than one) coordinate becomes cyclic. If we are able to integrate such a new dynamics, the problem is to invert the solution in order to get the evolution in the previous variables.

In the next two sections, we provide two examples and show in detail how the conditions (6.11) and (6.20) allow one to select the form of Lagrangians presenting symmetries and then reducible in cyclic coordinates.

6.2 Particle Lagrangian with Unknown Potential

Here we introduce the Noether Symmetry Approach by means of the application to particle Lagrangians with unknown potentials. Let us start by outlining the general scheme that will be used to select theories with symmetries, based on the Noether Theorem. We have to do the following:

1. Write the general form of the generator of symmetries, depending on the variables of the system;
2. Impose the condition $X^{[1]}\mathcal{L} + \mathcal{L}\partial_a\xi^a = \partial_a g^a$ (or $L_X\mathcal{L} = 0$), which leads to a system of differential equations;
3. Solve the system and find the infinitesimal generators η^i;
4. When necessary, change variables using Eq. (2.37) in order to reduce the dynamics.

Let us now consider a canonical mechanical Lagrangian depending on the coordinates through the potential $V(q^i)$, namely

$$\mathcal{L} = \frac{1}{2}(\dot{q}^i)^2 - V(q^i), \tag{6.25}$$

where the index i runs from 1 to 3. We aim to select a viable potential $V(q^i)$ by symmetry considerations. In this case, the first prolongation of the Noether vector takes the form

$$\begin{aligned} X^{[1]} = \xi(t, q^i, \dot{q}^i)\frac{\partial}{\partial t} + \eta^i(t, q^i, \dot{q}^i)\frac{\partial}{\partial q^i} \\ + \left(\dot{\eta}^i(t, q^i, \dot{q}^i) - \dot{q}^i\xi(t, q^i, \dot{q}^i)\right)\frac{\partial}{\partial \dot{q}^i}. \end{aligned} \tag{6.26}$$

Therefore, the application of the condition (2.24) to Lagrangian (6.25) yields

$$\eta^i\frac{\partial V(q^i)}{\partial q^i} + \left(\dot{\eta}^i - \dot{q}^i\dot{\xi}\right)\dot{q}_i + \frac{1}{2}\dot{\xi}(\dot{q}^i)^2 - \xi\, V(q^i) = \dot{g}. \tag{6.27}$$

By computing the total derivatives, Eq. (6.27) can be written as

$$\eta^i \frac{\partial V(q^i)}{\partial q^i} + \frac{\partial \eta^i}{\partial t}\dot{q}_i + \frac{\partial \eta^i}{\partial q^k}\dot{q}^k\dot{q}_i - \frac{1}{2}(\dot{q}^i)^2\frac{\partial \xi}{\partial t} - \frac{1}{2}(\dot{q}^i)^2\frac{\partial \xi}{\partial q^k}\dot{q}^k$$
$$-\frac{\partial \xi}{\partial t}V(q^i) - \frac{\partial \xi}{\partial q^k}V(q^i)\dot{q}^k = \frac{\partial g}{\partial t} + \frac{\partial g}{\partial q^k}\dot{q}^k. \tag{6.28}$$

Similarly to the previous applications, we split the preceding equation by imposing terms sharing the same time-derivatives to vanish. This procedure yields the system

$$\begin{cases} \eta^i \dfrac{\partial V(q^i)}{\partial q^i} - \dfrac{\partial \xi}{\partial t}V(q^i) = \dfrac{\partial g}{\partial t} \\ \dfrac{\partial \eta^i}{\partial t}\dot{q}_i - \dfrac{\partial \xi}{\partial q^k}V(q^i)\dot{q}^k = \dfrac{\partial g}{\partial q^k}\dot{q}^k \\ \dfrac{\partial \eta^i}{\partial q^k}\dot{q}^k\dot{q}_i - \dfrac{1}{2}(\dot{q}^i)^2\dfrac{\partial \xi}{\partial t} = 0 \\ \dfrac{1}{2}(\dot{q}^i)^2\dfrac{\partial \xi}{\partial q^k}\dot{q}^k = 0. \end{cases} \tag{6.29}$$

Since $g(t, q^i)$ is an arbitrary function that does not affect the solution of the system, it can be assumed to vanish. This is an ad hoc choice that allows one to simplify the treatment. Nevertheless, the gauge function g can be arbitrarily chosen according to the case considered. The last equation of the system suggests that the sum

$$\frac{\partial \xi}{\partial q^k}\dot{q}^k = \frac{\partial \xi}{\partial q^1}\dot{q}^1 + \frac{\partial \xi}{\partial q^2}\dot{q}^2 + \frac{\partial \xi}{\partial q^3}\dot{q}^3 \tag{6.30}$$

must vanish in order for the system to hold. Being the variables q^1, q^2, and q^3 independent of each other, the only possible solution is $\dfrac{\partial \xi}{\partial q^k} = 0$. Replacing such a constraint into the second equation of the system, we obtain

$$\frac{\partial \eta^i}{\partial t}\dot{q}_i = 0, \tag{6.31}$$

which leads to the condition $\dfrac{\partial \eta}{\partial t} = 0$. Therefore, it turns out that the system can be reduced to two equations constrained by two conditions on the infinitesimal generators, namely

$$\begin{cases} \eta^i \dfrac{\partial V(q^i)}{\partial q^i} - \dfrac{\partial \xi}{\partial t}V(q^i) = 0 \\ \dfrac{\partial \eta^i}{\partial q^k}\dot{q}^k\dot{q}_i - \dfrac{1}{2}(\dot{q}^i)^2\dfrac{\partial \xi}{\partial t} = 0 \\ \xi \equiv \xi(t), \qquad \eta^i \equiv \eta^i(q^k). \end{cases} \tag{6.32}$$

By setting $\eta^1 = \alpha$, $\eta^2 = \beta$, $\eta^3 = \gamma$, $q^1 = a$, $q^2 = b$, and $q^3 = c$, the system takes the form

$$\begin{cases} \alpha\partial_a V + \beta\partial_b V + \gamma\partial_c V - V\partial_t\xi = 0 \\ \partial_a\alpha\dot{a}^2 + \partial_b\alpha\dot{a}\dot{b} + \partial_c\alpha\dot{a}\dot{c} + \partial_a\beta\dot{a}\dot{b} + \partial_b\beta\dot{b}^2 + \partial_c\beta\dot{b}\dot{c} + \\ + \partial_a\gamma\dot{a}\dot{c} + \partial_b\gamma\dot{b}\dot{c} + \partial_c\gamma\dot{c}^2 - \frac{1}{2}\left(\dot{a}^2 + \dot{b}^2 + \dot{c}^2\right)\partial_t\xi = 0 \\ \xi \equiv \xi(t), \qquad \eta^i \equiv \eta^i(q^k). \end{cases} \tag{6.33}$$

The second equation can be further split in other six equations, after equaling to zero the coefficients of different time derivatives:

$$\begin{cases} \alpha\partial_a V + \beta\partial_b V + \gamma\partial_c V - V\partial_t\xi = 0 \\ \partial_a\alpha - \frac{1}{2}\partial_t\xi = 0 \\ \partial_b\alpha + \partial_a\beta = 0 \\ \partial_c\alpha + \partial_a\gamma = 0 \\ \partial_b\beta - \frac{1}{2}\partial_t\xi = 0 \\ \partial_c\beta + \partial_b\gamma = 0 \\ \partial_c\gamma - \frac{1}{2}\partial_t\xi = 0 \\ \xi \equiv \xi(t), \qquad \eta^i \equiv \eta^i(q^k). \end{cases} \tag{6.34}$$

The preceding system admits two different solutions for the infinitesimal generators η^i, that is,

$$\begin{cases} \xi = \xi_0 t + \xi_1 \\ \alpha = \frac{1}{2}\xi_0 a + \alpha_1 b + \alpha_2 c \\ \beta = \frac{1}{2}\xi_0 b - \alpha_1 a + \beta_2 c \\ \gamma = \frac{1}{2}\xi_0 c - \alpha_2 a - \beta_2 b, \end{cases} \qquad \begin{cases} \xi = \xi_0 t + \xi_1 \\ \alpha = \frac{1}{2}\xi_0 a + \alpha_0 b\, c \\ \beta = \frac{1}{2}\xi_0 b + \alpha_0 a\, c \\ \gamma = \frac{1}{2}\xi_0 c - \alpha_0 a\, b. \end{cases} \tag{6.35}$$

In addition, from the first equation of the system (6.32), four potentials $V(q^i)$ can be selected. Two of them lead to external symmetry generators of the form

$$\begin{cases} \xi = \xi_0 t + \xi_1 \\ \alpha = \frac{1}{2}\xi_0 a \\ \beta = \frac{1}{2}\xi_0 b \\ \gamma = \frac{1}{2}\xi_0 c \\ V(a,b,c) = f_0 a^2 + g_0 b^2 + h_0 c^2 = k_i (q^i)^2, \end{cases} \qquad \begin{cases} \xi = \xi_0 t + \xi_1 \\ \alpha = \frac{1}{2}\xi_0 a \\ \beta = \frac{1}{2}\xi_0 b \\ \gamma = \frac{1}{2}\xi_0 c \\ V(a,b,c) = V_0 a^k b^n c^\ell, \\ \text{with } k + n + \ell = 2. \end{cases} \tag{6.36}$$

Notice that, in order to analytically find the form of the potential, the infinitesimal generators (6.35) must be further constrained with respect to Eq. (6.35). Similarly, the solutions for internal symmetries with vanishing $\xi(t)$ are

$$\begin{cases} \xi = 0 \\ \alpha = \alpha_1 b + \alpha_2 c \\ \beta = -\alpha_1 a \\ \gamma = -\alpha_2 a \\ V(a,b,c) = f_0 a + g_0 b + h_0 c \equiv k_i q^i, \end{cases} \qquad \begin{cases} \xi = 0 \\ \alpha = \frac{1}{2}\xi_0 a + \alpha_1 b \\ \beta = \frac{1}{2}\xi_0 b - \alpha_1 a + \beta_2 c \\ \gamma = \frac{1}{2}\xi_0 c - \beta_2 b \\ V(a,b,c) = f_0 a + g_0 b + h_0 c \equiv k_i q^i, \end{cases}$$

$$\begin{cases} \xi = 0 \\ \alpha = \alpha_1 b + \alpha_2 c \\ \beta = -\alpha_1 a + \beta_2 c \\ \gamma = -\alpha_2 a - \beta_2 b \\ V(a,b,c) = V_0 e^{ka+nb+\ell c} \equiv V_0 e^{k_i q^i}, \end{cases} \qquad \begin{cases} \xi = 0 \\ \alpha = \alpha_0 b\, c \\ \beta = \alpha_0 a\, c \\ \gamma = -\alpha_0 a\, b \\ V(a,b,c) = f_0 a^2 + g_0 b^2 \\ \qquad + h_0 c^2 \equiv k_i (q^i)^2, \end{cases}$$

$$\begin{cases} \xi = 0 \\ \alpha = \alpha_0 b\, c \\ \beta = \alpha_0 a\, c \\ \gamma = -\alpha_0 a\, b \\ V(a,b,c) = V_0 e^{ka^2+nb^2+\ell c^2} \equiv V_0 e^{k_i (q^i)^2}. \end{cases} \tag{6.37}$$

Summing up, we discussed the approach capable of selecting the unknown potential-containing symmetries, by means of the simple case of free-particles Lagrangian. This is a straightforward example of the *Noether Symmetry Approach.* It is worth noticing that the harmonic potential also occurs as a solution of the system (6.34), and the symmetry generator turns out to be the same as that previously found in Section 3.5.2.

Another interesting result occurs when internal symmetries are considered. It relies on the application of the Noether vector X to the Lagrangian. The non-extended method, which imposes the Lie derivative of the Lagrangian to vanish, is not complete since it is based on the assumption $\xi(t) = 0$. Indeed, generally, several solutions arise when coordinates transformations are not neglected, but cannot be selected by the relation $X\mathcal{L} = 0$. For this reason, though the non-extended approach may often simplify the treatment, the entire set of symmetries can be found only by the application of the first prolongation of Noether vector. As better pointed out in this second part of the book, cosmological Lagrangians that provide only trivial solutions when the vanishing Lie derivative condition is considered can yield a suitable set of symmetries when the complete approach is dealt with.

6.3 Application to the Point like Canonical Lagrangian

As previously shown, from the condition $L_X\mathcal{L} = 0$, only internal symmetries can be selected, according to which the infinitesimal generator $\delta x^a = \xi^a$ does not contribute to the transformation. However, the function $\xi^a(x^a, \phi^i)$ often plays a crucial role in the search for symmetry. In the previous example, we considered a particle Lagrangian with potential $V(q)$ and showed that the Noether system solutions can be distinguished in two main categories, occurring when $\xi^a = 0$ and when $\xi^a \neq 0$, respectively. Specifically, in the latter case, the function ξ^a turned out to be independent of the configuration space variables. In light of this result, we wonder whether this condition holds for any canonical Lagrangian, or whether it is a particular solution relying on the Lagrangian (6.25) only. In order to find a general constraints for the function ξ, here we apply the Noether Symmetry Approach to the most general canonical Lagrangian, namely

$$\mathcal{L}\left(q^i, \dot{q}^i\right) = a_{ij}\left(q^i\right)\dot{q}^i\dot{q}^j - V\left(q^i\right) - A_i\left(q^i\right)\dot{q}^i, \tag{6.38}$$

with a_{ij} being general functions of the variables q^i, $V(q)$ the potential, and A_i a vector function of q^i. Assuming a two-particle system, the index i runs from 1 to 2 and the first prolongation of the Noether vector can be written as

$$X^{[1]} = \xi\partial_t + \alpha\partial_{q_1} + \beta\partial_{q_2} + (\dot{\alpha} - \dot{q}_1\dot{\xi})\partial_{\dot{q}_1} + (\dot{\beta} - \dot{q}_2\dot{\xi})\partial_{\dot{q}_2}, \tag{6.39}$$

where $\eta^i \equiv (\alpha, \beta)$ and $q^i \equiv (q_1(t), q_2(t))$. Notice that Eq. (6.39) is the two-dimensional representation of the vector (2.16). The application of the identity (2.24) yields the system of differential equations

$$\begin{cases}
\alpha\partial_{q_1}V + \beta\partial_{q_2}V + (\partial_t\alpha)A_1 + (\partial_t\beta)A_2 + (\partial_t\xi)V - \partial_t g = 0\\
-\alpha\partial_{q_1}A_1 - \beta\partial_{q_2}A_1 - A_1\partial_{q_1}\alpha + 2(\partial_t\alpha)a_{12}\\
-A_2\partial_{q_1}\beta + (\partial_t\beta)a_{12} + (\partial_t\beta)a_{21} - \partial_{q_1}g = 0\\
-\alpha\partial_{q_1}A_2 - \beta\partial_{q_2}A_2 - A_1\partial_{q_2}\alpha + (\partial_t\alpha)a_{12} + (\partial_t\alpha)a_{21}\\
-A_2\partial_{q_2}\beta + 2(\partial_t\beta)a_{22} - \partial_{q_2}g = 0\\
\alpha\partial_{q_1}a_{11} + \beta\partial_{q_2}a_{11} + 2\left(\partial_{q_1}\alpha\right)a_{11} - (\partial_t\xi)a_{11} + \left(\partial_{q_1}\beta\right)a_{12} + \left(\partial_{q_1}\beta\right)a_{21} = 0\\
\alpha\partial_{q_1}a_{22} + \beta\partial_{q_2}a_{22} + \left(\partial_{q_2}\alpha\right)a_{12} + \left(\partial_{q_2}\alpha\right)a_{21} + 2\left(\partial_{q_2}\beta\right)a_{22} - \left(\partial_q\xi\right)a_{22} = 0\\
\alpha\partial_{q_1}a_{12} + \alpha\partial_{q_1}a_{21} + \beta\partial_{q_2}a_{12} + \beta\partial_{q_2}a_{21} + 2\left(\partial_{q_2}\alpha\right)a_{11}\\
+\left(\partial_{q_1}\alpha\right)a_{12} + \left(\partial_{q_1}\alpha\right)a_{21} - (\partial_t\xi)a_{12} - (\partial_t\xi)a_{21} + \left(\partial_{q_2}\beta\right)a_{12}\\
+\left(\partial_{q_2}\beta\right)a_{21} + 2\left(\partial_{q_1}\beta\right)a_{22} = 0\\
-\left(\partial_{q_1}\xi\right)a_{12} - \left(\partial_{q_1}\xi\right)a_{21} - \left(\partial_{q_2}\xi\right)a_{11} = 0\\
-\left(\partial_{q_2}\xi\right)a_{12} - \left(\partial_{q_2}\xi\right)a_{21} - \left(\partial_{q_1}\xi\right)a_{22} = 0\\
-2\left(\partial_{q_1}\xi\right)a_{11} + \left(\partial_{q_1}\xi\right)a_{11} = 0\\
-2\left(\partial_{q_2}\xi\right)a_{22} + \left(\partial_{q_2}\xi\right)a_{22} = 0.
\end{cases} \tag{6.40}$$

The large amount of unknown functions does not allow one to find a unique solution for the system, which turns out to be overdetermined. However, only the last two equations are needed to put a general constraint on the function ξ. Indeed, assuming $a_{11} \neq 0$ and $a_{22} \neq 0$, the last two equations are satisfied if and only if $\partial_{q_2}\xi = \partial_{q_1}\xi = 0$, which means that the function ξ must be a function of time only. This general result can be imposed a priori in future applications, as often as the theory is labeled by a canonical Lagrangian.

6.4 The Noether Symmetry Approach for Theories of Gravity

As pointed out in the previous sections, the Noether Theorem can be recast as a method to select theories that are symmetric with respect to certain transformations and to find the corresponding generators of symmetries. This application of the theorem is useful in the context of electromagnetic, weak, and strong interactions, which can be well described by the Yang–Mills formalism. In the case of theories of gravity, it can be crucial because, for the above-mentioned reasons, there is no final candidate capable of describing the gravitational interaction at any scales.

In Chapter 4, we introduced modified/extended theories of gravity as possible candidates to describe the gravitational interaction, avoiding the shortcomings suffered by GR. However, such theories often lack some strong physical foundation because the corresponding actions involve unknown functions of some curvature/torsion invariant. One possible remedy, largely considered in the literature, is to use the Noether Symmetry Approach to select those actions containing symmetries. Specifically, in order to constrain the functional form of the starting action, it is possible to search for symmetries of gravitational Lagrangians, following the steps reported in Section 6.3.

As we will show, the condition $L_X\mathcal{L} = 0$ can be used as a starting point to reveal transformations leaving the Lagrangian unchanged. Moreover, this prescription can be extended by considering the identity in Eq. (2.24), by means of which symmetries of EL equations can be selected. As the former approach only involves internal transformations, the latter also include space-time coordinate variations and their n-derivatives. In this regard, we showed that Noether's vector can be extended up to the nth prolongation when the nth derivative is considered in the Lagrangian. Despite this fact, here we only focus on the first prolongation since higher-order derivatives in the gravitational Lagrangians can always be integrated out. This extended method can be easily addressed to the vanishing Lie derivative condition by imposing the infinitesimal generator δt to vanish.

As we will show, when internal symmetries are considered, the condition $L_X\mathcal{L} = 0$ is sometimes unable to provide any (nontrivial) transformations generating the symmetry. However, in most of the analyzed cases, both approaches permit one to find integrals of motion, by means of which the Lagrangian can be written in terms of cyclic variables. This procedure turns out to be very useful in classical and quantum cosmology, where cyclic variables allow one to recover the Hartle criterion and the corresponding wave function related to observable universes.

Now some remarks seem to be appropriate in view of the application of Noether Symmetry Approach to the gravitational interaction. For instance, it may seem that the most natural way to apply Noether's theorem is to consider the starting covariant action of the given modified theory of gravity, without choosing the background a priori. Though it would be better to select theories of gravity without focusing on particular metrics, Noether's theorem was demonstrated under the assumption of flat space-times. For this reason, we must consider the point-like Lagrangian instead of the full Lagrangian density. As a consequence, geometry is understood in the starting action as dynamical fields evolving in a flat background.

The first chapters of this part of the book are devoted to the application of the Noether Symmetry Approach to cosmological Lagrangians. The last chapter considers three different theories of gravity in a spherically symmetric space-time.

In the former case, the analysis is performed in a FLRW spatially flat line element, namely

$$ds^2 = dt^2 - a(t)^2 d|\mathbf{x}|^2, \tag{6.41}$$

with $\mathbf{x}$ being the spatial position. The extension to any spatially curved geometry is straightforward. Since the cosmological Lagrangians are usually dependent on the cosmic time, as pointed out in Section 6.1, the Noether identity (2.24) reduces to

$$X^{[1]}\mathcal{L} + \dot{\xi}\mathcal{L} = \dot{g}(t, q^i), \tag{6.42}$$

with the first prolongation of Noether's vector and conserved quantity being, respectively:

$$X^{[1]} = \xi\frac{\partial}{\partial t} + \eta^i\frac{\partial}{\partial q^i} + \eta^{i\,[1]}\frac{\partial}{\partial \dot{q}^i} = \xi\frac{\partial}{\partial t} + \eta^i\frac{\partial}{\partial q^i} + (\dot{\eta}^i - \dot{q}^i\dot{\xi})\frac{\partial}{\partial \dot{q}^i}, \tag{6.43}$$

$$j(t, q^i, \dot{q}^i) = \xi\left(\dot{q}^i\frac{\partial\mathcal{L}}{\partial\dot{q}^i} - \mathcal{L}\right) - \eta^i\frac{\partial\mathcal{L}}{\partial\dot{q}^i} + g(t, q^i). \tag{6.44}$$

Furthermore, when the infinitesimal generator ξ vanishes, internal symmetries occur, so that the Noether's vector and the conserved quantity take the form

$$\begin{cases} X = \eta^i\dfrac{\partial}{\partial q^i} + \dot{\eta}^i\dfrac{\partial}{\partial \dot{q}^i} \\[2ex] j(t, q^i, \dot{q}^i) = \eta^i\dfrac{\partial\mathcal{L}}{\partial\dot{q}^i} \\[2ex] X\mathcal{L} = 0 \;\rightarrow\; j \sim const. \end{cases} \tag{6.45}$$

Let us stress that when the condition $j = 0$ is realized, two possibilities can follow:

1. The conserved quantity vanishes identically, regardless of the field equations solution;
2. The conserved quantity vanishes after replacing the field equations solution.

The former case leads to trivial models since the conserved quantity vanishes independently of the cosmological solutions. In the latter case, which is not trivial, the conserved quantity is initially different from zero, but it vanishes when the cosmological solutions are considered.

In a cosmological background, the minisuperspace variables are functions of the cosmic time so that the Lagrangian density can be straightforwardly integrated over the hypersurface, providing:

$$S = \int \mathscr{L}\, d^D x = \int \mathscr{L}\, d^{D-1} x dt = \int \mathcal{L}\, dt, \tag{6.46}$$

with the definition $\mathcal{L} \equiv \int \sqrt{-g}\mathscr{L}\, d^{D-1}x$ and D being the number of dimensions. Moreover, some mathematical difficulties can be overcome by the assumption of a homogeneous and isotropic universe.

As we will show, several exact solutions occur analytically, in both cosmological and spherically symmetric cases, after performing the Noether Approach.

In all cases, the starting Lagrangian will be obtained by means of the Lagrange multipliers method. The Lagrange multipliers will be coupled to the cosmological (or spherically symmetric) expression of the geometric term considered in the starting action. To this purpose, the dependence of the geometric terms on the scale factor $a(t)$ will be manifestly written case by case. However, as usual in the formalism of Lagrange multipliers, geometric terms and scale factors will be treated as separated fields. The relations occurring between the two ones will thus be recovered by the EL equations. The latter provide the same solutions as the field equations and will be used in order to determine the corresponding dynamics.

Nevertheless, the equations of motion can be solved only after selecting the form of the function by the Noether symmetries; otherwise the system of differential EL equations turns out to be overdetermined. In some cases, we will also use Eq. (2.37) to find suitable changes of variables that allow one to recast the Lagrangian in terms of the conserved quantities j. This prescription will be further investigated in Chapter 17, where the application to Quantum Cosmology will be considered. In particular, it will turn out that a suitable change of variables is necessary in order to find the Hamiltonian and the related Wave Function of the Universe (see Chapter 5 for details).

A methodical procedure to change variables can be found only when internal symmetries are considered, that is, when the vanishing Lie derivative condition is applied. However, the vanishing Lie derivative condition, sometimes, yields only trivial results.

It is worth pointing out that the Lagrangian formalism turns out to be equivalent to the variational one only when the given system enjoys special symmetry properties. Indeed, field equations of modified theories of gravity are, in general, different from EL equations. This is due to the fact that the amount of field equations is independent of the minisuperspace dimension, while the amount of EL equations strictly relies on the minisuperspace considered. In the cosmological spatially flat background (6.41), for instance, only two linearly independent field equations occur after varying the action with respect to the metric. The only EL equation, instead, is the one with respect to the scale factor.

In cosmology, this apparent disagreement between the two approaches can be addressed by means of the *energy condition*, which must be considered along with the EL equations to provide the final set of equations of motion. In particular, it is a condition of zero energy, which, in the cosmological case, must be included to recover the first Friedman equation. More precisely, when the dynamics of a gravitational field is thought of as the time evolution of the minisuperspace variables in a flat space-time, the energy momentum tensor in Eq. (4.11) reduces to that in Eq. (3.18). Therefore, if we consider as matter Lagrangian the point-like Lagrangian $\mathcal{L}$ of the minisuperspace variables q^i evolving in a fixed spacetime, the variational principle gives

$$\mathcal{T}^0_0 = \dot{q}^i \frac{\partial \mathcal{L}}{\partial \dot{q}^i} - \mathcal{L} = 0. \tag{6.47}$$

The remaining field equations are automatically provided by the EL equations with respect to each minisuperspace variable. However, one can equivalently neglect the energy condition and introduce a lapse function $N(t)$ as the first component of the line element (6.41) so that the EL equation with respect to N can provide back the first Friedman equation when $N \to 1$. The same prescription cannot be applied, for example, to axial symmetry, where field equations are generally different from EL equations.

The key step in the implementation of this program consists of passing from the Lagrangian of the relevant fields to a point-like Lagrangian or, in other words, of going from a system with an infinite number of degrees of freedom to one with a finite number of degrees of freedom.

In spherical symmetry, with metric given by

$$g_{\mu\nu} = \text{diag}(P(r)^2, -Q(r)^2, -r^2, -r^2 \sin^2\theta), \tag{6.48}$$

the variational approach provides three equations, while the Lagrangian approach provides only two equations (the EL equations with respect to P and Q). The third equation cannot be recovered by the latter approach since the energy condition cannot be considered for dissipative Lagrangians. Nevertheless, EL equations are completely equivalent to the two (out of three) field equations. Moreover, in most cases, it turns out that the three field equations are not independent, with the consequence that only two of them are needed to determine the related dynamics. For this reason, the above ansatz on the shape of the metric is consistent even when adopting the Lagrangian approach. Conversely, one can also start from the most general line element:

$$g_{\mu\nu} = \mathrm{diag}(P(r)^2, -Q(r)^2, -M(r)^2, -M(r)^2 \sin^2\theta). \tag{6.49}$$

In such a case the dynamics is determined by three EL equations and by the energy condition since the Lagrangian is no longer dissipative. The EL equation with respect to $M(r)$, which is nothing else but the third field equation of the variational approach, will be exactly equivalent to the energy condition. After setting $M(r) = r$, one recovers the 2-sphere and it turns out that the three EL equations are not independent so that the dynamics is provided by two equations only.

Summing up, regardless of the dimensions considered, the variational approach turns out to be equivalent to the Lagrangian approach. One can either vary the action after adopting some ansatz or get the field equations before considering the minisuperspace. Fortunately, this is feasible in cosmology because most models of physical interest are spatially homogeneous Bianchi models and the observed universe is spatially homogeneous and isotropic to a high degree. In spherical symmetry, with line element (6.48), there are two unknown functions and the Lagrangian approach provides two EL equations. This is expected also for the variational approach, which, in turn, provides three equations that are not independent.

This part of the book is organized as follows: first we consider the most simple extensions of the EH action, including functions of the curvature, torsion, or both ($f(R, T)$ gravity). Then we also include second-order curvature invariants into the action as well as the Gauss–Bonnet term or its teleparallel equivalent. Subsequently we search for symmetries in higher-order theories of gravity containing, for example, terms like $\Box R$ or $\Box T$. All the previously mentioned theories will be investigated in a FLRW spatially flat background, and the matter Lagrangian will be neglected from the beginning. This is due to the fact that we are interested in searching for exact solutions, describing the cosmological evolution of the universe, either in its very early stages or in the late time. Thus, we will assume that cosmological solutions describe those stages of the Universe where geometric contributions are dominating with respect to the matter fields. Therefore, models selected by the Noether approach will be considered in vacuum, where de Sitter and AdS-like solutions can be found. In this way, it is possible to check whether modified theories of gravity can mimic the role of dark energy.

In Chapter 10, we apply the Noether Symmetry Approach to scalar-tensor theories of gravity by coupling the scalar field to curvature, torsion, and the Gauss–Bonnet term, respectively. In Chapter 11, we focus on nonlocal actions of curvature (or torsion) invariants. Finally, in Chapters 12 and 13, the Noether approach is applied to Bianchi and spherically symmetric universes, respectively.

It is worth pointing out that the physical viability of models with symmetries has to be checked through the comparison with phenomenology. To this purpose, in the third part of the book, we will test the physical meaning and consistency of these models by the observations.

7

The Extensions of GR, TEGR, and STEGR

Geometry will draw the soul toward truth
and create the spirit of philosophy.

— *Plato*
(*De Republica* [375 B.C])

This is the first chapter devoted to the application of Noether's theorem to Extended Gravity cosmology; specifically, here we are going to consider the most straightforward extensions of the GR, TEGR, and STEGR actions. In Section 7.1, we apply the Noether Symmetry Approach to $f(R)$ gravity, selecting those actions containing symmetries and the corresponding conserved quantities in a cosmological spatially flat background. In Section 7.2 we develop a similar approach for $f(T)$ gravity, introduced in Chapter 4. Even in this case, the application of the first prolongation of Noether's vector allows to select models mathematically easier to handle, from which analytic solutions occur. Moreover, in Section 7.2.1 the boundary term also is included in the starting action, so that standard $f(R)$ gravity is restored as a particular case of $f(T, B)$ gravity. In addition, by means of the boundary term, both curvature and torsion can be considered, as pointed out in Section 7.2.2. Finally, in Section 7.3, we apply the Noether Symmetry Approach to extensions of the STEGR action, containing different nonmetricity invariants. More precisely, we consider a general theory of all possible quadratic, first-order derivative terms of the nonmetricity tensor, classifying the models that are invariant under point transformations in a cosmological background and using the symmetries of these models to reduce dynamics.

7.1 GR Extension: The Case of $f(R)$ Gravity

Let us first consider $f(R)$ gravity, a straightforward extension of GR, whose main aspects are overviewed in Section 4.3. The starting action is

$$S = \int \sqrt{-g} f(R) \, d^4x. \tag{7.1}$$

Before pursuing the approach, it is worth discussing some cosmological aspects of the theory in the Lagrangian formalism. In order to find the point-like Lagrangian, we use the Lagrange multipliers method in a cosmological background. Specifically, in a spatially flat FLRW spacetime, given by the line element (6.41), the cosmological expression of the curvature scalar is

$$R = -6\left(\frac{\ddot{a}}{a} + \frac{\dot{a}^2}{a^2}\right), \tag{7.2}$$

while the square root of the metric determinant is $\sqrt{-g} = a^3$. By introducing the Lagrange multiplier λ, and considering that variables in the minisuperspace are only time-dependent, the action can be written as

$$S = \int \left[a^3 f(R) - \lambda\left(R + 6\frac{\ddot{a}}{a} + 6\frac{\dot{a}^2}{a^2}\right)\right] dt. \tag{7.3}$$

Since all variables are independent of the spatial coordinates, the hypersurface can be neglected and the general integral in Eq. (7.1) reduces to an integration over the time. By varying the action with respect to the scalar curvature it is possible to find the Lagrange multiplier λ:

$$\frac{\delta S}{\delta R} = a^3 f_R(R) - \lambda = 0, \qquad \lambda = a^3 f_R(R). \tag{7.4}$$

With the aim to get a canonical Lagrangian, the terms containing second derivatives must be integrated by parts and then the divergence theorem must be used. In this way, the point-like Lagrangian turns out to be [243]

$$\mathcal{L} = a^3\left[f(R) - Rf_R(R)\right] + 6a\dot{a}^2 f_R(R) + 6a^2\dot{a}\dot{R} f_{RR}(R). \tag{7.5}$$

Let us now define the Hessian matrix determinant

$$||H|| = \begin{vmatrix} \dfrac{\partial \pi_a}{\partial \dot{a}} & \dfrac{\partial \pi_a}{\partial \dot{R}} \\ \\ \dfrac{\partial \pi_R}{\partial \dot{a}} & \dfrac{\partial \pi_R}{\partial \dot{R}} \end{vmatrix}. \tag{7.6}$$

Considering the conjugate momenta

$$\begin{aligned} \pi_a &= \frac{\partial \mathcal{L}}{\partial \dot{a}} = 6a^2 f_{RR}(R)\,\dot{R} + 12a\dot{a}f_R(R), \\ \pi_R &= \frac{\partial \mathcal{L}}{\partial \dot{R}} = 6a^2\dot{a}f_{RR}(R), \end{aligned} \tag{7.7}$$

it turns out that Eq. (7.6) can be written as

$$||H|| = \begin{vmatrix} 12af_R(R) & 6a^2 f_{RR}(R) \\ 6a^2 f_{RR}(R) & 0 \end{vmatrix} = -36a^4 f_{RR}(R). \tag{7.8}$$

This means that, among the $f(R)$ extensions, the only degenerate Lagrangian, with the Hessian determinant equal to zero, is the EH one.

From Lagrangian (7.5), one gets a set of three differential equations of the form

$$\begin{cases} 6a^2\dot{a}\dot{R}f_{RR}(R) + 6a\dot{a}^2 f_R(R) - a^3[f(R) - Rf_R(R)] = 0 \\ \dot{R}^2 f_{RRR}(R) + \ddot{R}f_{RR}(R) + \dfrac{\dot{a}^2}{a^2} f_R(R) + 2\dfrac{\ddot{a}}{a} f_R(R) \\ -\dfrac{1}{2}[f(R) - Rf_R(R)] + 2\dfrac{\dot{a}}{a}\dot{R}f_{RR}(R) = 0 \\ R = -6\left(\dfrac{\ddot{a}}{a} + \dfrac{\dot{a}^2}{a^2}\right). \end{cases} \tag{7.9}$$

Notice that the third equation, namely the equation of motion with respect to R, provides the cosmological expression of the scalar curvature, used in Eq. (7.2) to construct the starting action. It is the equation corresponding to the Lagrange multiplier. This result is expected by construction and can be used as a check to verify the consistency of the treatment.

The preceding system can be solved after selecting the form of the unknown function $f(R)$ by the Noether symmetry. To this purpose, notice that the first prolongation of the Noether vector acts in the minisuperspace of variables $\mathcal{S} \equiv \{a, R\}$, so that it takes the form

$$\begin{aligned} X^{[1]} = \; & \xi(a,R,t)\partial_t + \alpha(a,R,t)\partial_a + \beta(a,R,t)\partial_R \\ & + \left(\dot{\alpha}(a,R,t) - \dot{\xi}(a,R,t)\dot{a}\right)\partial_{\dot{a}} + \left(\dot{\beta}(a,R,t) - \dot{\xi}(a,R,t)\dot{R}\right)\partial_{\dot{R}}. \end{aligned} \tag{7.10}$$

By recasting the time derivatives of the infinitesimal generators as

$$\begin{cases} \dot{\alpha}(a,R,t) = \partial_t\alpha(a,R,t) + \dot{a}\partial_a\alpha(a,R,t) + \dot{R}\partial_R\alpha(a,R,t) \\ \dot{\beta}(a,R,t) = \partial_t\beta(a,R,t) + \dot{a}\partial_a\beta(a,R,t) + \dot{R}\partial_R\beta(a,R,t) \\ \dot{\xi}(a,R,t) = \partial_t\xi(a,R,t) + \dot{a}\partial_a\xi(a,R,t) + \dot{R}\partial_R\xi(a,R,t), \end{cases} \tag{7.11}$$

a system of nine differential equations arises by application of the Noether identity (6.42) to the point-like Lagrangian (7.5) [243, 244]. It reads as follows:

$$\begin{cases} 3\alpha f(R) - 3\alpha Rf_R(R) - \beta a f_{RR}(R) - \dfrac{\partial_t g}{a^2} + \partial_t\xi\, a[f(R) - Rf_R(R)] = 0 \\ \partial_R\alpha\, f_{RR}(R) = 0 \\ \partial_R\xi\, f_{RR}(R) = 0 \\ 6\partial_t\alpha\, f_{RR}(R) + a[f(R) - Rf_R(R)]\partial_R\xi - \dfrac{\partial_R g}{a^2} = 0 \\ 2\alpha f_{RR}(R) + \beta a f_{RRR} - \partial_t\xi\, a f_{RR}(R) + 2\partial_R\alpha f_R(R) + \partial_R\beta\, a f_{RR}(R) \\ +\partial_a\alpha\, a f_{RR}(R) = 0 \\ f_R\partial_a\xi = 0 \\ 12\partial_t\alpha\, f_R(R) + 6a\partial_t\beta\, f_{RR}(R) + a^2[f(R) - Rf_R(R)]\partial_a\xi - \dfrac{\partial_a g}{a} = 0 \\ \alpha f_R(R) + \beta\, a f_{RR}(R) - a\partial_t\xi f_R(R) + 2a\partial_a\alpha\, f_R(R) + a^2\partial_a\beta\, f_{RR}(R) = 0 \\ \partial_R\xi\, f_R(R) + a\partial_a\xi\, f_{RR}(R) = 0, \end{cases} \tag{7.12}$$

where α and β are the components of the function η^i, considered in Eq. (6.2), namely $\eta^i \equiv (\alpha, \beta)$ with the index i running from 1 to 2. After applying the condition (6.42), the system is obtained by separately setting to zero all the terms sharing the same order in time derivatives. A first solution can be easily found by setting $f_{RR}(R) = 0$ and it reads as follows:

$$f(R) = f_0 R + \Lambda, \quad \alpha = \pm\sqrt{\alpha_0 a + \alpha_1}, \quad \xi = \xi_0 t + \xi_1, \quad \beta = \beta(a, R, t), \tag{7.13}$$

with α_0, α_1, ξ_0, ξ_1, and f_0 being integration constants. The conserved quantity related to such a symmetry vanishes identically, independently of the dependence of the minisuperspace variables on the cosmic time. Therefore it is a trivial solution without any physical meaning. On the other hand, when $f_{RR}(R) \neq 0$, the system can be simplified to three differential equations as follows:

$$\begin{cases} 3\alpha[f(R) - Rf_R(R)] - \beta\, aRf_{RR}(R) + a[f(R) - Rf_R(R)]\partial_t \xi = 0 \\ 2\alpha f_{RR}(R) + \beta\, af_{RRR}(R) - \partial_t \xi\, af_{RR}(R) \\ +\partial_R \beta\, af_{RR}(R) + a\partial_a \alpha\, f_{RR}(R) = 0 \\ \alpha f_R(R) + \beta\, af_{RR}(R) - \partial_t \xi\, af_R(R) + 2a\partial_a \alpha f_R(R) + a^2 \partial_a \beta f_{RR}(R) = 0 \\ \xi \equiv \xi(t), \quad \alpha \equiv \alpha(a), \quad \beta \equiv \beta(a, R). \end{cases} \tag{7.14}$$

Being Lagrangian (7.5) in a canonical form, the function ξ can only depend on the cosmic time, as already pointed out in Section 6.3. By the impositions $f_{RR} \neq 0$ and $\xi(t) \neq 0$, three other solutions can be found from the preceding system, namely

$$\begin{cases} I : \alpha = \dfrac{7\xi_0}{36} a, \quad \beta = -\dfrac{7\xi_0}{6} R, \quad \xi = \xi_0 t + \xi_1, \quad f(R) = f_0 R^{\frac{19}{14}} + \Lambda \\ II : \xi = \xi_0, \quad \alpha = \beta = 0, \quad f(R) = f_0 R^{\frac{19}{14}} + \Lambda \\ III : \xi = \xi_0 t + \xi_1, \ \alpha = \alpha_0 a, \ \beta = -2\xi_0 R, \ f(R) = f_0 R^k, \ k = 1 + \dfrac{3\alpha_0}{2\xi_0}, \end{cases} \tag{7.15}$$

so that the corresponding generators and conserved quantities are

$$\begin{cases} \mathcal{X}_1 = \xi_0 \partial_t \\ j_1 = a^2[f(R) - Rf_R(R)] + 6[a\dot{a}^2 f_R(R) + a^2 \dot{a}\dot{R} f_{RR}(R)] + \Lambda a^3 \\ \\ \mathcal{X}_2 = (\xi_0 t + \xi_1)\partial_t + \dfrac{7\xi_0}{36} a\partial_a - \dfrac{7\xi_0}{6} R\partial_R \\ j_2 = a^3 t[f(R) - Rf_R(R)] + 6t[a\dot{a}^2 f_R(R) + a^2 \dot{a}\dot{R} f_{RR}(R)] + \Lambda t a^3 \\ +7a^2 \left[-\dfrac{1}{3}\dot{a} f_R(R) - \dfrac{1}{12} a\dot{R} f_{RR}(R) + \dot{a} R f_{RR}(R)\right] \\ \\ \mathcal{X}_3 = (\xi_0 t + \xi_1)\partial_t + \alpha_0 a\partial_a - 2\xi_0 R\partial_R \\ j_3 = 2\xi_0 aR\left(6a\dot{a} f_{RRR}\dot{R} + 6\dot{a}^2 f_{RR} - a^2 R f_{RR}\right) \\ -\alpha_0 a\left[6a\dot{a}\dot{R} f_{RR} + 6\dot{a}^2 f_R + a\left(6\dot{a}\dot{R} f_{RR} + 2a\left(f - Rf_R\right)\right) + a^2\left(f - Rf_R\right)\right] \\ +\xi_0 t\left[6a^2 \dot{a}\dot{R} f_{RR} + a\dot{a}\left(12\dot{a} f_R + 6a\dot{R} f_{RR}\right)\right. \\ \left. - a\left(6a\dot{a}\dot{R} f_{RR} + 6\dot{a}^2 f_R + a^2\left(f - Rf_R\right)\right)\right], \end{cases} \tag{7.16}$$

Table 7.1 Noether solutions in $f(R)$ cosmology

$\xi(t)$	$\alpha(a)$	$\beta(a,R)$	$f(R)$
$\xi_0 t+\xi_1$	$\pm\sqrt{\alpha_0 a+\alpha_1}$	$\beta(a,R,t)$	$f_0 R+\Lambda$
$\xi_0 t+\xi_1$	$\frac{7\xi_0}{36}a$	$-\frac{7\xi_0}{6}R$	$f_0 R^{\frac{19}{14}}+\Lambda$
$\xi_0 t+\xi_1$	$\alpha_0 a$	$-2\xi_0 R$	$f_0 R^{1+\frac{3\alpha_0}{2\xi_0}}$
ξ_0	0	0	$f_0 R^{\frac{19}{14}}+\Lambda$
0	$\frac{\alpha_0}{a^2}$	$-3\alpha_0\frac{R}{a^3}$	$f_0 R$
0	$\alpha_0\sqrt{a}$	$-\frac{\alpha_0}{2}\frac{R}{\sqrt{a^3}}$	$f_0 R$
0	$\frac{\alpha_0}{a}$	$-2\alpha_0\frac{R}{a^2}$	$f_0 R^{\frac{3}{2}}$
0	0	$\frac{\beta_0}{a}$	$f_0 R+f_1 R^2$

with β_0 a real constant. Notice that the third solution depends on the free parameter k that will be constrained by observations in the third part of the book. Imposing $\xi(t)=0$, a set of four different solutions arise from the Noether system [245, 246]:

$$\begin{cases} \mathcal{X}_I=\frac{\alpha_0}{a^2}\partial_a-3\alpha_0\frac{R}{a^3}\partial_R \\ j=24\alpha_0 f_0\frac{\dot{a}}{a}, \quad f(R)=f_0 R \\ \mathcal{X}_{II}=\alpha_0\sqrt{a}\partial_a-\frac{\alpha_0}{2}\frac{R}{\sqrt{a^3}}\partial_R \\ j=24\alpha_0 f_0 a^{\frac{3}{2}}\dot{a}, \quad f(R)=f_0 R \\ \mathcal{X}_{III}=\frac{\alpha_0}{a}\partial_a-2\alpha_0\frac{R}{a^2}\partial_R \\ j=9\alpha_0 f_0(2R\dot{a}+a\dot{R})R^{-\frac{1}{2}}, \quad f(R)=f_0 R^{\frac{3}{2}} \\ \mathcal{X}_{IV}=\frac{\beta_0}{a}\partial_R \\ j=24\beta_0 f_1 a\dot{a}, \quad f(R)=f_0 R+f_1 R^2. \end{cases} \tag{7.17}$$

Since the preceding solutions describe internal symmetries, they can be equivalently obtained from the imposition of the vanishing Lie derivative of Lagrangian (7.5). In Table **7.1** the complete set of solutions coming from the system (7.12) is outlined.

It is worth pointing out that selected functions in Table **7.1** are solutions of Noether's system even when the matter Lagrangian $\mathcal{L}_m = \rho_0 a^s$ (with s constant) is included in Eq. (7.5). Specifically, though in such a case the infinitesimal generators are different, the shape of the function $f(R)$ is independent of the matter. This is because introducing the matter Lagrangian yields a further equation into the system (7.12), not providing any additional constraint to the function $f(R)$. In other words, the matter Lagrangian does not impose further restrictions on the starting action, though it can change the symmetry generators and, consequently, the conserved quantities.

In what follows, the functions selected by Noether's approach will be replaced in the equations of motion (7.9) in order to investigate the cosmological solutions of $f(R)$ gravity. First, the function $f(R) = f_0 R + \Lambda$ yields the well-known Einstein–de Sitter vacuum expansion. Let us then consider the function $f(R) = f_0 R^k$, whose Lagrangian generalizes that of the second model of system (7.15), when Λ vanishes. It reads

$$\mathcal{L} = f_0 a R^{k-2} \left(6k(k-1)a\dot{a}\dot{R} + 6kR\dot{a}^2 - (k-1)a^2R^2\right). \tag{7.18}$$

The EL equations yield the solution

$$a(t) = a_0 t^{\frac{(2k-1)(k-1)}{2-k}}, \quad R(t) = \frac{6k(5-4k)(2k-1)(k-1)}{(k-2)^2 t^2}, \tag{7.19}$$

which holds only for $k \neq 2$. When $k = 2$, the scalar curvature is constant and exponential solutions occur:

$$a(t) = a_0 e^{\ell t}, \quad R(t) = -12\ell^2, \tag{7.20}$$

with ℓ a real constant. Notice that Eq. (7.19), with $k = 19/14$, contains the scale factor solving the field equations related to the second model of the system (7.15), as expected. Moreover, as pointed out in Section 16.2.3, where the slow-roll inflation is discussed in the framework of $f(R)$ gravity, the scale factor can describe either an accelerating or a decelerating universe, depending on the value of k. Radiation, stiff matter, and dust matter fluids also can be mimicked by geometry based on the values of k:

- $k = \frac{2}{3} \pm \frac{\sqrt{10}}{6} \rightarrow$ Stiff Matter: $a(t) = a_0 t^{\frac{1}{3}}$
- $k = \frac{5}{4} \rightarrow$ Radiation: $a(t) = a_0 t^{\frac{1}{2}}$
- $k = \frac{7}{12} \pm \frac{\sqrt{73}}{12} \rightarrow$ Dust Matter: $a(t) = a_0 t^{\frac{2}{3}}$

Another class of solutions is provided by the function $f(R) = f_0 R^{\frac{3}{2}}$, whose Lagrangian turns out to be

$$\mathcal{L} = f_0 a R^{-\frac{1}{2}} \left(-a^2R^2 + 18R\dot{a}^2 + 9a\dot{a}\dot{R}\right) \tag{7.21}$$

and provides the following equations of motion:

$$\begin{cases} 36R^2\dot{a}^2 + 36aR(\dot{a}\dot{R} + 2R\ddot{a}) + 3a^2(2R^3 - 3\dot{R}^2 + 6R\ddot{R}) = 0 \\ R = -6\left(\dfrac{\ddot{a}}{a} + \dfrac{\dot{a}^2}{a^2}\right) \\ a^2R^2 + 18R\dot{a}^2 + 9a\dot{a}\dot{R} = 0 \end{cases} \tag{7.22}$$

The preceding system yields the solution

$$a(t) = a_0\left[c_4t^4 + c_3t^3 + c_2t^2 + c_1t + c_0\right]^{1/2}, \tag{7.23}$$

with c_i integration constants. Notice that by setting $c_0 = c_1 = c_2 = c_3 = 0$, $k{=}3/2$, Eqs. (7.23) and (7.19) are equivalent. However, with respect to Eq. (7.19), four other terms arise that cannot be analytically found when a general k is considered.

The constants c_i are free parameters that can be fixed through experimental observations at astrophysical and cosmological levels. Though not discussed in this book, a detailed study of solution (7.23) might provide the values of c_i best-fitting current experiments at low and high energies. For example, $c_4 \neq 0$ gives a power law inflation while, if the regime is dominated by the linear term in c_1, we get a radiation-dominated stage. See [247] for details. As discussed in [116], the case for $k = 3/2$ is the so-called Liouville field theory and it is one of the few cases where a fourth-order Lagrangian can be expressed, in the Einstein frame, in terms of elementary functions under a conformal transformation.

Finally, the last function containing Noether symmetries is $f(R) = f_0R + f_1R^2$ [119]. It is the Starobinsky model [122] that provides, as discussed, one of the most important inflationary scenarios. The corresponding Lagrangian reads

$$\mathcal{L} = 2a[-f_1a^2R^2 + 6(f_0 + 2f_1R)\dot{a}^2 + 12f_1a\dot{a}\dot{R}], \tag{7.24}$$

and the equations of motion

$$\begin{cases} 2(f_0 + 2f_1R)\dot{a}^2 + 4a[2f_1\dot{a}\dot{R} + (f_0 + 2f_1R)\ddot{a}] + f_1a^2(R^2 + 4\ddot{R}) = 0 \\ R = -6\left(\dfrac{\ddot{a}}{a} + \dfrac{\dot{a}^2}{a^2}\right) \\ f_1a^2R^2 + 6(f_0 + 2f_1R)\dot{a}^2 + 12f_1a\dot{a}\dot{R} = 0 \end{cases} \tag{7.25}$$

can be analytically solved only when either f_0 or f_1 vanish. In the latter case, the theory trivially reduces to GR, while in the former case the following solutions occur:

$$a(t) = a_0\sqrt{t},\ f(R) = f_1R^2, \quad \text{and} \quad a(t) = a_0e^{\ell t},\ f(R) = f_1R^2, \tag{7.26}$$

which mean that a radiation-dominated behavior and a de Sitter solution are immediately recovered. It is worth noticing that $a(t) \sim \sqrt{t}$ gives $R(t) = 0$, and then the field equations are identically satisfied, independently of the form of the function $f(R)$. However, among functions $f(R) = f_0R^k$, the case $k = 2$ leads to an exponential scale factor. This is related to the scale-invariance of $f(R) = f_0 R^2$ gravity, as shown in [248]. These features can be naively checked by expanding the action around a Gaussian fixed point and considering the coupling constant mass dimensions. Notice that conformal invariance naturally arises from the solutions of Noether's system, suggesting that the Noether Symmetry Approach is also related to the conformal symmetry [249].

Applications of the approach to $f(R)$ gravity in spherical symmetry are discussed, for example, in [250] and [251]. In the latter reference, the author uses a function containing symmetries to fit the mass-radius diagram of neutron stars (see Chapter 14 for details). Another application of Noether's symmetries for spherically symmetric configurations can be found in Ref. [252], where $f(R)$ gravity is used to constrain the Fundamental Plane of galaxy.

Specific applications of the Noether Symmetry Approach to spherically symmetric problems will be discussed in Chapter 13, while the comparison with observations will be considered in the third part of the book, where selected models will be investigated in the contexts of stellar and galactic astrophysics and cosmology.

7.2 Teleparallel Extension: The Case of $f(T)$ Gravity

In order to find a suitable point-like Lagrangian for $f(T)$ extension of TEGR action, we use the Lagrange multipliers method with the constraint (4.113). In this way, all dynamical variables depend only on the cosmic time, and the three-dimensional surface term can be easily integrated. The action therefore reads

$$S = \int \left[a^3 f(T) - \lambda \left(T + 6\frac{\dot{a}^2}{a^2} \right) \right] dt. \tag{7.27}$$

By varying the action with respect to the torsion, we get the form of the Lagrange multiplier λ

$$\frac{\delta S}{\delta T} = 0 \quad \rightarrow \quad \lambda = a^3 f_T(T). \tag{7.28}$$

Notice that, since $f(T)$ field equations are of second order, no higher-order derivative occurs in the action so that no term needs to be integrated out. Therefore, the point-like Lagrangian turns out to be

$$\mathcal{L} = a^3 [f(T) - T f_T(T)] - 6 a \dot{a}^2 f_T(T), \tag{7.29}$$

where the subscript T denotes the differentiation with respect to T. The EL equations and the energy condition for the two variables $\{a, T\}$ are, respectively,

$$\begin{cases} \dfrac{\partial \mathcal{L}}{\partial T} = 0 \quad \rightarrow \quad T = -6 \left(\dfrac{\dot{a}}{a} \right)^2 \\ \dfrac{d}{dt}\dfrac{\partial \mathcal{L}}{\partial \dot{a}} = \dfrac{\partial \mathcal{L}}{\partial a} \quad \rightarrow \quad \ddot{a} + \dfrac{\dot{a}^2}{2a} + \dot{a}\dot{T}\dfrac{f_{TT}(T)}{f_T(T)} - \dfrac{1}{4} a \dfrac{T f_T(T) - f(T)}{f_T(T)} = 0 \\ \dot{a}\dfrac{\partial \mathcal{L}}{\partial \dot{a}} = \mathcal{L} \quad \rightarrow \quad a^3 [f(T) - T f_T(T)] + 6 a \dot{a}^2 f_T(T) = 0 \end{cases} \tag{7.30}$$

and can be solved only after selecting the form of the function. This system of differential equations is dynamically equivalent to that provided by Eqs. (4.111) and (4.112), as expected by construction. Note that the Lagrangian is independent of $\dot{T}$ so that the related EL equation provides the constraint imposed on the torsion scalar. In order to apply the identity (6.42), we consider the following first prolongation of Noether's vector:

$$X^{[1]} = \xi\partial_t + \alpha\partial_a + \beta\partial_T + \left(\dot{\alpha} - \dot{\xi}\dot{a}\right)\partial_{\dot{a}}, \tag{7.31}$$

with $\alpha = \alpha(a, T)$ and $\beta = \beta(a, T)$ being the components of the infinitesimal generator η^i. A Noether system of differential equations can be constructed by imposing terms sharing the same time-derivative to vanish. Thus, the system reads as follows [253]:

$$\begin{cases} \partial_t\xi = \partial_T\xi = \partial_a\xi = 0 \\ a^3[Tf_T(T) - f(T)] = \partial_T g \\ (3\alpha a^2 + \partial_t\xi a^3)\,[Tf_T(T) - f(T)] + \beta a^3 Tf_{TT}(T) = \partial_t g \\ 12\partial_t\alpha a f_T(T) = \partial_a g \\ \alpha f_T(T) + \beta a T f_{TT}(T) + 2a\partial_a f_T(T) - \partial_t\xi a f_T(T) = 0. \end{cases} \tag{7.32}$$

Three possible solutions of the preceding system occur, but all of them admit the same form of the function, namely $f(T) = f_0 T^k$ with $k \in \mathbb{R}$. The generators of the symmetries together with the conserved quantities are as follows [254, 255]:

$$\begin{cases} \mathcal{X}_1 = \left[\dfrac{\xi_0}{2k-1}\right] t\partial_t + \left[\dfrac{\xi_0}{3} a + \alpha_0 a^{1-\frac{3}{2k}}\right]\partial_a \\ + \left[\dfrac{1}{k}\left(\beta_1 - 3\alpha_0 a^{-\frac{3}{2k}}\right) + \dfrac{\xi_0}{2k-1} + \beta_2\right] T\partial_T, \qquad k \neq \dfrac{3}{2}, \dfrac{1}{2} \\ j_1 = -12 f_0 k\left(\xi_0 a^2 + \alpha_0 a^{2-\frac{3}{2k}}\right) T^{k-1}\dot{a} \\ \\ \mathcal{X}_2 = \xi_0 t\partial_t + [\alpha_0 a + \alpha_1]\,\partial_a \\ + \left[\dfrac{138}{5}\alpha_0 - 14\xi_0 - \dfrac{\beta_0}{a} + \dfrac{2}{5}\left(\dfrac{84}{5}\alpha_0 - 2\xi_0\right)\right] T\partial_T, \qquad k = \dfrac{3}{2} \\ j_2 = -18 f_0\left[\alpha_0 a^2 + \beta_0 a\right] T^{\frac{1}{2}}\dot{a} \\ \\ \mathcal{X}_3 = \xi_0 t\partial_t + \left(-2\xi_0 + \alpha_0 a^{\frac{1}{4}}\right)\partial_a + \left(4\xi_0 + \beta_0 + \dfrac{3\alpha_0}{2} a^{-\frac{3}{4}}\right) T\partial_T, \qquad k = \dfrac{1}{2} \\ j_3 = -6 f_0\left(-2\xi_0 a + \alpha_0 a^{\frac{3}{4}}\right) T^{-\frac{1}{2}}\dot{a}, \end{cases} \tag{7.33}$$

with C, c_{1-7} constant parameters. Substituting the function into Lagrangian (7.29), we get

$$\mathcal{L} = f_0\left[(1-k)a^3 T - 6ka\dot{a}^2\right] T^{k-1}. \tag{7.34}$$

The EL equations can be analytically solved only for $k = \frac{1}{2}$ so that the dynamical scale factor and the torsion scalar turn out to be [256]

$$a(t) = a_0 t^{\frac{1}{3}}, \qquad T(t) = -\frac{2}{3}\frac{1}{t^2}$$

$$a(t) = e^{\ell t}, \qquad T(t) = -6k^2 \quad \forall \ell. \tag{7.35}$$

The function $f(T)$ in turn is constrained to

$$f(T) = f_0\sqrt{T}. \tag{7.36}$$

This means that the function $f(T) = f_0\sqrt{T}$ is the only one that simultaneously solves the equations of motion and contains extra Noether symmetries. The time power-law scale factor expression describes a stiff matter-dominated era, while the other epochs can be recovered by coupling Eq. (7.34) with the matter Lagrangian. Moreover, in the exponential solution, the free parameter ℓ can be constrained by observations [254].

As formerly mentioned, though TEGR and GR are exactly equivalent at the level of dynamics, $f(R)$ and $f(T)$ gravity are different for several reasons. Therefore, Noether or field equation solutions of $f(R)$ gravity are not expected to be recovered in this case [97].

7.2.1 TEGR Extension with Boundary Term: $f(T, B)$ Gravity

In the previous section, we considered the $f(T)$ extension of TEGR. $f(R)$ gravity cannot be directly recovered from $f(T)$ gravity since the boundary term (4.105), which relates the curvature scalar to the torsion scalar, cannot be neglected when TEGR action is extended. With the aim of comparing $f(R)$ gravity with the teleparallel equivalent, let us introduce a function of torsion scalar and boundary term, that is, $f(T, B)$. Therefore, $f(R)$ gravity can be recovered by the choice

$$f(T, B) = f(-T + B) \equiv f(R). \tag{7.37}$$

The action

$$S = \int h f(T, B)\, d^4x \tag{7.38}$$

leads to the following field equations [173, 174]:

$$\begin{aligned} &2hh_a^{\lambda}\Box f_B - 2hh_a^{\sigma}\nabla^{\lambda}\nabla_{\sigma} f_B + hBh_a^{\lambda} f_B + 4h\left[(\partial_\mu f_B) + (\partial_\mu f_T)\right] S_a^{\ \mu\lambda} \\ &+ 4\partial_\mu\left(hS_a^{\ \mu\lambda}\right) f_T - 4hf_T T^{\sigma}_{\ \mu a} S_\sigma^{\ \lambda\mu} - hfh_a^{\lambda} = 0. \end{aligned} \tag{7.39}$$

To apply Noether's approach to $f(T, B)$ gravity, we must consider the Lagrangian formalism and the related EL equations. To this purpose, we develop the Lagrange multipliers method by coupling the Lagrange multipliers with the cosmological expression of T and the boundary term B. The latter, in a FLRW spacetime, takes the following form:

$$B = -6\frac{\dot{a}^2}{a^2} - 6\left(\frac{\ddot{a}}{a} + \frac{\dot{a}^2}{a^2}\right) = -12\frac{\dot{a}^2}{a^2} - 6\frac{\ddot{a}}{a}. \tag{7.40}$$

Considering also Eq. (4.113), the action can be written as

$$S = 2\pi^2 \int \left\{ a^3 f(T, B) - \lambda_1 \left[T + 6\left(\frac{\dot{a}}{a}\right)^2 \right] \right.$$
$$\left. -\lambda_2 \left[B + 6\left(\frac{\ddot{a}}{a} + 2\left(\frac{\dot{a}}{a}\right)^2\right) \right] \right\} dt. \tag{7.41}$$

By varying this action with respect to T and B, we find the Lagrange multipliers λ_1 and λ_2:

$$\begin{cases} \dfrac{\delta S}{\delta T} = 0 \ \rightarrow \ \lambda_1 = a^3 f_T(T, B) \\ \\ \dfrac{\delta S}{\delta B} = 0 \ \rightarrow \ \lambda_2 = a^3 f_B(T, B). \end{cases} \tag{7.42}$$

Integrating out the terms containing second derivatives and neglecting the boundaries, the cosmological point-like Lagrangian turns out to be

$$\mathcal{L} = a^3 \left[f - T f_T - B f_B \right] - 6 a \dot{a}^2 f_T + 6 a^2 \dot{a} \left(f_{BT} \dot{T} + f_{BB} \dot{B} \right), \tag{7.43}$$

which provides the following equations of motion:

$$\begin{aligned} &3a^2 \left(2\ddot{B} f_{BB} + 4\dot{B}\dot{T} f_{TBB} + 2\dot{B}^2 f_{BBB} \right. \\ &\left. + 2\ddot{T} f_{TB} + 2\dot{T}^2 f_{TTB} + B f_B + T f_T - f \right) \\ &- 6\dot{a}^2 f_t - 12 \left[\ddot{a} f_T + \dot{a} \left(\dot{B} f_{TB} + \dot{T} f_{TT} \right) \right] = 0,. \end{aligned} \tag{7.44}$$

$$\left(6\frac{\ddot{a}}{a} + 12\frac{\dot{a}^2}{a^2} + B \right) f_{TB} + \left(6\frac{\dot{a}^2}{a^2} + T \right) f_{TT} = 0, \tag{7.45}$$

$$\left(6\frac{\ddot{a}}{a} + 12\frac{\dot{a}^2}{a^2} + B \right) f_{BB} + \left(6\frac{\dot{a}^2}{a^2} + T \right) f_{TB} = 0, \tag{7.46}$$

$$6 a \dot{a} \left(\dot{B} f_{BB} + \dot{T} f_{TB} \right) - 6\dot{a}^2 f_T + a^2 \left(B f_B + T f_T - f \right) = 0. \tag{7.47}$$

The first three are the equations with respect to a, T, and B, respectively. The last one, instead, is the energy condition. Also, the equations of motion with respect to the torsion and the boundary term provide the constraints (4.113) and (7.40) by construction. Notice that if we choose $f(T, B) = f(T)$, we recover the teleparallel $f(T)$ Lagrangian (7.29). In addition, when $f(T, B) = f(-T+B) = f(R)$, the point-like Lagrangian of $f(R)$ gravity (7.5) is restored.

As an example, let us investigate only the internal symmetries and select modified gravity models by setting the Lie derivative of the Lagrangian to zero. The general approach is applied in Section 8.2, where a function of T, B, and the Gauss–Bonnet teleparallel invariant will be analyzed. The minisuperspace considered consists of the three variables a, T, and B and yields the Noether's vector

$$X = \alpha \partial_a + \beta \partial_T + \gamma \partial_B + \dot{\alpha} \partial_{\dot{a}} + \dot{\beta} \partial_{\dot{T}} + \dot{\gamma} \partial_{\dot{B}}. \tag{7.48}$$

The application of Eq. (7.48) to Lagrangian (7.43) provides the following system of differential equations:

$$\begin{cases} f_T\,(2a\partial_a\alpha+\alpha)+f_{TB}\,(a\gamma-a^2\partial_a\beta)+af_{TT}\beta-a^2f_{BB}\partial_a\gamma=0\\ f_{TB}\,(a\partial_a\alpha+a\partial_T\beta+2\alpha)+af_{BB}\partial_T\gamma+a\,(f_{TTB}\beta+f_{TBB}\gamma)=0\\ f_{BB}\,(a\partial_a\alpha+a\partial_B\gamma+2\alpha)+af_{TB}\partial_B\beta+a\,(\beta f_{TBB}+\gamma f_{BBB})=0\\ 3\,(f-Bf_B-Tf_T)\,\alpha-a\,(Bf_{TB}+Tf_{TT})\,\beta-a\,(Bf_{BB}+Tf_{TB})\,\gamma=0\\ \alpha\equiv\alpha(a). \end{cases} \quad . \tag{7.49}$$

The preceding system cannot provide a unique set of solutions since the variables in the minisuperspace are more than those required to uniquely solve the system. Therefore, we impose specific models for $f(T,B)$ and search for relevant symmetries. From a physical viewpoint, this prescription allows one to straightforwardly study reliable theories. When a general product of the form $F_1(T)F_2(B)$ is substituted into (7.49), the infinitesimal generators that solve the system are

$$\alpha=\frac{\alpha_0}{a^2},\quad \beta=-\frac{6\alpha_0 T}{a^3},\quad \gamma=-\frac{3\alpha_0 B}{a^3},\quad f(T,B)=f_0B^kT^{\frac{1}{2}(1-k)}, \tag{7.50}$$

with conserved quantity

$$j=3\alpha_0f_0(k-1)B^{k-2}T^{-\frac{1}{2}(k+1)}\left(2B^2\frac{\dot a}{a}+2kT\dot B-kB\dot T\right). \tag{7.51}$$

Notice that only power-law models of the form $f(T,B)=f_0B^kT^{\frac{1}{2}(1-k)}$ are allowed by Noether symmetries. The corresponding Lagrangian can be found by substituting this function into Eq. (7.43):

$$\mathcal{L}=\frac{1}{2}f_0(k-1)aB^{k-2}T^{-\frac{1}{2}(k+1)}\left[6B^2\dot a^2-6ka\dot a(B\dot T-2\dot BT)-B^2Ta^2\right]. \tag{7.52}$$

It is easy to see that the trivial case $k=1$, which produces $f=f_0B$, gives the expected result where the field equations are identically zero. The EL equation for the scale factor $a(t)$ and the energy condition gives, for $k\neq1$ and $f_0\neq0$,

$$\begin{aligned}\frac{d}{dt}\frac{\partial\mathcal{L}}{\partial\dot a}=\frac{\partial\mathcal{L}}{\partial a}\ \rightarrow\ & 2B^3T\dot a^2-2aB^2\left[(k+1)B\dot a\dot T-2T(B\ddot a+k\dot a\dot B)\right]\\ &+a^2\left[4k(k-2)T^2\dot B^2+4kBT(T\dot B-(k-1)\dot B\dot T)\right.\\ &\left.+kB^2\left((k+1)\dot T^2-2T\ddot T\right)+B^3T^2\right]=0\end{aligned}$$

$$\dot q^i\frac{\partial\mathcal{L}}{\partial\dot q^i}-\mathcal{L}=0\ \rightarrow\ 6ka\dot a\,(2T\dot B-B\dot T)+6B^2\dot a^2+a^2B^2T=0, \tag{7.53}$$

while the other two equations provide the cosmological expressions of T and B, regardless of the form of the function. The preceding set of differential equations can be analytically solved, providing

$$a(t)=a_0t^{\frac{1+k}{3}},\qquad T(t)=-\frac{2}{3}\frac{(k-1)^2}{t^2},\qquad B(t)=-2\frac{k(k+1)}{t^2}. \tag{7.54}$$

When we set $k = 0$, the function turns out to be independent of B, so that Eq. (7.54) reduces to the previously discussed Eq. (7.35), namely

$$f(T) = f_0\sqrt{T}, \quad a(t) = a_0 t^{\frac{1}{3}}. \tag{7.55}$$

Stiff matter, radiation, and dust matter can be recovered from the vacuum solution (7.54) by varying the free parameter k. Specifically, stiff matter-dominated era occurs for $k = 0$, radiation for $k = 1/2$, and dust matter for $k = 1$. Notice that stiff matter only occurs in the extended teleparallel framework, namely when the function $f(T, B)$ reduces to $f(T, B) = f_0\sqrt{T}$. Similarly, a dust matter behavior is recovered when the function is linearly dependent on the boundary term.

Another solution of (7.49) can be found by assuming a function of the form $f(T, B) = F_1(T) + F_2(B)$. When substituted into the system (7.49), it yields:

$$\alpha = \frac{\alpha_0}{a^2}, \quad \beta \equiv \beta(T), \quad \frac{\alpha_0}{3a^3 B}, \quad f(T, B) = T + f_0 B + \frac{1}{3} B \ln(B) \tag{7.56}$$

and the corresponding Lagrangian reads

$$\mathcal{L} = -\frac{1}{3}a^3 B - a\dot{a}^2 + \frac{2a^2\dot{a}\dot{B}}{B}. \tag{7.57}$$

The EL equations with respect to the scale factor and the energy condition are

$$\frac{d}{dt}\frac{\partial \mathcal{L}}{\partial \dot{a}} = \frac{\partial \mathcal{L}}{\partial a} \quad \to \quad -6\dot{a}^2 - 12a\ddot{a} + a^2\frac{B^3 - 2\dot{B}^2 + 2B\ddot{B}}{B^2} = 0$$

$$\dot{q}^i\frac{\partial \mathcal{L}}{\partial \dot{q}^i} - \mathcal{L} = 0 \quad \to \quad \frac{1}{3}a^3 B - 6a\dot{a}^2 + \frac{2a^2\dot{a}\dot{B}}{B} = 0, \tag{7.58}$$

whose solution is

$$a(t) = \left[\frac{3c_0 e^{c_1 t}}{c_1^2} + 3c_2(t + c_3)\right]^{\frac{1}{3}}. \tag{7.59}$$

This solution reduces to a de Sitter exponential expansion by setting $c_2 = 0$. Also, time power-law solutions provided by $f(T)$ gravity can be recovered for $c_0 = c_3 = 0$. It is worth pointing out that the physically relevant cases analyzed in this section do not account for the whole set of solutions coming from the system (7.49). As an example, other solutions occur when T and B are combined such that $f(T, B) = f(-T + B, T) = f(R, T)$, as shown in the next subsection.

7.2.2 $f(R, T)$ Gravity

A subcase that deserves a separate discussion is the $f(R, T)$ theory of gravity, discussed in detail in Ref. [257]. Here we use the vanishing Lie derivative condition in order to find models with internal symmetries and the related dynamics. This section is mainly aimed at comparing $f(R)$ and $f(T, B)$ models and, specifically, at showing that the former can be compared with the latter even at the level of symmetries.

As discussed in Chapter 4, the curvature scalar can be written in terms of the metric tensor only if the antisymmetric part of the Levi–Civita connection is assumed to vanish. In this way, connections are metric-compatible and, in a spatially flat FLRW space-time, the cosmological expression (7.2) is recovered. On the other hand, the Weitzenböck connection can be recast in terms of non-trivial tetrad fields only when $R = 0$; otherwise the relation (4.113) does not hold. From this point of view, if one considers a function of R and T, expressions (7.2) and (4.113) cannot be applied simultaneously. However, here we pursue a different strategy, making use of the boundary term to properly recover the scalar curvature in the teleparallel context. Therefore, the system (7.49) can be reduced as the function $f(T, B)$ takes the form $f(T, B) = f(-T+B, T) = f(R, T)$. In this way, the dynamical degree of freedom $B(t)$ can be replaced by the scalar curvature so that the related minisuperspace turns out to be $\mathcal{S} \equiv \{a, R, T\}$ and the Noether vector can be written as

$$
\begin{aligned}
X = {} & \alpha(a, R, T)\partial_a + \beta(a, R, T)\partial_T + \gamma(a, R, T)\partial_R \\
& + \dot{\alpha}(a, R, T)\partial_{\dot{a}}\dot{\beta}(a, R, T)\partial_{\dot{T}} + \dot{\gamma}(a, R, T)\partial_{\dot{R}}.
\end{aligned} \tag{7.60}
$$

The point-like Lagrangian can be easily obtained from Eq. (7.43) with the substitution $B = R + T$, or, equivalently, by coupling the Lagrange multipliers to the new variables. It reads as

$$
\mathcal{L} = a^3 \left(f - Rf_R - Tf_T\right) + 6a\dot{a}^2 \left(f_R - f_T\right) + 6a^2\dot{a}\left(\dot{R}f_{RR} + \dot{T}f_{RT}\right). \tag{7.61}
$$

The first term in the bracket can be intended as an effective potential, which vanishes when TEGR or GR actions are considered. The EL equations with respect to the curvature R and the torsion T provide, respectively,

$$
\begin{cases}
T + 6\dfrac{\dot{a}^2}{a^2} = 0 \\
R + 6\dfrac{\ddot{a}}{a} + 6\dfrac{\dot{a}^2}{a^2} = 0,
\end{cases} \tag{7.62}
$$

which are automatically satisfied when the cosmological expressions of R and T are replaced. Dynamics is provided by the equation of motion with respect to a and the energy condition, which read, respectively,

$$
\begin{aligned}
& 6\left(f_R - f_T\right)\left(2\dot{a}^2 - a^2 + 2a\ddot{a}\right) - 3a^2\left(f - Tf_T - Rf_R\right) \\
& - 12a\dot{a}\left(\dot{T}f_{TT} - \dot{R}f_{RR}\right) - 12a\dot{a}\left(\dot{R} - \dot{T}\right)f_{RT} \\
& + 6a^2\left(\ddot{T}f_{RT} + \ddot{R}f_{RR} + \dot{T}^2 f_{RTT} + 2\dot{R}\dot{T}f_{RRT} + \dot{R}^2 f_{RRR}\right) = 0,
\end{aligned}
$$

$$
\begin{aligned}
& 6a\dot{a}^2\left(f_R - f_T\right) + a^3\left(f - Tf_T - Rf_R\right) \\
& - 6a^2\dot{a}\left(\dot{T}f_{RT} + \dot{R}f_{RR}\right) = 0.
\end{aligned} \tag{7.63}
$$

By setting $L_X\mathcal{L} = 0$ and neglecting trivial solutions such as $f_R = 0$ or $f_T = 0$, a system of nine differential equation occurs:

$$\begin{cases} 6\alpha\,(f_R - f_T) + 6\gamma a\,(f_{RR} - f_{RT}) + 6\beta a\,(f_{RT} - f_{TT}) \\ +12a\partial_a\alpha(f_R - f_T) + 6a^2\partial_a\gamma f_{RR} + 6a^2\partial_a\beta f_{RT} = 0 \\ 12\alpha a f_{RT} + 6\gamma a^2 f_{RRT} + 6\beta a^2 f_{RTT} + 6a^2\partial_a\alpha f_{RT} \\ +12a\partial_T\alpha(f_R - f_T) + 6a^2\partial_T\gamma f_{RR} + 6a^2\partial_T\beta f_{RT} = 0 \\ 12\alpha a f_{RR} + 6\gamma a^2 f_{RRR} + 6\beta a^2 f_{RRT} + 6a^2\partial_a\alpha f_{RR} \\ +12a\partial_R\alpha(f_R - f_T) + 6a^2\partial_R\gamma f_{RR} + 6a^2\partial_R\beta f_{RT} = 0 \\ 6a^2\partial_R\alpha(f_{RT} + f_{RR}) = 0 \\ 3\alpha a^2\,(f - Rf_R - Tf_T) - \gamma a^3\,(Tf_{RT} + Rf_{RR}) - \beta a^3\,(Tf_{TT} + Rf_{RT}) = 0 \\ \partial_T\alpha f_{RT} = \partial_R\alpha f_{RR} = 0. \end{cases} \tag{7.64}$$

The preceding system can be also obtained from Eq. (7.49) as a particular limit. However, as previously pointed out, it is overdetermined and, if solvable, enables one to assign α, β, γ, and $f(R, T)$, once a specific ansatz for the function is replaced. By considering a product between a function of R and a function of T, the system provides the constraint $f(R, T) = f_0 R^k T^{1-k}$. The infinitesimal generators are the same as the first solution of the previous section, under given limits. The preceding function yields the equations of motion

$$\begin{aligned} & k(k-1)(k-2)a^2T^2\dot{R}^2 + (k-1)R^3\left[-2ka\dot{a}\dot{T} + T(\dot{a}^2 + 2a\ddot{a})\right] \\ & + k(k-1)aRT\left[-2(k-1)a\dot{T}\dot{T} + T(2\dot{a}\dot{R} + a\ddot{R})\right] \\ & + kR^2\left[k(k-1)a^2\dot{T}^2 + T^2(\dot{a}^2 + 2a\ddot{a})\right] \\ & + (k-1)aT\left[2\dot{a}(\dot{R} - \dot{T}) - a\ddot{T}\right] = 0, \end{aligned}$$

$$(k-1)R^2\dot{a} + k(k-1)aT\dot{R} - kR\left[(k-1)a\dot{T} - T\dot{a}\right] = 0, \tag{7.65}$$

whose solutions are

$$\begin{cases} a(t) = a_0 t^{\frac{k-1}{4k-3}}, \quad T(t) - 6\left(\dfrac{k-1}{4k-3}\right)^2 \dfrac{1}{t^2}, \quad R(t) = \dfrac{6k(k-1)}{(4k-3)^2}\dfrac{1}{t^2}, \quad k \neq \dfrac{3}{4} \\ \\ a(t) = a_0 e^{nt}, \quad T(t) = -\dfrac{6n^2}{t^2}, \quad R(t) = -\dfrac{6n(3n-1)}{t^2}, \quad k = \dfrac{3}{4}, \end{cases} \tag{7.66}$$

with n being a real constant. As we can see, the second solution constrains the parameter k to be $k = 3/4$ so that the only function compatible with the de Sitter expansion is

$$f(R, T) = f_0\sqrt[4]{\frac{R^3}{T}}. \tag{7.67}$$

In Section 7.3, we shall apply the Noether Symmetry Approach to extended gravitational actions including different nonmetricity invariants. We do not consider the case of $f(Q)$ gravity since the cosmological expression of the nonmetricity scalar Q is exactly the same as the torsion scalar T. Therefore, the Noether Symmetry Approach yields exactly the same solutions, which, in turn, provide the same dynamics. This holds in a spatially flat cosmological universe, but cannot be applied to a spherically symmetric background, where Q and T are generally different. Cosmological implications of $f(Q)$ gravity are discussed, for example, in Ref. [258], where the authors investigate the dynamics provided by the function $f(Q) = f_0(Q + c_1 Q^2)$, with c_1 being a dimensional parameter.

7.3 Geometric Extensions of STEGR

In this section, sketching the steps outlined in Ref. [188], we want to show how conserved quantities coming from symmetries can be used to reduce the dynamics and find exact solutions in the context of STEGR. To this purpose we consider a starting Lagrangian that contains all possible combinations of nonmetricity tensor.

Let us consider the following definitions:

$$\begin{aligned} &A \equiv Q_{\alpha\mu\nu} Q^{\alpha\mu\nu}\,, \quad B \equiv Q_{\alpha\mu\nu} Q^{\mu\alpha\nu}\,, \quad C \equiv Q_\alpha Q^\alpha\,, \\ &D \equiv \tilde{Q}_\alpha \tilde{Q}^\alpha\,, \quad E \equiv \tilde{Q}_\alpha Q^\alpha\,, \end{aligned} \tag{7.68}$$

and a gravitational action of the form

$$S = \int d^4x \sqrt{-g}\, f(A, B, C, D, E)\,. \tag{7.69}$$

In the cosmological spatially flat background given by the line element $ds^2 = N(t)^2 dt^2 - a(t)^2 d\mathbf{x}^2$, with $N(t)$ being the lapse function, the action can be written in terms of Lagrange multipliers λ. Here, as better pointed out in what follows, introducing the lapse function is needed in order to develop a nontrivial prescription. Considering the cosmological expressions of the invariants in Eq. (7.68), Eq. (7.69) can be recast as

$$\begin{aligned} S = \int &\{a^3 N f(A, B, C, D, E) - \lambda_A \left[A + 4N^{-2}(3H^2 + T^2)\right] \\ &- \lambda_B (B + 4N^{-2} T^2) - \lambda_C \left[C + 4N^{-2}(3H + T)^2\right] \\ &- \lambda_D (D + 4N^{-2} T^2) - \lambda_E \left[E + 4N^{-2}(3H + T)\, T\right]\}\, dt\,, \end{aligned} \tag{7.70}$$

where H is the Hubble constant and we used the following quadratic invariants for the FLRW metric:

$$
\begin{aligned}
T &\equiv \dot{N}/N \\
A &\equiv Q_{\alpha\mu\nu}Q^{\alpha\mu\nu} = -4N^{-2}\left(3H^2 + T^2\right), \\
B &\equiv Q_{\alpha\mu\nu}Q^{\mu\alpha\nu} = -4N^{-2}T^2, \\
C &\equiv Q_\alpha Q^\alpha = -4N^{-2}\left(3H + T\right)^2, \\
D &\equiv \tilde{Q}_\alpha \tilde{Q}^\alpha = -4N^{-2}T^2, \\
E &\equiv \tilde{Q}_\alpha Q^\alpha = -4N^{-2}\left(3H + T\right)T.
\end{aligned} \tag{7.71}
$$

Notice that, in the absence of the lapse function, we have $B = D = E = 0$ and $A = C$. Therefore, in such a case, the treatment would turn out to be equivalent to that of Section 7.2 since the action would simply reduce to the $f(Q)$ one. The standard form of nonmetricity scalar is recovered as a linear combination of the previously defined quantities, that is,

$$
Q = -6N^{-2}H^2 = -\frac{1}{4}A + \frac{1}{2}B + \frac{1}{4}C - \frac{1}{2}E. \tag{7.72}
$$

Equation (7.70) can be written in a compact form by denoting the five invariants $A, B, C, D,$ and E collectively with χ, that is, $\chi \in \{A, B, C, D, E\}$. Varying with respect to each χ, we obtain the five equations

$$
a^3 N f_\chi - \lambda_\chi = 0, \tag{7.73}
$$

so that the point-like Lagrangian, coming from action (7.70), reads

$$
\begin{aligned}
\mathcal{L} = {} & a^3 N f(A, B, C, D, E) - \frac{a^3}{N}\sum_\chi f_\chi \left(N^2\chi + 4T^2\right) \\
& - \frac{12a^3}{N}\left[\left(f_A + 3f_C\right)H^2 + \left(2f_C + f_E\right)HT\right].
\end{aligned} \tag{7.74}
$$

There are two EL equations that follow from the preceding Lagrangian, namely the equation with respect to the lapse function and the equation with respect to the scale factor. The former is the (0,0) component of the field equation and is equivalent to the energy condition in the limit $N \to 1$, while the latter is the second Friedman equation. They read, respectively, as

$$
\begin{aligned}
& \frac{N^2}{2}f + 6\left(2f_C + f_E\right)\dot{H} + 6\left(2f_A + 12f_C + 3f_E\right)H^2 \\
& + 4\sum_\chi f_\chi \dot{T} + 6\left(2f_A + 4f_C + 2f_B + 2f_D + 3f_E\right)HT + \\
& + \sum_{\chi_1}\left[6\left(2f_{C\chi_1} + f_{E\chi_1}\right)H + 4\sum_{\chi_2} f_{\chi_1\chi_2}T\right]\dot{\chi}_1 = 0, \\
& \frac{N^2}{2}f + 4\left(f_A + 3f_C\right)\left(\dot{H} + 3H^2 - HT\right) + 2\left(2f_C + f_E\right)\left(\dot{T} - T^2 + 3HT\right) \\
& + 4\sum_\chi \left[2\left(f_{A\chi} + 3f_{C\chi}\right)H + \left(2f_{C\chi} + f_{E\chi}\right)T\right]\dot{\chi} = 0.
\end{aligned} \tag{7.75}
$$

Notice that $f(Q)$ cosmological field equations are recovered when the invariants χ are combined as in Eq. (7.72).

Of course, the Noether Symmetry Approach cannot be directly applied to the Lagrangian (7.74) since the Noether system of differential equations would end up being overdetermined, due to the large amount of variables in the minisuperspace. Therefore, we can reduce the minisuperspace $\mathcal{S} \equiv \{a, A, B, C, D, E, N\}$ by considering viable functions selected by physical criteria. For instance, we can focus on some more-specific classes of theories that enjoy enhanced gauge invariance, such as models that are invariant under conformal transformations. There are three quadratic combinations that transform covariantly under conformal rescalings, namely [259]

$$\hat{A} \equiv A + 4D - 2E, \quad \hat{B} \equiv B - D, \quad \hat{C} \equiv C + 16D - 8E. \tag{7.76}$$

The general class of quadratic theories, which describes a general scale-invariant scalar nonmetricity theory in a fixed gauge, contains all the three quantities in Eq. (7.76) and two additional terms. The latter come from the kinetic term and from the quartic potential that is allowed by the scale invariance. In light of these considerations, the scale-invariant function reads

$$f = b_1 \hat{A} + b_2 \hat{B} + b_3 \hat{C} + b_4 C + b_5 \,. \tag{7.77}$$

Our purpose now is to apply the Noether Symmetry Approach to Lagrangian (7.74), with function (7.77), and constrain the free parameters b_i such that the model contains Noether symmetries. Before doing so, it is worth noticing that the point-like Lagrangian (7.74) does not depend on the derivative of the χ configuration variables. This means that, in principle, one could solve the EL equations and substitute back to the Lagrangian to make it canonical. However, this would uselessly complicate the calculations, especially considering that the Noether condition (6.42) is still valid for singular Lagrangians. The point-like Lagrangian (7.74), with the function f given by Eq. (7.77), takes the form

$$\begin{aligned} \mathcal{L} = {} & 24\,(b_1 + 3b_3 - b_4)\,\frac{a^2}{N^2}\dot{a}\dot{N} - 12\,[b_1 + 3\,(b_3 + b_4)]\,\frac{a}{N}\dot{a}^2 \\ & - 4\,(3b_1 + 9b_3 + b_4)\,\frac{a^3}{N^3}\dot{N}^2 + b_5 a^3 N . \end{aligned} \tag{7.78}$$

Apparently, there is no explicit dependence on the nonmetricity scalars, and the configuration space simply reduces to $\mathcal{S} \equiv \{a, N\}$. Consequently, the symmetry generator takes the form

$$\mathcal{X} = \xi(t, a, N)\partial_t + \alpha(t, a, N)\partial_a + \mathcal{N}(t, a, N)\partial_N, \tag{7.79}$$

but, being the Lagrangian in a canonical form, we can set $\xi \equiv \xi(t)$ from the beginning (see Section 6.3). By applying the condition (6.42) to the Lagrangian (7.74), we get a system of 10 equations, which, of course, are not all independent.

Assuming the gauge function to be constant, we get six different classes of theories that show symmetries. The corresponding generators and functions (written in terms of the old variables A, B, C, D, and E) are as follows:

$$
\begin{aligned}
\mathcal{X}_1 &= \xi(t)\partial_t - \left(\frac{a\mathcal{N}(t,a,N)}{3N} + \frac{1}{3}a\xi'(t)\right)\partial_a + \mathcal{N}(t,a,N)\partial_N \\
f_1(A,B,C,D,E) &= -3b_3A + b_2B + b_3C - (b_2 - 4b_3)\,D - 2b_3E + b_5 \\
\\
\mathcal{X}_2 &= \xi_0\partial_t \\
f_2(A,B,C,D,E) &= -3A(b_3+b_4) + b_2B + (b_3+b_4)\,C \\
&\quad - (b_2 - 4b_3 + 12b_4)\,D - (2b_3 - 6b_4)\,E + b_5 \\
\\
\mathcal{X}_3 &= (c_1t + c_2)\partial_t - \frac{8\sqrt{2}c_3a^{1/4}\left[(b_1+3b_3)\,N\right]^{7/4}}{3N}\partial_a \\
&\quad + \left\{\frac{8\sqrt{2}c_3a^{1/4}\left[(b_1+3b_3)\,N\right]^{7/4}}{3N} - c_1N\right\}\partial_N \\
f_3(A,B,C,D,E) &= b_1A + b_2B + b_3C + (4b_1 - b_2 + 16b_3)\,D \\
&\quad - 2\,(b_1 + 4b_3)\,E + b_5 \\
\\
\mathcal{X}_4 &= (c_1t + c_2)\partial_t - c_1N\partial_N \\
f_4(A,B,C,D,E) &= b_1A + b_2B + (b_3+b_4)C + (4b_1 - b_2 + 16b_3)\,D \\
&\quad - 2\,(b_1 + 4b_3)\,E + b_5 \\
\\
\mathcal{X}_5 &= (c_1t + c_2)\partial_t - c_1N\partial_N \\
f_5(A,B,C,D,E) &= -3b_3A + b_2B + (b_3+b_4)\,C - (b_2 - 4b_3)\,D - 2b_3E + b_5 \\
\\
\mathcal{X}_6 &= \xi(t)\partial_t + \frac{1}{6}a\left[-6a^3N\frac{d\alpha(a^3N)}{d(a^3N)} + 3\alpha\,(a^3N) + \dot{\xi}\right]\partial_a \\
&\quad + \frac{1}{2}a\left[6a^3N\frac{d\alpha(a^3N)}{d(a^3N)} + 3\alpha\,(a^3N) - \dot{\xi}\right]\partial_N \\
f_6(A,B,C,D,E) &= -3b_3A + b_2B + (b_3+b_4)\,C - (b_2 - 4b_3)\,D - 2b_3E,
\end{aligned}
\tag{7.80}
$$

with c_i being integration constants. Let us find the cosmological solutions related to the second function, and let us set for simplicity $b_5 = 0$. The cosmological point-like Lagrangian (7.78) becomes

$$
\mathcal{L} = \frac{32b_4a^2}{N^3}\left(a\dot{N}^2 - 3N\dot{a}\dot{N}\right) \tag{7.81}
$$

and the EL equations read

$$\frac{96b_4 a^2}{N^3}\left(N\ddot{N} - \dot{N}^2\right) = 0\,, \tag{7.82}$$

$$-\frac{32b_4 a}{N^4}\left[2aN(3\dot{a}\dot{N} + a\ddot{N}) - 3N^2\left(a\ddot{a} + 2\dot{a}^2\right) - 3a^2\dot{N}^2\right] = 0\,, \tag{7.83}$$

whose solution is

$$N(t) = N_1 e^{N_2 t} \quad \text{and} \quad a(t) = a_1\left(3(t - a_2)\right)^{1/3} e^{\frac{N_2}{3}(t-3a_2)}\,, \tag{7.84}$$

where a_1, a_2, N_1, and N_2 are constants of integration. We see that, for specific choices of the integration constants, we can have both power-law and de Sitter-like solutions. The same approach to find out exact cosmological solutions can be worked out for the other preceding cases [188].

To conclude, we want to point out that modifications of the alternative formulations of GR, that is, TEGR and STEGR, lead to different phenomenologies due to the differences in the underlying geometries. We considered a general theory containing all five even-parity quadratic scalars of nonmetricity, $f(A, B, C, D, E)$. In order to select specific models of the theory, apart from observations, we can apply some theoretical constraints as well. We focused on a scale-invariant theory, given by Eq. (7.77), and showed that selected models can give rise to physically relevant cosmological solutions. It is worth pointing out the difference between the approach considered in this section and the scheme pursued in the rest of the book. Here, given the complexity of dynamics, we chose a scale-invariant action depending on five free parameters b_1, b_2, b_3, b_4, and b_5. Then we used Noether's approach to find relations among the parameters leading to models with symmetries.

8

Higher-Order Extensions with the Gauss–Bonnet Invariant

The laws of nature are but the mathematical thoughts of God.

— *Euclid*

(*Elements* [300 B.C.])

Apart from introducing combinations of the Ricci scalar and its derivatives, extended actions can also contain other curvature invariants. GR assumes that the Riemann tensor must be contracted with the metric tensor such that the starting action is

$$S_{E-H} = \frac{1}{2}\int \sqrt{-g}\, g^{\mu\rho} g^{\nu\sigma} R_{\mu\nu\rho\sigma}\, d^4x. \tag{8.1}$$

In this way, other curvature invariants are neglected, which in principle may contribute to the dynamics as the curvature scalar. For instance, the Weyl tensor $C_{\mu\nu\rho\sigma}$ can be defined as

$$\begin{aligned} C_{\mu\nu\rho\sigma} &= R_{\mu\nu\rho\sigma} + \frac{1}{D-2}\left(R_{\mu\sigma}g_{\nu\rho} - R_{\mu\rho}g_{\nu\sigma} + R_{\nu\rho}g_{\mu\sigma} - R_{\nu\sigma}g_{\mu\rho}\right) \\ &+ \frac{1}{(D-1)(D-2)} R\left(g_{\mu\rho}g_{\nu\sigma} - g_{\mu\sigma}g_{\nu\rho}\right), \end{aligned} \tag{8.2}$$

where D is the dimension of the manifold. Interestingly, it is worth noticing that $C^{\mu}{}_{\nu\rho\sigma}$ is the only rank-4 tensor invariant under conformal transformations. The Weyl scalar C^2 can be defined by the following contraction of the Weyl tensor $C^2 \equiv C^{\mu\nu\rho\sigma}C_{\mu\nu\rho\sigma}$, whose four-dimensional representation is [260, 261]

$$C^{\mu\nu\rho\sigma}C_{\mu\nu\rho\sigma} = \frac{1}{3}R^2 - 2R^{\mu\nu}R_{\mu\nu} + R^{\mu\nu\rho\sigma}R_{\mu\nu\rho\sigma}. \tag{8.3}$$

The preceding quantity vanishes identically in a spatially flat FLRW space-time. Because of the vanishing cosmological expression of the Weyl tensor, the following three quantities are equivalent in cosmology:

$$\mathcal{G} = R^2 - 4R^{\mu\nu}R_{\mu\nu} + R_{\mu\nu\rho\sigma}R^{\mu\nu\rho\sigma}, \tag{8.4}$$

$$\mathcal{G} + C^2 = \frac{4}{3}R^2 - 6R^{\mu\nu}R_{\mu\nu} + 2R_{\mu\nu\rho\sigma}R^{\mu\nu\rho\sigma}, \tag{8.5}$$

$$\mathcal{G} - C^2 = \frac{2}{3}R^2 - 2R^{\mu\nu}R_{\mu\nu}. \tag{8.6}$$

Another important second-order invariant is the Kretschmann scalar, defined as

$$K = R_{\mu\nu\rho\sigma} R^{\mu\nu\rho\sigma}. \tag{8.7}$$

The expression of K in a cosmological background is

$$K(t) = 12 \left(\frac{\ddot{a}^2}{a^4} + \frac{\dot{a}^4}{a^4} \right). \tag{8.8}$$

As we can see from Eq. (8.8), the impossibility of integrating out second derivatives does not allow one to develop a suitable Lagrangian formalism and therefore to apply the Noether Symmetry Approach. A detailed classification of second-order curvature invariants can be found in [262].

In this chapter, we consider actions containing the Gauss–Bonnet term, defined in Eq. (4.61). It is one of the most important second-order invariants in physics, due to several reasons. On the one hand, since in four dimensions $\mathcal{G}$ turns out to be a topological surface term, it can contribute in reducing the dynamics and providing exact solutions. On the other hand, a function of the Gauss–Bonnet invariant can be reduced to $f(R)$ gravity under given limits. As a linear contribution of $\mathcal{G}$ does not change the dynamics, the $f(\mathcal{G})$ extension also is trivial in three dimensions or less. It is worth noticing the appearance of $\mathcal{G}$ in the formulation of QFT in curved space-time, as discussed in [263].

In general, $f(R, \mathcal{G})$ gravity is taken into account to recover GR, in a given limit, assuming as starting action

$$S = \int \sqrt{-g} \left(\frac{1}{2} R + f(\mathcal{G}) \right) d^4x. \tag{8.9}$$

GR is thus safely recovered when $f(\mathcal{G}) = 0$ so that the function plays the role of an effective cosmological constant generated by $\mathcal{G}$. The preceding action has been extensively studied. In [264], cosmologically viable models are considered by studying the stability of a late-time de Sitter solution and the existence of radiation and matter fluids. In [265], possible power-law scaling solutions have been taken into account by developing the scalar-tensor equivalent of the preceding theory. In particular, in [266], the authors show that density perturbations cause instabilities. In [267], the author shows that the preceding theory is ruled out as a possible explanation of the late-time acceleration by Solar System tests. In [268], the Gauss–Bonnet term is added to a $f(R)$ five-dimensional Lagrangian and a static spherically symmetric solution is studied. In [269], the authors study the energy bounds for Gauss–Bonnet gravity in an AdS_7 background. Finally, in [270], a mimetic version of the preceding theory is considered and they find, besides accelerating behaviors, solutions that unify the inflationary era together with dark energy. In addition, dark matter can be described in the framework of this model. In [271], the Newtonian and PN limits of these models are studied in detail.

An important consideration, related to the Gauss–Bonnet term, is in order here. For almost half a century, higher-dimensional theories of gravity have been studied in many different contexts, see, for example, [272]. The aforementioned puzzling phenomena in gravity can sometimes be explained by invoking extra dimensions [273–276]. Braneworlds and other higher-dimensional modifications of Einstein's theory, for example, Lovelock theory [277], have been considered as possible extensions in the hunt for a self-consistent theory of gravity.

All of the preceding researches deal with a theory that safely recovers GR in the background or in some limit. This means that if one switches off the effect of the Gauss–Bonnet contribution, that is, $f(\mathcal{G}) \to 0$, then the action reduces to the EH one and GR is recovered. This happens because GR has to be restored in view of observations and experimental tests.

Observational and theoretical constraints have also been obtained for other forms of $f(R, \mathcal{G})$, but pure $f(\mathcal{G})$ theories are not, in general, considered because GR seems excluded. However, it turns out that GR can be restored as a particular case of $f(\mathcal{G})$ gravity and the further degrees of freedom related to $R_{\mu\nu}$ and $R^{\alpha}{}_{\lambda\mu\nu}$ can be neglected with respect to R. This happens if particular symmetries are adopted like in homogeneous and isotropic cosmology or in other specific cases. The action containing only a function of $\mathcal{G}$ can be helpful in reducing the dynamics and analytically solve the equations of motion. For instance, in cosmological contexts, the above second-order curvature invariants P and $\mathcal{Q}$ are comparable with R^2 so that $\sqrt{\mathcal{G}} \sim \sqrt{R^2}$, up to a constant factor. Therefore, $\mathcal{G}$ and R^2 can be considered dynamically equivalent on the solutions (up to a constant factor) if homogeneity and isotropy hold [107]. In a spherically symmetric configuration, as pointed out in Chapter 13, exact solutions can be found in any dimensions.

Moreover, as shown in [127], the action (4.59) yields new spin-2 polarization modes in GWs. It is worth noticing that the linearization of the metric tensor in the $f(\mathcal{G})$ theory yields the same dispersion relation as GR [278]. It means that the $f(\mathcal{G})$ Gauss–Bonnet gravity is the only modified action where no ghost mode occurs, but only the spin-2 massless modes of GR and the scalar mode.

However, as pointed out in [279], $f(R, \mathcal{G})$ theories could suffer the presence of ghosts even at the level of cosmological perturbations, where superluminal modes can occur. Specifically, in the $f(R, \mathcal{G})$ gravity case, this is a serious issue because superluminal modes are present in the master equation that governs the evolution of primordial curvature perturbations. This is mainly due to the fact that the Gauss–Bonnet invariant introduces higher derivatives with respect to the metric. As a consequence, the Lagrangian cannot be recast in a canonical form, with the result that the corresponding Hamiltonian turns out to be linearly unstable.

Specifically, instability stems from the coordinates that the given Lagrangian depends on, which are fewer than the initial parameters needed to specify the problem. Even when adopting appropriate gauge choices, higher-order derivatives with respect to the metric cannot be deleted from the field equations of $f(R, \mathcal{G})$ gravity, and therefore ghost modes might appear.

This shortcoming was addressed in Ref. [270, 280], where the authors use Lagrange multipliers formalism to provide a remedy to ghost instability. Such a formalism leads to the elimination of the ghost degrees of freedom in both $f(\mathcal{G})$ and $f(R, \mathcal{G})$ gravity theories, and thus the resulting models can, in principle, produce ghost-free primordial curvature perturbations. To demonstrate that the theories can be rendered ghost-free, the authors perform a conformal transformation of the metric and work in the Jordan frame. With the aim of developing a ghost-free theory, they couple the Gauss–Bonnet invariant with a scalar field and choose an ad hoc form for the potential. Thus the metric is expanded around the flat space-time and field equations are then evaluated. As a result, in the Jordan frame, field equations contain only first and second derivatives of $\delta g_{\mu\nu}$ with respect to the space-time coordinates, but the scalar field turns out not to be dynamical so that no further degree of freedom occurs with respect to GR.

In the following sections, we provide cosmological solutions for extended theories of gravity containing the Gauss–Bonnet term, selected by the Noether Symmetry Approach. We do not discuss that, when working in the Jordan frame, these selected models can be directly related to ghost-free theories. However, given the large amount of free parameters, we expect that integration constants can be properly set in order to avoid ghost instability, with the result that these models can be relevant both at large and at small scales.

8.1 $f(R, \mathcal{G})$ Gravity

We start by considering the general $f(R, \mathcal{G})$ theory of gravity. In this section and in Section 8.1.2 we consider both the vanishing Lie derivative condition $L_X\mathcal{L} = 0$ and the identity (6.42), with the aim to compare the selected models. As a result, it turns out that the latter approach yields a set of nontrivial solutions, which cannot be recovered from the former. Once again, this prescription shows the importance of dealing with the first prolongation of the Noether vector, when searching for models with symmetries. The application of the vanishing Lie derivative to $f(R, \mathcal{G})$ gravity is discussed in detail in Ref. [281], while the first prolongation of the Noether vector is applied to the cosmological Lagrangian in Ref. [282]. In both cases, the minisuperspace is three-dimensional and leads to an overdetermined Noether system of differential equations. For this reason, solutions can be found only after performing some physical ansatz on the shape of the function. The starting action is

$$\int \sqrt{-g} f(R, \mathcal{G})\, d^4x, \tag{8.10}$$

whose variation with respect to the metric tensor provides the field equations

$$\begin{aligned} f_R G_{\mu\nu} &= \frac{1}{2} g_{\mu\nu} (f - R f_R) + D_\mu D_\nu f_R - g_{\mu\nu} \Box f_R \\ &+ f_{\mathcal{G}} \left(-2RR_{\mu\nu} + 4R_{\mu\rho} R^\rho_\nu - 2R_\mu^{\ \rho\lambda\sigma} R_{\nu\rho\lambda\sigma} + 4g^{\rho\lambda} g^{\sigma\alpha} R_{\mu\rho\nu\sigma} R_{\lambda\alpha}\right) \\ &+ 2\left(D_\mu D_\nu f_{\mathcal{G}}\right) R - 2g_{\mu\nu}\left(\Box f_{\mathcal{G}}\right) R + 4\left(\Box f_{\mathcal{G}}\right) R_{\mu\nu} - 4\left(D_\rho D_\mu f_{\mathcal{G}}\right) R^\rho_\nu \\ &- 4\left(D_\rho D_\nu f_{\mathcal{G}}\right) R^\rho_\mu + 4g_{\mu\nu}\left(D_\rho D_\lambda f_{\mathcal{G}}\right) R^{\rho\lambda} - 4\left(D_\lambda D_\alpha f_{\mathcal{G}}\right) g^{\rho\lambda} g^{\sigma\alpha} R_{\mu\rho\nu\sigma}, \end{aligned} \tag{8.11}$$

which are a particular case of those in Eq. (4.60). The cosmological expression of the Gauss–Bonnet scalar in a FLRW cosmology,

$$\mathcal{G} \equiv R^2 - 4R^{\mu\nu} R_{\mu\nu} + R^{\mu\nu\rho\sigma} R_{\mu\nu\rho\sigma} = 24\left(\frac{\dot{a}^2 \ddot{a}}{a^3}\right) = \frac{8}{a^3} \frac{d}{dt}\left(\dot{a}^3\right), \tag{8.12}$$

can be used to obtain the point-like Lagrangian by the Lagrange multipliers method. Specifically, by means of Eq. (8.12), the action can be recast as

$$\begin{aligned} S = \int \Bigg\{ & a^3 f(R, \mathcal{G}) - a^3 f_R(R, \mathcal{G}) \left[R + 6\left(\frac{\ddot{a}}{a} + \frac{\dot{a}^2}{a^2}\right)\right] \\ & - a^3 f_{\mathcal{G}}(R, \mathcal{G}) \left[\mathcal{G} - 24\left(\frac{\ddot{a}\dot{a}^2}{a^3}\right)\right]\Bigg\} dt, \end{aligned} \tag{8.13}$$

which, after a straightforward integration by parts of the second derivatives, yields

$$\begin{aligned} \mathcal{L} = \; & 6a\dot{a}^2 f_R + 6a^2 \dot{a}\left(f_{RR}\dot{R} + f_{R\mathcal{G}}\dot{\mathcal{G}}\right) \\ & - 8\dot{a}^3\left(f_{R\mathcal{G}}\dot{R} + f_{\mathcal{G}\mathcal{G}}\dot{\mathcal{G}}\right) + a^3\left(f - Rf_R - \mathcal{G}f_{\mathcal{G}}\right). \end{aligned} \tag{8.14}$$

The equations of motion with respect to $\mathcal{G}$ and R provide back the cosmological expressions of the two variables. Therefore the only two equations that determine dynamics are the EL equation with respect to the scale factor and the energy condition, namely

$$\begin{aligned} & 2\dot{a}^2 f_R + 4a\ddot{a} f_R - a^2\left(f - Rf_R - \mathcal{G}f_{\mathcal{G}}\right) \\ & - 16\dot{a}\ddot{a}\left(\dot{\mathcal{G}}f_{\mathcal{G}\mathcal{G}} + \dot{R}f_{R\mathcal{G}}\right) + 4a\dot{a}\left(\dot{\mathcal{G}}f_{R\mathcal{G}} + \dot{R}f_{RR}\right) \\ & + 2a^2\left(\ddot{\mathcal{G}}f_{R\mathcal{G}} + \dot{\mathcal{G}}^2 f_{R\mathcal{G}\mathcal{G}} + \ddot{R}f_{RR} + 2\dot{\mathcal{G}}\dot{R}f_{RR\mathcal{G}}\dot{R}^2 f_{RRR}\right) \\ & - 8\dot{a}^2\left(\ddot{\mathcal{G}}f_{\mathcal{G}\mathcal{G}} + \dot{\mathcal{G}}^2 f_{\mathcal{G}\mathcal{G}\mathcal{G}} + \ddot{R}f_{R\mathcal{G}} + 2\dot{\mathcal{G}}\dot{R}f_{R\mathcal{G}\mathcal{G}} + \dot{R}^2 f_{RR\mathcal{G}}\right) = 0; \end{aligned}$$

$$\begin{aligned} & 6a\dot{a}^2 f_R - a^3(f - Rf_R - \mathcal{G}f_{\mathcal{G}}) + \\ & - 24\dot{a}^3(\dot{\mathcal{G}}f_{\mathcal{G}\mathcal{G}} + \dot{R}f_{R\mathcal{G}}) + 6a^2\dot{a}(\dot{\mathcal{G}}f_{R\mathcal{G}} + \dot{R}f_{RR}) = 0. \end{aligned} \tag{8.15}$$

The Noether vector X, in the minisuperspace $\mathcal{S} \equiv \{a, R, \mathcal{G}\}$, can be written as

$$X = \alpha(a, R, \mathcal{G})\partial_a + \beta(a, R, \mathcal{G})\partial_R + \gamma(a, R, \mathcal{G})\partial_{\mathcal{G}} + \dot{\alpha}(a, R, \mathcal{G})\partial_{\dot{a}} + \dot{\beta}(a, R, \mathcal{G})\partial_{\dot{R}} + \dot{\gamma}(a, R, \mathcal{G})\partial_{\dot{\mathcal{G}}}. \tag{8.16}$$

Thus, by applying the vanishing Lie derivative condition to Lagrangian (8.14) and neglecting trivial solutions, we get the following system of differential equations [281]:

$$\begin{cases} \alpha f_R + \beta a f_{RR} + \gamma a f_{R\mathcal{G}} + 2a\partial_a \alpha f_R + a^2\partial_a \beta f_{RR} + a^2 \partial_{\mathcal{G}}\gamma f_{R\mathcal{G}} = 0 \\ 2\alpha f_{RR} + \beta a f_{RRR} + \gamma a f_{RR\mathcal{G}} + a\partial_a \alpha f_{RR} + 2\partial_R \alpha f_R \\ + a\partial_R \beta f_{RR} + a\partial_R \gamma f_{R\mathcal{G}} = 0 \\ \beta f_{R\mathcal{G}\mathcal{G}} + \gamma f_{\mathcal{G}\mathcal{G}\mathcal{G}} + 3\partial_a \alpha f_{\mathcal{G}\mathcal{G}} + \partial_{\mathcal{G}}\beta f_{R\mathcal{G}} + \partial_{\mathcal{G}}\gamma f_{\mathcal{G}\mathcal{G}} = 0 \\ \partial_R \alpha f_{RR} = \partial_{\mathcal{G}}\alpha f_{\mathcal{G}\mathcal{G}} = \partial_R \alpha f_{R\mathcal{G}} = \partial_{\mathcal{G}}\alpha f_{R\mathcal{G}} = 0 \\ \partial_R \alpha f_{\mathcal{G}\mathcal{G}} + \partial_{\mathcal{G}}\alpha f_{R\mathcal{G}} = 0 \\ \partial_R \alpha f_{R\mathcal{G}} + \partial_{\mathcal{G}}\alpha f_{RR} = 0 \\ 2\alpha f_{R\mathcal{G}} + \beta a f_{RR\mathcal{G}} + \gamma a f_{R\mathcal{G}\mathcal{G}} + a\partial_a \alpha f_{R\mathcal{G}} + 2\partial_{\mathcal{G}}\alpha f_R \\ + a\partial_{\mathcal{G}}\beta f_{RR} + a\partial_{\mathcal{G}}\gamma f_{R\mathcal{G}} = 0 \\ \partial_a \left(\beta f_{R\mathcal{G}} + \gamma f_{\mathcal{G}\mathcal{G}}\right) = 0 \\ \beta f_{RR\mathcal{G}} + \gamma f_{R\mathcal{G}\mathcal{G}} + \partial_R \beta f_{R\mathcal{G}} + \partial_R \gamma f_{\mathcal{G}\mathcal{G}} + 3\partial_a \alpha f_{R\mathcal{G}} = 0 \\ 3\alpha \left(f - Rf_R - \mathcal{G}f_{\mathcal{G}}\right) - \beta a \left(Rf_{RR} + \mathcal{G}f_{R\mathcal{G}}\right) - \gamma a \left(\mathcal{G}f_{\mathcal{G}\mathcal{G}} + Rf_{R\mathcal{G}}\right) = 0. \end{cases} \tag{8.17}$$

Assuming that $f(R, \mathcal{G})$ can be recast as $f(R, \mathcal{G}) = F_1(R) + F_2(\mathcal{G})$, with F_1 and F_2 unknown analytic functions, the only possible solution provided by the system is $f(R, \mathcal{G}) = f_0 R + f_1 \mathcal{G}$. However, due to the topological features of $\mathcal{G}$, it trivially reduces to standard GR. On the other hand, when the function is assumed to have the form $f(R, \mathcal{G}) = F_1(R)F_2(\mathcal{G})$, Noether's system provides the constraint $f(R, \mathcal{G}) = f_0 R^k \mathcal{G}^{1-k}$ with $k \in \mathbb{R}$. Nonetheless, the latter solution is trivial since it yields an identically null conserved quantity. The generator of the symmetry, indeed, is

$$\mathcal{X} = \beta(a, R, \mathcal{G}, t)\partial_R + \frac{\mathcal{G}}{R}\beta(a, R, \mathcal{G}, t)\partial_{\mathcal{G}}. \tag{8.18}$$

A larger set of solutions can be found by considering condition (6.42), with the first prolongation of the Noether vector being of the form

$$X^{[1]} = \xi(a, R, \mathcal{G}, t)\partial_t + \alpha(a, R, \mathcal{G}, t)\partial_a + \beta(a, R, \mathcal{G}, t)\partial_R + \gamma(a, R, \mathcal{G}, t)\partial_{\mathcal{G}} + \dot{\alpha}(a, R, \mathcal{G}, t)\partial_{\dot{a}} + \dot{\beta}(a, R, \mathcal{G}, t)\partial_{\dot{R}} + \dot{\gamma}(a, R, \mathcal{G}, t)\partial_{\dot{\mathcal{G}}}.$$

The application of the above vector to Lagrangian (8.14) provides the following system [282]:

$$\begin{cases} 6a\left(af_{RR}\partial_t\beta + af_{\mathcal{G}R}\partial_t\gamma\right) + \partial_a g = 0 \\ f_{\mathcal{G}R}\partial_t\beta + f_{\mathcal{G}\mathcal{G}}\partial_t\gamma = 0 \\ f_{\mathcal{G}RR}\beta + f_{\mathcal{G}\mathcal{G}R}\gamma + f_{\mathcal{G}\mathcal{G}}\partial_R\gamma + f_{\mathcal{G}R}\left(3\partial_a\alpha + \partial_R\beta - 3\partial_t\xi\right) = 0 \\ f_{\mathcal{G}\mathcal{G}R}\beta + f_{\mathcal{G}\mathcal{G}\mathcal{G}}\gamma + f_{\mathcal{G}R}\partial_{\mathcal{G}}\beta + f_{\mathcal{G}\mathcal{G}}\left(3\partial_a\alpha + \partial_{\mathcal{G}}\gamma - 3\partial_t\xi\right) = 0 \\ f_R\left(\frac{\alpha}{a} + 2\partial_a\alpha - \partial_t\xi\right) + f_{RR}\beta + f_{\mathcal{G}R}\gamma + a\left(f_{RR}\partial_a\beta + f_{\mathcal{G}R}\partial_a\gamma\right) = 0 \\ f_{RR}\left(2\frac{\alpha}{a} + \partial_a\alpha + \partial_R\beta - \partial_t\xi\right) + f_{RRR}\beta + f_{RR\mathcal{G}}\gamma + f_{\mathcal{G}R}\partial_R\gamma = 0 \\ f_{\mathcal{G}R}\left(2\frac{\alpha}{a} + \partial_a\alpha + \partial_{\mathcal{G}}\gamma - \partial_t\xi\right) + f_{RR\mathcal{G}}\eta^2 + f_{\mathcal{G}\mathcal{G}R}\gamma + f_{RR}\partial_{\mathcal{G}}\beta = 0 \\ 3a^2\left(Rf_R + \mathcal{G}f_{\mathcal{G}} - f\right)\alpha + a^3\left(Rf_{RR} + \mathcal{G}f_{R\mathcal{G}}\right)\beta \\ +a^3\left(Rf_{R\mathcal{G}} + \mathcal{G}f_{\mathcal{G}\mathcal{G}}\right)\gamma + a^3\left(Rf_R + Gf_G - f\right)\partial_t\xi + \partial_t g = 0 \\ f_{\mathcal{G}R}\partial_a\beta + f_{\mathcal{G}\mathcal{G}}\partial_a\gamma = 0, \end{cases} \tag{8.19}$$

with the infinitesimal generators ξ and α being further constrained to $\xi = \xi(t)$ and $\alpha = \alpha(a, t)$. The former constraint naturally arises from the system (8.19), but cannot be imposed a priori. This is due to the fact that $f(R, \mathcal{G})$ gravity cannot be addressed to those cases discussed in Section 6.3 since the corresponding Lagrangian cannot be written in a canonical form. Linear solutions in the Ricci scalar have been neglected in the development of the system since a detailed treatment of the subcase $f_{RR} = 0$ is considered in Sections 8.1.1 and 8.1.2 of this chapter. In addition, notice that (8.19) comprehends (8.17) as a particular limit, namely $\xi = g = 0$.

As previously done, let us start by assuming $f(R, \mathcal{G})$ to be made of a sum between a function of R and a function of $\mathcal{G}$, namely $f(R, \mathcal{G}) = F_1(R) + F_2(\mathcal{G})$. Substituting f into the system, the latter yields the constraint $f(R, \mathcal{G}) = f_0 R^k + f_1 \mathcal{G}^z$; thus Lagrangian (8.14) becomes

$$\begin{aligned} \mathcal{L} = \; & a^3[f_1(1-z)\mathcal{G}^z + f_0(1-k)R^k] + 6f_0 kaR^{k-1}\dot{a}^2 \\ & -8f_1 z(z-1)\mathcal{G}^{z-2}\dot{a}^3\dot{\mathcal{G}} + 6f_0 k(k-1)a^2R^{k-2}\dot{a}\dot{R} \end{aligned} \tag{8.20}$$

and Eqs. (8.15) take the form

$$\begin{cases} 4f_0 kaR^{k-2}[(k-1)\dot{a}^2\dot{R} + R\ddot{a}] + 2\dot{a}[f_0 kR^{k-1}\dot{a} \\ -4f_1 z(z-1)(z-2)\mathcal{G}^{z-3}\dot{a}\dot{\mathcal{G}}^2 - 4f_1 z(z-1)\mathcal{G}^{z-2} + (2\dot{\mathcal{G}}\ddot{a} + \dot{a}\ddot{\mathcal{G}})] \\ +f_1 a^2(z-1)\mathcal{G}^z + f_0(k-1)a^2R^{k-3}[R^3 + 2k(k-2)\dot{R}^2 + 2kR\ddot{R}] = 0 \\ \\ a^3[f_1(z-1)\mathcal{G}^z + f_0(k-1)R^k] + 6f_0 kaR^{k-1}\dot{a}^2 \\ -24f_1 z(z-1)\mathcal{G}^{z-2}\dot{a}^3\dot{\mathcal{G}} + 6f_0 k(k-1)a^2R^{k-2}\dot{a}\dot{R} = 0. \end{cases} \tag{8.21}$$

Here k and z are real constants, with $k \neq 1, 0$.[1] Symmetry generators strongly depend on the free parameters. Therefore, in what follows, after setting the values

[1] The cases $k = 1$ and $k = 0$ are discussed in Sections 8.1.1 and 8.1.2 of this chapter.

of k and z, we outline the corresponding generators and conserved quantities. First, by setting $z = 1$ and $k = \frac{3}{2}$, the function turns out to be

$$f(R, \mathcal{G}) = f_0 R^{3/2} + f_1 \mathcal{G} \tag{8.22}$$

and three different generators and conserved quantities arise. They read as follows:

$$\begin{cases} \mathcal{X}_1 = \dfrac{1}{a}\partial_a - \dfrac{2R}{a^2}\partial_R \\ j_1 = -9f_0 a\sqrt{R}\left(\dfrac{\dot{a}}{a} + \dfrac{\dot{R}}{2R}\right) \\ \\ \mathcal{X}_2 = 3t\partial_t + 2a\partial_a - 6R\partial_R \\ j_2 = -9f_0 a^3\sqrt{R}\left(\dfrac{\dot{a}}{a} + \dfrac{\dot{R}}{R}\right) \\ \\ \mathcal{X}_3 = \dfrac{t}{a}\partial_a - \dfrac{2R\,t}{a^2}\partial_R \\ j_3 = -9f_0 a\sqrt{R}\left(\dfrac{\dot{a}}{a} + \dfrac{\dot{R}}{2R}\right)t + 9f_0 a\sqrt{R}. \end{cases} \tag{8.23}$$

Interestingly, it is worth noticing that though the dynamics provided by function (8.22) is the same as that of $f(R) = R^{3/2}$ gravity, the set of Noether symmetries is different. This is because the topological invariance of $\mathcal{G}$ does not provide any contribution to the equations of motion, but can play an important role in the search for symmetries. From this point of view, the Gauss–Bonnet scalar can be used to obtain different classes of symmetries without changing the field equations. On the other hand, cosmological solutions are totally equivalent to those in Eq. (7.23).

When k and z are set such that the function is $f(R, \mathcal{G}) = f_0 R^{\frac{7}{8}} + f_1 \mathcal{G}$, the EL equations are the same as those for $f(R)$ gravity, but the generators and the conserved quantities are different. They read as follows:

$$\begin{cases} \mathcal{X}_1 = 4t\partial_t + a\partial_a - 8R\partial_R \\ j_1 = \dfrac{21}{4}f_0 R^{-\frac{1}{8}}\left(-3a^2\dot{a} + a^3\dfrac{\dot{R}}{8R}\right) \\ \\ \mathcal{X}_2 = 2t^2\partial_t + ta\partial_a - 8tR\partial_R \\ j_2 = \dfrac{21}{4}f_0 R^{-\frac{1}{8}}\left(-3a^2\dot{a} + a^3\dfrac{\dot{R}}{8R}\right)t + \dfrac{21}{4}f_0 a^3 R^{-\frac{1}{8}}. \end{cases} \tag{8.24}$$

Also here, dynamics is provided by Eq. (7.19). For all other values of $k \neq 3/2, 7/8$ we have

$$
\begin{cases}
\mathcal{X} = 3t\partial_t + (2k-1)a\partial_a - 6R\partial_R \\
f(R,\mathcal{G}) = f_0 R^k + f_1\mathcal{G} \\
j = 6f_0 k a^3 R^{k-1}\left[2(k-2)\dfrac{\dot{a}}{a} - (k-1)(2k-1)\dfrac{\dot{R}}{R}\right],
\end{cases}
\tag{8.25}
$$

with cosmological solutions given by Eq. (7.19). Finally, setting $k = 1/2$ and $z = 1/4$, the function takes the form

$$
f(R,\mathcal{G}) = f_0 R^{\frac{1}{2}} + f_1 \mathcal{G}^{\frac{1}{4}}
\tag{8.26}
$$

and provides the following symmetry:

$$
\begin{cases}
\mathcal{X} = t\partial_t - 2R\partial_R - 4G\partial_{\mathcal{G}} \\
\\
j = -6\left(\frac{f_0}{2}R^{-\frac{1}{2}}a^2\dot{a} + g_0 G^{-\frac{3}{4}}\dot{a}^3\right).
\end{cases}
\tag{8.27}
$$

In this case, cosmological solutions cannot be addressed to $f(R)$ gravity since the Gauss–Bonnet term does not contribute linearly in the field equations. By substituting the function (8.26) into the system (8.19) we get

$$
a(t) = a_0 t^2, \quad R(t) \sim \frac{1}{t^2}, \quad \mathcal{G}(t) \sim \frac{1}{t^4}.
\tag{8.28}
$$

Therefore, power-law solutions are admitted in $f(R,\mathcal{G})$ gravity and can describe the Universe's acceleration in specific cosmological epochs.

Other symmetries can be found when the product $f(R,\mathcal{G}) = F_1(R)F_2(\mathcal{G})$ is considered. First, by substituting the function into the system (8.17), F_1 and F_2 turn out to be the power-law functions of R and $\mathcal{G}$, respectively. Therefore, in this case the only function containing symmetries is

$$
f(R,\mathcal{G}) = f_0 R^k \mathcal{G}^z.
\tag{8.29}
$$

The generator and conserved quantity occurring as a solution of the system are

$$
\begin{cases}
\mathcal{X} = 3t\partial_t + (4z+2k-1)a\partial_a - 6R\partial_R - 12\mathcal{G}\partial_{\mathcal{G}} \\
j = 6f_0 R^k \mathcal{G}^z a^3 \left\{2(2z+k-2)\dfrac{\dot{a}}{a}\left(\dfrac{k}{R} + \dfrac{4z\dot{a}^2}{\mathcal{G}a^2}\right)\right. \\
\left. -(4z+2k-1)\left[\dfrac{k}{R}\left((k-1)\dfrac{\dot{R}}{R} + z\dfrac{\dot{\mathcal{G}}}{\mathcal{G}}\right) + \dfrac{4z\dot{a}^2}{\mathcal{G}a^2}\left(k\dfrac{\dot{R}}{R} + (z-1)\dfrac{\dot{\mathcal{G}}}{\mathcal{G}}\right)\right]\right\}.
\end{cases}
$$

Substituting Eq. (8.29) into Eq. (8.14), the Lagrangian can be written as

$$
\begin{aligned}
\mathcal{L} = {} & -f_0\mathcal{G}^{z-2}R^{k-2}\left\{(z+k-1)a^2\mathcal{G}^2R^2 - 6ka\mathcal{G}^2R\dot{a}^2\right. \\
& \left. -6ka^2\mathcal{G}\dot{a}[zR\dot{\mathcal{G}} + (k-1)\mathcal{G}\dot{R}] + 8zR\dot{a}^3[(z-1)R\dot{\mathcal{G}} + k\mathcal{G}\dot{R}]\right\}.
\end{aligned}
\tag{8.30}
$$

The equations of motion cannot be analytically solved for any k and z, but by imposing the relation $z = 1-k$ exact solutions occur. In such a case, the scale factor that solves the equations of motion (8.15) is

$$a(t) = a_0 t^{2k-1}. \tag{8.31}$$

Notice that the function $f(R,\mathcal{G}) = R^k \mathcal{G}^{1-k}$ is the same selected by the vanishing Lie derivative condition, considered at the beginning of this section. However, as internal transformations only select a trivial conserved quantity, when the first prolongation of Noether's vector is taken into account, the function turns out to be physically interesting.

The cosmological solutions provided in this section can be further investigated by a dynamical system analysis, as pointed out in Ref. [283], where the authors consider the function in Eq. (8.29). Specifically, they found eight fixed points in the dynamical analysis. Two of them admit an accelerating expansion of the Universe and can be considered as possible candidates able to model the dark energy era. Moreover, a radiation-dominated era occurs when $k = 1 - z < \frac{1}{5}\left(2 + \sqrt{14}\right)$, matter-dominated epochs are recovered when k is close to 1, while, for $n > 1$, the universe accelerates forever without the possibility of structure formation.

Similarly, in [108], the authors analyze a fully geometric approach to dark energy in the framework of $f(R,\mathcal{G})$ gravity by considering the cosmological solution (8.31). Confronting these models with data, the authors show that they predict a shift in the position of the peaks of the Cosmic Microwave Background (CMB) temperature power spectrum and an increase in the matter power spectrum amplitude with respect to the Cosmological Standard Model. The observational viability of this approach is also tested by comparing its theoretical predictions with joint data of CMB, Supernovae Type Ia, local measurement of the Hubble constant, and the matter power spectrum. It turns out that both the anisotropy temperature data and the matter distribution are well fitted, with the price of a high total density of the geometric contribution. For this reason, the authors conclude that power law solutions in $f(R,\mathcal{G})$ gravity can describe the current observational data without any further dark energy contributions.

This result is in agreement with results provided by Ref. [284], where the authors recast the extra terms occurring in $f(R,\mathcal{G})$ gravity as an effective perfect fluid of geometric origin. They show that this approach can improve the perspective of interpreting the dark side of the universe as fully geometric effects.

In the following subsections we analyze two subcases of $f(R,\mathcal{G})$ gravity, which deserve a separate discussion. In the first one, the function is supposed to be linearly dependent on the scalar curvature, namely $f(R,\mathcal{G}) = R + f(\mathcal{G})$. In the second case, the function is assumed to be independent of R, namely $f(R,\mathcal{G}) = f(\mathcal{G})$.

In the former case, GR can be recovered in the limit $f(\mathcal{G}) = 0$, and the function plays the role of an effective cosmological constant [285]. The main aspects of $R + f(\mathcal{G})$ gravity are also discussed in [286–289]. The latter case is the "pure" $f(\mathcal{G})$ gravity, where GR limit is not imposed as a requirement. As pointed out in [107,128], $f(\mathcal{G})$ gravity provides several important results in the framework of cosmology and higher-dimensional black holes.

8.1.1 $R + f(\mathcal{G})$ Gravity

We start from the general action

$$S = \int \sqrt{-g}\left[\frac{R}{2} + f(\mathcal{G})\right] d^4x, \tag{8.32}$$

by means of which we will use the Lagrange multipliers method to get the point-like Lagrangian. Noether's approach can be pursued either by assuming R and a as independent variables or by replacing the cosmological expression of the scalar curvature into the action. The latter approach reduces the minisuperspace to two dimensions and can lead to interesting models. For this reason, in this subsection, we consider the cosmological expression of R from the beginning so that the only dynamical degrees of freedom are the Gauss–Bonnet term and the scale factor. In this way, after integrating out second derivatives, the Lagrangian can be written as

$$\mathcal{L} = 3a\dot{a}^2 + a^3(f - \mathcal{G}f_{\mathcal{G}}) - 8\dot{a}^3\dot{\mathcal{G}}f_{\mathcal{G}\mathcal{G}}. \tag{8.33}$$

Notice that the minisupersace reduction can be pursued only because the scalar curvature appears linearly into the action; otherwise higher derivatives cannot be straightforwardly integrated and a point-like Lagrangian cannot be found. From Eq. (8.33) three equations of motion follow, but the only ones that determine the dynamics are the EL equation with respect to a and the energy condition, namely

$$\begin{cases} 2a\ddot{a} + \dot{a}^2\left(1 - 8\dot{\mathcal{G}}^2 f_{\mathcal{G}\mathcal{G}\mathcal{G}} - 8\ddot{\mathcal{G}}f_{\mathcal{G}\mathcal{G}}\right) - 16\dot{a}\ddot{a}\dot{\mathcal{G}}f_{\mathcal{G}\mathcal{G}} + a^2\left(\mathcal{G}f_{\mathcal{G}} - f\right) = 0 \\ 3a\dot{a}^2 - a^3(f - \mathcal{G}f_{\mathcal{G}}) - 24\dot{a}^3\dot{\mathcal{G}}f_{\mathcal{G}\mathcal{G}} = 0. \end{cases} \tag{8.34}$$

As previously mentioned, the EL equation with respect to $\mathcal{G}$ provides back the cosmological expression of the Gauss–Bonnet term by construction. Since the minisuperspace is independent of the curvature scalar, the generator of symmetry can be recast as

$$\mathcal{X} = \xi\partial_t + \alpha\partial_a + \gamma\partial_{\mathcal{G}}. \tag{8.35}$$

Considering the first prolongation of Noether's vector and the Lagrangian (8.33), the identity (6.42) provides the system

$$\begin{cases} \dfrac{\alpha}{a} + 2\partial_a\alpha - \partial_t\xi = 0 \\ \gamma f_{\mathcal{G}\mathcal{G}\mathcal{G}} + f_{\mathcal{G}\mathcal{G}}\left(3\partial_a\alpha + \partial_{\mathcal{G}}\gamma - 3\partial_t\xi\right) = 0 \\ (f - \mathcal{G}f_{\mathcal{G}})(3\alpha + a\partial_t\xi) - \gamma a\mathcal{G}f_{\mathcal{G}\mathcal{G}} = 0 \\ \xi = \xi(t), \quad \alpha = \alpha(a), \quad \gamma = \gamma(\mathcal{G}), \end{cases} \tag{8.36}$$

whose only solution is

$$\begin{cases} \mathcal{X} = t\partial_t + \frac{1}{3}a\partial_a - 4\mathcal{G}\partial_\mathcal{G} \\ f(R, \mathcal{G}) = \dfrac{R}{2} + f_0\sqrt{\mathcal{G}} + f_1\mathcal{G}. \end{cases} \tag{8.37}$$

Therefore, the application of Noether's identity to $R + f(\mathcal{G})$ gravity only selects $f(\mathcal{G}) = f_0\sqrt{\mathcal{G}}$. In homogeneous and isotropic cosmology, it is $R^2 - 4R^{\mu\nu}R_{\mu\nu} + R^{\mu\nu\rho\sigma}R_{\mu\nu\rho\sigma} \sim R^2$ so that the function $f(\mathcal{G}) = f_0\sqrt{\mathcal{G}}$ turns out to be dynamically equivalent to the Ricci curvature scalar.

Neglecting the linear contribution of the Gauss–Bonnet term, the only function selected by symmetries leads to the Lagrangian

$$\mathcal{L} = \frac{1}{2}f_0 a^3\sqrt{\mathcal{G}} + 3a\dot{a}^2 + 2f_0\dot{a}^3\dot{\mathcal{G}}\mathcal{G}^{-\frac{3}{2}}, \tag{8.38}$$

which provides the scale factor

$$a(t) = a_0 t^{\frac{4f_0^2+3\pm\sqrt{3}\sqrt{16f_0^2+3}}{2(3-2f_0^2)}}. \tag{8.39}$$

This solution will be studied in detail in Section 16.3.1, where energy conditions and slow-roll inflation will be considered. Also notice that among all possible actions

$$S = \int \sqrt{-g}\left[\frac{R}{2} + f_0\mathcal{G}^k\right] d^4x, \tag{8.40}$$

which depend on the free parameter k, the only one yielding a power-law scale factor occurs for $k = 1/2$. In fact, the generalization of the EL equations (8.34) to a function $f(\mathcal{G}) = f_0\mathcal{G}^k$ provides the general solution

$$a(t) = a_0 \exp\left\{\left[\frac{24^k f_0(1-k)}{3}\right]^{\frac{1}{2-4k}} t\right\}. \tag{8.41}$$

The preceding holds for any k except for $k = 1/2$, for which the action (8.40) admits extra Noether symmetries. Therefore, when $k \neq 1/2$, the Gauss–Bonnet term acts like a cosmological constant and provides exponential solutions for the scale factor. This solution is analyzed in the context of energy conditions and slow-roll inflation, respectively, in Sections 16.2.2 and 16.3.

8.1.2 $f(\mathcal{G})$ Gravity

Let us generalize the Gauss–Bonnet cosmology to any $d+1$-dimensional spacetime, with d labeling the number of spatial dimensions. This is due to the fact that the Gauss–Bonnet term might provide important contributions also for $d \neq 3$. For instance, as pointed out in Section 5.2, it naturally arises in higher-dimensional Lovelock Lagrangians or in Chern–Simons AdS_5 Lagrangians. In this context, the study of the higher-dimensional AdS-invariant Lagrangians is relevant for several reasons. On the one hand, unlike the Poincaré group, the AdS group is semisimple, which means that it can be recast as a direct sum of simple Lie algebras. Moreover, according to Cartan's criteria, semisimple groups are closely related to nondegenerate Killing forms, which in turn can be used to define kinetic terms for the gauge fields and make AdS Lagrangians physically worthwhile for gravity at small scales. Moreover, five-dimensional (or even higher-dimensional) theories of gravity are largely considered with the purpose of developing a theory that unifies the four fundamental interactions. This is the case, for example, for the previously mentioned String Theory, M-Theory, and Kaluza–Klein theory.

The existence of higher dimensions was firstly conjectured by Einstein himself, who suggested that Electromagnetism could be the result of a gravitational field polarized in the fifth dimension. This prescription, however, requires a modification of GR, as Minkowski space-time and Maxwell's equations are supposed to be embedded in a five-dimensional Riemann curvature tensor.

Searching for extra dimensions can provide a wide support to the theories we have just mentioned and might hold the explanation for several open questions such as the fine-tuning of physical constants. The main idea is that there might be additional dimensions of space that, for some reasons, we do not yet perceive, maybe because they are so small that they have so far remained undetected.

Though several attempts to detect the presence of extra dimensions have been pursued, so far there is no evidence for their occurrence. One option is to compare energies and momenta before and after the collision of particles, with the hope that such collisions can eject some of the energy from our three dimensions, forcing it to enter into the other ones. If the total amount of energy (or momentum) after the collision is lower than before, this would be evidence that energy (and momentum) has drifted away toward another path, reasonably leading to other dimensions.

Another possibility is to consider the higher-dimensional spherically symmetric solution of GR, which provides a potential deviating from the Newtonian one (see Section 13.1 for details). Deviations from Newton's law at small spatial scales can be considered in this context to reveal the presence of additional dimensions.

Another option is to search for indirect indications of the existence of higher dimensions. This is the case, for example, with the holographic principle, according to which the extra dimensions are manifestations of curvature effects in a spacetime with one fewer dimension.

Here, we consider the modified $f(\mathcal{G})$ theory of gravity in a space-time with d spatial dimensions. As we have mentioned, in four dimensions (i.e. $d=3$), a linear term $\mathcal{G}$ in the action is trivial because, as a topological invariant, it turns into a surface term and the related integral is null. Despite this, cosmological and spherically symmetric solutions can be valid even in four dimensions, if the Gauss–Bonnet coupling constant diverges when $d=3$. This idea was recently proposed, for example, in Refs. [290, 291], where the authors deal with a four-dimensional theory in which the Gauss–Bonnet term contributes to dynamics. Nonetheless, as in four dimensions GR perfectly fits the current observations at Solar System scales, so in more dimensions the Gauss–Bonnet term might provide a description for a higher-dimensional universe. Therefore, the Gauss–Bonnet gravity we are going to discuss here is given by the action

$$S = \int \sqrt{|g|} f(\mathcal{G})\, d^{d+1}x + S^m , \tag{8.42}$$

with S^m being the matter action. Here we define the determinant $\sqrt{|g|}$ in modulus because its signature depends on the number of spatial dimensions d. The preceding action starts being topologically trivial in three dimensions or less, so that $f(\mathcal{G})$ gravity may represent a valuable alternative to Einstein GR to address issues at small and large scales. By varying Eq. (8.42) with respect to the metric, we find the following field equations:

$$\begin{aligned} &2RD_\mu D_\nu f_{\mathcal{G}}(\mathcal{G}) - 2g_{\mu\nu}R\Box f_{\mathcal{G}}(\mathcal{G}) - 4R^\lambda_\mu D_\lambda D_\nu f_{\mathcal{G}}(\mathcal{G}) + 4R_{\mu\nu}\Box f_{\mathcal{G}}(\mathcal{G}) \\ &+ 4g_{\mu\nu}R^{\rho\sigma}D_\rho D_\sigma f_{\mathcal{G}}(\mathcal{G}) + 4R_{\mu\nu\rho\sigma}D^\rho D^\sigma f_{\mathcal{G}}(\mathcal{G}) + \frac{1}{2}g_{\mu\nu}[f(\mathcal{G}) - \mathcal{G}f_{\mathcal{G}}(\mathcal{G})] = \mathcal{T}_{\mu\nu} , \end{aligned} \tag{8.43}$$

where $\mathcal{T}_{\mu\nu}$ is the energy-momentum tensor of matter fields. In addition, the trace equation provides

$$\left(\frac{d+1}{2}\right) f(\mathcal{G}) - 2\mathcal{G}f_{\mathcal{G}}(\mathcal{G}) - 2(d-2)\left(R\Box - 2R^{\mu\nu}D_\mu D_\nu\right) f_{\mathcal{G}}(\mathcal{G}) = \mathcal{T} , \tag{8.44}$$

with $\mathcal{T}$ being the trace of the energy-momentum tensor $\mathcal{T}_{\mu\nu}$. Equation (8.43) can be obtained from Eq. (4.60) by the choice $f(R, P, Q) = f(R^2 - 4P + Q)$. In order to get the form of the Gauss–Bonnet scalar in cosmology, we have to calculate the $(d+1)$-dimensional Riemann tensor, Ricci tensor, and Ricci scalar in FLRW metric. We choose a spatially flat FLRW line element of the form

$$ds^2 = dt^2 - a(t)^2 \delta_{ij} dx^i dx^j , \tag{8.45}$$

where the indexes i, j label all the spatial dimensions and run from 1 to d. The nonvanishing curvature components are

$$\begin{aligned} &R_{00} = -d\frac{\ddot{a}}{a} , \qquad R_{ij} = \left[(d-1)\dot{a}^2 + a\ddot{a}\right]\delta_{ij} , \\ &R = -2d\frac{\ddot{a}}{a} - d(d-1)\frac{\dot{a}^2}{a^2} , \qquad R_{0i0j} = a\ddot{a}\delta_{ij} , \qquad R_{ijm\ell} = a^2\dot{a}^2\delta_{im}\delta_{j\ell} . \end{aligned} \tag{8.46}$$

By properly contracting the above quantities, the $(d+1)$ dimensional Gauss–Bonnet term turns out to be

$$\mathcal{G} = \frac{p(d)\left[(d-3)\dot{a}^4 + 4a\dot{a}^2\ddot{a}\right]}{a^4} \equiv p(d)\frac{\left[(d-3)\dot{a}^4 + 4a\dot{a}^2\ddot{a}\right]}{a^4} , \tag{8.47}$$

with $p(d) = d(d-1)(d-2)$. As we can see, in less than four dimensions (i.e. $d < 3$) it vanishes regardless of the value of the scale factor, while in four-dimensions ($d = 3$) it turns into the preceding topological surface term of the form

$$\mathcal{G} = 24\left(\frac{\dot{a}^2\ddot{a}}{a^3}\right). \tag{8.48}$$

As in the previous cases, dynamics can be derived either starting from field Eqs. (8.43) or from the EL equations provided by a point-like Lagrangian. Because of our further considerations related to the Noether Theorem, we conform to the previous section and pursue the latter approach. The point-like Lagrangian can be found, thanks to the Lagrange multipliers method, with constraint (8.47), as follows:

$$S = \int\left[\sqrt{-g}f(\mathcal{G}) - \lambda\left\{\mathcal{G} - \frac{p(d)\left[(d-3)\dot{a}^4 + 4a\dot{a}^2\ddot{a}\right]}{a^4}\right\} + \mathcal{L}_m\right] d^{d+1}x\,, \tag{8.49}$$

with $\mathcal{L}_m$ being the matter Lagrangian. Considering the cosmological volume element in $d+1$-dimensions, the action can be written as

$$S = \int\left[a^d f(\mathcal{G}) - \lambda\left\{\mathcal{G} - \frac{p(d)\left[(d-3)\dot{a}^4 + 4a\dot{a}^2\ddot{a}\right]}{a^4}\right\} + \mathcal{L}_m\right] d^{d+1}x\,. \tag{8.50}$$

By varying the action with respect to $\mathcal{G}$, we can find the Lagrange multiplier λ:

$$\delta S = \frac{\partial S}{\partial \mathcal{G}}\delta\mathcal{G} = a^d f_{\mathcal{G}}(\mathcal{G}) - \lambda = 0, \qquad \lambda = a^d f_{\mathcal{G}}(\mathcal{G})\,. \tag{8.51}$$

Substituting the result into Eq. (8.50) and integrating out the second derivative, the Lagrangian finally takes the following form:

$$\mathcal{L} = \frac{1}{3}a^{d-4}\left[(3-d)p(d)\dot{a}^4 f_{\mathcal{G}}(\mathcal{G}) + 3a^4[f(\mathcal{G}) - \mathcal{G}f_{\mathcal{G}}(\mathcal{G})] - 4ap(d)\dot{a}^3\dot{\mathcal{G}}f_{\mathcal{G}\mathcal{G}}(\mathcal{G})\right], \tag{8.52}$$

where we neglected the matter Lagrangian contribution. The dynamical system is given by the two EL equations coming from Lagrangian (8.52), with respect to the scale factor a and the Gauss–Bonnet scalar $\mathcal{G}$, respectively. The system is completed by the Energy condition

$$E_{\mathcal{L}} = \left(\dot{a}\partial_{\dot{a}} + \dot{\mathcal{G}}\partial_{\dot{\mathcal{G}}} - 1\right)\mathcal{L} = 0\,, \tag{8.53}$$

so that the final set of equations of motion consists of the following three partial differential equations:

$$\begin{cases} d\,a^4[f(\mathcal{G}) - \mathcal{G}f_{\mathcal{G}}(\mathcal{G})] + 4(d-3)p(d)\,a\dot{a}^2[\ddot{a}f_{\mathcal{G}}(\mathcal{G}) + \dot{a}\dot{\mathcal{G}}f_{\mathcal{G}}(\mathcal{G})] \\ +\,4a^2p(d)\dot{a}[2\dot{\mathcal{G}}\ddot{a}f_{\mathcal{G}\mathcal{G}}(\mathcal{G}) + \dot{a}\ddot{\mathcal{G}}f_{\mathcal{G}\mathcal{G}}(\mathcal{G}) + \dot{a}\dot{\mathcal{G}}^2 f_{\mathcal{G}\mathcal{G}\mathcal{G}}(\mathcal{G})] \\ +\,(d-3)(d-4)p(d)\dot{a}^4 f_{\mathcal{G}}(\mathcal{G}) = 0 \\ \\ \mathcal{G} = p(d)\dfrac{\left[(d-3)\dot{a}^4 + 4a\dot{a}^2\ddot{a}\right]}{a^4} \\ \\ (3-d)p(d)\dot{a}^4 f_{\mathcal{G}}(\mathcal{G}) - a^4[f(\mathcal{G}) - \mathcal{G}f_{\mathcal{G}}(\mathcal{G})] - 4ap(d)\dot{a}^3\dot{\mathcal{G}}f_{\mathcal{G}\mathcal{G}}(\mathcal{G}) = 0. \end{cases} \tag{8.54}$$

It is worth noticing that the equation for $\mathcal{G}$ provides exactly the cosmological constraint on the Gauss–Bonnet scalar (8.47). Now we adopt the Noether Symmetry condition (6.42), starting from the generator

$$\mathcal{X} = \xi(a, \mathcal{G}, t)\partial_t + \alpha(a, \mathcal{G}, t)\partial_a + \beta(a, \mathcal{G}, t)\partial_{\mathcal{G}} \,. \tag{8.55}$$

After applying Eq. (6.42) and setting terms with derivative powers of a and $\mathcal{G}$ equal to zero, we get a system of four differential equations plus the constraints on the infinitesimal generators α, β, and ξ. It reads

$$\begin{cases} d\, a^2\alpha(f - \mathcal{G}f_{\mathcal{G}}) - 4p(d)\dot{a}^2\dot{\mathcal{G}}f_{\mathcal{G}\mathcal{G}}\partial_t\alpha - a^3\left[\beta f'_{\mathcal{G}\mathcal{G}} - (f - \mathcal{G}f_{\mathcal{G}})\partial_t\xi\right] = 0 \\ (d-3)\left[f_{\mathcal{G}}\partial_t\alpha + af_{\mathcal{G}\mathcal{G}}\partial_t\beta\right] = 0 \\ (d-3)\alpha f_{\mathcal{G}\mathcal{G}} + a\beta f_{\mathcal{G}\mathcal{G}\mathcal{G}} + af_{\mathcal{G}\mathcal{G}}\left(3\partial_a\alpha + \partial_{\mathcal{G}}\beta - 3\partial_t\xi\right) = 0 \\ (d-3)(d-4)\alpha f_{\mathcal{G}} + (d-3)a\beta f_{\mathcal{G}\mathcal{G}} - (d-3)af_{\mathcal{G}}\left(3\partial_t\xi - 4\partial_a\alpha\right) \\ + 4a^2 f_{\mathcal{G}\mathcal{G}}\partial_a\beta = 0 \,, \end{cases} \tag{8.56}$$

with $\xi = \xi(t)$, $\alpha = \alpha(a, t)$, and $g = g_0$. Here, we neglect a priori the possibility $p(d) = 0$. Only three solutions satisfy the whole system. All of them provide the same dependence of the infinitesimal generators on the variables, namely

$$\alpha = \alpha_0 a \,, \quad \beta = \beta_0\mathcal{G} \,, \quad \xi = \xi_0 t + \xi_1, \tag{8.57}$$

but with different values of the constants α_0, β_0, and ξ_0. The final solutions with the corresponding infinitesimal generators are

$$\begin{aligned} &1):\ \alpha = \alpha_0 a \,, \quad \beta = \beta_0\mathcal{G} \,, \quad \xi = \alpha_0\left(\frac{d}{3}\right)t + \xi_1 \,, \qquad f(\mathcal{G}) = f_0\mathcal{G}; \\ &2):\ \alpha = \alpha_0 a \,, \quad \beta = -4\xi_0\mathcal{G} \,, \quad \xi = \xi_0 t + \xi_1 \,,\ f(\mathcal{G}) = \frac{4f_0\xi_0}{\alpha_0 d + \xi_0}\mathcal{G}^{\frac{\alpha_0 d + \xi_0}{4\xi_0}}; \\ &3):\ \alpha = 0 \,, \quad \beta = \beta_0\mathcal{G} \,, \quad \xi = 0 \,, \qquad f(\mathcal{G}) = f_0\mathcal{G} + f_1, \end{aligned} \tag{8.58}$$

where the exponent of the second function must be different from 1. The first and the third solutions are nontrivial only in more than four dimensions, while the second provides contributions to the equations of motion even for $d = 3$. Without loss of generality, in order to find the dynamics of the scale factor, we choose the function $f(\mathcal{G}) = f_0\mathcal{G}^k$, where we define

$$\frac{\alpha_0 d + \xi_0}{4\xi_0} = k \tag{8.59}$$

and incorporate the coefficient of $\mathcal{G}^k$ into f_0. In this way, the point-like Lagrangian can be written as

$$\mathcal{L} = -\frac{1}{3}a^{d-4}\mathcal{G}^{k-2}\left[3(k-1)a^4\mathcal{G}^2 + k(d-3)\mathcal{G}p(d)\dot{a}^4 + 4k(k-1)ap(d)\dot{a}^3\dot{\mathcal{G}}\right] . \tag{8.60}$$

EL equations (8.54) can now be exactly solved, providing

$$a(t) = a_0 e^{qt} \,, \qquad \mathcal{G}(t) = (d+1)p(d)\, q^4 \,, \qquad k = \frac{d+1}{4} \tag{8.61}$$

and

$$
a(t) = a_0 t^{-4\frac{(k-1)(4k-1)}{4k-d-1}},
$$
$$
\mathcal{G}(t) = \frac{256\left[(k-1)(4k-1)\right]^3\left[4+(d+1)(4k-5)\right]p(d)}{((d+1)t-4kt)^4}, \tag{8.62}
$$

with q constant. It is worth noticing that the de Sitter-like expansion only holds in more than four dimensions, unlike the power-law solutions, which are valid even for $d=3$. However, the $d=3$ case deserves a separate discussion. In this case, the only solution reads

$$
a(t) = a_0 t^{1-4k}, \qquad \mathcal{G}(t) = -96k(1-4k)^3 t^{-4} \equiv \mathcal{G}_0 t^{-4}, \tag{8.63}
$$

which is a power-law expansion. From Eq. (8.63), we can distinguish radiation, stiff matter, and dust matter universes even in vacuum, as depending on the geometric contributions

$$
\begin{aligned}
k &= \frac{1}{8} \rightarrow a(t) \sim t^{\frac{1}{2}}, \quad \mathcal{G} = -\frac{3}{2}t^{-4} \quad : \text{radiation},\\
k &= \frac{1}{6} \rightarrow a(t) \sim t^{\frac{1}{3}}, \quad \mathcal{G} = -\frac{16}{27}t^{-4} \quad : \text{stiff matter},\\
k &= \frac{1}{12} \rightarrow a(t) \sim t^{\frac{2}{3}}, \quad \mathcal{G} = -\frac{64}{27}t^{-4} \quad : \text{dust matter}.
\end{aligned} \tag{8.64}
$$

Cosmological solutions (8.63) are, therefore, in agreement with the FLRW solutions of GR but are recovered without imposing the Ricci scalar in the gravitational action. It is worth noticing that, in all cases, the Gauss–Bonnet term turns out to be negative so that the function $f(\mathcal{G}) = f_0\mathcal{G}^k$ may lead to some problems for fractional values of k. To avoid these kinds of singularities, we want to stress that the function $f(\mathcal{G})$ is still a solution of Noether's system, even including the modulus of the Gauss–Bonnet term, that is, $f(\mathcal{G}) = f_0|\mathcal{G}|^k$. The same happens in several other modified theories; for example, as pointed out in Section 7.1, the Noether approach in $f(R)$ gravity provides $f(R) \sim R^{3/2}$, whose time power-law solution $a(t) \sim t^p$ leads to a complex function for $p < 0$ and $p > 1/2$. Hence, without loss of generality and in agreement with Noether's approach, we can always require the function in the action to be positive. However, as shown in [247] for $f(R) \sim |R|^{3/2}$, some exact solutions can imply transitions from decelerating/accelerating behaviors, that is, dust/dark energy behaviors according to the values of solution parameters. In the present case, however, we are discussing only exact solutions emerging from Noether's symmetries, where there is no change of concavity in the evolution of scale factor and then no transition from decelerated to accelerating behaviors (and vice versa). To conclude, let us impose the vanishing Lie derivative condition on the four-dimensional limit of Lagrangian (8.60), showing that only trivial solutions can be selected in the search for internal symmetries. The application of condition $L_X\mathcal{L} = 0$ to Lagrangian (8.60), when $d = 3$, yields

$$
\begin{aligned}
X\mathcal{L} &= 3\alpha a^2[f(\mathcal{G}) - \mathcal{G}f_{\mathcal{G}}(\mathcal{G})] - 24\partial_a\alpha f_{\mathcal{G}\mathcal{G}}(\mathcal{G})\dot{a}^3\dot{\mathcal{G}} - 24\partial_{\mathcal{G}}\alpha f_{\mathcal{G}\mathcal{G}}(\mathcal{G})\dot{a}^2\dot{\mathcal{G}}^2\\
&\quad - \beta a^3\mathcal{G}f_{\mathcal{G}\mathcal{G}}(\mathcal{G}) - 8\beta f_{\mathcal{G}\mathcal{G}\mathcal{G}}(\mathcal{G})\dot{a}^3\dot{\mathcal{G}} - 8\partial_a\beta f_{\mathcal{G}\mathcal{G}}(\mathcal{G})\dot{a}^4 - 8\partial_{\mathcal{G}}\beta f_{\mathcal{G}\mathcal{G}}(\mathcal{G})\dot{a}^3\dot{\mathcal{G}},
\end{aligned} \tag{8.65}
$$

from which the following system occurs:

$$\begin{cases} 3\alpha[f(\mathcal{G}) - \mathcal{G}f_{\mathcal{G}}(\mathcal{G})] - \beta a \mathcal{G} f_{\mathcal{G}\mathcal{G}}(\mathcal{G}) = 0 \\ 3\partial_a \alpha f_{\mathcal{G}\mathcal{G}}(\mathcal{G}) + \partial_{\mathcal{G}}\beta f_{\mathcal{G}\mathcal{G}}(\mathcal{G}) + \beta f_{\mathcal{G}\mathcal{G}\mathcal{G}}(\mathcal{G}) = 0 \\ \alpha = \alpha(a), \quad \beta = \beta(\mathcal{G}). \end{cases} \tag{8.66}$$

By differentiating the second equation with respect to the scale factor, one obtains

$$\frac{\partial^2 \alpha}{\partial a^2} = 0 \qquad \to \qquad \alpha = \alpha_0 a + \alpha_1, \tag{8.67}$$

so that the first equation can be written as

$$\{3\alpha_0[f(\mathcal{G}) - \mathcal{G}f_{\mathcal{G}}(\mathcal{G})] - \beta\mathcal{G}f_{\mathcal{G}\mathcal{G}}(\mathcal{G})\}\, a = -3\alpha_1[f(\mathcal{G}) - \mathcal{G}f_{\mathcal{G}}(\mathcal{G})], \tag{8.68}$$

from which it automatically follows that $\alpha_1 = 0$. Moreover, a linear combination of the first two equations yields $\alpha = 0$, $\beta = 0$, and $f_{\mathcal{G}\mathcal{G}\mathcal{G}}(\mathcal{G}) = 0$. This means that the only possible functions solving the system (8.66) are at most quadratic with respect to the Gauss–Bonnet term. Therefore, the condition $L_X\mathcal{L} = 0$ yields solutions that are totally included in Eq. (8.58). Moreover, since the infinitesimal generators and the conserved quantity vanish identically, internal symmetries are physically trivial.

8.2 Teleparallel Equivalent of Gauss–Bonnet Gravity

This final section is devoted to the study of the general action

$$S = \int \sqrt{-g} f(T, B, T_{\mathcal{G}})\, d^4x, \tag{8.69}$$

with $T_{\mathcal{G}}$ being the teleparallel equivalent of the Gauss–Bonnet term. In cosmology the latter takes the same form as the standard Gauss–Bonnet term, while in general, they differ from each other for a pure divergence. Therefore, the relation $\mathcal{G} = -T_{\mathcal{G}} + B_{\mathcal{G}}$ holds, with $B_{\mathcal{G}}$ being a boundary term. Specifically, $T_{\mathcal{G}}$ can be written as a function of the Contorsion Tensor (4.79) as

$$\begin{aligned} T_{\mathcal{G}} = \Big(& K_{a\ e}^{\ g} K_{b}^{\ eh} K_{c\ f}^{\ r} K_{d}^{\ fs} - 2K_{a}^{\ gh} K_{b\ e}^{\ r} K_{c\ f}^{\ e} K_{d}^{\ fs} + 2K_{a}^{\ gh} K_{b\ e}^{\ r} K_{f}^{\ es} K_{d\ c}^{\ f} \\ & + 2K_{a}^{\ gh} K_{b\ e}^{\ r} \partial_c K_{d}^{\ es} \Big) \delta_{g}^{a} \delta_{h}^{b} \delta_{r}^{c} \delta_{s}^{d}. \end{aligned} \tag{8.70}$$

A detailed discussion about foundations and applications of the teleparallel equivalent of the Gauss–Bonnet term can be found in [292–297]. Notice that the function considered in this section generalizes all the previously discussed cases. Indeed, $f(R)$ gravity can be recovered from $f(T, B)$ gravity by the choice $f(T, B) = f(-T + B)$, as pointed out in Section 7.2. Due to the equivalence between $\mathcal{G}$ and $T_{\mathcal{G}}$, $f(R, \mathcal{G})$ gravity can be obtained from $f(-T + B, T_{\mathcal{G}})$. Finally, pure $f(\mathcal{G})$ cosmology is exactly equivalent to $f(T_{\mathcal{G}})$.

It is worth noticing that the system is overdetermined due to the presence of four variables in the minisuperspace. For this reason, here we only write the Noether system of differential equations, but search for solutions with symmetries only in $f(T, T_{\mathcal{G}})$ cosmology.

The general action

$$S = \int h f(T, B, T_{\mathcal{G}})\, d^4x \tag{8.71}$$

can be recast in terms of Lagrange multipliers as

$$\begin{aligned} S = \int & \left[a^3 f(T, B, T_{\mathcal{G}}) - \lambda_1 \left(T + 6\frac{\dot{a}^2}{a^2} \right) \right. \\ & \left. -\lambda_2 \left(T_{\mathcal{G}} - 24\frac{\dot{a}^2 \ddot{a}}{a^3} \right) - \lambda_3 \left(B + 12\frac{\dot{a}^2}{a^2} + 6\frac{\ddot{a}}{a} \right) \right] d^4x, \end{aligned}$$

where we used the cosmological expressions of T, B, and $T_{\mathcal{G}}$ provided in Eqs. (4.113), (7.40), and (8.12), respectively. Moreover, from the variation of the action with respect to T, B, and $T_{\mathcal{G}}$, one can find the Lagrange multipliers λ_1, λ_2, and λ_3. After integrating second derivatives, the Lagrangian turns out to be

$$\begin{aligned} \mathcal{L} = & \, 6a\dot{a}^2 f_T - 6a^2 \dot{a} \left(f_{BT}\dot{T} + f_{BB}\dot{B} + f_{BT_{\mathcal{G}}}\dot{T}_{\mathcal{G}} \right) \\ & + 8\dot{a}^3 \left(f_{TT_G}\dot{T} + f_{BT_G}\dot{B} + f_{T_{\mathcal{G}}T_G}\dot{T}_{\mathcal{G}} \right) \\ & + a^3 \left(f - Tf_T - Bf_B - T_G f_{T_G} \right). \end{aligned} \tag{8.72}$$

Dynamics is provided by the EL equation with respect to the scale factor, and the energy condition, which in turn represents the first Friedman equation. The remaining three equations are identically satisfied by construction, when the cosmological expressions of T, B, and $T_{\mathcal{G}}$ are considered. Specifically, the whole set of equations of motion reads

$$\begin{cases} f - Tf_T - Bf_B - T_G f_{T_G} - 2f_T\dfrac{\dot{a}^2}{a^2} - 4\left(\dfrac{\dot{a}}{a}\dot{f}_T + f_T\dfrac{\ddot{a}}{a} \right) \\ -16\dot{f}_{T_G}\dfrac{\dot{a}\ddot{a}}{a^2} - 8\ddot{f}_{T_G}\dfrac{\dot{a}^2}{a^2} + 2\ddot{f}_B = 0 \\ 6f_T a\dot{a}^2 - 6a^2\dot{a}\dot{f}_B + 24\dot{a}^3\dot{f}_{T_G} - a^3\left(f - Tf_T - Bf_B - T_G f_{T_G} \right) = 0 \\ T = -6\dfrac{\dot{a}^2}{a^2}, \qquad B = -12\dfrac{\dot{a}^2}{a^2} - 6\dfrac{\ddot{a}}{a}, \qquad T_{\mathcal{G}} = 24\dfrac{\dot{a}^2\ddot{a}}{a^3}. \end{cases} \tag{8.73}$$

Noether's identity (6.42) can be applied to Lagrangian (8.72) in the minisuperspace $\mathcal{S} \equiv \{a, T, B, T_{\mathcal{G}}\}$, where the generator of the symmetry is

$$\mathcal{X} = \xi\partial_t + \alpha\partial_a + \beta\partial_T + \gamma\partial_B + \zeta\partial_{T_{\mathcal{G}}}. \tag{8.74}$$

Neglecting trivial solutions, the system consists of the following 30 differential equations [298]:

$$
\begin{cases}
12af_T\partial_t\alpha - 6a^2\left(f_{BT}\partial_t\beta + f_{BB}\partial_t\gamma + f_{BT_G}\partial_t\zeta\right) - \partial_a g = 0 \\
6a^2 f_{BT}\partial_t\alpha + \partial_T g = 0 \\
6a^2 f_{BB}\partial_t\alpha + \partial_B g = 0 \\
6a^2 f_{BT_{\mathcal{G}}}\partial_t\alpha + \partial_{T_{\mathcal{G}}} g = 0 \\
f_{TT_G}\partial_t\alpha = 0 \\
f_{BT_G}\partial_t\alpha = 0 \\
f_{T_GT_G}\partial_t\alpha = 0 \\
f_{TT_G}\partial_t\beta + f_{BT_G}\partial_t\gamma + f_{T_GT_G}\partial_t\zeta = 0 \\
f_{BT_G}\partial_B\alpha = 0 \\
f_{T_GT_G}\partial_{T_{\mathcal{G}}}\alpha = 0 \\
f_{BT}\partial_B\alpha + f_{BB}\partial_T\alpha = 0 \\
f_{BT}\partial_T\alpha = 0 \\
f_{BB}\partial_B\alpha = 0 \\
f_{BT_G}\partial_T\alpha = 0 \\
f_{TT_G}\partial_T\alpha = 0 \\
f_T\left(\alpha + 2a\partial_a\alpha - a\partial_t\xi\right) + af_{TT}\beta + af_{TB}\gamma + af_{TT_G}\zeta \\
-a^2\left(f_{BT_G}\partial_a\beta + f_{BB}\partial_a\gamma + f_{BT_G}\partial_a\zeta\right) = 0 \\
f_{BT}\left(\alpha + a\partial_a\alpha + a\partial_T\beta - a\partial_t\xi\right) + af_{BTT}\beta + af_{BBT}\gamma \\
+af_{BTT_G}\zeta - 2f_T\partial_T\alpha + af_{BB}\partial_T\gamma + af_{BT_G}\partial_T\zeta = 0 \\
f_{BB}\left(2\alpha + a\partial_a\alpha + a\partial_B\gamma - a\partial_t\xi\right) + af_{BBT}\beta + af_{BBB}\gamma \\
+af_{BBT_G}\zeta - 2f_T\partial_B\alpha + af_{BT}\partial_B\beta + af_{BT_G}\partial_B\zeta = 0 \\
f_{BT_G}\left(2\alpha + a\partial_a\alpha + a\partial_{T_{\mathcal{G}}}\beta - a\partial_t\xi\right) + af_{BTT_G}\beta + af_{BBT_G}\gamma \\
+af_{BT_GT_G}\zeta - 2f_T\partial_{T_{\mathcal{G}}}\alpha + af_{BT_G}\partial_{T_{\mathcal{G}}}\beta + af_{BB}\partial_{T_{\mathcal{G}}}\gamma = 0 \\
f_{BT}\partial_{T_{\mathcal{G}}}\alpha + f_{BT_G}\partial_T\alpha = 0 \\
f_{TT_G}\partial_B\alpha + f_{BT_G}\partial_T\alpha = 0 \\
f_{BT_G}\partial_{T_{\mathcal{G}}}\alpha + f_{T_GT_G}\partial_B\alpha = 0 \\
f_{BB}\partial_{T_{\mathcal{G}}}\alpha + f_{BT_G}\partial_B\alpha = 0 \\
f_{TT_G}\partial_{T_{\mathcal{G}}}\alpha + f_{T_GT_G}\partial_T\alpha = 0 \\
f_{TT_G}\partial_a\beta + f_{BT_G}\partial_a\gamma + f_{T_GT_G}\partial_a\zeta = 0 \\
f_{TT_G}\left(3\partial_a\alpha + \partial_T\beta - 3\partial_t\xi\right) + f_{TTT_G}\beta + f_{TBT_G}\gamma \\
+f_{TT_GT_G}\zeta + f_{BT_G}\partial_T\gamma + f_{T_GT_G}\partial_T\zeta = 0 \\
f_{BT_G}\left(3\partial_B\alpha + \partial_B\gamma - 3\partial_t\xi\right) + f_{TBT_G}\beta + f_{BBT_G}\gamma \\
+f_{BT_GT_G}\zeta + f_{TT_G}\partial_B\beta + f_{T_GT_G}\partial_B\zeta = 0 \\
f_{T_GT_G}\left(3\partial_a\alpha + \partial_{T_{\mathcal{G}}}\zeta - 3\partial_t\xi\right) + f_{TT_GT_G}\beta + f_{BT_GT_G}\gamma \\
+f_{T_GT_GT_G}\zeta + f_{TT_G}\partial_{T_{\mathcal{G}}}\beta + f_{BT_G}\partial_{T_{\mathcal{G}}}\gamma = 0 \\
3\alpha a^2\left(Tf_T + Bf_B + T_Gf_{T_G} - f\right) + \beta a^3\left(Tf_{TT} + Bf_{TB} + T_{TG}f_{TT_G}\right) \\
+\gamma a^3\left(Tf_{TB} + Bf_{BB} + T_{BG}f_{BT_G}\right) \\
\zeta a^3\left(Tf_{TT_{\mathcal{G}}} + Bf_{BT_{\mathcal{G}}} + T_Gf_{T_GT_{\mathcal{G}}}\right) + a^3\partial_t\xi\left(Tf_T + Bf_B + T_Gf_{T_G} - f\right) + \partial_t\xi = 0 \\
\xi = \xi(t)
\end{cases}
\tag{8.75}
$$

In this section we do not aim to solve the preceding system; rather, we only focus on the teleparallel equivalent of $f(R, \mathcal{G})$ gravity, namely $f(T, T_{\mathcal{G}})$. Therefore, we set $f_B = 0$ and start from the action [183]

$$
S = \int \left[a^3 f(T, T_{\mathcal{G}}) - \lambda_2\left(T + 6\frac{\dot{a}^2}{a^2}\right) - \lambda_2\left(T_{\mathcal{G}} - 24\frac{\dot{a}^2\ddot{a}}{a^3}\right)\right] dt, \tag{8.76}
$$

whose point-like Lagrangian is

$$
\mathcal{L} = a^3\left(f - T_{\mathcal{G}}f_{T_{\mathcal{G}}} - Tf_T\right) - 8\dot{a}^3\left(T_{\mathcal{G}}f_{T_{\mathcal{G}}T_{\mathcal{G}}} + \dot{T}f_{TT_{\mathcal{G}}}\right) - 6f_T a\dot{a}^2. \tag{8.77}
$$

Notice that by setting $f_T = 0$, Eq. (8.77) reduces to the $d = 3$ limit of Eq. (8.52), while for $f_{T_{\mathcal{G}}} = 0$ it reduces to Eq. (7.29). The EL equations with energy condition yield the following system of differential equations:

$$
\begin{cases}
f - Tf_T - T_G f_{T_G} - 2f_T\frac{\dot{a}^2}{a^2} - 4\left(\frac{\dot{a}}{a}\dot{f}_T + f_T\frac{\ddot{a}}{a}\right) - 16\dot{f}_{T_G}\frac{\dot{a}\ddot{a}}{a^2} - 8\ddot{f}_{T_G}\frac{\dot{a}^2}{a^2} = 0\\
6\dot{f}_T a\dot{a}^2 + 24\dot{a}^3\dot{f}_{T_G} - a^3\left(f - Tf_T - T_G f_{T_G}\right) = 0\\
\left(a^3 T_{\mathcal{G}} - 24\dot{a}^2\ddot{a}\right) f_{T_{\mathcal{G}}T_{\mathcal{G}}} + \left(a^3 T + 6a\dot{a}^2\right) f_{TT_{\mathcal{G}}} = 0\\
\left(a^2 T + 6\dot{a}^2\right) a f_{TT} - \left(a^3 T_{\mathcal{G}} - 24\dot{a}^2\ddot{a}\right) f_{T_{\mathcal{G}}T} = 0.
\end{cases}
\tag{8.78}
$$

The last two equations gives $T = -6\left(\frac{\dot{a}^2}{a^2}\right)$ and $T_{\mathcal{G}} = 24\left(\frac{\dot{a}^2\ddot{a}}{a^3}\right)$ as expected by construction. In the minisuperspace $\mathcal{S} \equiv \{a, T, T_{\mathcal{G}}\}$, where the symmetry generator is

$$
\mathcal{X} = \alpha(a, T, T_{\mathcal{G}})\partial_a + \beta(a, T, T_{\mathcal{G}})\partial_T + \gamma(a, T, T_{\mathcal{G}})\partial_{T_{\mathcal{G}}},
\tag{8.79}
$$

Noether's system of differential equations takes the following form:

$$
\begin{cases}
\partial_a\gamma f_{T_{\mathcal{G}}T_{\mathcal{G}}} + \partial_a\beta f_{TT_{\mathcal{G}}} = 0\\
\gamma f_{T_{\mathcal{G}}T_{\mathcal{G}}T_{\mathcal{G}}} + \beta f_{TT_{\mathcal{G}}T_{\mathcal{G}}} + 3\partial_a\alpha f_{T_{\mathcal{G}}T_{\mathcal{G}}} + \partial_{T_{\mathcal{G}}}\gamma f_{T_{\mathcal{G}}T_{\mathcal{G}}} + \partial_{T_{\mathcal{G}}}\beta f_{TT_{\mathcal{G}}} = 0\\
\gamma f_{TT_{\mathcal{G}}T_{\mathcal{G}}} + \beta f_{TTT_{\mathcal{G}}} + 3\partial_a\alpha f_{TT_{\mathcal{G}}} + \partial_T\gamma f_{T_{\mathcal{G}}T_{\mathcal{G}}} + \partial_T\beta f_{TT_{\mathcal{G}}} = 0\\
\alpha f_T + \gamma f_{TT_{\mathcal{G}}} a + \beta f_{TT} a + 2f_T a\partial_a\alpha = 0\\
3\alpha\left(f - Tf_T - T_{\mathcal{G}}f_{T_{\mathcal{G}}}\right) - a\gamma\left(T_{\mathcal{G}}f_{T_{\mathcal{G}}T_{\mathcal{G}}} + Tf_{TT_{\mathcal{G}}}\right)\\
-a\beta\left(T_{\mathcal{G}}f_{TT_{\mathcal{G}}} + Tf_{TT}\right) = 0\\
\alpha = \alpha(a),
\end{cases}
\tag{8.80}
$$

and can be easily obtained from (8.75) by setting $\xi = f_B(T, T_{\mathcal{G}}, B) = 0$. As well as $f(R, \mathcal{G})$ gravity, Noether's system can be solved only after imposing some ansatz on the form of the function. After substituting the ansatz $f(T, T_{\mathcal{G}}) = F_1(T) + F_2(T_{\mathcal{G}})$ into Eq. (8.80), the Noether system provides

$$
f(T, T_{\mathcal{G}}) = f_0 T^k + f_1 T_{\mathcal{G}}.
\tag{8.81}
$$

Being a topological term, the preceding function yields the same dynamics as $f(T)$ cosmology, discussed in Section 7.2. On the other hand, when $f(T, T_{\mathcal{G}})$ is assumed to be a product between $F_1(T)$ and $F_2(T_{\mathcal{G}})$, the system provides

$$
f(T, T_{\mathcal{G}}) = f_0 T_{\mathcal{G}}^k T^{1-k}.
\tag{8.82}
$$

Notice that Eq. (8.82) can be addressed to the function in Eq. (8.29), by a proper redefinition of the constant k. This is because T and R differ only for a boundary term, while $\mathcal{G}$ and $T_{\mathcal{G}}$ are exactly equivalent in a cosmological background. Two symmetry generators correspond to the function (8.82), namely

$$\begin{cases} \mathcal{X} = \alpha_0 a^{1-\frac{3}{2k}}\partial_a - \dfrac{3\alpha_0}{k} T a^{-\frac{3}{2k}}\partial_T + \gamma(a, \mathcal{G}, T)\partial_{T_{\mathcal{G}}} \\ \mathcal{X} = \beta(a, T, T_{\mathcal{G}})\dfrac{T}{T_{\mathcal{G}}}\partial_T + \gamma(a, T, T_{\mathcal{G}})\partial_{T_{\mathcal{G}}}. \end{cases} \tag{8.83}$$

With the aim of studying cosmological solutions of $f(T, T_{\mathcal{G}}) = f_0 T_{\mathcal{G}}^k T^{1-k}$ gravity, we substitute Eq. (8.82) into Lagrangian (8.77) so that the latter takes the form

$$\mathcal{L} = f_0(k-1)\dot{a}^2 T_{\mathcal{G}}^{k-2} T^{-k}\left[4k\dot{a}(T_{\mathcal{G}}\dot{T} - T\dot{T}_{\mathcal{G}}) + 3aT_{\mathcal{G}}^2\right] \tag{8.84}$$

and EL equations (8.78) admit the following solution:

$$a(t) = a_0 t^{2k+1}, \quad T = -6\frac{(2k+1)^2}{t^2}, \quad T_{\mathcal{G}}(t) = 24\frac{2k(2k+1)^3}{t^4}. \tag{8.85}$$

It is worth comparing Eq. (8.31) with cosmological solutions coming from $f(R, \mathcal{G}) = R^k \mathcal{G}^{1-k}$ gravity (8.85). First notice that the two solutions are different because $f(R)$ cosmology cannot be directly recovered from the teleparallel equivalent, when the boundary term is not included. Nevertheless, though they depend on the free parameter k in a different way, both lead to power-law scale factors.

9

Extensions with Higher Derivatives of R and T

There are two possible outcomes:
if the result confirms the hypothesis,
then you've made a measurement.
If the result is contrary to the hypothesis,
then you've made a discovery.

— *Enrico Fermi*
(as quoted in *Nuclear Principles in Engineering* [2005])

In this chapter, we apply the Noether Symmetry Approach to Lagrangians depending on general functions of higher-order terms, such as $\Box R$ or $\Box T$. First we introduce the main features of higher-derivative gravity, considering the application of the d'Alembert operator to the scalar curvature. Obtained results can be easily generalized to teleparallel equivalent gravity, where the operator $\Box$ acts on the torsion scalar.

The main reason to consider these higher-derivative terms in the action is due to the need of finding renormalizable approaches to gravity. Indeed, as one can naively infer from the power-counting renormalization procedure (see Section 18.3 for details), renormalizable theories of gravity require higher-order corrections or non-minimal coupling between scalar fields and geometry to make the UV divergences cancel out. The application of the d'Alembert operator to geometric terms leads to a positive mass dimension of the coupling constant when the action is expanded around a Gaussian fixed point. Moreover, these quantities naturally arise as quantum corrections of EH action [299–301]. As an example, at the one-loop level, the energy-momentum tensor reads as follows [263]:

$$\begin{aligned}< \mathcal{T}_{\mu\nu} > &= k\left(2D_\mu D_\nu R - 2g_{\mu\nu}\Box R + 2RR_{\mu\nu} - \frac{1}{2}g_{\mu\nu}R^2\right) \\ &+k_1\left(R^\sigma_\mu R_{\nu\sigma} - \frac{2}{3}RR_{\mu\nu} - \frac{1}{2}g_{\mu\nu}R^{\sigma\tau}R_{\sigma\tau} + \frac{1}{4}g_{\mu\nu}R^2\right),\end{aligned} \tag{9.1}$$

whose trace is

$$< \mathcal{T} >= \left[k\left(M-\frac{2}{3}\Box R\right)+k_1\mathcal{G}\right], \tag{9.2}$$

where

$$\begin{aligned} M &= R^{\alpha\beta\gamma\delta}R_{\alpha\beta\gamma\delta}-2R^{\alpha\beta}R_{\alpha\beta}+\frac{1}{3}R^2, \\ \mathcal{G} &= R^2-4R^{\alpha\beta}R_{\alpha\beta}+R^{\alpha\beta\gamma\delta}R_{\alpha\beta\gamma\delta}, \end{aligned} \tag{9.3}$$

and k, k_1 are real constants. In what follows we consider the general functions $f(R,\Box R)$ and $f(T,\Box T)$. Notice that among all possible choices, most of them are physically trivial. This is the case, for example, with $(\Box^i R)$, which is a boundary term unable to provide a nontrivial dynamics. More precisely, it turns out that any higher-derivative theory can be related to $f(R,\Box^i R)$ gravity [302–304]. In our analysis, we set $i=1$ and consider $\Box R$ as an independent degree of freedom, so that the cosmological minisuperspace consists of the three variables $\{a, R, \Box R\}$. Hereafter we set $\Box R\equiv\chi$ and $\Box T\equiv Y$.

9.1 $f(R,\Box R)$ Gravity

Let us apply the Noether Symmetry Approach to a higher-derivative theory of gravity with action

$$S=\int\sqrt{-g}f(R,\Box R)\,d^4x\equiv\int\sqrt{-g}f(R,\chi)\,d^4x. \tag{9.4}$$

Field equations can be obtained from the $k=1$ limit of Eq. (4.55) and read[1]

$$\begin{aligned} G_{\alpha\beta}\left(f_R+\Box f_\chi\right) &= \frac{1}{2}g_{\alpha\beta}\left(f-Rf_R-R\Box f_\chi\right) \\ &+ D_\alpha D_\beta\left(\Box f_\chi+f_R\right)-g_{\alpha\beta}\Box^2 f_\chi-g_{\alpha\beta}\Box f_R+\frac{1}{2}D_{\{\alpha}f_\chi D_{\beta\}}R \\ &- \frac{1}{2}g_{\alpha\beta}\left(f_\chi\Box R+g^{\mu\nu}D_\mu f_\chi D_\nu R\right). \end{aligned} \tag{9.5}$$

In order to find the point-like cosmological Lagrangian, we use again the Lagrange multipliers method with constraints:

$$R=-6\left(\frac{\ddot a}{a}+\frac{\dot a^2}{a^2}\right), \tag{9.6}$$

$$\chi=\Box R=\ddot R+3\frac{\dot a}{a}\dot R. \tag{9.7}$$

[1] See Appendix A for details about the variational principle.

By means of the preceding equations, the action can be written in terms of the Lagrange multipliers λ_1 and λ_2 as

$$S = \int \left[a^3 f(R,\chi) - \lambda_1 \left(R + 6\frac{\ddot{a}}{a} + 6\frac{\dot{a}^2}{a^2} \right) \right. \\ \left. -\lambda_2 \left(\chi - \ddot{R} - 3\frac{\dot{a}}{a}\dot{R} \right) \right] dt, \tag{9.8}$$

where we integrated the three-dimensional surface and neglected the matter Lagrangian, in agreement with the previous chapters. Multipliers λ_1 and λ_2 can be found by varying the action with respect to R and χ respectively and equaling the result to zero. After doing so, second-order derivatives arise into the action, but they can be straightforwardly integrated out. This procedure finally yields the Lagrangian:

$$\mathcal{L} = a^3 \left[f(R,\chi) - R f_R(R,\chi) - \chi f_\chi(R,\chi) \right] + 6a\dot{a}^2 f_R(R,\chi) \\ + 6a^2 \dot{a}\dot{R} f_{RR}(R,\chi) + 6a^2 \dot{a}\dot{\chi} f_{R\chi}(R,\chi) - a^3 \dot{R}^2 f_{R\chi}(R,\chi) \\ - a^3 \dot{R}\dot{\chi} f_{\chi\chi}(R,\chi), \tag{9.9}$$

which, in the limit $f_\chi(R,\chi) = 0$, provides Eq. (7.5). The EL equations and the energy condition read as follows:

$$\left[R f_R(R,\chi) + \chi f_\chi(R,\chi) - f(R,\chi) \right] \\ +2\left(\frac{\dot{a}^2}{a^2} + 2\frac{\ddot{a}}{a} \right) f_R(R,\chi) + 2\left(\chi - \frac{\dot{a}}{a}\dot{R} \right) f_{RR}(R,\chi) \\ +\dot{R}(\dot{\chi}) f_{\chi\chi}(R,\chi) + 2\dot{R}^2 f_{RRR}(R,\chi) \\ +2(\dot{\chi})^2 f_{R\chi\chi}(R,\chi) + 4\dot{R}\dot{\chi} f_{RR\chi}(R,\chi) \\ + \left[2(\chi)^2 - 2\frac{\dot{a}}{a}\dot{\chi} + \dot{R}^2 \right] f_{R\chi}(R,\chi) = 0;$$

$$\Box f_\chi = 0;$$

$$a^2 \left[f_{R\chi} \left(R + 6\frac{\ddot{a}}{a} + 6\frac{\dot{a}^2}{a^2} \right) + f_{\chi\chi} \left(\chi - \ddot{R} - 3\frac{\dot{a}}{a}\dot{R} \right) \right] = 0;$$

$$\frac{\dot{a}^2}{a^2} f_R(R,\chi) + \frac{\dot{a}}{a}\frac{d}{dt}\left[f_R(R,\chi) \right] + \frac{1}{6}\left[R f_R(R,\chi) \right. \\ \left. + \chi f_\chi(R,\chi) - f(R,\chi) - \dot{R}\frac{d}{dt}\left(f_\chi(R,\chi) \right) \right] = 0. \tag{9.10}$$

Notice that the equation with respect to χ identically vanishes, when the cosmological expressions of R and χ are replaced. The preceding system is solved after selecting the form of the function containing the Noether symmetry. To this purpose, we consider the Noether vector

$$X = \alpha(a,R,\chi)\partial_a + \beta(a,R,\chi)\partial_R + \gamma\partial_\chi + \dot{\alpha}(a,R,\chi)\partial_{\dot{a}} + \dot{\beta}(a,R,\chi)\partial_{\dot{R}} + \dot{\gamma}(a,R,\chi)\partial_{\dot{\chi}}, \tag{9.11}$$

whose application to Lagrangian (9.9) leads to the following Noether system of differential equations [245, 305]:

$$\begin{cases} f_R\left(\alpha + 2a\partial_a\alpha\right) + af_{RR}\left(\beta + a\partial_a\beta\right) + af_{R\chi}\left(\gamma + a\partial_a\gamma\right) = 0 \\ -6f_{RR}\partial_R\alpha + f_{R\chi}\left(3\alpha + 2a\partial_R\beta\right) + \beta af_{RR\chi} + \gamma af_{R\chi\chi} + a\partial_R\gamma f_{\chi\chi} = 0 \\ 6f_{R\chi}\partial_\chi\alpha - af_{\chi\chi}\partial_\chi\beta = 0 \\ 12f_R\partial_R\alpha + 6f_{RR}\left(2\alpha + a\partial_a\alpha + a\partial_R\beta\right) + af_{R\chi}\left(6\partial_R\gamma - 2a\partial_a\beta\right) \\ -a^2\partial_a\gamma f_{\chi\chi} + 6\beta af_{RRR} + +6\gamma af_{RR\chi} = 0 \\ 12f_R\partial_\chi\alpha + f_{R\chi}\left(12\alpha + 6a\partial_a\alpha + 6a\partial_\chi\gamma\right) + 6af_{RR}\partial_\chi\beta \\ -a^2f_{\chi\chi}\partial_a\beta + 6\gamma af_{R\chi\chi} + 6\beta af_{RR\chi} = 0 \\ f_{\chi\chi}\left(3\alpha + a\partial_\chi\gamma + a\partial_R\beta\right) + f_{R\chi}\left(2a\partial_R\beta - 6\partial_R\alpha\right) \\ -6f_{RR}\partial_\chi\alpha + \beta af_{R\chi\chi} + \gamma af_{\chi\chi\chi} = 0 \\ 3\alpha\left(f - Rf_R - \chi f_\chi\right) - \beta aRf_{RR} \\ -\gamma a\chi f_{\chi\chi} - af_{R\chi}(\beta\chi + \gamma R) = 0. \end{cases} \tag{9.12}$$

As standard when cosmology is ruled by three-dimensional minisuperspaces, the system is overdetermined and cannot provide a set of unique solutions. Therefore, we mainly analyze those solutions of physical interest, as previously done, for example, for $f(R, \mathcal{G})$ or $f(T, B)$ gravities. In the search for symmetries, we neglect the possibility $f_\chi = 0$, which only provides the same internal solutions as Section 7.1. Two solutions can be found by substituting the function $f(R,\chi) = F_1(R) + F_2(\chi)$ into Eq. (9.12), that is,

$$\begin{cases} f(R,\chi) = f_0R + f_1(\chi)^k \\ f(R,\chi) = f_0R + f_1R^2 + f_2\chi. \end{cases} \tag{9.13}$$

On the other hand, replacing the product between $F_1(R)$ and $F_2(\chi)$, the only selected function is

$$f(R,\chi) = f_0\sqrt{R\chi}. \tag{9.14}$$

Two other functions of physical interest solve the preceding system, but they cannot be recast in terms of sum or product between F_1 and F_2. They are

$$\begin{cases} f(R,\chi) = f(R,\chi) = f_0R + f_1R^2 + f_2R\chi \\ f(R,\chi) = f_0R + f_1R^2 + f_2(R\chi)^{\frac{1}{2}}. \end{cases} \tag{9.15}$$

The preceding solutions are physically nontrivial, since the corresponding conserved quantities are finite and nonvanishing. Several other solutions of system

(9.12) can be obtained, but most of them have no physical meaning because the integral of motion vanishes identically. Summing up, we can investigate the dynamics of five theories containing symmetries. We outline them here, with the corresponding symmetry generators and conserved quantities:

$$\begin{cases} \text{I:} & \mathcal{X} = \beta_0 \partial_R, \quad f_0 R + f_1 (\chi)^k \\ & j_0 = -\beta_0 f_1 k(k-1) a^3 (\chi)^{k-2} \dot{\chi} \\ \text{II:} & \mathcal{X} = \dfrac{\beta_0}{a} \partial_R + \gamma(a, R, \chi) \partial_\chi, \quad f(R,\chi) = f_0 R + f_1 R^2 + f_2 \chi \\ & j_0 = 12 \beta_0 f_1 a \dot{a} \\ \text{III:} & \mathcal{X} = \beta_0 \partial_R + \beta_0 \dfrac{\chi}{R} \partial_\chi, \quad f(R,\chi) = f_0 \sqrt{R\chi} \\ & j_0 = 4 (R\chi)^{-\frac{3}{2}} \left[\beta_0 f_0 a^3 R \left(R \dot{\chi} - \dot{R} \chi \right) \right] \\ \text{IV:} & \mathcal{X} = \beta_0 \partial_R + \beta_0 \dfrac{\chi}{R} \partial_\chi, \quad f(R,\chi) = f_0 \sqrt{R\chi} + f_1 R \\ & j_0 = 4 (R\chi)^{-\frac{3}{2}} \left[\beta_0 f_0 a^3 R \left(R \dot{\chi} - \dot{R} \chi \right) \right]. \end{cases} \tag{9.16}$$

Let us begin by considering the first function. In this case, Lagrangian (9.9) becomes

$$\mathcal{L} = 6 f_0 a \dot{a}^2 - f_1 (k-1) a^3 (\chi)^{k-2} [(\chi)^2 + k \dot{\chi} \dot{R}], \tag{9.17}$$

whose only solution of the corresponding equations of motion is

$$a(t) = a_0 e^{\ell t}, \qquad \text{with} \qquad f_0 = 0 \tag{9.18}$$

where ℓ is a real constant. When the scale factor has an exponential form, the scalar curvature is constant and the term χ vanishes identically. This means that higher derivatives act like a cosmological constant that describes an expanding accelerated universe. Therefore, χ can be considered as an auxiliary scalar field. Also notice that solution (9.18) holds only for $f_0 = 0$, so that the function reduces to $f(R, \chi) = f_1 (\chi)^k$. Moreover, a phase-space view and conformal analysis, as that in [306], gives the conditions for the onset and duration of inflation, which, specifically, depends on the sign and the value of f_1. This fact restricts the set of initial conditions capable of providing satisfactory inflationary cosmology as discussed in [306].

The second function in Eq. (9.16) yields exactly the Starobinsky Lagrangian, that is,

$$\mathcal{L} = 6(f_0 + 2 f_1 R) a \dot{a}^2 + 12 f_1 a^2 \dot{a} \dot{R} - f_1 a^3 R^2, \tag{9.19}$$

since the linear contribution $f_1 \chi$ is a boundary term that does not affect the equations of motion. Cosmological dynamics provided by Eq. (9.19) is discussed in Section 7.1.

The fourth solution of (9.12) can be easily addressed to the third one by setting $f_1 = 0$. As a consequence of this imposition, the only solution provided is

$$a(t) = a_0\sqrt{t}, \quad R(t) = 0, \tag{9.20}$$

which is particularly worth noting in the context of String Theory, since it represents a part of the so-called a_3 anomaly [307]. Finally, no solution occurs when $f_1 \neq 0$, so that Eq. (9.20) concludes the analysis of the cosmological dynamics provided by Noether symmetries. Notice that, generally, in the limit of the classical theory, higher-derivative Lagrangians give rise to superfluous degrees of freedom. Some authors [308] discuss the possibility that only solutions that are expandable in powers of the Planck constant $\hbar$ are self-consistent, others [309] consider such superfluous degrees of freedom as oscillation phases around the Friedman background. In any case, solutions provided by Noether's approach are consistent at the cosmological level, since extra terms in the effective gravitational action can be recovered by asking for symmetries of Lagrangian (9.9).

9.2 $f(T, \Box T)$ Gravity

A further extension of TEGR can come from considering higher-order derivative terms in the torsion scalar as $\Box T$. In principle, the expression of the operator $\Box$ is different from the one in GR, due to the different form of connections. In this case, the Weitzenböck connection introduced in Section 4.2 yields

$$\begin{aligned}\Box T \equiv Y = \nabla_\mu \nabla^\mu T = \nabla_\mu \partial^\mu T = \partial_\mu \partial^\mu T + \Gamma^\rho_{\mu\rho}\partial^\mu T \\ = \ddot{T} + \Gamma^\rho_{0\rho}\dot{T} = \ddot{T} + 3\left(\frac{\dot{a}}{a}\right)\dot{T}\,.\end{aligned} \tag{9.21}$$

Note that despite the different form of the connection, the d'Alembert operator has the same expression as in the corresponding metric theory.

Thanks to Eq. (9.21), we can use the Lagrange multipliers method in order to get the point-like Lagrangian. To develop the approach, we treat the torsion T and its d'Alembertian $Y = \Box T$ as separated fields, although they are related through Eq. (9.21). Let us start by writing the general action

$$S = \int h f(T, Y)\, d^4x \tag{9.22}$$

so that, after integrating the three-dimensional hypersurface, the Lagrange multiplier gives

$$S = 2\pi^2 \int \left\{ a^3 f(T, Y) - \lambda_1 \left(T + 6\frac{\dot{a}^2}{a^2} \right) + \lambda_2 \left(Y - \ddot{T} - 3\frac{\dot{a}}{a}\dot{T} \right) \right\} dt\,. \tag{9.23}$$

By varying the action with respect to T and Y, we find that

$$\lambda_1 = a^3 \frac{\partial f(T, Y)}{\partial T} \tag{9.24}$$

and

$$\lambda_2 = a^3 \frac{\partial f(T, Y)}{\partial Y} . \tag{9.25}$$

Substituting Eqs. (9.24) and (9.25) into the action (9.23) and integrating the second-order time derivatives, we finally get

$$\begin{aligned}\mathcal{L} = & \, a^3 \left(f - T\frac{\partial f}{\partial T} - Y\frac{\partial f}{\partial Y} \right) \\ & - 6a\dot{a}^2 \frac{\partial f}{\partial T} - a^3 \dot{T}\dot{Y}\frac{\partial^2 f}{\partial Y^2} - a^3 \dot{T}^2 \frac{\partial^2 f}{\partial T \partial Y} .\end{aligned} \tag{9.26}$$

Note that by setting $\frac{\partial f}{\partial Y} = 0$, we recover the Lagrangian (7.29), discussed in Section 7.2. The EL equation with respect to Y provides the cosmological expressions of T and Y, while the EL equation related to the torsion gives the constraint $\Box f_Y = 0$. The equation with respect to the scale factor, together with the energy condition, provide the dynamics of the variables in the minisuperspace $\mathcal{S} \equiv \{a, T, Y\}$ whose tangent space is $\mathcal{TS} \equiv \{a, \dot{a}, T, \dot{T}, Y, \dot{Y}\}$. The complete system of differential equations read as

$$\begin{aligned}& 4a\ddot{a}f_T(T, Y) + 4a\dot{a}\dot{Y}f_{TY}(T, Y) + 4a\dot{a}\dot{T}f_{TT}(T, Y) \\ & -a^2 \dot{Y}\dot{T}f_{YY}(T, Y) - a^2 \dot{T}^2 f_{TY}(T, Y) - a^2 Y f_Y(T, Y) \\ & +a^2 f(T, Y) + 2\dot{a}^2 f_T(T, Y) - a^2 T f_T(T, Y) = 0, \\ & \\ & \Box f_Y = 0, \\ & \\ & \frac{\partial^2 f(T, Y)}{\partial Y^2} \left(Y - 3\frac{\dot{a}}{a}\dot{T} - \ddot{T} \right) + \frac{\partial^2 f(T, Y)}{\partial T \partial Y} \left(T + 6\frac{\dot{a}^2}{a^2} \right) = 0, \\ & \\ & 6\dot{a}^2 f_T(T, Y) + a^2 \dot{Y}\dot{T}f_{YY}(T, Y) + a^2 \dot{T}^2 f_{TY}(T, Y) \\ & -a^2 T f_T(T, Y) + a^2 f(T, Y) - a^2 Y f_Y(T, Y) = 0 .\end{aligned} \tag{9.27}$$

As before, the preceding equations cannot be solved until the function $f(T, Y)$ is selected. The application of the condition $L_X\mathcal{L} = 0$ to the Lagrangian (9.26) allows one to find the symmetries of the higher-order theory $f(T, Y)$. The minisuperspace of configurations contains three variables, so that Noether's vector has the form

$$X = \alpha\partial_a + \beta\partial_T + \gamma\partial_Y + \dot{\alpha}\partial_{\dot{a}} + \dot{\beta}\partial_{\dot{T}} + \dot{\gamma}\partial_{\dot{Y}}. \tag{9.28}$$

Imposing the condition $X\mathcal{L} = 0$ and equating to zero the coefficients of $\dot{a}^2$, $\dot{T}\dot{Y}$, $\dot{T}^2$, $\dot{a}\dot{T}$, $\dot{a}\dot{Y}$, and $\dot{Y}^2$, we get a system of seven differential equations:

$$\begin{cases} 3\alpha\left(f - T\dfrac{\partial f}{\partial T} - Y\dfrac{\partial f}{\partial Y}\right) - \beta a T\dfrac{\partial^2 f}{\partial T^2} \\ -\beta a Y\dfrac{\partial^2 f}{\partial T\partial Y} - \gamma a T\dfrac{\partial^2 f}{\partial T\partial Y} - \gamma a Y\dfrac{\partial^2 f}{\partial Y^2} = 0, \\ \left(\alpha + 2a\dfrac{\partial\alpha}{\partial a}\right)\dfrac{\partial f}{\partial T} + \beta a\dfrac{\partial^2 f}{\partial T^2} + \gamma a\dfrac{\partial^2 f}{\partial T\partial Y} = 0, \\ \left(3\alpha + a\dfrac{\partial\gamma}{\partial Y} + a\dfrac{\partial\beta}{\partial T}\right)\dfrac{\partial^2 f}{\partial Y^2} + \beta a\dfrac{\partial^3 f}{\partial T\partial Y^2} + \gamma a\dfrac{\partial^3 f}{\partial Y^3} = 0, \\ \left(3\alpha + 2a\dfrac{\partial\beta}{\partial T}\right)\dfrac{\partial^2 f}{\partial T\partial Y} + \beta a\dfrac{\partial^3 f}{\partial T^2\partial Y} \\ +\gamma a\dfrac{\partial^3 f}{\partial T\partial Y^2} + a\dfrac{\partial\gamma}{\partial T}\dfrac{\partial^2 f}{\partial Y^2} = 0, \\ 12\dfrac{\partial\alpha}{\partial T}\dfrac{\partial f}{\partial T} + a^2\dfrac{\partial\gamma}{\partial a}\dfrac{\partial^2 f}{\partial Y^2} + 2a^2\dfrac{\partial\beta}{\partial a}\dfrac{\partial^2 f}{\partial T\partial Y} = 0, \\ 12\dfrac{\partial\alpha}{\partial T}\dfrac{\partial f}{\partial T} + a^2\dfrac{\partial\beta}{\partial a}\dfrac{\partial^2 f}{\partial Y^2} = 0, \\ \dfrac{\partial\beta}{\partial Y} = 0\,, \end{cases} \tag{9.29}$$

whose nontrivial solutions are

$$\begin{cases} \alpha = 0 \\ \beta = \beta_0 \\ \gamma = \beta_0\dfrac{Y}{T} \\ f_I(T, Y) = f_0\, T^{1-k}Y^k, \end{cases} \qquad \begin{cases} \alpha = -\dfrac{1}{3}\gamma_0 k Y a^{-\frac{1}{2}} \\ \beta = \beta_0 a^{-\frac{3}{2}} \\ \gamma = \gamma_0(Y)^2 a^{-\frac{3}{2}} \\ f_{II}(T, Y) = f_0(Y)^k + f_1\, T, \end{cases}$$

$$\begin{cases} \alpha = -\dfrac{1}{3}\gamma_0 k a Y\chi(a) \\ \beta = \beta_0\chi(a) \\ \gamma = \gamma_0(Y)^2 \\ f_{III}(T, Y) = f_0(Y)^k, \end{cases} \qquad \begin{cases} \alpha = \alpha_0 a^{1-\frac{3}{2k}} \\ \beta = -\dfrac{3\alpha_0}{k}Ta^{-\frac{3}{2k}} \\ \gamma = \gamma(a, T, Y) \\ f_{IV}(T, Y) = f_0\, T^k + f_1\, Y. \end{cases} \tag{9.30}$$

The first function is the only one made of a product of T and Y. It is particularly interesting since the product between the two variables can be seen as a modified $f(T)$ theory of gravity non-minimally coupled to an effective scalar field. Moreover, it is the generalization of Eq. (9.14) to any k. In the limit $k = 1/2$, it reduces to

$$f(T, Y) = f_0\sqrt{TY}, \tag{9.31}$$

which is formally equivalent to Eq. (9.14) but dynamically different. For general k the point-like Lagrangian can be found by replacing the function $f_I(T, Y)$ into Eq. (9.26), obtaining

$$\mathcal{L} = f_0(k-1)\left\{6a\dot{a}^2\left(\frac{Y}{T}\right)^k - ka^3\dot{T}\dot{Y}\left(\frac{Y}{T}\right)^{k-2}\frac{1}{T}\right.$$
$$\left. + ka^3\dot{T}^2\left(\frac{Y}{T}\right)^{k-1}\frac{1}{T}\right\}. \tag{9.32}$$

The EL equations can be analytically solved for $k \geq 2$, providing a de Sitter-like cosmological solution of the form

$$a(t) = a_0 e^{nt}, \quad T(t) = -6n^2, \quad Y(t) = 0, \tag{9.33}$$

with n being an arbitrary real number. In the second and fourth solutions of (9.30), the functions consist of a sum between powers of the two variables T and $\Box T$. Specifically, the latter leads to the same Lagrangian as $f(T)$ gravity, since the additive contribution of the higher-order term $\Box T$ does not introduce any further degree of freedom: in such a case, also the generator coefficients are, in turn, the same as those found in Section 7.2.

It is worth pointing out the difference with respect to $f(R,\Box R)$ gravity. Noether's system (9.29) provides a more general class of solutions with respect to the corresponding metric case. Consequently, the teleparallel equivalent is easier to handle, and analytic solutions can be found. This is due to the deep difference occurring between $f(R)$ and $f(T)$ gravity: as the former is a fourth-order theory with respect to the metric, the latter yields second-order field equations with respect to tetrad fields. This allows one to simplify the dynamics and obtain more physically viable models.

10

Scalar-Tensor Theories of Gravity

Not only is the Universe stranger than we think,
it is stranger than we can think.

— *Werner Karl Heisenberg*
(*Across the Frontiers, Harper & Row* [1974])

Despite the need for extending/modifying GR, self-interaction potentials, couplings, and kinetic terms give rise to infinite choices, which can lead to a frustrating indetermination in fitting observations and addressing conceptual problems. One can adjust models and parameters to match single datasets and phenomena, but a theory in agreement with the whole phenomenology seems far from being achieved. In other words, any single theory loses its general predictive power and cannot be used to reproduce a self-consistent cosmic history, starting from UV to IR scales. Therefore, some selection criteria, based on physical requirements, are needed to discriminate among the plethora of modified scalar-tensor gravities. These criteria can be based, for example, on symmetries, conservation laws, and on general physical motivations.

Here, we want to consider scalar fields non-minimally coupled with different geometric invariants, in particular the Ricci scalar R, the torsion scalar T, and the Gauss–Bonnet scalar $\mathcal{G}$. The aim is to demonstrate that all these non-minimally coupled invariants can give rise to consistent cosmological dynamics once we know how their conservation laws. Furthermore, these scalar-tensor theories can be dealt under the standard of the Noether Symmetry Approach, which allows one to fix couplings, potentials, and kinetic terms by requiring the existence of symmetries and related conserved quantities. In other words, the purpose is to compare, by the Noether symmetries, the dynamics of different actions, containing scalar fields non-minimally coupled to different geometric invariants, pointing out the possible equivalence of the various representations of gravity.

Specifically, in the starting action, we have the coupling $f(\phi)$, the kinetic term coefficient $u(\phi)$ and the potential $V(\phi)$. The function $f(\phi)$ can be non-minimally coupled to R, T, and $\mathcal{G}$ scalars. It is worth pointing out that analogue (or identical) features emerge if couplings, kinetic terms, and potentials are selected by the Noether symmetries. The final result of this study is that all the previously mentioned theories provide the same dynamics and the same solutions. This is a further proof of the equivalence between the Gauss–Bonnet models and $f(R)$ gravity, where GR is a particular case.

10.1 The Curvature Scalar Coupled to a Scalar Field

A scalar-tensor action, written in terms of the scalar curvature, reads as

$$S = \int \sqrt{-g}\left\{f(\phi)\,R + u(\phi)g^{\mu\nu}\phi_{;\mu}\phi_{;\nu} - V(\phi)\right\}d^4x. \tag{10.1}$$

The variation of the action, with respect to the metric tensor $g_{\mu\nu}$, provides the field equations [310]

$$\begin{aligned} &R_{\mu\nu}g(\phi) - \frac{1}{2}g_{\mu\nu}\left[Rf(\phi) + u(\phi)\phi_{;\alpha}\phi^{;\alpha} - V(\phi)\right] \\ &-f(\phi)_{;\mu;\nu} + g_{\mu\nu}\Box f(\phi) + u(\phi)\phi_{;\mu}\phi_{;\nu} = 0\,, \end{aligned} \tag{10.2}$$

that clearly reduce to the Einstein equations when $f(\phi) = \text{const.}$ and $V(\phi) = u(\phi) = 0$. Action (10.1) is the paradigm for a very large class of theories. For example, $f(R)$ gravity in metric formalism can be easily recovered from (10.1) if $u(\phi)$ is set to zero and [91, 155]

$$V(\phi) = \frac{1}{2}\frac{f(R) - Rf_R}{[f_R]^2}\,. \tag{10.3}$$

As mentioned in Section 4.2 of Chapter 4, it turns out that a second-order theory of gravity, non-minimally coupled to a scalar field, can always be recast as a fourth-order theory. Cosmological solutions considered in this chapter, therefore, can be addressed to $f(R)$, $f(T)$, and $f(\mathcal{G})$ cosmology, respectively,

Let us start our analysis by considering the action (10.1) where the form of unknown functions will be fixed by the Noether symmetries. After selecting these functions, we shall find analytic cosmological solutions thanks to the reduction of dynamics.

From the variation of the action with respect to the scalar field ϕ, we get the Klein–Gordon equation

$$u_\phi(\phi)\phi^{;\alpha}\phi_{;\alpha} + 2u(\phi)\Box\phi - RF_\phi(\phi) + V_\phi(\phi) = 0, \tag{10.4}$$

where the subscript ϕ denotes the derivative with respect to ϕ. Together with Eqs. (10.2), it completes the set of field equations. The action can be simplified by focusing on a spatially flat FLRW metric and integrating over the 3-D surface term. Moreover, replacing the cosmological expression of the Ricci scalar into the action and integrating out second-order derivatives, we get the cosmological point-like Lagrangian

$$\mathcal{L} = 6f(\phi)a\dot{a}^2 + 6f_\phi(\phi)a^2\dot{a}\dot{\phi} + a^3u(\phi)\dot{\phi}^2 - a^3V(\phi). \tag{10.5}$$

As standard when the Lagrangian approach is considered, field equations and the Klein–Gordon equation result in the EL equations along with the energy condition $E_{\mathcal{L}} = 0$, that is, the 00 equation. Therefore, the dynamical system turns out to be

$$\begin{cases} 2f\dot{a}^2 + 4a\left(f_\phi\dot{a} + f\ddot{a}\right) + a^2\left(V - u\dot{\phi}^2 + 2f_{\phi\phi}\dot{\phi}^2 + 2f_\phi\ddot{\phi}\right) = 0 \\ 6\dot{a}^2 f_\phi + 6au\dot{a}\dot{\phi} + a\left[6f_\phi\ddot{a} + a\left(V_\phi + \dot{\phi}^2 w_\phi + 2w\ddot{\phi}\right)\right] = 0 \\ 6f\dot{a}^2 + 6af_\phi\dot{\phi}\dot{a} + a^2\left(V + w\dot{\phi}^2\right) = 0 \end{cases} \tag{10.6}$$

and can be solved after the three functions of ϕ are selected through the Noether symmetries. The approach can be developed in the two-dimensional minisuperspace $\mathcal{S} \equiv \{a, \phi\}$ whose corresponding symmetry generator is

$$\mathcal{X} = \xi(a,\phi,t)\partial_t + \alpha(a,\phi,t)\partial_a + \beta(a,\phi,t)\partial_\phi. \tag{10.7}$$

After equating to zero terms containing the same time derivatives of variables, the application of Noether's identity (6.42) to Lagrangian (10.5) provides a system of 10 differential equations. Nevertheless, by imposing a priori the condition $\xi = \xi(t)$ (holding for Lagrangians in canonical forms) and neglecting redundant equations, the system reduces to four differential equations plus the condition on the infinitesimal generators:

$$\begin{cases} \alpha - \beta V_\phi + V\partial_t\xi = 0 \\ \beta\left[2f_\phi V_\phi^2 + V^2 u_\phi - V\left(uV_\phi + 2f_\phi V_{\phi\phi}\right)\right] \\ -2V\left[Vu\left(\partial_t\xi - \partial_\phi\beta\right) + f_\phi V_\phi\partial_\phi\beta\right] = 0 \\ \beta\left[2fV_\phi^2 + 3V^2 f_{\phi\phi} - V(3f_\phi V_\phi + 2fV_{\phi\phi})\right] \\ +V^2\left[f_\phi\left(-6\partial_t\xi + 3\partial_\phi\beta\right) + au\partial_a\beta\right] - VV_\phi\left(2f\partial_\phi\beta + af_\phi\partial_a\beta\right) = 0 \\ 3\beta\left(Vf_\phi - fV_\phi\right) + 3aVf_\phi\partial_a\beta - 2f\left(3V\partial_t\xi + uV_\phi\partial_a\beta\right) = 0 \\ \alpha = \alpha(a,\phi), \quad \beta = \beta(a,\phi), \quad \xi = \xi(t). \end{cases} \tag{10.8}$$

Such a system is clearly overdetermined and cannot provide any explicit form without imposing some constraint. Since we want to investigate functions with physical meaning for cosmology, we substitute into the system both power-law and exponential potentials, so that it provides the following solutions [311]:

$$
\begin{cases}
\mathcal{X} = (\xi_0 t + \xi_1)\,\partial_t - \dfrac{\xi_0}{3} a \left(\dfrac{k+c}{k-c}\right)\partial_a + \dfrac{2\xi_0}{k-c}\phi\partial_\phi \\
f(\phi) = f_0\phi^k, \quad u(\phi) = u_0\phi^{k-2}, \quad V(\phi) = V_0\phi^c, \quad k \neq c \\
\\
\mathcal{X} = \dfrac{k\alpha_0}{3} a^{-\frac{1}{5}}\partial_a + \alpha_0 a^{-\frac{6}{5}}\partial_\phi \\
f(\phi) = f_0\phi^k, \quad u(\phi) = u_0\phi^{k-2}, \quad V(\phi) = V_0\phi^k, \quad k = \pm\sqrt{\dfrac{2u_0}{3f_0}} \\
\\
\mathcal{X} = (\xi_0 t + \xi_1)\,\partial_t - \dfrac{\xi_0}{3} a \left(\dfrac{k+c}{k-c}\right)\partial_a + \dfrac{2\xi_0}{k-c}\partial_\phi \\
f(\phi) = f_0 e^{k\phi}, \quad u(\phi) = u_0 e^{k\phi}, \quad V(\phi) = V_0 e^{c\phi}, \quad k \neq c \\
\\
\mathcal{X} = -\dfrac{k\alpha_0}{3} a^{-\frac{1}{5}} e^{-\frac{k u_0 \phi}{f_0 k^2 + u_0}}\partial_a + \alpha_0 a^{-\frac{6}{5}} e^{-\frac{k u_0 \phi}{f_0 k^2 + u_0}}\partial_\phi \\
f(\phi) = f_0 e^{k\phi}, \quad u(\phi) = u_0 e^{k\phi}, \quad V(\phi) = V_0 e^{k\phi}, \quad k = \pm\sqrt{\dfrac{2u_0}{3f_0}},
\end{cases}
\tag{10.9}
$$

with $\alpha_0, \beta_0, \xi_0, k, c, s, u_0, f_0, V_0$ real constants. We neglect trivial solutions, such as constant coupling or vanishing potential. It is worth noticing that the exponential potential also leads to exponential coupling and exponential kinetic terms like in string–dilaton cosmology [312–314]. This means that the string–dilaton Lagrangian can be naturally obtained from Noether symmetries, as pointed out in Section 18.1. From this point of view, solutions occurring in Eq. (10.9) can be considered more general than those provided in [313], since both the solutions outlined there by the authors are contained in the last two of Eq. (10.9). Inserting the above functions into the EL equations, the dynamics is reduced and equations of motion can be analytically solved.

Let us first solve the EL equations (10.6) for cases corresponding to the first and the third solutions of Eq. (10.9). In the former case, Lagrangian (10.5) is

$$
\mathcal{L} = 6f_0\phi^k a\dot{a}^2 + 6f_0 k\phi^{k-1} a^2\dot{a}\dot{\phi} + a^3 u_0\phi^{k-2}\dot{\phi}^2 - a^3 V_0\phi^c, \tag{10.10}
$$

and the corresponding equations of motion can be analytically solved with the constraint $k = c = 2$, providing a de Sitter-like expansion of the following form:

$$
a(t) = a_0 e^{qt}, \quad \phi(t) = \phi_0 \exp\left\{\frac{1}{2}\left[-3q \pm \sqrt{\frac{-48f_0 q^2 - 4V_0 + 9q^2 u_0}{u_0}}\right] t\right\}, \tag{10.11}
$$

with $q, a_0, \phi_0 \in \mathbb{R}$. Considering the third case of (10.9), the point-like Lagrangian (10.5) can be written as

$$
\mathcal{L} = 6f_0 e^{k\phi} a\dot{a}^2 + 6f_0 k e^{k\phi} a^2\dot{a}\dot{\phi} + a^3 u_0 e^{k\phi}\dot{\phi}^2 - a^3 V_0 e^{c\phi} \tag{10.12}
$$

and, even in this case, the equations of motion set the value of the parameter k and c, introducing the further constraint $k = c$. Therefore, discarding the solutions with minimal coupling, we find

$$a(t) = a_0 e^{qt}, \quad \phi(t) = \frac{3f_0 kq \pm \sqrt{(3f_0 kq)^2 - 6f_0 u_0 q^2 - V_0 u_0}}{u_0}, \quad q \in \mathbb{R}. \tag{10.13}$$

The values of the constants f_0, u_0, V_0 can be fixed according to cosmological observations [315].

In what follows we focus on the particular cases admitting a constant function $u(\phi)$. Besides being physically relevant, several other solutions of Noether's system can be found. This is due to the fact that the system is further reduced and analytic solutions occur even without performing any additional ansatz on the coupling and the potential. We start by setting $u(\phi) = 1/2$ in the action (10.1), so that the latter is [316]

$$S = \int \sqrt{-g} \left[f(\phi) R + \frac{1}{2} g^{\mu\nu} \partial_\mu \phi \partial_\nu \phi - V(\phi) \right] d^4x. \tag{10.14}$$

Notice that the general action (10.1) can be recast in terms of Eq. (10.14) by an appropriate change of coordinates. Specifically, the scalar field, the coupling function, and the potential can be always reparameterized such that the kinetic term coefficient is constant. Though cosmological solutions coming from Eq. (10.1) can be related to those of Eq. (10.14), Noether's system provides a different set of symmetries. Substituting the cosmological expression of the scalar curvature into the action (10.14) and integrating out the boundary term, we get the point-like Lagrangian

$$\mathcal{L} = 6a\dot{a}^2 f(\phi) + 6a^2 \dot{a}\dot{\phi} f_\phi(\phi) + a^3 \left[\frac{\dot{\phi}^2}{2} - V(\phi) \right]. \tag{10.15}$$

The EL equations and the energy condition can be obtained from Eq. (10.6) by setting $u(\phi) = 1/2$. They read

$$\begin{cases} 4f(\phi)\dot{a}^2 + 8a \left[\dot{a}\dot{\phi} f_\phi(\phi) + f(\phi)\ddot{a} \right] \\ +a^2 \left[2V(\phi) - \dot{\phi}^2 + 4\dot{\phi}^2 f_{\phi\phi}(\phi) \right] + 4\ddot{\phi} f_\phi(\phi) = 0 \\ 6\dot{a}^2 f_\phi(\phi) + 3a\dot{a}\dot{\phi} + a \left[6f_\phi(\phi)\ddot{a} + aV_\phi(\phi) + a\ddot{\phi} \right] = 0 \\ 6f(\phi)\dot{a}^2 + 6a\dot{a}\dot{\phi} f_\phi(\phi) + a^2 \left[V(\phi) + \frac{1}{2}\dot{\phi}^2 \right] = 0. \end{cases} \tag{10.16}$$

The two-dimensional minisuperspace $\mathcal{S} \equiv \{a, \phi\}$ leads to the symmetry generator

$$\mathcal{X} = \xi(t)\partial_t + \alpha(a, \phi, t)\partial_a + \beta(a, \phi, t)\partial_\phi, \tag{10.17}$$

where the infinitesimal generator ξ is assumed to depend on the cosmic time only. Noether's identity (6.42) can be applied to canonical Lagrangian (10.15), providing the following system of differential equations [317]:

$$
\begin{cases}
f(\phi)\alpha + a\beta f_\phi(\phi) + 2a\partial_a\alpha f(\phi) + a^2\partial_a\beta f_\phi(\phi) - af(\phi)\partial_t\xi = 0 \\
\frac{3}{2}a^2\alpha + 6a^2 f_\phi(\phi)\partial_\phi\alpha + a^3\partial_\phi\beta - \frac{1}{2}a^3\partial_t\xi = 0 \\
12a\alpha f_\phi(\phi) + 6a^2\beta f_{\phi\phi}(\phi) + 12af(\phi)\partial_\phi\alpha \\
+ 6a^2 f_\phi(\phi)\partial_a\alpha + 6a^2 f_\phi(\phi)\left(\partial_\phi\beta - \partial_t\xi\right) + a^3\partial_a\beta = 0 \\
2af(\phi)\partial_t\alpha + f_\phi(\phi)a^2\partial_t\beta = \partial_a g \\
6a^2 f_\phi(\phi)\partial_t\alpha + a^3\partial_t\beta = \partial_\phi g \\
-3a^2\alpha V(\phi) - a^3\left[\beta V_\phi(\phi) + V(\phi)\partial_t\xi\right] = \partial_t g
\end{cases}
\tag{10.18}
$$

whose solutions are

$$
\begin{cases}
I\colon \mathcal{X} = -\frac{2}{3}\alpha_0 a\partial_a + \alpha_0\phi\partial_\phi, \qquad g = 0 \\
f(\phi) = f_0\phi^2, \quad V(\phi) = V_0\phi^2 \\
j = \alpha_0\left[4f_0\phi^2 a^2\dot{a} + (1 - 8f_0)a^3\phi\dot{\phi}\right] \\
\\
II\colon \mathcal{X} = -\frac{2(s+1)}{2s+3}\beta_0 a^{s+1}\phi^{\frac{2s^2+4s}{2s+3}}\partial_a + \beta_0 a^s\phi^{\frac{2s^2+6s+3}{2s+3}}\partial_\phi, \qquad g = 0 \\
f(\phi) = \ell(s)\phi^2, \quad V(\phi) = V_0\phi^{\frac{6(s+1)}{2s+3}}, \qquad \ell(s) = \frac{(2s+3)^2}{48(s+1)(s+2)} \\
j = a^{s+2}\phi^{\frac{2s^2+6s+3}{2s+3}}\left[(2s+3)\phi\dot{a} + 2(s+1)a\dot{\phi}\right], \quad s \neq 0 \\
\\
III\colon \mathcal{X} = (\alpha_0 t + \xi_1)\partial_t + \alpha_0\frac{s+2}{3(s-2)}a\partial_a - \frac{2\alpha_0}{s-2}\phi\partial_\phi, \qquad g = g_0 \\
f(\phi) = f_0\phi^2, \quad V(\phi) = V_0\phi^s \\
j = -g_0(s-2) + 4f_0(s-4)a^2\phi^2\dot{a} + 2a^3\phi\dot{\phi}\left[2f_0(s+2) - 1\right].
\end{cases}
\tag{10.19}
$$

Notice that two internal solutions occur [316]. They can also arise by imposing the vanishing Lie derivative condition to Lagrangian (10.15). In order not to lose generality, we consider the coupling $f(\phi) = f_0\phi^2$ and the potential $V(\phi) = V_0\phi^k$, since all the listed solutions can be related to this case. In particular, solution I can be recovered for $k = 2$, solution II for $f_0 = \frac{(2s+3)^2}{48(s+1)(s+2)}$, and $k = \frac{6(s+1)}{2s+3}$, solution III for $k = s$. See [316] for details. Therefore, when the coupling and the potential are chosen as $f(\phi) = f_0\phi^2$ and $V(\phi) = V_0\phi^k$, the Lagrangian reads as

$$
\mathcal{L} = a\left[6f_0\phi^2\dot{a}^2 + 12f_0 a\phi\dot{a}\dot{\phi} + \frac{1}{2}a^2\left(-2V_0\phi^k + \dot{\phi}^2\right)\right]
\tag{10.20}
$$

and yields the EL equations

$$
\begin{cases}
4f_0\phi^2\dot{a}^2 + 8f_0 a\phi\left(2\dot{a}\dot{\phi} + \phi\ddot{a}\right) + a^2\left[2V_0\phi^k + (8f_0 - 1)\dot{\phi}^2 + 8f_0\phi\ddot{\phi}\right] = 0 \\
12f_0\phi\dot{a}^2 + 3a\left(\dot{a}\dot{\phi} + 4f_0\phi\ddot{a}\right) + a^2\left(kV_0\phi^{k-1} + \ddot{\phi}\right) = 0 \\
6f_0 a\phi^2\dot{a}^2 + 12f_0 a^2\phi\dot{a}\dot{\phi} + \frac{1}{2}a^3\left(2V_0\phi^k + \dot{\phi}^2\right) = 0.
\end{cases}
\tag{10.21}
$$

The above system of differential equations can be analytically solved, providing two different sets of cosmological solutions:

$$a(t) = a_0 t^{\frac{k+2}{3(k-2)}}, \qquad \phi = \phi_0 t^{-\frac{2}{k-2}};$$
$$a(t) = a_0 e^{c_0(t-t_0)}, \qquad \phi = \phi_0. \tag{10.22}$$

However, Eq. (10.22) holds for $k \neq 2, 4$. For the cases $k = 2$ and $k = 4$, two other sets of solutions arise from EL equations (10.21). The former ($k = 2$) is

$$a(t) = (c_0 t + c_1)^{\frac{2}{3}} e^{\frac{2}{3}(4V_0 t^2 - c_2 t - c_3)}, \qquad \phi(t) = \frac{e^{-4V_0 t^2 + c_2 t + c_3}}{\sqrt{c_0 t + c_1}}. \tag{10.23}$$

In Ref. [317] the authors consider Eq. (10.23) and investigate the corresponding inflationary model by properly setting the free parameters. Furthermore, in Ref. [318], Eq. (10.23) is used to investigate the relation between the Hubble parameter and the cosmic time.

Eqs. (10.21) for $k = 4$ yield [318]

$$a(t) = a_0 e^{-\frac{c_1}{3}(t-t_0)}(e^{c_1(t-t_0)} + c_2)^{\frac{2}{3}}, \qquad \phi(t) = \phi_0 \sqrt{\frac{e^{c_0(t-t_0)}}{e^{c_1(t-t_0)} - 1}}. \tag{10.24}$$

Thus, once the values of the free parameters by observations have been set, the ϕ^4 potential can provide either an exponentially accelerating expansion of the universe or a bouncing cosmological model.

10.1.1 Generalization to Higher Dimensions

Let us generalize now the preceding considerations to scalar-tensor theories of gravity in D dimensions, applying the Noether Symmetry Approach and finding the corresponding cosmological solutions [314]. We also show that when the four-dimensional limit is considered, results provided in Section 10.1 are recovered. Starting from the line element (8.45), we use Noether's prescriptions to select suitable models that may represent realizations for higher-dimensional theories, such as Kaluza–Klein and String Theory.

Moreover, in the Einstein frame, the three-dimensional EH action is invariant under the local Poincaré group, thus studying the corresponding Jordan frame can be useful in view of a possible quantum description of the gravitational interaction. On the one hand, increasing the number of dimensions can be a fundamental step toward a unitary scheme for the fundamental interactions. Furthermore, scalar-tensor theories give rise to reliable inflationary behaviors that can be confronted with observations.

Let us then consider the following scalar-tensor action in $D = d+1$ dimensions:

$$S = \int \sqrt{-g}\left[Rf(\phi) + \frac{1}{2}\partial_\mu\phi\partial^\mu\phi - V(\phi)\right] d^{d+1}x. \tag{10.25}$$

Taking into account the D-dimensional expression of the Ricci scalar (8.46) in FLRW spatially flat interval (8.45), the D-dimensional action can be written as

$$S = \int a^d \left\{ -f(\phi)\left[2d\frac{\ddot{a}}{a} + d(d-1)\frac{\dot{a}^2}{a^2}\right] + \frac{1}{2}\dot{\phi}^2 - V(\phi)\right\} d^{d+1}x. \tag{10.26}$$

The point-like Lagrangian can be easily computed by integrating out the second derivatives and the d-dimensional hypersurface:

$$\mathcal{L} = d(d-1)a^{d-2}\dot{a}^2 f(\phi) + 2da^{d-1}\dot{a}\dot{\phi}f_\phi(\phi) + a^d\left[\frac{1}{2}\dot{\phi}^2 - V(\phi)\right], \tag{10.27}$$

which is the generalization of Eq. (10.15) to $D = d+1$ dimensions. The EL equations, both with the energy condition, read as

$$\begin{cases} 2(d-1)\frac{\ddot{a}}{a}f(\phi) + (d-1)(d-2)\left(\frac{\dot{a}}{a}\right)^2 f(\phi) + 2(d-1)\frac{\dot{a}}{a}\dot{\phi}f_\phi(\phi) \\ +2\ddot{\phi}f_\phi(\phi) + 2f_{\phi\phi}(\phi)\dot{\phi}^2 - \left[\frac{1}{2}\dot{\phi}^2 - V(\phi)\right] = 0 \\ \ddot{\phi} + d(d-1)\left(\frac{\dot{a}}{a}\right)^2 f_\phi(\phi) + 2d\frac{\ddot{a}}{a}f_\phi(\phi) + d\frac{\dot{a}}{a}\dot{\phi} + V_\phi(\phi) = 0 \\ d(d-1)\left(\frac{\dot{a}}{a}\right)^2 f(\phi) + 2d\frac{\dot{a}}{a}\dot{\phi}f_\phi(\phi) + \frac{\dot{\phi}^2}{2} + V(\phi) = 0. \end{cases} \tag{10.28}$$

Here we only search for internal symmetries by imposing the condition $L_X\mathcal{L} = 0$ to Lagrangian (10.27). Independently of the number of space-time dimensions, the minisuperspace is the same as Section 10.1, and the corresponding generator (10.17) can be applied to Lagrangian (10.27), providing the system

$$\begin{cases} (d-1)\left[(d-2)\alpha + 2a\partial_a\alpha\right]f + af_\phi\left[(d-1)\beta + 2a\partial_a\beta\right] = 0 \\ d\alpha + 4df_\phi\partial_\phi\alpha + 2a\partial_\phi\beta = 0 \\ a\beta f_{\phi\phi} + \left[(d-1)\alpha + a\partial_a\alpha + a\partial_\phi\beta\right]f_\phi + (d-1)\partial_\phi\alpha f + \frac{a^2}{2d}\partial_a\beta = 0 \\ d\alpha V + a\beta V_\phi = 0. \end{cases} \tag{10.29}$$

Note that, with respect to Eq. (10.18), two fewer equations occur. This is due to the fact that in the four-dimensional case, we considered the first prolongation of the Noether vector and the related identity (6.42), while here we are searching for internal symmetries only. Therefore, equations including derivatives with respect to the cosmic time disappear identically. Nonetheless, by setting $d = 3$ in Eq. (10.29), we fully recover the system (10.18) with $\xi = 0$. The general solution admits a coupling function proportional to an elliptical integral of the second kind, which generalizes the first two solutions of (10.19). In addition, three other solutions occur, one of which leads to a degenerate Lagrangian, that is:

$$f(\phi) = \frac{(d-1)}{8d}\phi^2. \tag{10.30}$$

Indeed, by equaling the Hessian determinant to zero, one gets

$$|He| = \left|\frac{\partial^2 \mathcal{L}}{\partial \dot{a}\partial \dot{\phi}}\right| = (d-1)f(\phi) - df_\phi(\phi)^2 = 0 \to f(\phi) = \frac{(d-1)}{8d}\phi^2. \tag{10.31}$$

The other solutions provide the following generators, couplings, and potentials:

$$\begin{cases} \mathcal{X}_I = -\frac{2}{d}p(d,s)\beta_0 a^{s+1}\phi^{m(d,s)-1}\partial_a + \beta_0 a^s \phi^{m(d,s)}\partial_\phi \\ j_0 = -\beta_0 a^{d+s-1}\phi^{m(s,d)} \\ \times \{4D(d,s)[1+(d+1)(p(d,s)-1) \\ \qquad -2p(d,s)]\phi\dot{a} + [8D(d,s)p(d,s)-1]a\dot{\phi}\} \\ f(\phi) = D(d,s)\phi^2, \qquad V(\phi) = V_0\phi^{2p(d,s)} \\ \\ \mathcal{X}_{II} = -\frac{2}{3}\beta_0 a\partial_a + \beta_0\phi\partial_\phi \\ j_0 = \beta_0\left[4f_0\phi^2 a^2\dot{a} + (1-8f_0)a^3\phi\dot{\phi}\right] \\ f(\phi) = f_0\phi^2, \qquad V(\phi) = V_0\phi^2, \end{cases} \tag{10.32}$$

where we have defined

$$p(d,s) = \frac{d(2s+d-1)}{(d-1)(d+2s)}, \qquad m(d,s) = \frac{4ds+4s^2+d^2-d-2}{(d-1)(2s+d)},$$

$$D(s,d) = \frac{(d-1)(2s+d)^2}{8d(2s+d-1)(2s+d+1)}. \tag{10.33}$$

Notice that both solutions are equivalent to the first and second of Eq. (10.19), for $d=3$.

Let us first consider solution I and find the related dynamics. Substituting coupling and potential into Eq. (10.28), we get a nonlinear system of differential equations, which cannot be solved without properly adopting some ansatz. In order to reduce dynamics, we consider Eq. (2.37) and introduce a cyclic variable z into the starting Lagrangian. By means of the first set of infinitesimal generators (10.32), the change of variables can be written as

$$\begin{cases} -\frac{2}{d}p(d,s)\beta_0 a^{s+1}\phi^{m(d,s)-1}\partial_a z + \beta_0 a^s\phi^{m(d,s)}\partial_\phi z = 1 \\ -\frac{2}{d}p(d,s)a\phi^{m(d,s)-1}\partial_a w + a\phi^{m(d,s)}\partial_\phi w = 0, \end{cases} \tag{10.34}$$

whose solution is

$$z = \frac{d}{\beta_0\chi(d,s)}a^{-s}\phi^{1-m(d,s)}, \qquad w = w_0 a^d\phi^{2p(d,s)}, \tag{10.35}$$

with w_0 being an integration constant and $\chi(d, s)$ defined as

$$\chi(d, s) = \frac{-4ds}{(d-1)(2s+d)}. \tag{10.36}$$

Lagrangian (10.27), written in terms of the new variables, turns out to be

$$\begin{aligned}
\mathcal{L} &= \gamma(d, s) w^{\frac{s}{d}} \dot{z}\dot{w} - V_0 w, \\
\gamma(d, s) &= \left[\frac{\zeta^2 \xi^d}{\chi^2}\right] \left[\frac{-2ds}{(2s+d+1)(2s+d-1)}\right], \\
\zeta &= \left(\frac{\beta_0^d \chi}{d w_0^s}\right)^{\frac{1}{\chi}}, \quad \xi = \left[\frac{1}{w_0 \zeta^{2p}}\right]^{\frac{1}{d}},
\end{aligned} \tag{10.37}$$

where cyclicity of z is evident. Eqs. (10.28) become

$$\begin{cases} \gamma(d, s) w^{\frac{s}{d}} \ddot{z} + V_0 = 0 \\ \gamma(d, s) w^{\frac{s}{d}} \dot{w} = j_0 \\ \gamma(d, s) w^{\frac{s}{d}} \dot{z}\dot{w} + V_0 w = 0. \end{cases} \tag{10.38}$$

Equations of motion can be straightforwardly solved, providing

$$\begin{aligned}
z(t) &= L_1 \left[k_1 t + k_2\right]^{\frac{s+2d}{s+d}} + L_0, \quad w(t) = \left[k_1 t + k_2\right]^{\frac{d}{s+d}}, \\
k_1 &= \left[\frac{s+d}{\gamma(d, s) d}\right] j_0, \quad k_2 = w_0^{\frac{s+d}{d}}, \\
L_0 &= z_0 - \frac{V_0 \gamma(d, s) d w_0^{2d+s+2}}{(2d+s+2) j_0^2}, \quad L_1 = \left(\frac{\lambda \gamma(d, s)(d+1)}{s+2d+2}\right) \frac{1}{j_0^2},
\end{aligned} \tag{10.39}$$

with z_0 integration constant and j_0 the conserved quantity. Coming back to the old variables, we finally obtain

$$\begin{aligned}
a(t) &= \xi \left[k_1 t + k_2\right]^{\frac{d+2s+1}{2(s+d)}} \left[L_1 \left(k_1 t + k_2\right)^{\frac{2d+s}{s+d}} + L_0\right]^{\frac{2s+d-1}{2s}}, \\
\phi(t) &= \zeta \left\{ \left[k_1 t + k_2\right]^{\frac{1}{(s+d)}} \left[L_1 \left(k_1 t + k_2\right)^{\frac{s+2d}{s+d}} + L_0\right]^{\frac{1}{s}} \right\}^{-\frac{(d+1)(2s+d)}{4}}.
\end{aligned} \tag{10.40}$$

Notice that Eq. (10.40) generalizes Eq. (17.73) to any dimensions and yields a time power-law scale factor at the late time, where the related scalar field converges only if $s \geq -(d+1)/2$. On the other hand, the $s = 0$ limit yields the string–dilaton cosmology action, by which the scale factor duality is recovered (see Section 18.1 for further details). Finally, by setting $s = -d-1$ from the beginning, EL equations provide another solution that, in terms of old variables, reads as

$$\begin{aligned}
a(t) &= a_0 e^{-\left(\frac{d+1}{2d}\right) k_1 t} \left[k_3 \left(e^{k_1 t} - 1\right) + k_4\right]^{\frac{c+1}{2d}}, \\
\phi(t) &= \left[-\frac{j_0^2}{V_0}\right]^{\frac{d-1}{4}} \left[\frac{1}{1 - e^{-j_0 t}}\right]^{\frac{d-1}{4}}.
\end{aligned} \tag{10.41}$$

Here, the Noether Symmetry Approach is applied to a spatially flat background. In Ref. [314], the generalization to open and closed universes is considered. As a consequence, Noether's system contains one further differential equation, whose solution provides power-law coupling and potential.

In the next subsection, we discuss the link between internal symmetries and conformal transformations in order to check whether the latter can be addressed to the former, preserving the Lie derivative along the flux of a given vector field X.

10.1.2 Conformal Transformations

As mentioned in Chapter 4, dynamics can be equivalently described either in the Jordan frame or in the Einstein frame. Here we want to show that the structure of the Lie derivative also is preserved by Noether's symmetry in the two frames. This means that symmetries in the Einstein frame can be reduced to symmetries in the Jordan frame, when conformal transformations are performed. As a consequence, the Lie derivative written in terms of the cosmic time can be related to that written in terms of the conformal time. To this purpose, let us consider the minisuperspace $\mathcal{S} \equiv \{\tilde{a}(\eta), \tilde{\phi}(\eta)\}$, with η being the conformal time $\eta = \int \frac{dt}{a}$. The related Noether vector is

$$\tilde{X}_\eta = \tilde{\alpha}\partial_{\tilde{a}} + \tilde{\beta}\partial_{\tilde{\phi}} + \tilde{\alpha}'\partial_{\tilde{a}'} + \tilde{\beta}'\partial_{\tilde{\phi}'}, \tag{10.42}$$

where the prime denotes the derivative with respect to η. The application of the Lie derivative to Lagrangian (10.15) yields [249]

$$\begin{aligned}
L_{\tilde{X}_\eta}\tilde{\mathcal{L}}_\eta &= 6\left[2f(\tilde{\phi})\partial_{\tilde{a}}\tilde{\alpha} + \left(\tilde{\beta} + \tilde{a}\partial_{\tilde{a}}\tilde{\beta}\right) f_{\tilde{\phi}}(\tilde{\phi})\right]\tilde{a}'^2 \\
&+ 6\left[\tilde{a}\tilde{\beta}f_{\tilde{\phi}\tilde{\phi}}(\tilde{\phi}) + \left(\tilde{\alpha} + \tilde{a}\partial_{\tilde{a}}\tilde{\alpha} + \tilde{a}\partial_{\tilde{\phi}}\tilde{\beta}\right) f_{\tilde{\phi}}(\tilde{\phi}) + 2f(\tilde{\phi})\partial_{\tilde{\phi}}\tilde{\alpha} + \frac{\tilde{a}^2}{6}\partial_{\tilde{a}}\tilde{\beta}\right]\tilde{a}'\tilde{\phi}' \\
&+ \tilde{a}\left[\tilde{\alpha} + 6f_{\tilde{\phi}}(\tilde{\phi})\partial_{\tilde{\phi}}\tilde{\alpha} + \tilde{a}\partial_{\tilde{\phi}}\tilde{\beta}\right]\tilde{\phi}'^2 - \tilde{a}^3\left[4\tilde{\alpha}\,V(\tilde{\phi}) + \tilde{a}\tilde{\beta}\,V_\phi(\tilde{\phi})\right].
\end{aligned} \tag{10.43}$$

Coming back to the cosmic time, the preceding equation can be recast as

$$\begin{aligned}
L_{X_\eta}\mathcal{L}_\eta &= 6a\left[\alpha f(\phi) + 2af(\phi)\partial_a\alpha + a\left(\beta + a\partial_a\beta\right) f_\phi(\phi)\right]\dot{a}^2 \\
&+ 6a^2\left[a\beta f_{\phi\phi}(\phi) + \left(2\alpha + a\partial_a\alpha + a\partial_\phi\beta\right) f_\phi(\phi) + 2f(\phi)\partial_\phi\alpha + \frac{a^2}{6}\partial_a\beta\right]\dot{a}\dot{\phi} \\
&+ a^3\left[\frac{3}{2}\alpha + 6f_\phi(\phi)\partial_\phi\alpha + a\partial_\phi\beta\right]\dot{\phi}^2 - a^3\left[3\alpha\,V(\phi) + a\beta\,V_\phi(\phi)\right] \\
&- 6\alpha f(\phi)a\dot{a}^2 - \frac{1}{2}\alpha a^3\dot{\phi}^2 + -6\alpha f_\phi(\phi)a^2\dot{a}\dot{\phi} - \alpha a^3\,V(\phi).
\end{aligned} \tag{10.44}$$

On the other hand, when the Lie derivative is directly applied to Lagrangian (10.15) in terms of the cosmic time, we have

$$L_{X_t}\mathcal{L}_t = 6\left[\alpha f(\phi) + 2af(\phi)\partial_a\alpha + a\left(\beta + a\partial_a\beta\right)f_\phi(\phi)\right]\dot{a}^2$$
$$+ 6a\left[a\beta f_{\phi\phi}(\phi) + \left(2\alpha + a\partial_a\alpha + a\partial_\phi\beta\right)f_\phi + 2f(\phi)\partial_\phi\alpha + \frac{a^2}{6}\partial_a\beta\right]\dot{a}\dot{\phi}$$
$$+ a^2\left[\frac{3}{2}\alpha + 6f_\phi(\phi)\partial_\phi\alpha + a\partial_\phi\beta\right]\dot{\phi}^2 - a^2\left[3\alpha V(\phi) + a\beta V_\phi(\phi)\right]. \tag{10.45}$$

By comparing Eq. (10.44) with Eq. (10.45), the following condition occurs:

$$L_{X_\eta}\mathcal{L}_\eta = aL_{X_t}\mathcal{L}_t - \left(L_{X_t}a\right)E_{\mathcal{L}(t)}, \tag{10.46}$$

with E_t defined as

$$E_{\mathcal{L}(t)} = \dot{a}\partial_{\dot{a}}\mathcal{L} + \dot{\phi}\partial_{\dot{\phi}}\mathcal{L} - \mathcal{L}. \tag{10.47}$$

The same relation holds in the Einstein frame, which means that conformal transformations preserve the condition of Eq. (10.46). In cosmology, where the energy condition is a component of the equations of motion, the last term of Eq. (10.46) vanishes, so that the existence of symmetries in the Einstein frame implies the existence of symmetries in the Jordan frame (and vice versa). In general, the quantities $E_{\mathcal{L}(\eta)}$ and $E_{\mathcal{L}(t)}$ satisfy the condition $E_{\mathcal{L}(\eta)} = aE_{\mathcal{L}(t)}$, which, when substituted into Eq. (2.60), provides

$$L_{X_\eta}\mathcal{L}_\eta = a\dot{g} + \left(a\dot{\xi} - L_{X_t}a\right)E_{\mathcal{L}(t)}. \tag{10.48}$$

See Ref. [249] for a detailed discussion.

10.2 Hybrid Gravity

In Section 4.5, we introduced the Palatini formalism, showing that the metric can be disentangled from the connection and the variational principle yields two sets of field equations. We also showed that the action (4.78) can be of particular interest in view of cosmological applications.[1] Indeed, the function $f(\mathcal{R})$ can be intended as an effective term mimicking dark energy, while GR is restored for $f(\mathcal{R}) = 0$. Considering the prescription outlined in Section 4.5, $\mathcal{R}$ can be recast in terms of a scalar field after performing a conformal transformation. As a final result, one gets the relation

$$\mathcal{R} = R + \frac{3}{2\phi}\partial_\alpha\phi\partial_\alpha\phi - \frac{3}{2}\Box\phi. \tag{10.49}$$

A suitable redefinition of the potential yields the so-called *Hybrid gravity* action, that is:

$$S = \frac{1}{2}\int\sqrt{-g}\left[(\phi+1)R + \frac{3}{2\phi}\partial^\alpha\phi\partial_\alpha\phi - V(\phi)\right]d^4x. \tag{10.50}$$

[1] R denotes the curvature scalar written in terms of the Levi–Civita connection, while $\mathcal{R}$ is the curvature scalar in terms of a general affine connection.

It is called *Hybrid* gravity because the affine and the metric terms are mixed through the scalar field ϕ. The purpose of this section is to apply the Noether symmetry condition to the preceding action, selecting suitable forms of the potential. Considering a FLRW metric and integrating the hypersurface, the point-like cosmological Lagrangian turns out to be

$$\mathcal{L} = 6a\dot{a}^2(1+\phi) + 6a^2\dot{a}\dot{\phi} + \frac{3}{2\phi}a^3\dot{\phi}^2 + a^3 V(\phi), \tag{10.51}$$

whose equations of motion are

$$\begin{cases} \ddot{a} + \dfrac{1-\phi}{2a}\dot{a}^2 - \dfrac{1}{2}\dot{a}\dot{\phi} - \dfrac{a}{3\phi}\dot{\phi}^2 - \dfrac{1}{12}a\left(3V - 2\phi V_\phi\right) = 0 \\ \ddot{\phi} + \dfrac{\phi(\phi+1)}{a^2}\dot{a}^2 + \dfrac{\phi+3}{a}\dot{a}\dot{\phi} + \dfrac{\phi-2}{4\phi}\dot{\phi}^2 + \dfrac{\phi}{6}\left[3V(\phi) - 2(\phi+1)V_\phi\right] = 0 \\ 6a\dot{a}^2(1+\phi) + 6a^2\dot{a}\dot{\phi} + \dfrac{3}{2\phi}a^3\dot{\phi}^2 - a^3 V(\phi) = 0. \end{cases} \tag{10.52}$$

In the two-dimensional minisuperspace $\mathcal{S} \equiv \{a, \phi\}$, the symmetry generator takes the form

$$\mathcal{X} = \xi(t)\partial_t + \alpha(a,\phi,t)\partial_a + \beta(a,\phi,t)\partial_\phi, \tag{10.53}$$

so that the identity (6.42) leads to the system

$$\begin{cases} 12a\partial_t\alpha\,(1+\phi) + 6a^2\partial_t\beta - \partial_a g = 0 \\ 4\phi\alpha + 4\phi(1+\phi)\partial_\phi\alpha + 2a\phi\left(-\partial_t\xi + \partial_\phi\beta + \partial_a\alpha\right) + a^2\partial_a\beta = 0 \\ \alpha(1+\phi) + a\left[\beta + (1+\phi)(2\partial_a\alpha\partial_t\xi) + a\partial_a\beta\right] = 0 \\ 6a^2\phi\partial_t\alpha - \phi\partial_\phi g + 3a^3\partial_\phi\beta = 0 \\ 3\alpha\phi + 4\phi^2\partial_\phi\alpha - a\beta - a\phi\left(\partial_t\xi - 2\partial_\phi\beta\right) = 0 \\ 3\alpha a^2 + \partial_t g + a^3\left[\beta V_\phi(\phi) + \partial_t\xi\, V(\phi)\right] = 0 \\ \xi = \xi(t). \end{cases} \tag{10.54}$$

The preceding equations only provide $V(\phi) = V_0$, with generator and conserved quantity of the form

$$\begin{cases} \mathcal{X} = \dfrac{\sqrt{\phi}}{a}\partial_\phi \\ j = 3\dfrac{a}{\sqrt{\phi}}(2\phi\dot{a} + a\dot{\phi}). \end{cases} \tag{10.55}$$

The latter can be used to properly introduce a cyclic variable z into the Lagrangian. To this purpose, we define

$$a = w^{\frac{2}{3}}, \qquad \phi = z^2 w^{-\frac{4}{3}}, \tag{10.56}$$

so that the Lagrangian takes the form

$$\mathcal{L}(w, \dot{w}, \dot{z}) = \frac{8}{3}\dot{w}^2 + 6w^{\frac{2}{3}}\dot{z}^2 + V_0 w^2 \tag{10.57}$$

and the equation of motion provides

$$\int \frac{dw}{\sqrt{\frac{3}{8}V_0 w^2 - \frac{9}{4}j'^2 w^{-\frac{2}{3}}}} = \int dt, \tag{10.58}$$

with j' being the conserved quantity in terms of new variables: $j' = \dot{z}w^{\frac{2}{3}}$. Notice that, since (10.55) is an internal solution, the change of variables can be equivalently pursued by following the methodical procedure in Eq. (2.37). An analytic solution of Eq. (10.58) can be found in the Einstein frame, as pointed out in Ref. [319]. Specifically, the authors consider a conformal transformation of the action (10.50) and use the ADM formalism to find the Wave Function of the Universe and recover classical trajectories.

10.3 The Torsion Scalar Coupled to a Scalar Field

In this section, we develop similar considerations as in Section 10.1 for non-minimally coupled TEGR. We shall show that dynamics and solutions, derived from Noether symmetries, are equivalent to those obtained in Section 10.1. This purpose can be pursued even without introducing the boundary term B, namely without imposing the $f(R)$ limit as a requirement.

Let us consider the teleparallel equivalent of action (4.64), that is:

$$S = \int h\left[Tf(\phi) + u(\phi)\partial_\mu\phi\partial^\mu\phi - V(\phi)\right] d^4x, \tag{10.59}$$

whose Klein–Gordon equation reads as

$$Tf_\phi(\phi) + u_\phi(\phi)\partial_\mu\phi\partial^\mu\phi - V_\phi(\phi) - 2u(\phi)\Box\phi = 0. \tag{10.60}$$

Unlike the previous case, the cosmological expression of torsion does not contain second derivatives that must be integrated out; therefore, the point-like Lagrangian can be easily found only by substituting the relation (4.113) into the action and integrating the three-dimensional surface:

$$\mathcal{L} = -6af(\phi)\dot{a}^2 + a^3u(\phi)\dot{\phi}^2 - a^3V(\phi). \tag{10.61}$$

Note that this Lagrangian is already canonical and the equations of motion are simplified with respect to Eqs. (10.6). They are

$$\begin{cases} -2f\dot{a}^2 + a^2(V - u\dot{\phi}^2) - 4a(f_\phi\dot{a}\dot{\phi} + f\ddot{a}) = 0 \\ 6\dot{a}^2 f_\phi + 6au\dot{a}\dot{\phi} + a^2(V_\phi + u_\phi\dot{\phi}^2 + 2u\ddot{\phi}) = 0 \\ -6af\dot{a}^2 + a^3(V + u\dot{\phi}^2) = 0. \end{cases} \tag{10.62}$$

The minisuperspace considered is two-dimensional as in the previous case ($\mathcal{S} \equiv \{a, \phi\}$), and the generator of symmetry is, in turn:

$$\mathcal{X} = \xi(a,\phi,t)\partial_t + \alpha(a,\phi,t)\partial_a + \beta(a,\phi,t)\partial_\phi . \tag{10.63}$$

Notice that, since the Lagrangian is in a canonical form, the condition $\xi = \xi(t)$ immediately holds. The application of the extended Noether vector to the point-like Lagrangian (10.61) provides a system of 12 equations, which can be reduced to four equations with the constraints on the infinitesimal generators, that is [311, 320]:

$$\begin{cases} 6f\partial_\phi\alpha - 2ua^2\partial_a\beta = 0 \\ \alpha f + \beta a f_\phi + af\partial_t\xi - 2af\partial_a\alpha = 0 \\ 3\alpha V + \beta a V_\phi - aV\partial_t\xi = 0 \\ 3\alpha u + \beta a u_\phi - au\partial_t\xi + 2au\partial_\phi\beta = 0 \\ \alpha = \alpha(a,\phi), \quad \beta = \beta(a,\phi), \quad \xi = \xi(t). \end{cases} \tag{10.64}$$

After some manipulations, the system can be recast in two differential equations containing the three functions $f(\phi)$, $u(\phi)$, $V(\phi)$ and two unknown infinitesimal generators. It is thus clear that the system cannot provide a unique solution, and a choice must be adopted in order to fix the related dynamics. Therefore, we substitute in (10.64) power-law and exponential potentials, which are of cosmological interest. However, the assumption is not excessively strict, since only the form of the potential is needed in order to solve exactly the system. Solutions containing power-law and exponential potentials are

$$\begin{cases} \mathcal{X} = (\xi_0 t + \xi_1)\,\partial_t - \dfrac{\xi_0}{3} a \left(\dfrac{k+c}{k-c}\right)\partial_a + \dfrac{2\xi_0}{k-c}\phi\partial_\phi \\ f(\phi) = f_0\phi^k, \quad u(\phi) = u_0\phi^{k-2}, \quad V(\phi) = V_0\phi^c, \quad k \neq c \\ \\ \mathcal{X} = (\xi_0 t + \xi_1)\,\partial_t - \dfrac{\xi_0}{3} a \left(\dfrac{k+c}{k-c}\right)\partial_a + \dfrac{2\xi_0}{k-c}\partial_\phi \\ f(\phi) = f_0 e^{k\phi}, \quad u(\phi) = u_0 e^{k\phi}, \quad V(\phi) = V_0 e^{c\phi}, \quad k \neq c \\ \\ \mathcal{X} = -\dfrac{k}{3} a\beta(\phi)\partial_a + \beta(\phi)\partial_\phi \\ f(\phi) = f_0 e^{k\phi}, \quad u(\phi) = u_0 e^{k\phi}, \quad V(\phi) = V_0 e^{k\phi}. \end{cases} \tag{10.65}$$

Also in this case, the exponential solutions of the Noether system allow us to find the teleparallel equivalent of string–dilaton cosmology, namely the string–dilaton action with torsion instead of curvature.

Let us now solve Eqs. (10.62) for two different set of couplings, potentials, and kinetic terms. We choose the most general solutions among those in (10.65), namely the first and the second. In the former case the Lagrangian (10.61) turns out to be

$$\mathcal{L} = -6f_0 a\phi^k \dot{a}^2 + u_0 a^3\phi^{k-2}\dot{\phi}^2 - V_0 a^3\phi^c . \tag{10.66}$$

Assuming the condition $k = c$ and the de Sitter-like expansion for the scale factor, we have

$$a(t) = a_0 e^{qt}\,, \quad \phi(t) = \phi_0 \exp\left\{ \pm\sqrt{\frac{6f_0 q^2 - V_0}{u_0}}\, t \right\}, \quad q = \sqrt{\frac{V_0 u_0}{6f_0 u_0 - 4k^2 f_0^2}}\,. \tag{10.67}$$

By taking into account the second set of functions, the Lagrangian takes the form

$$\mathcal{L} = -6f_0 a e^{k\phi} \dot{a}^2 + u_0 a^3 e^{k\phi} \dot{\phi}^2 - V_0 a^3 e^{c\phi}, \tag{10.68}$$

leading to the exponential solutions constrained by the relation $k = c$:

$$a(t) = a_0 e^{qt}\,, \quad \phi(t) = \pm\sqrt{\frac{6f_0 q^2 - V_0}{u_0}}\, t, \quad q = \sqrt{\frac{V_0 u_0}{6f_0 u_0 - 4k^2 f_0^2}}\,. \tag{10.69}$$

It is worth stressing the difference between the scalar field coupled to the curvature scalar and to the torsion scalars. The Noether approach performed in Section 10.1 allows one to find exact expressions for the scalar field and for the scale factor, but the analytic relations between the free parameters cannot be obtained analytically. In the case treated here, instead, such a relation can be analytically found, so that an exact solution of Eqs. (10.62) occurs. This is due to the cosmological expression of T, which, not containing second derivatives, leads immediately to a canonical Lagrangian.

Let us now evaluate solutions with a constant kinetic term coefficient by setting $u(\phi) = 1/2$ in the starting Lagrangian (10.61). Thus the latter takes the form

$$\mathcal{L} = -6f(\phi) a \dot{a}^2 + \left[\frac{1}{2}\dot{\phi}^2 - V(\phi)\right] a^3. \tag{10.70}$$

By imposing the Noether identity onto Lagrangian (10.70), we find the same system of four differential equations as Eq. (10.64) in the limit $u(\phi) = 1/2$:

$$\begin{cases} \alpha f(\phi) + \beta a f_\phi(\phi) + 2f(\phi) a \partial_a \alpha - a f(\phi) \partial_t \xi = 0 \\ 3\alpha + 2a\partial_\phi \beta . a\partial_t \xi = 0 \\ a^2 \partial_a \beta - 12 f(\phi) \partial_\phi \alpha = 0 \\ (3\alpha - a\partial_t \xi)\, V(\phi) + \beta a V_\phi(\phi) = 0\,, \end{cases} \tag{10.71}$$

with the further conditions $\alpha = \alpha(a, \phi)$, $\beta = \beta(a, \phi)$, $\xi = \xi(t)$. Nontrivial solutions are provided by three internal symmetries and an external one, which are not included in (10.65). They read

$$
\begin{cases}
I: \mathcal{X} = -\dfrac{2\beta_0}{2\ell+3} a^{\ell+1} \phi^{-\frac{2\ell}{2\ell+3}} \partial_a + \beta_0 a^{\ell} \phi^{\frac{3}{2\ell+3}} \partial_\phi \\
f(\phi) = \dfrac{(2\ell+3)^2}{48}\phi^2, \qquad V(\phi) = V_0 \phi^{\frac{6}{2\ell+3}} \\
j_0 = \dfrac{1}{2}\beta_0 (2\ell+3) a^{\ell+2} \phi^{2-\frac{2\ell}{2\ell+3}} \dot{a} + \beta_0 a^{\ell+3} \phi^{\frac{3}{2\ell+3}} \dot{\phi} \\
\\
II: \mathcal{X} = -\dfrac{2q\beta_0}{3} \dfrac{e^{q\phi}}{\sqrt{a}} \partial_a + \beta_0 \dfrac{e^{q\phi}}{\sqrt{a^3}} \partial_\phi \\
f(\phi) = \dfrac{3}{16 q^2}, \qquad V(\phi) = V_0 e^{2q\phi} \\
j_0 = \dfrac{\beta_0}{2q} e^{q\phi} \sqrt{a} \left(3\dot{a} + 2qa\dot{\phi}\right) \\
\\
III: \mathcal{X} = -\dfrac{2}{3} a^{\frac{1}{4}} (c_2 + 2c_3\phi) \partial_a + a^{-\frac{3}{4}} (c_1 + c_2\phi + c_3\phi^2) \partial_\phi \\
f(\phi) = \dfrac{3}{64 c_3} (c_1 + c_2\phi + c_3\phi^2), \qquad V(\phi) = V_0 (c_1 + c_2\phi + c_3\phi^2)^2 \\
j_0 = \dfrac{1}{8c_3} a^{\frac{5}{4}} (c_1 + c_2\phi + c_3\phi^2) \left[3\dot{a}(c_2 + 2c_3\phi) + 8c_3 a\dot{\phi}\right] \\
\\
IV: \mathcal{X} = (-\dfrac{\xi_0}{3} t + \xi_1)\partial_t - \dfrac{2}{3}\beta_0 \sqrt{V_0} a \partial_a + \beta_0 \sqrt{V_0} \phi \partial_\phi \\
f(\phi) = f_0 \phi^2, \qquad V(\phi) = V_0 \phi^2 \\
j_0 = \beta_0 \sqrt{V_0} a^2 \phi (8 f_0 \phi \dot{a} + a\dot{\phi}).
\end{cases}
\tag{10.72}
$$

The first and the second functions do not admit analytic cosmological solutions. In particular, the latter reduces to standard TEGR when the potential vanishes, so that the corresponding trajectories are the same as Einstein GR. From the third function of (10.72), analytic solutions occur only by setting $c_1 = c_2 = 0$. By means of such an imposition, Solution III can be addressed to IV, after we properly redefine the constant parameters. Let us then focus on the third one, with the aim of investigating the cosmological implications. The corresponding quantum treatment will be discussed in Section 17.5. The point-like Lagrangian that arises, once the functions $f_{IV}(\phi)$ and $V_{IV}(\phi)$ are substituted into Eq. (10.70), takes the form

$$
\mathcal{L} = -6 f_0 a \phi^2 \dot{a}^2 + a^3 \left(- V_0 \phi^2 + \frac{1}{2} \dot{\phi}^2 \right), \tag{10.73}
$$

and the EL equations are in turn

$$
\begin{cases}
4 f_0 \phi^2 \dot{a}^2 + a^2 (-2 V_0 \phi^2 + \dot{\phi}^2) + 8 f_0 a \phi (2\dot{a}\dot{\phi} + \phi \ddot{a}) = 0 \\
12 f_0 \phi \dot{a} + 3 a \dot{a} \dot{\phi} + a^2 (2 V_0 \phi + \ddot{\phi}) = 0 \\
-6 f_0 a \phi^2 \dot{a}^2 + \dfrac{1}{2} a^3 (2 V_0 \phi^2 + \dot{\phi}^2) = 0.
\end{cases}
\tag{10.74}
$$

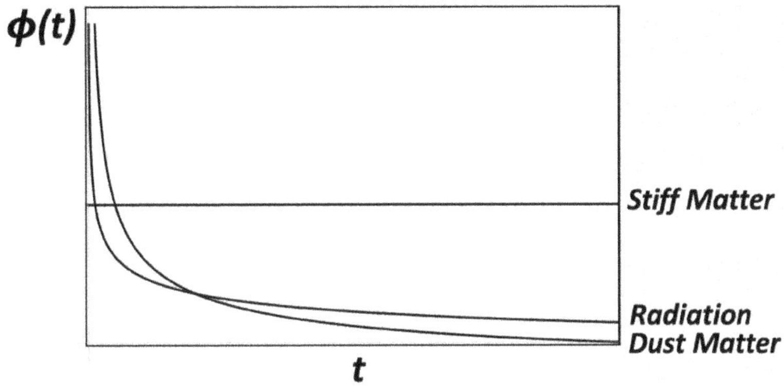

Figure 10.1 Behavior of the scalar field in the three cosmological eras as a function of time

Equations of motion (10.74), when $f \neq 0$, $V \neq 0$, only provide exponential solutions of the form

$$a(t) = a_0 e^{kt}, \quad \phi = \phi_0 e^{8f_0 k(t-t_0)}, \quad V(\phi) = 2(3f_0 k^2 - 16 f_0^2 k^2)\phi^2. \tag{10.75}$$

By neglecting the contribution of the potential, the EL equations also yield power-law solutions, which further constrain $f(\phi)$ to be

$$\begin{aligned} a(t) &= a_0 t^k, \quad \phi(t) = \phi_0 t^{\frac{1}{2}(1-3k)}, \\ f(\phi) &= \frac{(3k-1)^2}{48k^2}\phi^2 \, . \end{aligned} \tag{10.76}$$

From the preceding expressions for the scale factor, we can distinguish the three different Friedman eras:

$$\begin{aligned} \text{Radiation}: &\to a(t) = a_0 t^{\frac{1}{2}}, \quad \phi(t) = \phi_0 t^{-\frac{1}{4}}; \\ \text{Stiff Matter}: &\to a(t) = a_0 t^{\frac{1}{3}}, \quad \phi(t) = \phi_0; \\ \text{Dust Matter}: &\to a(t) = a_0 t^{\frac{2}{3}}, \quad \phi(t) = \phi_0 t^{-\frac{1}{2}}. \end{aligned}$$

The evolution of the scalar field is reported in Fig. **10.1**. See [320] for details. Similar results occur considering the Gauss–Bonnet topological term non minimally coupled to a scalar field, as we are going to discuss in the forthcoming section.

10.4 The Gauss–Bonnet Invariant Coupled to a Scalar Field

Another interesting scalar-tensor theory of gravity can be obtained by coupling a function of the Gauss–Bonnet invariant to a scalar field ϕ. Following the same scheme as in previous sections, we search for symmetries of a general scalar-tensor action, written in terms of the coupling $f(\phi)$, the potential $V(\phi)$, and the kinetic term coefficient $u(\phi)$. Thus, we consider the subcase $u(\phi) = 1/2$ to finally compare Noether symmetries with the results provided in Sections 10.1 and 10.3 of this chapter. As pointed out in Section 4.3, of Chap. 4, in some cosmological contexts, the terms $\mathbf{Riem}^2$ and $\mathbf{Ricci}^2$ are comparable to R^2. For this reason, unlike Sections 10.1 and 10.3 of this chapter, here the Gauss–Bonnet invariant is included in the initial action through the power-law function $f(\mathcal{G}) = \mathcal{G}^n$. Let us start by considering the action

$$S = \int \sqrt{-g}\left[\mathcal{G}^n f(\phi) + u(\phi)\partial_\mu\phi\partial^\mu\phi - V(\phi)\right] d^4x, \qquad n \in \mathbb{R}, \tag{10.77}$$

whose Klein–Gordon equation and field equations read, respectively

$$\mathcal{G}^n f_\phi(\phi) + u_\phi(\phi)\partial^\mu\phi\partial_\mu\phi - 2u(\phi)\Box\phi - V_\phi(\phi) = 0 \tag{10.78}$$

and

$$\begin{aligned}
&-2nf(\phi)\left(RR_{\mu\nu} - 2R_{\mu\alpha}R^\alpha{}_\nu + R_\mu{}^{\alpha\beta\gamma}R_{\nu\alpha\beta\gamma} - 2R^{\alpha\beta}R_{\mu\alpha\nu\beta}\right)\mathcal{G}^{n-1} \\
&+ nf(\phi)\Big[2RD_\mu D_\nu + 4G_{\mu\nu}\Box - 4(R^\rho{}_\nu D_\mu + R^\rho{}_\mu D_\nu)D_\rho + 4g_{\mu\nu}R^{\rho\sigma}D_\rho D_\sigma \\
&- 4R_{\mu\alpha\nu\beta}D^\alpha D^\beta\Big]\mathcal{G}^{n-1} - \frac{1}{2}g_{\mu\nu}u(\phi)D_\alpha\phi D^\alpha\phi - D_\mu D_\nu f(\phi) + g_{\mu\nu}\Box f(\phi) \\
&+ u(\phi)\partial_\mu\phi\partial_\nu\phi + \frac{1}{2}g_{\mu\nu}V(\phi) + \frac{1}{2}g_{\mu\nu}\mathcal{G}^n f(\phi) = 0.
\end{aligned} \tag{10.79}$$

Note that the minisuperspace is three-dimensional, that is, $\mathcal{S} \equiv \{a, \phi, \mathcal{G}\}$. By replacing the cosmological expression of $\mathcal{G}$ into the action, we obtain second-order derivatives that cannot be eliminated through a simple integration. In order to find the point-like Lagrangian, we have to define a Lagrange multiplier λ, which must be introduced into the action through the constraint (8.48). After integrating the surface term, the action turns out to be

$$S = \int a^3\left\{\left[f(\phi)\mathcal{G}^n + u(\phi)\dot{\phi}^2 - V(\phi)\right] - \lambda\left[\mathcal{G} - 24\left(\frac{\dot{a}^2\ddot{a}}{a^3}\right)\right]\right\} dt. \tag{10.80}$$

The Lagrange multiplier can be found by varying the action with respect to the Gauss–Bonnet invariant. Thus, substituting the result into the action and integrating out the second derivatives, the point-like Lagrangian becomes

$$\begin{aligned}
\mathcal{L} = (1-n)a^3\mathcal{G}^n f(\phi) - 8n\dot{a}^3\dot{\phi}f_\phi(\phi)\mathcal{G}^{n-1} + a^3u(\phi)\dot{\phi}^2 \\
-a^3V(\phi) - 8n(n-1)\mathcal{G}^{n-2}\dot{a}^3\dot{\mathcal{G}}f(\phi).
\end{aligned} \tag{10.81}$$

Clearly the Gauss–Bonnet contribution does not disappear for $n = 1$, like in the case of minimal coupling, since the surface term, thanks to the scalar field, provides a nontrivial dynamics. In this case, we have three EL equations and the energy condition; the further equation is the one for $\mathcal{G}$, which provides the cosmological expression of the Gauss–Bonnet surface term by construction. The equations of motion therefore read

$$\begin{cases} 3a^2\left[(n-1)f\mathcal{G}^n + V - u\dot{\phi}^2\right] \\ -24nG[t]^{n-3}\dot{a}\left\{(n-1)f\left[2\mathcal{G}\dot{\mathcal{G}}\ddot{a} + (n-2)\dot{a}\dot{\mathcal{G}}^2 + \dot{a}\mathcal{G}\ddot{\mathcal{G}}\right]\right. \\ \left. +2\mathcal{G}^2 f_\phi\dot{\phi}\ddot{a} + \dot{a}\mathcal{G}\left[\mathcal{G}\dot{\phi}^2 f_{\phi\phi} + 2f_\phi(n-1)\dot{\mathcal{G}}\dot{\phi} + \mathcal{G}f_\phi\ddot{\phi}\right]\right\} = 0 \\ 6a^2u\dot{a}\dot{\phi} - 24n\mathcal{G}^{n-1}\dot{a}^2 f_\phi\ddot{a} + a^3\left[(n-1)\mathcal{G}^n f_\phi + V_\phi + u_\phi\dot{\phi}^2 + 2u\ddot{\phi}\right] = 0 \\ a^3\mathcal{G} - 24\dot{a}^2\ddot{a} = 0 \\ -8n\mathcal{G}^{n-2}\dot{a}^3\left[2(n-1)f\dot{\mathcal{G}} + 3\mathcal{G}f_\phi\dot{\phi}\right] + a^3\left[(n-1)f\mathcal{G}^n + V + u\dot{\phi}^2\right] = 0. \end{cases} \tag{10.82}$$

Notice that the third EL equation provides back the expression of $\mathcal{G}$ in a FLRW background. The generator of the symmetry in the three-dimensional minisuperspace is

$$\mathcal{X} = \xi(a,\phi,\mathcal{G},t)\partial_t + \alpha(a,\phi,\mathcal{G},t)\partial_a + \beta(a,\phi,\mathcal{G},t)\partial_\phi + \gamma(t,a,\phi,\mathcal{G})\partial_{\mathcal{G}}, \tag{10.83}$$

with γ being the new infinitesimal generator associated to $\mathcal{G}$. The application of Eq. (6.42) to Lagrangian (10.81) provides the following system of four differential equations [311]:

$$\begin{cases} (n-1)\left(\gamma f_\phi + f\partial_\phi\gamma\right) + \beta\mathcal{G}f_{\phi\phi} + f_\phi\mathcal{G}\left(-3\partial_t\xi + \partial_\phi\beta + 3\partial_a\alpha\right) = 0 \\ \beta\mathcal{G}f_\phi + f\left[(n-2)\gamma + \mathcal{G}\left(-3\partial_t\xi + \partial_{\mathcal{G}}\gamma + 3\partial_a\alpha\right)\right] = 0 \\ 3\alpha u + a\left[\beta u_\phi - u(\partial_t\xi - 2\partial_\phi\beta)\right] = 0 \\ 3\alpha\mathcal{G}\left[(n-1)f\mathcal{G}^n + V\right] + a(n-1)f\mathcal{G}^n\left(n\gamma + \mathcal{G}\partial_t\xi\right) \\ +a\mathcal{G}\beta\left[(n-1)\mathcal{G}^n f_\phi + V_\phi\right] + a\mathcal{G}V\partial_t\xi = 0 \\ \alpha = \alpha(a), \quad \beta = \beta(\phi), \quad \gamma = \gamma(a,\phi,\mathcal{G}), \quad \xi = \xi(t). \end{cases} \tag{10.84}$$

Notice that the constraint $\xi = \xi(t)$ cannot be imposed from the beginning, due to the noncanonical form of the Lagrangian. As in the previous cases, the system is overdetermined and admits an infinite class of solutions depending on the form of the unknown coupling, namely

$$\begin{aligned} &\alpha = \alpha_0 a, \quad \beta = -\frac{(3\alpha_0 + \xi_0 - 4n\xi_0)f(\phi)}{f_\phi(\phi)}, \quad \gamma = -4\xi_0\mathcal{G}, \quad \xi = \xi_0 t + \xi_1, \\ &u = \frac{f(\phi)^{\frac{-3\alpha_0+(8n-3)\xi_0}{3\alpha_0+\xi_0-4n\xi_0}} f_\phi(\phi)^2}{(3\alpha_0 + \xi_0 - 4n\xi_0)^2}, \qquad V = V_0 f(\phi)^{\frac{3\alpha_0+\xi_0}{3\alpha_0+\xi_0-4n\xi_0}}. \end{aligned} \tag{10.85}$$

By substituting into (10.85) potentials of physical interest, namely exponential and power-law potentials, the infinitesimal generators, couplings, and kinetic term coefficients take the form

$$
\begin{cases}
\mathcal{X}_I = (\xi_0 t + \xi_1)\partial_t + \alpha_0 a\partial_a - \dfrac{3\alpha_0 + \xi_0 - 4n\xi_0}{k}\partial_\phi - 4\xi_0 \mathcal{G}\partial_{\mathcal{G}}, \\
f(\phi) = f_0 e^{k\phi}, \;\; u(\phi) = u_0 e^{\frac{k(3\alpha_0 - \xi_0)\phi}{3\alpha_0 + \xi_0 - 4n\xi_0}}, \;\; V = V_0 f_0^{\frac{3\alpha_0 + \xi_0}{3\alpha_0 + \xi_0 - 4n\xi_0}} e^{\frac{k(3\alpha_0 + \xi_0)\phi}{3\alpha_0 + \xi_0 - 4n\xi_0}} \\
\\
\mathcal{X}_{II} = (\xi_0 t + \xi_1)\partial_t + \alpha_0 a\partial_a - \left(\dfrac{3\alpha_0 + \xi_0 - 4n\xi_0}{k}\right)\phi\partial_\phi - 4\xi_0 \mathcal{G}\partial_{\mathcal{G}}, \\
f(\phi) = f_0 \phi^k, \;\; u(\phi) = u_0 \phi^{\frac{k(3\alpha_0 - \xi_0)}{3\alpha_0 + \xi_0 - 4n\xi_0} - 2}, \;\; V = V_0 f_0^{\frac{3\alpha_0 + \xi_0}{3\alpha_0 + \xi_0 - 4n\xi_0}} \phi^{\frac{k(3\alpha_0 + \xi_0)}{3\alpha_0 + \xi_0 - 4n\xi_0}} \\
\\
\mathcal{X}_{III} = (\xi_0 t + \xi_1)\partial_t + \dfrac{3\xi_0}{k}\phi\partial_\phi, \\
f(\phi) = -\dfrac{1}{3V_0^2}\phi^k, \quad V(\phi) = V_0 \phi^{-\frac{k}{3}}, \quad u(\phi) = u_0 \phi^{\frac{k}{3} - 2} \\
\\
\mathcal{X}_{IV} = (\xi_0 t + \xi_1)\partial_t + \dfrac{3\xi_0}{k}\partial_\phi, \\
f(\phi) = \dfrac{1}{k} e^{k\phi}, \quad V(\phi) = V_0 e^{-\frac{k}{3}\phi}, \quad u(\phi) = u_0 e^{\frac{k}{3}\phi}.
\end{cases}
\tag{10.86}
$$

As before, $\xi_0, \alpha_0, \beta_0, \gamma_0, k, n$ are real constants. While the first two solutions hold for any $n \neq 1$, solutions *III* and *IV* set the value of n to 1. The $n = 1$ limit is topologically trivial only when $k = 1$, where the contribution of the geometry in the corresponding action turns into a topological surface term. For this reason, there is no interest in investigating cosmological solutions occurring for $k = n = 1$ and, in what follows, we will focus on the $n = 1/2$ limit, which represents the Gauss–Bonnet equivalent of GR in the cosmological framework.

Let us now derive cosmological solutions for the Noether symmetry (10.85). We will solve the related Eqs. (10.82) for $n = 1/2$ in order to compare the results with the above curvature and torsion cases. For $f(\mathcal{G}) = \mathcal{G}^{1/2}$, the first two Noether solutions (10.86) can be written as

$$
\begin{aligned}
\mathcal{X} &= \left[\frac{\ell}{2}(z - k)t + \xi_1\right]\partial_t + \frac{\ell}{6}(z + k)a\partial_a - \ell\partial_\phi - 2\ell(z - k)\mathcal{G}\partial_{\mathcal{G}}, \\
u(\phi) &= u_0 e^{k\phi}, \qquad V(\phi) = \tilde{V}_0 e^{z\phi}, \qquad f(\phi) = f_0 e^{k\phi};
\end{aligned}
\tag{10.87}
$$

$$
\begin{aligned}
\mathcal{X} &= \left[\frac{\ell}{2}(z - k)t + \xi_1\right]\partial_t + \frac{\ell}{6}(z + k)a\partial_a - \ell\phi\partial_\phi - 2\ell(z - k)\mathcal{G}\partial_{\mathcal{G}}, \\
u &= u_0 \phi^{k-2}, \qquad V = \tilde{V}_0 \phi^z, \qquad f(\phi) = f_0 \phi^k,
\end{aligned}
\tag{10.88}
$$

where we have defined

$$
\ell \equiv \frac{3\alpha_0 - \xi_0}{k}, \qquad V_0 f_0^{1 + \frac{2\xi_0}{k\ell}} \equiv \tilde{V}_0, \qquad z \equiv k\left(\frac{3\alpha_0 + \xi_0}{3\alpha_0 - \xi_0}\right).
\tag{10.89}
$$

Let us start by analyzing the action containing a power-law coupling, namely

$$S=\int\sqrt{-g}\left[f_0\sqrt{\mathcal{G}}\phi^k+\frac{1}{\ell^2}\phi^{k-2}\dot{\phi}^2+\tilde{V}_0\phi^z\right]d^4x. \tag{10.90}$$

After solving the system (10.82), we obtain a de Sitter-like solution that fixes the values of k and z to $k=z=\frac{\sqrt{6}}{2\ell^2 f_0}$. It reads as follows:

$$a(t)=a_0 e^{-\frac{\sqrt{-6\tilde{V}_0}}{2\ell f_0}t},\quad \phi(t)=\phi_0 e^{\ell\sqrt{-\tilde{V}_0}t},\quad \mathcal{G}(t)=\frac{54\tilde{V}_0^2}{\ell^4 f_0^4}. \tag{10.91}$$

This means that, by merging the result provided by the EL equations with those coming from the Noether approach, the only generator associated to this case is

$$\mathcal{X}=\frac{1}{\sqrt{6}\ell f_0}a\partial_a-\ell\phi\partial_\phi, \tag{10.92}$$

which describes an internal gauge symmetry. Let us now analyze the second solution with exponential coupling, potential, and kinetic terms; the corresponding action takes the following form:

$$S=\int\sqrt{-g}\left[f_0\sqrt{\mathcal{G}}e^{k\phi}+\frac{1}{\ell^2}\dot{\phi}^2e^{k\phi}+\tilde{V}_0e^{z\phi}\right]d^4x. \tag{10.93}$$

By substituting Eq. (10.87) into the equations of motion (10.82), it turns out that the latter can be analytically solved by imposing the constraint $k=z$, so that the scale factor and the scalar field behave like

$$\begin{aligned} a(t)&=a_0\exp\left\{\frac{k\ell}{3}\sqrt{\tilde{V}_0}(1+\sqrt{2})\,t\right\}, &\qquad \phi(t)&=-\ell\sqrt{\tilde{V}_0}t,\\ \mathcal{G}(t)&=\frac{8k^4\ell^4}{27}\tilde{V}_0^2(1+\sqrt{2})^4, & k&=\frac{3}{\ell^2}\sqrt{\frac{1}{F_0\sqrt{21-12\sqrt{2}}}}. \end{aligned} \tag{10.94}$$

As a final remark, it is worth noticing that the non-minimal couplings with the invariants R, T, and $\mathcal{G}$ all admit de Sitter solutions that can be easily compared with each other.

However, the remaining free parameters should be fixed by observations in order to provide physical constraints that can be used to match large-scale structure and cosmic evolution at early and late times.

10.5 Equivalence among Scalar-Tensor Theories by Noether Symmetries

Scalar-tensor theories with curvature, torsion, and Gauss–Bonnet scalars can be compared by Noether's symmetries. It is possible to show that, if the symmetries are the same, the different theories are equivalent. Specifically, we use solutions provided in Sections 10.1, 10.3, and 10.4 of this chapter to show that the three descriptions of scalar-tensor cosmology are dynamically equivalent. Furthermore, we start from a general scalar-tensor action and find the transformation to pass from the Einstein to the Jordan frames. Specifically, it turns out that such a transformation preserves Noether symmetries.

Let us take into account the second symmetry of (10.19), the first of (10.72), and that of (10.88), namely

$$R :\rightarrow \begin{cases} \mathcal{X} = -\dfrac{2(s+1)}{2s+3}\beta_0 a^{s+1}\phi^{\frac{2s^2+4s}{2s+3}}\partial_a + \beta_0 a^s \phi^{\frac{2s^2+6s+3}{2s+3}}\partial_\phi \\ f(\phi) = \dfrac{(2s+3)^2}{48(s+1)(s+2)}\phi^2, \quad V(\phi) = V_0\phi^{\frac{6(s+1)}{2s+3}}, \end{cases} \tag{10.95}$$

$$T :\rightarrow \begin{cases} \mathcal{X} = -\dfrac{2\beta_0}{2s+3}a^{s+1}\phi^{-\frac{2s}{2s+3}}\partial_a + \beta_0 a^s \phi^{\frac{3}{2s+3}}\partial_\phi \\ f(\phi) = \dfrac{(2s+3)^2}{48}\phi^2, \quad V(\phi) = V_0\phi^{\frac{6}{2s+3}}, \end{cases} \tag{10.96}$$

$$\mathcal{G} :\rightarrow \begin{cases} \mathcal{X} = \dfrac{k\ell}{3}a\partial_a - \ell\phi\partial_\phi \\ f(\phi) = f_0\phi^k, \quad V = V_0\phi^k, \quad u(\phi) = \dfrac{1}{\ell^2}\phi^{k-2}. \end{cases} \tag{10.97}$$

In the last solution, to obtain the condition $\xi = 0$, we assume $k = z$, $\xi_1 = 0$. With the aim of comparing the three solutions, we set the coefficient of the kinetic term in the Gauss–Bonnet case (10.97) to be constant, as naturally provided by the Noether approach in the other two cases, (10.95) and (10.96). Therefore, for $s = 0$ and $k = 2$, (10.95), (10.96), and (10.97) become

$$R :\rightarrow \begin{cases} \mathcal{X} = -\dfrac{2}{3}\beta_0 a\partial_a + \beta_0\phi\partial_\phi \\ f(\phi) = \dfrac{3}{32}\phi^2, \quad V(\phi) = V_0\phi^2, \end{cases} \tag{10.98}$$

$$T :\rightarrow \begin{cases} \mathcal{X} = -\dfrac{2\beta_0}{3}a\partial_a + \beta_0\phi\partial_\phi \\ f(\phi) = \dfrac{3}{16}\phi^2, \quad V(\phi) = V_0\phi^2, \end{cases} \tag{10.99}$$

$$\mathcal{G} :\rightarrow \begin{cases} \mathcal{X} = \dfrac{2\ell}{3}a\partial_a - \ell\phi\partial_\phi \\ f(\phi) = f_0\phi^2, \quad V = V_0\phi^2. \end{cases} \tag{10.100}$$

The preceding results suggest that symmetries fix the equivalence among the three representations of gravity when a scalar field is coupled with R, T, and $\sqrt{\mathcal{G}}$, respectively.

In this sense, symmetries can be a criterion capable of comparing theories coming from different representations of gravity. More precisely, as provided by Eqs. (10.98), (10.99), and (10.100), $f(R)$, $f(T)$, and $f(\mathcal{G})$ gravity yield the same symmetries when $f(R) \sim R$, $f(T) \sim T$, and $f(\mathcal{G}) \sim \sqrt{\mathcal{G}}$. This equivalence has been theoretically discussed in the previous sections, where we showed that GR and TEGR are equivalent up to a boundary term and that GR and $f(\mathcal{G})$ gravity (with $f(\mathcal{G}) \sim \sqrt{\mathcal{G}}$) are dynamically equivalent in cosmology. The result of this section, on the other hand, shows that such an equivalence can also be inferred by considering the symmetry generators that arise after applying the Noether Symmetry Approach. In principle, one can start from a general action (depending on unknown functions of some second-order invariant), find a class of symmetries by means of Noether's approach, and select those actions leading to the same generators. This prescription cannot be proven in general; however, it provides a strong indication on the dynamical equivalence among different representations of gravity.

In what follows, we use the Klein–Gordon equations to recast actions (10.1), (10.59), and (10.77) in the Einstein frame, showing that Noether's symmetry is independent of the chosen frame. To this purpose, let us consider the Gauss–Bonnet action non-minimally coupled to a scalar field (10.77), with the imposition $u(\phi) = 0$. In this way, the Klein–Gordon equation (10.78), when $n = 1/2$, takes the form

$$\sqrt{\mathcal{G}} f_\phi(\phi) - V_\phi(\phi) = 0 \quad \rightarrow \quad \sqrt{\mathcal{G}} = \frac{V_\phi(\phi)}{f_\phi(\phi)} \equiv \mathcal{M}(\phi) \tag{10.101}$$

and leads to the relation

$$\phi = \mathcal{M}^{-1}(\sqrt{\mathcal{G}}). \tag{10.102}$$

Therefore, the action can be recast in terms of $\mathcal{M}$ as

$$\begin{aligned} S = \int \sqrt{-g} \Big\{ & \sqrt{\mathcal{G}} f[\mathcal{M}^{-1}(\sqrt{\mathcal{G}})] \\ & - \int \sqrt{\mathcal{G}} f'[\mathcal{M}^{-1}(\sqrt{\mathcal{G}})] d(\mathcal{M}^{-1}(\sqrt{\mathcal{G}})) \Big\} \, d^4x, \end{aligned} \tag{10.103}$$

where we set $u(\phi) = 0$. Integrating by parts and assuming that $\mathcal{M}^{-1}(\sqrt{\mathcal{G}}) \in \mathcal{C}^1$, the action turns out to be

$$S = \int \sqrt{-g} \left\{ \sqrt{\mathcal{G}} \int f[\mathcal{M}^{-1}(\sqrt{\mathcal{G}})] \, d(\sqrt{\mathcal{G}}) \right\} d^4x. \tag{10.104}$$

Replacing the power-law coupling and potential (10.97) in the definition of $\mathcal{M}(\phi)$ (10.101), the action (10.104) takes the form

$$\begin{aligned} & S = \int \sqrt{-g} \left[\sqrt{\mathcal{G}} f_0 e^{k\phi} - V_0 e^{c\phi} \right] d^4x \\ \rightarrow\ & S = \int \sqrt{-g} \left[f_0 \left(\frac{k}{c} \frac{f_0}{V_0} \right)^{\frac{k}{c-k}} - V_0 \left(\frac{k}{c} \frac{f_0}{V_0} \right)^{\frac{c}{c-k}} \right] \sqrt{\mathcal{G}}^{\frac{c}{c-k}} \, d^4x. \end{aligned} \tag{10.105}$$

On the other hand, when exponential coupling and potentials selected by Noether's symmetry are chosen, the action can be recast as

$$\begin{aligned} S &= \int \sqrt{-g}\left[\sqrt{\mathcal{G}} f_0 \phi^k - V_0 \phi^c\right] d^4x \\ \to S &= \int \sqrt{-g}\left[f_0 \left(\frac{k}{c}\frac{f_0}{V_0}\right)^{\frac{k}{c-k}} - V_0 \left(\frac{k}{c}\frac{f_0}{V_0}\right)^{\frac{c}{c-k}} \right] \sqrt{\mathcal{G}}^{\frac{c}{c-k}}\, d^4x. \end{aligned} \tag{10.106}$$

Therefore, it turns out that the actions (10.90) and (10.93) are dynamically equivalent to (10.105) and (10.106). This procedure allows to pass from the Jordan to the Einstein frame without using conformal transformations, rather considering the Klein–Gordon equation only. In addition, in the Einstein frame, the actions (10.105) and (10.106) have the same form as that selected by the Noether Symmetry Approach in Section 8.1.2. This result can be considered as an application of the general prescription outlined in Section 10.1.2, according to which Noether symmetries are preserved regardless of the frame chosen.

Notice that, to obtain Eqs. (10.105) and (10.106), no property of the Gauss–Bonnet invariant has been used. It is worth noticing that the procedure holds for other geometric invariants, including Ricci and torsion scalars.

10.6 Horndeski Gravity

Adopting Noether point symmetries, it is possible to classify and integrate dynamical systems coming from Horndeski cosmologies. Starting from these models, it is possible to derive several modified theories of gravity like $f(R)$ gravity, Brans–Dicke gravity, string-inspired gravity, and so on. In Section 4.4, we introduced the Horndeski action within the context of scalar-tensor theories. Recently, Horndeski gravity has been reconsidered [147, 321] and significant progress has been done at cosmological and smaller scales, for example, in the context of charged black holes [322–324], neutron stars [325, 326], cosmological perturbations [327], inflation [328], and so on. As previously pointed out, the Horndeski theory contains a lot of degrees of freedom encoded in the arbitrary functions of the action: $G_i(\phi, X)$, where $i = 2, \ldots, 5$, ϕ is the scalar field and $X = -1/2(\partial_\mu\phi\partial^\mu\phi)$ its kinetic term. Here, we want to select the functions $G_i(\phi, X)$ by the Noether Symmetry Approach.

Let us start by considering the Horndeski action (4.73), that is:

$$S_{Horndeski} = \sum_{i=2}^{5} \int d^4x\sqrt{-g}\,\mathcal{L}_i\,, \tag{10.107}$$

where

$$
\begin{aligned}
\mathcal{L}_2 &\equiv G_2(\phi, X)\,, \\
\mathcal{L}_3 &\equiv -G_3(\phi, X)\,\Box\phi\,, \\
\mathcal{L}_4 &\equiv G_4(\phi, X)R + G_{4X}\left[(\Box\phi)^2 - (D_\mu D_\nu \phi)^2\right], \\
\mathcal{L}_5 &\equiv G_5(\phi, X)\, G_{\mu\nu} D^\mu D^\nu \phi \\
&\quad - \frac{1}{6}\, G_{5X}\left[(\Box\phi)^3 - 3\,\Box\phi\,(D_\mu D_\nu\phi)^2 + 2\,(D_\mu D_\nu \phi)^3\right].
\end{aligned}
$$

In a spatially flat FLRW universe, the Horndeski Lagrangian can be obtained by substituting the cosmological expressions of $X, \Box\phi, \left(D_\mu D_\nu \phi\right)^2$, and $\left(D_\mu D_\nu \phi\right)^3$, namely

$$
\begin{aligned}
&X = \frac{1}{2}\dot{\phi}^2\,, \quad \Box\phi = \ddot{\phi} + 3\frac{\dot{a}}{a}\dot{\phi}, \\
&\left(D_\mu D_\nu \phi\right)^2 = \ddot{\phi}^2 + 3\frac{\dot{a}^2}{a^2}\dot{\phi}^2\,, \quad \left(D_\mu D_\nu \phi\right)^3 = \ddot{\phi}^3 + \frac{3\dot{a}^3}{a^3}\dot{\phi}^3\,,
\end{aligned}
\tag{10.108}
$$

into Eq. (10.107). We have

$$
\begin{aligned}
\mathcal{L} = {} & a^3 G_2 - 3a^2 G_3 \dot{a}\dot{\phi} - a^3 G_3 \ddot{\phi} + 6a G_{4X}\dot{a}^2\dot{\phi}^2 - 6a G_4 \dot{a}^2 \\
& + 6a^2 G_{4X}\dot{a}\dot{\phi}\ddot{\phi} - 6a^2 G_4 \ddot{a} + 3a G_5 \dot{a}^2\ddot{\phi} + 6a G_5 \dot{a}\dot{\phi}\ddot{a} + 3 G_5 \dot{a}^3\dot{\phi} \\
& - G_{5X}\dot{a}^3\dot{\phi}^3 - 3a G_{5X}\dot{a}^2\dot{\phi}^2\ddot{\phi}\,.
\end{aligned}
\tag{10.109}
$$

Notice that the Lagrangian is not in a canonical form, since higher-order time-derivatives of the scalar field occur. Though most of the higher-order terms can be integrated out, Eq. (10.109) cannot be rendered totally canonical. This represents an important obstacle in view of applying the identity (6.42), which is valid for a Lagrangian not containing second derivatives. In order to get a suitable Lagrangian, we must assume the second derivative of G_3 with respect to the kinetic term X, that is, G_{3XX}, to vanish. This ansatz implies $G_3(\phi, X) = g(\phi)X + h(\phi)$, with g and h arbitrary functions. In this way, all terms in Eq. (10.109) can be integrated by parts, and the canonical point-like Lagrangian turns out to be

$$
\begin{aligned}
\mathcal{L} = {} & a^3 G_2 - a^2 g(\phi)\dot{a}\dot{\phi}^3 + \frac{1}{6}a^3 g'(\phi)\dot{\phi}^4 + a^3 h'(\phi)\dot{\phi}^2 + 6a G_4 \dot{a}^2 \\
& + 3a\left(2G_{4X} - G_{5\phi}\right)\dot{a}^2\dot{\phi}^2 + 6a^2 G_{4\phi}\dot{a}\dot{\phi} - G_{5X}\dot{a}^3\dot{\phi}^3\,.
\end{aligned}
\tag{10.110}
$$

It is straightforward to see that the minisuperspace is two-dimensional, that is, $\mathcal{S} \equiv \{a, \phi\}$, and, therefore, the symmetry generator takes the form

$$
\mathcal{X} = \xi(t, a, \phi)\partial_t + \alpha(t, a, \phi)\partial_a + \beta(t, a, \phi)\partial_\phi. \tag{10.111}
$$

By applying Eq. (6.42) to Lagrangian (10.110), we get a system of 28 differential equations, whose solutions select the explicit form of the functions G_i. The only generator allowed by Noether's approach is

$$
\mathcal{X} = (\xi_1 t + \xi_2)\partial_t + \alpha(a)\partial_a + (\xi_1\phi + \phi_1)\partial_\phi\,, \tag{10.112}
$$

Table 10.1 Horndeski functions selected by the Noether Symmetry Approach

	$G_2(\phi, X)$	$G_3(\phi, X)$	$G_4(\phi, X)$	$G_5(\phi, X)$
I	$g_2(X)$	$c_1 + c_2 X + c_3\phi$	$g_4(X)$	$g_5(X) + c_4\phi$
II	$e^{-\frac{3\alpha_1\phi}{\phi_1}} g_2(X)$	$c_1 + e^{-\frac{3\alpha_1\phi}{\phi_1}}\left(c_2 X - \frac{c_3\phi_1}{3\alpha_1}\right)$	$e^{-\frac{3\alpha_1\phi}{\phi_1}} g_4(X)$	$c_4 - \frac{\phi_1 e^{-\frac{3\alpha_1\phi}{\phi_1}} g_5(X)}{3\alpha_1}$
III	$g_2(X)\left(\xi_1\phi + \phi_1\right)^{-\frac{3\alpha_1}{\xi_1}-1}$	$\left(\xi_1\phi + \phi_1\right)^{-\frac{3\alpha_1}{\xi_1}}\left(c_1 X - \frac{c_2}{3\alpha_1}\right) + c_3$	$g_4(X)\left(\xi_1\phi + \phi_1\right)^{1-\frac{3\alpha_1}{\xi_1}}$	$c_4 + \frac{g_5(X)(\xi_1\phi+\phi_1)^{2-\frac{3\alpha_1}{\xi_1}}}{2\xi_1 - 3\alpha_1}$
IV	$\frac{g_2(X)}{(\xi_1\phi+\phi_1)^3}$	$c_1 - \frac{c_2 - 2c_3\xi_1 X}{2\xi_1(\xi_1\phi+\phi_1)^2}$	$\frac{g_4(X)}{\xi_1\phi+\phi_1}$	$\frac{c_4 \ln(\xi_1\phi+\phi_1)}{\xi_1} + g_5(X) + c_4$
V	$\frac{g_2(X)}{\xi_1\phi+\phi_1}$	$\frac{c_1 \ln(\xi_1\phi+\phi_1)}{\xi_1} + c_2 X + c_3$	$\left(\xi_1\phi + \phi_1\right) g_4(X)$	$c_4 + \frac{g_5(X)(\xi_1\phi+\phi_1)^2}{2\xi_1}$
VI	$\frac{g_2(X)}{(\xi_1\phi+\phi_1)^3}$	$c_1 - \frac{c_2}{2\xi_1(\xi_1\phi+\phi_1)^2}$	$\frac{g_4(X)}{(\xi_1\phi+\phi_1)}$	$\frac{c_3 \ln(\xi_1\phi+\phi_1)}{\xi_1} + g_5(X)$
VII	$g_2(X)\left(\xi_1\phi + \phi_1\right)^{-\frac{\alpha_1}{\xi_1}-3}$	$c_1 - \frac{c_2(\xi_1\phi+\phi_1)^{-\frac{\alpha_1}{\xi_1}-2}}{\alpha_1 + 2\xi_1}$	$g_4(X)\left(\xi_1\phi + \phi_1\right)^{-\frac{\alpha_1}{\xi_1}-1}$	$c_3 - \frac{g_5(X)(\xi_1\phi+\phi_1)^{-\frac{\alpha_1}{\xi_1}}}{\alpha_1}$
VIII	$\frac{g_2(X)}{\xi_1\phi+\phi_1}$	$c_1 + \frac{c_2 \log(\xi_1\phi+\phi_1)}{\xi_1}$	$\frac{g_4(X)(\xi_1\phi+\phi_1)}{\xi_1}$	$c_3 + \frac{g_5(X)(\xi_1\phi+\phi_1)^2}{2\xi_1}$
IX	$e^{-\frac{\alpha_1\phi}{\phi_1}} g_2(X)$	$c_1 - \frac{c_2\phi_1 e^{-\frac{\alpha_1\phi}{\phi_1}}}{\alpha_1}$	$e^{-\frac{\alpha_1\phi}{\phi_1}} g_4(X)$	$c_3 - \frac{\phi_1 e^{-\frac{\alpha_1\phi}{\phi_1}} g_5(X)}{\alpha_1}$
X	$g_2(X)$	$c_1 + c_2\phi$	$g_4(X)$	$c_3\phi + g_5(X)$

with ξ_1, ξ_2, ϕ_1 integration constants. Depending on whether the function $g(\phi)$ vanishes or not, there are different solutions. In the class of solutions with $g(\phi) \neq 0$, the α coefficient becomes $\alpha(a) = \alpha_1 a$, with α_1 being constant. In the other case, where $g(\phi) = 0$, we get $\alpha(a) = \frac{1}{3}(\alpha_1 + 2\xi_1)a$. The whole set of solutions for the functions G_i is reported in Table **10.1** [149].

The arbitrariness of the functions $g_i(X)$ makes this classification broad enough, as far as the restrictions are concerned. By choosing specific classes (and thus symmetries) and playing with the form of the functions g_i, we can map modifications of GR to the Horndeski theory and see if they are invariant or not under the action of Noether point symmetries. In this perspective, the Noether Symmetry Approach is a selection criterion discriminating among integrable models. Clearly, in order to solve the EL equations of this theory, the function g_i must be assigned. This prescription is considered in Ref. [149], where the authors use the second-order invariants coming from the Noether Symmetry Approach and choose specific forms of the arbitrary functions g_i to recover $f(R)$ gravity, Brans–Dicke gravity, cubic galileon models, and other modified theories.

11
Nonlocal Gravity

Beati monoculi in terra caecorum.

— *Desiderius Erasmus Roterodamus*
(*Adagiorum collectanea* [1500])

11.1 Nonlocality in Physics

Locality is one of the fundamental principles of classical and relativistic physics, according to which any point of a given system interacts solely with its immediate surroundings. A theory including the principle of locality is said to be a "local theory," while a "nonlocal theory" admits interactions at a distance. More precisely, an action is "local" if and only if it is a functional of local fields only, that is, algebraic functions of fields or their derivatives evaluated at a single point of the manifold. The theory is said to be "nonlocal" if and only if the action is a functional of nonlocal fields (at least one), that is, transcendental functions of fields evaluated at more than one point.

As a matter of fact, while classical theories are *local theories*, Quantum Mechanics is kinematically nonlocal. It is important to distinguish kinematical and dynamical locality/nonlocality: while the former refers to the states describing a theory, the latter relies on the interactions, so that dynamical locality (nonlocality) is only due to the local (nonlocal) form of the action.

In general, any transcendental function of fields can be represented by the integral kernels of differential operators, namely the operator $\Box^{-1}$, which accounts for long-range nonlocal effects. It can be written in terms of the associated Green function $G(x, x')$ as

$$\Box^{-1}\phi\,(x) \equiv \int d^4x'\; G(x,x')\phi(x')\,.$$

Any classical theory can be localized at any time of its evolution, so that it is possible to perform a measure with arbitrary precision without perturbing the state. Thus, any classical field theory is kinematically local.

On the contrary, Quantum Mechanics is nonlocal by construction. For example, according to the Heisenberg Principle, a given particle cannot be localized with arbitrary precision. Due to this property, a particle starting its motion from a given position x_1 and moving toward a final position x_2 follows any possible path linking the two points. Unlike what happens in Classical Mechanics, where the actual path between two fixed endpoints is uniquely selected by the initial conditions, in Quantum Mechanics all the paths are simultaneously allowed.

It is worth pointing out that the kinematical nonlocality of the theory does not necessarily imply dynamical nonlocality; in order for a theory to be dynamically nonlocal, the action must have a nonlocal form as well. Another manifestation of nonlocality in Quantum Mechanics is given by the entanglement, which exhibits a kinematical nonlocality due to the interaction at a distance between particles.

In general, in the formalism of QFT, all the fundamental interactions exhibit dynamical nonlocality as soon as their one-loop effective actions are considered [329]. This is the case, for example, of the Euler–Heisenberg Lagrangian:

$$\mathcal{L}_{EH} = -\frac{1}{4}\mathcal{F}^2 - \frac{e^2}{32\pi^2}\int_0^\infty \frac{ds}{s}\, e^{i\varepsilon s} e^{-m^2 s}\left[\frac{\mathrm{Re}\cosh(esX)}{\mathrm{Im}\cosh(esX)}\mathcal{F}_{\mu\nu}\tilde{\mathcal{F}}^{\mu\nu} - \frac{4}{e^2 s^2} - \frac{2}{3}\mathcal{F}^2\right], \tag{11.1}$$

where $\mathcal{F}_{\mu\nu}$ is the Electromagnetic tensor, defined through the potential A_μ as $\mathcal{F}_{\mu\nu} = \partial_\mu A_\nu - \partial_\nu A_\mu$. The scalars $\mathcal{F}$ and X are defined respectively as $\mathcal{F} = \frac{1}{2}\left(|\mathbf{E}|^2 - |\mathbf{B}|^2\right)$ and $X = \mathcal{F} + i\mathbf{E}\cdot\mathbf{B}$, while the constant e and m are the charge and the mass, respectively. The above Lagrangian is the renormalized one-loop effective Lagrangian that arises after integrating out a massive fermion from the Quantum Electrodynamics full Lagrangian. Eq. (11.1) is of particular interest, since the nonlocality is due to the intrinsic nonlocal nature of an integration, which can be seen as the inverse of a differential operator.

Another example is given by the low-energy limit of the Yukawa theory with a massive scalar field ϕ, whose effective Lagrangian reads

$$\begin{aligned} &\mathcal{L}_Y = i\bar{\psi}\partial\!\!\!/\psi - \frac{1}{2}\phi(\Box + m^2)\phi + \lambda\,\phi\bar{\psi}\psi \\ \to\ &\mathcal{L}_{eff} = i\bar{\psi}\partial\!\!\!/\psi + \frac{\lambda^2}{2}\bar{\psi}\psi(\Box + m^2)^{-1}\bar{\psi}\psi\,. \end{aligned} \tag{11.2}$$

The nonlocality here is provided by the operator $(\Box + m^2)^{-1}$. After performing a straightforward Taylor expansion, the Lagrangian can be written as

$$\mathcal{L}_{eff} = i\bar{\psi}\partial\!\!\!/\psi + \frac{\lambda^2}{2m^2}\bar{\psi}\psi\bar{\psi}\psi - \frac{\lambda^2}{2m^4}\bar{\psi}\psi\Box\bar{\psi}\psi + \dots\ , \tag{11.3}$$

which coincides with the 4-Fermi theory when the scalar field ϕ accounts for the W and Z bosons.

Although nonlocal terms introduce extra degrees of freedom, inserting them into the field equations can represent a fundamental step toward the construction of a full theory of Quantum Gravity. In order to merge GR with Quantum Mechanics, the former should be dealt under the standard of nonlocal formalism. However, Nonlocal Theories of Gravity mainly aim to provide solutions for the high-energy shortcomings suffered by the EH action, preserving the unitarity and Lorentz invariance. Indeed, perturbed nonlocal actions turn out to be renormalizable at any level of perturbation [330–332].

Here, we discuss some cosmological results provided by Nonlocal Theories of Gravity, considering nonlocal terms in the starting actions. As pointed out in [333,334] nonlocality can play a crucial role also in the cosmological framework connecting early- and late-time evolution of the universe.

In what follows we briefly overview the main features of Nonlocal Gravity. The main difference with respect to the corresponding local modified theories, is that the effective action contains nonlocal operators of different form, which aim to merge the gravitational interaction with the quantum formalism. We will show that also Nonlocal Gravity can be investigated by the Noether Symmetry Approach.

Depending on the type of nonlocality, nonlocal modified theories can be split into two main classes: Infinite Derivative Theories of Gravity (IDGs) and Integral Kernel Theories of Gravity (IKGs). The former involve entire analytic transcendental functions of the covariant d'Alembert operator $\Box$. An example is given by the theory proposed in Ref. [331], which provides a solution for classical black hole and Big Bang singularities [330,335]. The latter mainly employs the inverse operator $\Box^{-1}$. It was firstly considered in [336], where the authors started from the following nonlocal extension of GR:

$$S = \frac{1}{2}\int d^4x\sqrt{-g}\,R\left[1 + F\left(\Box^{-1}R\right)\right] + S^{(m)}\,, \tag{11.4}$$

with $F\left(\Box^{-1}R\right)$ being an arbitrary function of the nonlocal term $\Box^{-1}R$. They also showed that the application of the nonlocal operator $\Box^{-1}$ to the curvature scalar R gives rise to a late-time cosmic expansion of the universe without invoking any dark energy. Action (11.4) is called the *Deser–Woodard action.*

In view of merging gravity with the other fundamental interactions, IDGs provide renormalizable and unitary Quantum Gravity theories [337], while IKGs deal with IR quantum corrections emerging from the formulation of QFT on curved space-time [329]. However, no Nonlocal Theory of Gravity formulated so far is capable of solving consistently all large-scale structure and cosmological evolution problems.

11.2 Infinite Derivatives Theories of Gravity

In order to introduce the main aspects of IDGs, let us consider the following infinite-derivatives Lorentz-invariant action:

$$S = \frac{1}{2}\int d^4x d^4y \phi(x)\mathcal{K}(x-y)\phi(y) - \int V(\phi)\, d^4x, \tag{11.5}$$

where ϕ is the scalar field depending on the coordinates and $\mathcal{K}(x-y)$ an operator depending on the distance $x-y$ through a generic function of the d'Alembert operator $\Box$, as

$$\mathcal{K}(x-y) = F(\Box)\delta^{(4)}(x-y). \tag{11.6}$$

A straightforward factorization of $F(\Box)$ yields

$$F(\Box) = e^{-\gamma(\Box)} \prod_{i=1}^{N} (\Box - m_i^2), \tag{11.7}$$

with $\gamma(\Box)$ being an entire function. By means of a Fourier transformation it is possible to show that bad ghosts appear when $N > 1$. For this reason, hereafter we focus on the $N = 1$ limit, where unitarity is preserved. The most general action made of functions of the d'Alembert operator, which is ghost-free and quadratic in the curvature, must contain infinite covariant derivatives. It reads [337–340]:

$$S = \frac{1}{2}\int d^4x\sqrt{-g}\left[R + \alpha\Big(RF_1(\Box_s)R + R_{\mu\nu}F_2(\Box_s)R^{\mu\nu} + R_{\mu\nu\rho\sigma}F_3(\Box_s)R^{\mu\nu\rho\sigma}\Big)\right], \tag{11.8}$$

where M_s is a mass/length scale and $F_i(\Box_s)$ a transcendental entire analytic function of $\Box_s \equiv \Box/M_s^2$. Bad ghosts are prevented by the nature of the entire function, which manifests no pole on the whole complex plane.

In Ref. [331], a subcase of Eq. (11.8) is proposed, which provides GR at the zero order. It is

$$S = \frac{1}{2}\int d^4x\sqrt{-g}\left(R - G_{\mu\nu}\frac{e^{H(-\Box_s)} - 1}{\Box}R^{\mu\nu}\right), \tag{11.9}$$

where $H(-\Box_s)$ is an entire analytic function of $\Box_s$. The field equation, up to the order $O(R^2)$, reads

$$G_{\mu\nu} + O(R^2) = e^{-H(-\Box_s)}\,\mathcal{T}^{(m)}_{\mu\nu}. \tag{11.10}$$

Notice that at the lowest order of Taylor expansion, the field equations are equivalent to those of GR. In spherical symmetry, Eq. (11.10) provides regular black holes without singularities [330], while, in an homogeneous and isotropic universe, it admits bouncing cosmological solutions [335]. The mechanism aimed at avoiding classical singularities is discussed in Refs. [341,342]. It basically consists of a nonlocal smearing of the point-like source of the Schwarzschild background, which automatically implies that the metric can no longer be considered as a vacuum solution. In the nonlocal region, where $r < 2/M_s$, the effects of nonlocality start being relevant, while, as soon as the radius starts increasing, the solution approaches the Schwarzschild one. As mentioned previously, the theory does not contain singularity for any value of r, including also $r = 0$.

Another example is provided by Ref. [343]. This theory turns out to be unitary and renormalizable and represents the maximal UV-completion of Starobinsky gravity. The corresponding action reads as

$$S = \frac{1}{2}\int d^4x\sqrt{-g}\left[R - G_{\mu\nu}\frac{V_2^{-1} - 1}{\Box}R^{\mu\nu} + \frac{1}{2}R\,\frac{V_0^{-1} - V_2^{-1}}{\Box}\,R\right], \tag{11.11}$$

where

$$V_2^{-1} \equiv e^{H_2(-\Box_s)}p^{(n_2)}(-\Box_s), \quad V_0^{-1} - V_2^{-1} \equiv \frac{1}{3}\left[e^{H_0(-\Box_s)}(1 + \Box_s) - e^{H_2(-\Box_s)}\right],$$

with H_i and $p^{(n_2)}$ being entire analytic functions. Other IDGs models are possible, and most of them solve some particular problem.

11.3 Integral Kernel Theories of Gravity

The other class of Nonlocal Theories of Gravity is represented by IKGs. They bring UV quantum corrections by the expansion around $s = 0$ of a Schwinger proper time integral [263]. IR corrections need an expansion around $s \to \infty$, which may represent a problem for several reasons. First, the Schwinger proper time integral can be settled only when masses of matter fields are greater than the potential. Second, in the massless limit, the proper time integration starts increasing for late times, up to infinity. These issues come directly from the perturbative approach used to calculate the Schwinger proper time integral, then a non-perturbative technique is needed in order to obtain both UV ($s = 0$) and IR ($s \to \infty$) corrections. The quantum effective action coming from such a non-perturbative technique reads [329]

$$W_0 = -\int d^4x\sqrt{-g}\left[V(x) + V(x)(\Box - V)^{-1}V(x)\right] + \frac{1}{6}\Sigma\,, \tag{11.12}$$

where $V(x)$ is a generic potential and Σ is a surface term defined through the inverse of the d'Alembert operator as [329]

$$\begin{aligned}\Sigma = \frac{1}{2}\int d^4x\sqrt{-g}\Big\{&R - R_{\mu\nu}\Box^{-1}G^{\mu\nu} + 2^{-1}R\,(\Box^{-1}R^{\mu\nu})\Box^{-1}R_{\mu\nu}\\ &- R^{\mu\nu}(\Box^{-1}R_{\mu\nu})\Box^{-1}R + (\Box^{-1}R^{\alpha\beta})(D_\alpha\Box^{-1}R)D_\beta\Box^{-1}R\\ &- 2\,(D^\mu\Box^{-1}R^{\nu\alpha})(D_\nu\Box^{-1}R_{\mu\alpha})\Box^{-1}R\\ &- 2\,(\Box^{-1}R^{\mu\nu})(D_\mu\Box^{-1}R^{\alpha\beta})D_\nu\Box^{-1}R_{\alpha\beta} + O[R^4_{\mu\nu}]\Big\}\,.\end{aligned} \tag{11.13}$$

In this case, the nonlocality is due to the integral operator $\Box^{-1}$, which is responsible for quantum corrections. However, the action (11.12) provides non-linear higher-order field equations, which may represent an obstacle toward the search for exact solutions. For this reason, several subcases have been treated in literature, among which are the Deser–Woodard action defined in Eq. (11.4).

It is possible to show that the presence of the operator $\Box^{-1}$ naturally explains late-time cosmic acceleration of the universe. To this purpose, let $t_0 \sim 10^{10}y$ be the present time, $t_{eq} \sim 10^5 y$ the time of the equivalence, and assume a matter-dominated era between t_0 and t_{eq}. In a spatially flat homogeneous and isotropic universe, simple computations show that the contribution of nonlocal causal effects, acting within the considered interval, are of the order

$$(\Box^{-1}R)(t_0) \sim 14.0\,, \tag{11.14}$$

which is the large number required by the current cosmic acceleration to avoid the fine-tuning of parameters.

In next sections, we will focus on higher-order IKGs, where nonlocality is induced by the propagator operator $\Box^{-1}$. They are generalizations of the IKGs considered, for example, in [333, 336, 344]. Even though we will not discuss in detail the UV quantum corrections of these theories, the presence of nonlocal operators is somehow useful in order to recover renormalizability and unitarity. In the search for exact cosmological solutions, we reduce the dynamics by Noether's approach.

Specifically, we will focus on three different nonlocal actions, finding exact cosmological solutions after selecting theories containing symmetries. In Section 11.4, the action $S = \int \sqrt{-g} f[R, \Box^{-1} h(R)] d^4x$ is considered. Notice that, with respect to the previously mentioned IKGs, here the propagator acts on a function of R that will be selected by symmetries. In Section 11.5, with the aim of reducing the dynamics, the Gauss–Bonnet term is included in the starting action, that is: $S = \int \sqrt{-g} f[\mathcal{G}, \Box^{-1} h(\mathcal{G})] d^4x$. Finally, in Section 11.6 we consider the teleparallel equivalent action, depending on a nonlocal function of the torsion scalar.

In all cases, in order to apply the condition (6.42), the Lagrangian must be localized by the introduction of local scalar fields, defined by means of the operator $\Box^{-1}$. As a general result, it is worth noticing that the presence of symmetries selects physically interesting models where the above features of renormalizability and unitarity naturally emerge.

11.4 Nonlocality with Curvature

In this section, we will study, in metric formalism, a nonlocal IKG theory whose action is

$$S = \int \sqrt{-g} f(R, \Box^{-1} h(R)) \, d^4x \,. \tag{11.15}$$

It is a straightforward extension of the Deser–Woodard action (11.4), which involves a function of the curvature scalar. It reduces to $f(R)$ gravity as soon as the function is independent of the nonlocal term.

In order to construct the point-like Lagrangian, we make use of the auxiliary *local* scalar field ϕ defined as

$$\Box^{-1} h(R) \equiv \phi(t), \qquad \text{so that} \qquad h(R) = \Box \phi(t). \tag{11.16}$$

By means of this definition, the theory reduces to a scalar-tensor theory with action

$$S = \int d^4x \sqrt{-g} \, f(R, \phi), \tag{11.17}$$

Adopting the cosmological expression of R and $\Box$ in a FLRW space-time, the action can be written in terms of the Lagrange multipliers as

$$\begin{aligned} S = \int & \left\{ a^3 f(R, \phi) - a^3 \lambda_1 \left[\ddot{\phi} + 3\frac{\dot{a}}{a}\dot{\phi} - h(R) \right] \right. \\ & \left. -\lambda_2 \left[R + 6\left(\frac{\ddot{a}}{a} + \frac{\dot{a}^2}{a^2} \right) \right] \right\} dt \,. \end{aligned} \tag{11.18}$$

The relation between λ_1 and λ_2 can be found by varying the action with respect to R:

$$\begin{aligned} \frac{\delta S}{\delta R} &= \int \left\{ a^3 f_R(R, \phi) - a^3 \epsilon(t) h_R(R) - \lambda_2 \right\} dt = 0 \\ & \rightarrow \lambda_2 = a^3 \left[f_R(R, \phi) - \epsilon(t) h_R(R) \right] \,, \end{aligned} \tag{11.19}$$

while the variation with respect to ϕ cannot provide any additional constraint. For this reason, following the same procedure as in, for example, [333, 334], the constant λ_1 can be promoted to a time-dependent scalar field by setting $\lambda_1 \equiv \epsilon(t)$. In this way, Eq. (11.18) can be rewritten as

$$S = \int \left\{ a^3 f(R,\phi) + a^3\epsilon(t)\left[\ddot{\phi} + 3\frac{\dot{a}}{a}\dot{\phi} - h(R)\right] \right. \\ \left. -\lambda_2 \left[R + 6\left(\frac{\ddot{a}}{a} + \frac{\dot{a}^2}{a^2}\right)\right]\right\} dt\,. \tag{11.20}$$

Furthermore, the variation of the action with respect to the scalar field ϕ provides

$$\begin{aligned} \frac{\delta S}{\delta\phi} &= \int \sqrt{-g}\left\{f_\phi(R,\phi) - \frac{\delta}{\delta\phi}\left[\epsilon(t)(\Box\phi - h(R))\right]\right\} d^4x \\ &= \int \sqrt{-g}\left\{f_\phi(R,\phi) - \frac{\delta}{\delta\phi}\left[\epsilon(t)\Box(\phi - \Box^{-1}h(R))\right]\right\} d^4x\,. \end{aligned} \tag{11.21}$$

Evaluating the last term and using the divergence theorem, we have

$$\begin{aligned} &\int \sqrt{-g}\left\{\frac{\delta}{\delta\phi}\left[\epsilon(t)\Box(\phi - \Box^{-1}h(R))\right]\right\} d^4x \\ &= \frac{\delta}{\delta\phi}\int \sqrt{-g}\Box\epsilon(t)(\phi - \Box^{-1}h(R))\,d^4x \\ &= \int \sqrt{-g}\,\Box\epsilon(t)\frac{\delta}{\delta\phi}(\phi - \Box^{-1}h(R)\,d^4x = \int \sqrt{-g}\,\Box\epsilon(t)\,d^4x\,, \end{aligned} \tag{11.22}$$

so that finally

$$\frac{\delta S}{\delta\phi} = 0 \qquad \rightarrow \qquad \Box\epsilon(t) = f_\phi(R,\phi)\,. \tag{11.23}$$

As standard, when dealing with Lagrange multipliers, we consider ϕ and R as independent fields. By varying Eq. (11.20) with respect to ϵ, we get the Klein–Gordon equation

$$\Box\phi = h(R). \tag{11.24}$$

Integrating out the second derivatives of the scalar fields, the point-like Lagrangian written in the configuration space $S \equiv \{a(t), R(t), \phi(t), \epsilon(t)\}$ reads as

$$\begin{aligned} \mathcal{L} &= a^3\left(f(R,\phi) - Rf_R(R,\phi) - \epsilon h(R) + \epsilon R h_R(R)\right) - a^3\dot{\epsilon}\dot{\phi} - 6a\epsilon\dot{a}^2 h_R(R) \\ &\quad - 6a^2\dot{a}\dot{\epsilon}h_R(R) - 6a^2\epsilon\dot{a}\dot{R}h_{RR}(R) + 6a\dot{a}^2 f_R(R,\phi) + 6a^2\dot{a}\dot{\phi}f_{R\phi}(R,\phi) \\ &\quad + 6a^2\dot{a}\dot{R}f_{RR}(R,\phi)\,, \end{aligned} \tag{11.25}$$

with EL equations

$$
\begin{cases}
2\dot{a}^2\left(f_R(R,\phi)-\epsilon h_R(R)\right)-4a\left[\ddot{a}\left(\epsilon h_R(R)-f_R(R,\phi)\right)\right.\\
\left.+\dot{a}\left(\dot{\epsilon}h_R(R)+\epsilon\dot{R}h_{RR}(R)-\dot{\phi}f_{R\phi}(R,\phi)-\dot{R}f_{RR}(R,\phi)\right)\right]\\
+a^2\left[-f(R,\phi)+\dot{\epsilon}\dot{\phi}-2h_R(R)\ddot{\epsilon}-4\dot{R}\dot{\epsilon}h_{RR}+Rf_R(R,\phi)\right.\\
+\epsilon\left(h(R)-Rh_R(R)-2\ddot{R}h_{RR}-2\dot{R}^2h_{RRR}(R)\right)\\
+2\ddot{\phi}f_{R\phi}(R,\phi)+2\dot{\phi}^2f_{R\phi\phi}(R,\phi)+2\ddot{R}f_{RR}(R,\phi)\\
\left.+4\dot{R}\dot{\phi}f_{RR\phi}(R,\phi)+2\dot{R}^2f_{RR}(R,\phi)\right]=0\\
\\
\Box\epsilon=f_\phi(R,\phi)\\
\\
R=-6\left(\dfrac{\ddot{a}}{a}+\dfrac{\dot{a}^2}{a^2}\right)\\
\\
\Box\phi=h(R)\\
\\
12\dot{a}^3\left(f_R(R,\phi)-\epsilon h_R(R)\right)-a^3\left[f(R,\phi)-\epsilon\left(h(R)-Rh_R(R)\right)\right]\\
+a^3\dot{\epsilon}\dot{\phi}-a^3\ddot{\epsilon}\dot{\phi}-a^3\dot{\epsilon}\ddot{\phi}+a^3Rf_R(R,\phi)+6a\dot{a}\left[\epsilon\left(-2h_R(R)\ddot{a}+\dot{a}h_R(R)\right.\right.\\
\left.-4\dot{a}\dot{R}h_{RR}(R)\right)+2\ddot{a}f_R(R,\phi)-\dot{a}\left(6\dot{\epsilon}h_R(R)+f_R(R,\phi)-6\dot{\phi}f_{R\phi}(R,\phi)\right.\\
\left.\left.-4\dot{R}f_{RR}(R,\phi)\right)\right]+6a^2\ddot{a}\left(\dot{\phi}f_{R\phi}(R,\phi)-\dot{\epsilon}h_R(R)\right)+6a^2\dot{a}\left[-\ddot{\epsilon}h_R(R)\right.\\
+\epsilon\dot{R}h_{RR}(R)+\dot{\epsilon}(h_R(R)-\dot{\phi}-3\dot{R}h_{RR}(R))-\epsilon\ddot{R}h_{RR}(R)-\epsilon\dot{R}^2h_{RRR}(R)\\
-\dot{\phi}f_{R\phi}(R,\phi)+\ddot{\phi}f_{R\phi}(R,\phi)+2\dot{\phi}^2f_{R\phi\phi}(R,\phi)-\dot{R}f_{RR}(R,\phi)+\ddot{R}f_{RR}(R,\phi)\\
\left.+3\dot{\phi}\dot{R}f_{RR\phi}(R,\phi)+\dot{R}^2f_{RR}(R,\phi)\right]=0\,.
\end{cases}
\tag{11.26}
$$

Notice that not all the EL equations provide new contributions to the dynamics. The equations with respect to ϵ and ϕ give back the Klein–Gordon equations (11.23) and (11.24). The equation with respect to the curvature scalar provides the cosmological expression of R by construction. The energy condition and the EL equation with respect to the scale factor are the only dynamical equations of motion by means of which analytic solutions will be found. They correspond to the "0,0" and "1,1" components of the field equations.

In the considered minisuperspace, the generator of the symmetry is

$$
\begin{aligned}
\mathcal{X} = {} & \xi(t,a,\phi,R,\epsilon)\partial_t+\alpha(t,a,\phi,R,\epsilon)\partial_a+\beta(t,a,\phi,R,\epsilon)\partial_\phi\\
& +\gamma(t,a,\phi,R,\epsilon)\partial_R+\delta(t,a,\phi,R,\epsilon)\partial_\epsilon\,.
\end{aligned}
\tag{11.27}
$$

The application of identity (6.42) to Lagrangian (11.25) yields a system of 28 partial differential equations, but most of them are linearly dependent. Neglecting linear combinations and using the constraint $\xi=\xi(t)$ (see Section 6.3), the system reduces to six differential equations with some additional constraints on the form of the infinitesimal generators [345]:

$$
\begin{cases}
3\alpha\,(f(R,\phi) - Rf_R(R,\phi) - \epsilon h(R) + \epsilon R h_R(R)) + a\,[\delta(Rh_R(R) - h(R)) \\
+\beta f_\phi(R,\phi) - \beta R f_{R\phi}(R,\phi) - \gamma R f_{RR}(R,\phi) + \partial_t\xi\,(f(R,\phi) - Rf_R(R,\phi)) \\
+\epsilon\gamma R h_{RR}(R) - \epsilon\partial_t\xi(h(R) - Rh_R(R))] = 0 \\
\\
2\alpha(f_{RR}(R,\phi) - \epsilon h_{RR}(R)) + a\left[-\delta h_{RR}(R) + \beta f_{RR\phi}(R,\phi) + \gamma f_{RRR}(R,\phi)\right. \\
-\partial_t\xi f_{RR}(R,\phi) + \partial_R\gamma f_{RR}(R,\phi) + \partial_a\alpha f_{RR}(R,\phi) \\
\left.-\epsilon(\gamma h_{RRR}(R) - \partial_t\xi h_{RR}(R) + \partial_R\gamma h_{RR}(R) + \partial_a\alpha h_{RR}(R))\right] = 0 \\
\\
12\alpha f_{R\phi}(R,\phi) + 6a\left[\beta f_{R\phi\phi}(R,\phi) + \gamma f_{RR\phi}(R,\phi)\right. \\
\left.-f_{R\phi}(R,\phi)(\partial_t\xi - \partial_\phi\beta - \partial_a\alpha)\right] + \partial_a\delta\; a^2 = 0 \\
\\
\alpha(\epsilon h_R(R) - f_R(R,\phi)) + a\left[h_R(R)\delta - \beta f_{R\phi}(R,\phi) - \gamma f_{RR}(R,\phi)\right. \\
+\partial_t\xi f_R(R,\phi) + \gamma\epsilon h_{RR}(R) - \partial_t\xi\epsilon h_R(R) + 2\partial_a\alpha\epsilon h_R(R) \\
\left.-2\partial_a\alpha f_R(R,\phi)\right] - \partial_a\beta a^2 f_{R\phi} + \partial_a\delta\; a^2 h_R(R) = 0 \\
\\
12\alpha h_R(R) + 6a[\gamma h_{RR} + h_R\,(-\partial_t\xi + \partial_\epsilon\delta + \partial_a\alpha)] - \partial_a\beta a^2 = 0 \\
\\
3\alpha - a\left(\partial_t\xi - \partial_\epsilon\delta - \partial_\phi\beta\right) = 0 \\
\\
\alpha = \alpha(a), \quad \beta = \beta(a,\phi), \quad \gamma = \gamma(R), \quad \delta = \delta(a,\epsilon), \quad \xi = \xi(t), \quad g = g_0\,.
\end{cases}
\tag{11.28}
$$

The preceding system leads to three different classes of internal symmetries, whose generators, functions, and conserved quantities are

$$
\begin{cases}
\mathcal{X} = (-3\alpha_0 t + \xi_1)\partial_t + \alpha_0 a\partial_a - 6\dfrac{\alpha_0}{n-1}R\partial_R + 6\alpha_0\epsilon\partial_\epsilon \\
h(R) = h_0 R, \qquad f(R,\phi) = f_0 R^n F(\phi) = f_0 R^n F(\Box^{-1}R) \\
j = \dfrac{6a^2}{R^2}\left\{\alpha_0\epsilon R^2\left(8h_0\dot{a} + a\dot{\phi}\right) + \alpha_0 aR\left[h_0 R\dot{\epsilon} - f_0 nR^n\dot{\phi}F_\phi(\phi)\right]\right. \\
\left.-f_0 nF(\phi)R^n\left[-4R\dot{a} + (n-1)a\dot{R}\right]\right\} \\
\\
\mathcal{X} = \gamma_0 R\partial_R \\
h(R) = h_0 R, \qquad f(R,\phi) = f_0 R^n + f_1\phi^k = f_0 R^n + f_1(\Box^{-1}R)^k\,. \\
j = 0 \\
\\
\mathcal{X} = (-3\alpha_0 t + \xi_1)\partial_t + \alpha_0 a\partial_a - 6\dfrac{\alpha_0}{n-1}R\partial_R + \beta_0\phi + 6\alpha_0\epsilon\partial_\epsilon \\
h(R) = h_0 R, \qquad f(R,\phi) = f_0 R^n e^{k\phi} = f_0 R^n e^{\Box^{-1}R} \\
j = a^2\left\{-a\left[6\alpha_0 h_0 + \beta_0\phi\right)\dot{\epsilon} - 6\alpha_0\epsilon\left(8h_0\dot{a} + a\dot{\phi}\right)\right. \\
-6f_0 ne^{k\phi}R^{n-1}\left[(4\alpha_0 - \beta_0 k\phi)\dot{a} - \alpha_0 ka\dot{\phi}\right] \\
\left.+6\alpha_0 f_0 n(n-1)e^{k\phi}aR^{n-2}\dot{R}\right\},
\end{cases}
\tag{11.29}
$$

where only the last one admits analytic cosmological solutions. Substituting the form of the last functions into Eqs. (11.26), we obtain the following de Sitter-like solution:

$$a(t) = a_0 e^{mt}, \quad \phi(t) = -4h_0 mt, \quad R(t) = -12m^2,$$

$$\epsilon(t) = \frac{e^{-3mt}\left[\frac{3^{1+n}4^n e^{(3-4h_0k)mt} f_0(-m^2)^n}{h_0(3-4h0k)}\right]}{(12m^2)}, \tag{11.30}$$

with the constraints

$$m = \frac{-3+10h_0k-8h_0^2k^2}{2(-3+18h_0k-31h_0^2k^2+20h_0^3k^3)}, \qquad n = \frac{3}{1+4h_0k}. \tag{11.31}$$

This result shows that nonlocal curvature cosmology is consistent with an accelerating expansion of the scale factor of the universe.

Also, external symmetries can be found from the system (11.28), whose generator reads

$$\mathcal{X} = (\xi_0 t + \xi_1)\partial_t + \frac{\xi_0}{3}(2n-1)\partial_a - 2\xi_0 R\partial_R + \frac{2\xi_0(1-\ell)}{n}\partial_\phi + (2\xi_0(1-n)\epsilon + \delta_1)\partial_\epsilon \tag{11.32}$$

with $\xi_0, \xi_1, n, \ell, \delta_0, \delta_1$ constants. Two functions correspond to the preceding generator, namely

$$f_I(R,\phi) = \frac{\delta_1}{2\xi_0(n-1)}R + [2\xi_0 R]^n \mathcal{F}\left(\phi + \frac{(1-n)}{\ell}\ln[2\xi_0 R]\right), \tag{11.33}$$

$$f_{II}(R,\phi) = \frac{\delta_1}{2\xi_0(n-1)}R + G(R)\, e^{k\phi}, \tag{11.34}$$

where $\mathcal{F}(\phi + \frac{(1-n)}{\ell}\ln[2\xi_0 R])$ is an arbitrary integration function of $\left(\phi + \frac{(1-n)}{\ell}\ln[2\xi_0 R]\right)$ and $G(R)$ a function of the curvature scalar. The latter can be addressed to the third solution of (11.29) by an appropriate choice of $G(R)$. In the search for cosmological solutions, we focus on f_I and assume the function to be linearly dependent on its argument, namely

$$\mathcal{F}_1\left(\phi + \frac{(1-n)}{\ell}\ln[2\xi_0 R]\right) \equiv \phi + \frac{(1-n)}{\ell}\ln[2\xi_0 R] + q,$$

with q constant. Under this assumption we get

$$f_1(R,\phi) = \frac{\delta_1}{2\xi_0(n-1)}R + (2\xi_0 R)^n(q+\phi) + (2\xi_0 R)^n\frac{(1-n)}{\ell}\ln[2\xi_0 R]. \tag{11.35}$$

Note that by choosing $n = 2$, the Starobinsky gravity, non-minimally coupled to a scalar field ϕ, is recovered and the function reduces to

$$f_1(R,\phi)\Big|_{n=2} = \frac{\delta_1}{2\xi_0(n-1)}R + 4\,\xi_0^2 R^2(q+\phi) - \frac{4\xi_0^2}{\ell}R^2\ln[2\xi_0 R]\,. \tag{11.36}$$

The solution of the equations of motion for arbitrary n yields three different scale factors. The first one is a de Sitter-like expansion that holds for $n = 3$:

$$a(t) = a_0 e^{\Lambda t}, \quad R(t) = -12\Lambda^2, \quad \phi(t) = -\frac{1}{3}(40 + 3q) - 4\Lambda t,$$
$$\epsilon(t) = 576\,(2\xi_0)^3\,\Lambda^5 t - \frac{C_3 e^{-3\Lambda t}}{3\Lambda} + \frac{\delta_1}{4\xi_0}, \tag{11.37}$$

with the constraint

$$\Lambda = \sqrt{-\frac{1}{24\xi_0 e}}\,(\xi_0 < 0)\,. \tag{11.38}$$

In this case, the function is further constrained by EL equations to be

$$f_1^{(1)}(R,\phi) = \frac{\delta_1}{4\xi_0} R + (\phi + q)\,(2\xi_0 R)^3 - \frac{16\xi_0^3}{\ell} R^3 \ln\left[2\xi_0 R\right]. \tag{11.39}$$

A second solution occurs for vanishing curvature scalar and leads to a power-law scale factor of the form

$$a(t) = a_0 t^{\frac{1}{2}}, \quad R(t) = 0, \quad \phi(t) = C_2, \quad \epsilon(t) = \frac{\delta_1}{2\xi_0(n-1)} - \frac{2C_3}{\sqrt{t}}. \tag{11.40}$$

Because of the vanishing Ricci scalar, in this case the theory turns out to be equivalent to GR minimally coupled to a scalar field, described by the function

$$f_1^{(2)}(R,\phi) = \frac{\delta_1}{2\xi_0(n-1)} R + \phi. \tag{11.41}$$

This represents a radiation-dominated universe. Finally, in the last case, we obtain

$$a(t) = a_0 t^{-10}, \quad R(t) \sim t^{-2}, \quad \phi(t) \sim C_2 + \ln(t),$$
$$\epsilon(t) = \frac{\delta_1}{2\xi_0(n-1)} + C_3 t^{31} + C_4\,(2\xi_0)^3\,t^{-4}, \tag{11.42}$$

where the corresponding function is constrained with respect to the general one in Eq. (11.35) to be

$$f_1^{(3)}(R,\phi) = \frac{\delta_1}{2\xi_0(n-1)} R + (\phi + q)\,(2\xi_0 R)^3 - \frac{16\xi_0^3}{\ell} R^3 \ln\left[2\xi_0 R\right]. \tag{11.43}$$

Notice that, regardless of the restriction induced by the EL equations' solutions, the above function is still a generalization of $f_1^{(1)}$.

The models described in this section contain both higher-order curvature invariants and local scalar fields capable of triggering the inflation and the late-time cosmic acceleration, respectively. In order to investigate in detail whether these solutions fit cosmological and astrophysical observations, a complete analysis is needed. Here we have only provided a starting point aimed at selecting the best nonlocal candidates to solve both UV and IR issues suffered by GR.

11.5 Nonlocality with Gauss–Bonnet Scalar

Let us now consider two nonlocal actions, containing functions of the Gauss–Bonnet term. Specifically, first we search for symmetries of a nonlocal Lagrangian only containing a function of $\mathcal{G}$ and $\Box^{-1}h(\mathcal{G})$, with $h(\mathcal{G})$ being a general function of $\mathcal{G}$. Then we restore GR limit by adding the curvature scalar into the action. Thus we compare results provided by the Noether Symmetry Approach, showing that the former can be addressed to the latter in given limits.

11.5.1 $f(\mathcal{G}, \Box^{1}\mathcal{G})$ Gravity

Similar to what discussed in Section 11.4, we consider a nonlocal IKG with action

$$S = \int \sqrt{-g} f(\mathcal{G}, \Box^{-1}h(\mathcal{G}))\, d^4x\,. \tag{11.44}$$

With respect to the previous case, the action depends only on the Gauss–Bonnet term, and a function $h(\mathcal{G})$ is considered in the starting action. The curvature scalar can be recovered as a limit of $f(\mathcal{G})$ gravity. As previously mentioned, in a FLRW background, the function $f(\mathcal{G}) = \sqrt{\mathcal{G}}$ provides the same solutions as Einstein GR. For this reason, the propagator $\Box^{-1}h(\mathcal{G})$ can be compared to $\Box^{-1}R$ for $h(\mathcal{G}) = \sqrt{\mathcal{G}}$.

Starting from this consideration, we want to derive the cosmological point-like Lagrangian and then search for Noether symmetries according to the lines sketched in the previous section. The first issue is to define a suitable localization of the nonlocal field. As before, we can define

$$\Box^{-1}h(\mathcal{G}) \equiv \phi(t), \qquad \text{so that} \qquad h(\mathcal{G}) = \Box\phi(t)\,. \tag{11.45}$$

According to Eq. (8.12) the Gauss–Bonnet scalar can be expressed as a function of the scale factor, so that the cosmological expression of $\mathcal{G}$ can be used as a Lagrange multiplier. Therefore, considering also the localization (11.45), we can define the further scalar field $\epsilon(t)$ such that action (11.44) can be written as:

$$S = \int \left\{ a^3 f(\mathcal{G},\phi) + a^3\epsilon(t)\left[\ddot{\phi} + 3\frac{\dot{a}}{a}\dot{\phi} - h(\mathcal{G})\right] - \lambda\left[\mathcal{G} - 24\left(\frac{\dot{a}^2\ddot{a}}{a^3}\right)\right]\right\} dt\,. \tag{11.46}$$

Also here, in order to write the cosmological Lagrangian and apply the Noether Approach, ϕ and $\mathcal{G}$ must be treated as separated fields. The variation of the action with respect to the Gauss–Bonnet term allows one to find the Lagrange multiplier λ:

$$\begin{aligned}\frac{\delta S}{\delta \mathcal{G}} &= \int \left\{a^3 f_{\mathcal{G}}(\mathcal{G},\phi) - a^3\epsilon(t)h_{\mathcal{G}}(\mathcal{G}) - \lambda\right\} dt = 0 \\ &\to \lambda = a^3\left[f_{\mathcal{G}}(\mathcal{G},\phi) - \epsilon(t)h_{\mathcal{G}}(\mathcal{G})\right]\,.\end{aligned} \tag{11.47}$$

Furthermore, it is easy to verify that once the action is varied with respect to $\epsilon(t)$, one recovers the definition $\Box\phi - h(\mathcal{G}) = 0$. From the variation with respect to the scalar field ϕ, we get

$$\frac{\delta S}{\delta\phi} = \int\sqrt{-g}\left\{f_\phi(\mathcal{G},\phi) - \frac{\delta}{\delta\phi}\left[\epsilon(t)(\Box\phi - h(\mathcal{G}))\right]\right\}d^4x \tag{11.48}$$
$$= \int\sqrt{-g}\left\{f_\phi(\mathcal{G},\phi) - \frac{\delta}{\delta\phi}\left[\epsilon(t)\Box(\phi - \Box^{-1}h(\mathcal{G}))\right]\right\}d^4x\,.$$

Using the divergence theorem, the last term in Eq. (11.49) can be written as

$$\int\sqrt{-g}\left\{\frac{\delta}{\delta\phi}\left[\epsilon(t)\Box(\phi - \Box^{-1}h(\mathcal{G}))\right]\right\}d^4x$$
$$= \frac{\delta}{\delta\phi}\int\sqrt{-g}\Box\epsilon(t)(\phi - \Box^{-1}h(\mathcal{G}))\,d^4x \tag{11.49}$$
$$= \int\sqrt{-g}\,\Box\epsilon(t)\frac{\delta}{\delta\phi}(\phi - \Box^{-1}h(\mathcal{G}))\,d^4x = \int\sqrt{-g}\,\Box\epsilon(t)\,d^4x\,,$$

and thus the variation with respect to the scalar field ϕ provides the following Klein–Gordon equation:

$$\frac{\delta S}{\delta\phi} = 0 \quad\rightarrow\quad \Box\epsilon(t) = f_\phi(\mathcal{G},\phi)\,. \tag{11.50}$$

After introducing the Lagrange multipliers and integrating out the higher derivatives, the point-like Lagrangian reads

$$\mathcal{L} = a^3\left[f(\mathcal{G},\phi) - \mathcal{G}f_{\mathcal{G}}(\mathcal{G},\phi) - \epsilon h(\mathcal{G}) + \epsilon\mathcal{G}h_{\mathcal{G}}(\mathcal{G})\right] - a^3\dot{\phi}\dot{\epsilon}$$
$$+ 8\epsilon\dot{a}^3\dot{\mathcal{G}}h_{\mathcal{G}\mathcal{G}}(\mathcal{G}) - 8\dot{a}^3\dot{\mathcal{G}}f_{\mathcal{G}\mathcal{G}}(\mathcal{G},\phi) + 8\dot{a}^3\dot{\epsilon}h_{\mathcal{G}}(\mathcal{G}) - 8\dot{a}^3\dot{\phi}f_{\mathcal{G}\phi}(\mathcal{G},\phi)\,, \tag{11.51}$$

and yields the following equations of motion:

$$\begin{cases} 8\dot{a}\left[2\ddot{a}\left(-\dot{\mathcal{G}}f_{\mathcal{G}\mathcal{G}}(\mathcal{G},\phi) - \dot{\phi}f_{\mathcal{G}\phi}(\mathcal{G},\phi) + \epsilon\dot{\mathcal{G}}h_{\mathcal{G}\mathcal{G}}(\mathcal{G})\right) + \dot{a}\left(-\ddot{\mathcal{G}}f_{\mathcal{G}\mathcal{G}}(\mathcal{G},\phi)\right.\right. \\ -2\dot{\mathcal{G}}\dot{\phi}f_{\mathcal{G}\mathcal{G}\phi}(\mathcal{G},\phi) - \dot{\mathcal{G}}^2 f_{\mathcal{G}\mathcal{G}\mathcal{G}}(\mathcal{G},\phi) - \ddot{\phi}f_{\mathcal{G}\phi}(\mathcal{G},\phi) - \dot{\phi}^2 f_{\mathcal{G}\phi\phi}(\mathcal{G},\phi) + \ddot{\epsilon}h_{\mathcal{G}}(\mathcal{G}) \\ \left.\left. +\epsilon\ddot{\mathcal{G}}h_{\mathcal{G}\mathcal{G}}(\mathcal{G}) + \epsilon\dot{\mathcal{G}}^2 h_{\mathcal{G}\mathcal{G}\mathcal{G}}(\mathcal{G})\right) + 2\dot{\epsilon}\left(\ddot{a}h_{\mathcal{G}}(\mathcal{G}) + \dot{a}\dot{\mathcal{G}}h_{\mathcal{G}\mathcal{G}}(\mathcal{G})\right)\right] \\ +a^2\left[\mathcal{G}f_{\mathcal{G}}(\mathcal{G},\phi) - f(\mathcal{G},\phi) + \dot{\epsilon}\dot{\phi} + \epsilon\left(h(\mathcal{G}) - \mathcal{G}h_{\mathcal{G}}(\mathcal{G})\right)\right] = 0 \\ \\ \Box\epsilon = f_\phi(\mathcal{G},\phi) \\ \\ \mathcal{G} = 24\dfrac{\dot{a}^2\ddot{a}}{a^3} \\ \\ \Box\phi = h(\mathcal{G}) \\ \\ a^3\left(f(\mathcal{G},\phi) - \epsilon h(\mathcal{G}) + \epsilon\mathcal{G}h_{\mathcal{G}}(\mathcal{G}) + \dot{\phi}\dot{\epsilon} - \mathcal{G}f_{\mathcal{G}}(\mathcal{G},\phi)\right) \\ -24\dot{a}^3\left(\dot{\epsilon}h_{\mathcal{G}}(\mathcal{G}) + \epsilon\dot{\mathcal{G}}h_{\mathcal{G}\mathcal{G}}(\mathcal{G}) - \dot{\phi}f_{\mathcal{G}\phi}(\mathcal{G},\phi) - \dot{\mathcal{G}}f_{\mathcal{G}\mathcal{G}}(\mathcal{G},\phi)\right) = 0\,. \end{cases} \tag{11.52}$$

Once the forms of the functions $h(\mathcal{G})$ and $f(\mathcal{G},\phi)$ are specified, the preceding system can provide exact cosmological solutions. In such a case, the minisuperspace is defined on the configuration space $\mathcal{S} \equiv \{a, \phi, \mathcal{G}, \epsilon\}$, so that the symmetry generator takes the explicit form

$$\begin{aligned}\mathcal{X} = \xi(t,a,\phi,\mathcal{G},\epsilon)\partial_t + \alpha(t,a,\phi,\mathcal{G},\epsilon)\partial_a + \beta(t,a,\phi,\mathcal{G},\epsilon)\partial_\phi \\ + \gamma(t,a,\phi,\mathcal{G},\epsilon)\partial_{\mathcal{G}} + \delta(t,a,\phi,\mathcal{G},\epsilon)\partial_\epsilon \,. \end{aligned} \tag{11.53}$$

The system of differential equations, coming from the preceding generator, is made of 37 equations but, after deleting all the linear combinations, it reduces to a system of five equations plus the conditions on the generator coefficients [346]:

$$\begin{cases} -\partial_t g + 3\alpha a^2 f(\mathcal{G},\phi) - \delta a^3 - 3\alpha a^2 \epsilon h(\mathcal{G}) + \delta a^3 \mathcal{G} h_{\mathcal{G}}(\mathcal{G}) + 3\alpha a^2 \epsilon \mathcal{G} h_{\mathcal{G}}(\mathcal{G}) \\ +\gamma a^3 \mathcal{G}\epsilon h_{\mathcal{G}\mathcal{G}}(\mathcal{G}) + \beta a^3 f_\phi(\mathcal{G},\phi) - 3\alpha a^2 \mathcal{G} f_{\mathcal{G}}(\mathcal{G},\phi) - \beta a^3 \mathcal{G} f_{\mathcal{G}\phi}(\mathcal{G},\phi) \\ -\gamma a^3 \mathcal{G} f_{\mathcal{G}\mathcal{G}}(\mathcal{G},\phi) + a^3 \partial_t \xi \left[f(\mathcal{G},\phi) - \epsilon h(\mathcal{G}) + \epsilon \mathcal{G} h_{\mathcal{G}}(\mathcal{G}) - \mathcal{G} f_{\mathcal{G}}(\mathcal{G},\phi)\right] = 0 \\ \\ \gamma h_{\mathcal{G}\mathcal{G}}(\mathcal{G}) - 3\partial_t\xi h_{\mathcal{G}}(\mathcal{G}) + \partial_\epsilon \delta\, h_{\mathcal{G}}(\mathcal{G}) + 3\partial_a \alpha h_{\mathcal{G}}(\mathcal{G}) \\ +\partial_\epsilon\gamma \left[\epsilon h_{\mathcal{G}\mathcal{G}}(\mathcal{G}) - f_{\mathcal{G}\mathcal{G}}(\mathcal{G},\phi)\right] = 0 \\ \\ f_{\mathcal{G}\phi}(\mathcal{G},\phi)\left(3\partial_t\xi - \partial_\phi\beta\right) + \partial_\phi\gamma\left[\epsilon h_{\mathcal{G}\mathcal{G}}(\mathcal{G}) - f_{\mathcal{G}\mathcal{G}}(\mathcal{G},\phi)\right] \\ -3\partial_a\alpha f_{\mathcal{G}\phi}(\mathcal{G},\phi) - \beta f_{\mathcal{G}\phi\phi} - \gamma f_{\mathcal{G}\mathcal{G}\phi}(\mathcal{G},\phi) = 0 \\ \\ \beta f_{\mathcal{G}\mathcal{G}\phi}(\mathcal{G},\phi) - \delta h_{\mathcal{G}\mathcal{G}}(\mathcal{G}) + \gamma f_{\mathcal{G}\mathcal{G}\mathcal{G}}(\mathcal{G},\phi) - 3\partial_t\xi f_{\mathcal{G}\mathcal{G}}(\mathcal{G},\phi) \\ +\partial_{\mathcal{G}}\gamma f_{\mathcal{G}\mathcal{G}}(\mathcal{G},\phi) + 3\partial_a\alpha f_{\mathcal{G}\mathcal{G}}(\mathcal{G},\phi) - \gamma\epsilon h_{\mathcal{G}\mathcal{G}\mathcal{G}}(\mathcal{G}) + 3\partial_t\xi\epsilon h_{\mathcal{G}\mathcal{G}} \\ -\partial_{\mathcal{G}}\gamma\epsilon h_{\mathcal{G}\mathcal{G}}(\mathcal{G}) - 3\partial_a\alpha\epsilon h_{\mathcal{G}\mathcal{G}}(\mathcal{G}) = 0 \\ \\ 3\alpha - a\left(\partial_t\xi - \partial_\epsilon\delta - \partial_\phi\beta\right) = 0 \\ \\ \alpha \equiv \alpha(a), \quad \beta \equiv \beta(\phi), \quad \gamma \equiv \gamma(\phi,\mathcal{G},\epsilon), \quad \delta \equiv \delta(\epsilon), \quad \xi \equiv \xi(t), \quad g \equiv g(t)\,. \end{cases} \tag{11.54}$$

The system admits five different solutions: in two of them the nonlocal function $f(\mathcal{G},\phi)$ is given by a sum of a function of ϕ and a function of $\mathcal{G}$, namely $f(\mathcal{G},\phi) = f_1(\mathcal{G}) + f_2(\phi)$. In the other three cases, solutions are products between the two functions, namely $f(\mathcal{G},\phi) = g_1(\mathcal{G})g_2(\phi)$ The entire set of solutions with the corresponding generators reads

$$
\begin{cases}
I: & \mathcal{X} = (\xi_0 t + \xi_1)\partial_t + \alpha_0 a \partial_a + (\beta_0\phi + \beta_1)\partial_\phi - 4\xi_0 \mathcal{G}\partial_\mathcal{G} + \delta_0\epsilon\partial_\epsilon \\
& h(\mathcal{G}) = h_0 \mathcal{G}^{\frac{1}{2}+\frac{n}{k}}, \quad f(\mathcal{G},\phi) = f_0\mathcal{G}^n + f_1\mathcal{G} + f_2\left(\beta_0\phi + \beta_1\right)^k \\
& j = 32 f_0 k^2 n(n-1)\xi_0 \mathcal{G}^{n-1}\dot{a}^3 - 24\alpha_0 f_0 k^2 n(n-1) a \mathcal{G}^{n-2}\dot{a}^2\dot{\mathcal{G}} \\
& \quad +4h_0(k+2n)\mathcal{G}^{\frac{n}{k}-\frac{1}{2}}\dot{a}^2\left\{\left[\delta_0 k + 2(k-2n)\xi_0\right]\epsilon\dot{a} + 3\alpha_0 k a\dot{\epsilon}\right\} \\
& \quad -6\alpha_0 h_0(k^2-4n^2) a\epsilon\mathcal{G}^{\frac{n}{k}-\frac{3}{2}}\dot{a}^2\dot{\mathcal{G}} - k^2 a^3\left[(\beta_1+\beta_0\phi)\dot{\epsilon} + \delta_0\epsilon\dot{\phi}\right] \\
\\
II: & \mathcal{X} = (\xi_0 t + \xi_1)\partial_t + \alpha_0 a \partial_a + (\beta_0\phi + \beta_1)\partial_\phi - 4\xi_0 \mathcal{G}\partial_\mathcal{G} + (\delta_0\epsilon + \delta_1)\partial_\epsilon \\
& h(\mathcal{G}) = h_0\mathcal{G}, \quad f(\mathcal{G},\phi) = f_0\mathcal{G}^n + f_1\mathcal{G} + f_2(\beta_0\phi+\beta_1)^{2n} \\
& j = \mathcal{G}^{-2}\left\{32 f_0\xi_0 n(n-1)\mathcal{G}^{n+3}\dot{a}^3 + 8h_0\mathcal{G}^2(\delta_0\epsilon+\delta_1)\dot{a}^3\right. \\
& \quad +24\alpha_0 a\dot{a}^2\left[h_0\mathcal{G}^2\dot{\epsilon} - f_0 n(n-1)\mathcal{G}^n\dot{\mathcal{G}}\right] \\
& \quad \left.-a^3\mathcal{G}^2\left[\beta_0\phi+\beta_1) + (\delta_0\epsilon+\delta_1)\dot{\phi}\right]\right\} \\
\\
III: & \mathcal{X} = (\xi_0 t + \xi_1)\partial_t + \alpha_0 a \partial_a + (\beta_0\phi + \beta_1)\partial_\phi - 4\xi_0 \mathcal{G}\partial_\mathcal{G} + \delta_0\epsilon\partial_\epsilon \\
& h(\mathcal{G}) = h_0\mathcal{G}^z, \quad f(\mathcal{G},\phi) = f_0\mathcal{G}^n(\beta_0\phi+\beta_1)^k \\
& j = \mathcal{G}^{-2}(\beta_1+\beta_0\phi)^{-1}\left\{-32\xi_0\mathcal{G}(\beta_1+\beta_0\phi)\left[h_0 z(z-1)\epsilon\mathcal{G}^z\right.\right. \\
& \quad \left.-f_0 n(n-1)\mathcal{G}^n(\beta_1+\beta_0\phi)^k\right]\dot{a}^3 - f_0 n(n-1)\mathcal{G}^n(\beta_1+\beta_0\phi)^{k+1}\dot{\mathcal{G}} \\
& \quad +(\beta_1+\beta_0\phi)^2\mathcal{G}^2\left[-8\beta_0 f_0 k n\mathcal{G}^{n-1}(\beta_1+\beta_0\phi)^{k-1}\dot{a}^3 - a^3\dot{\epsilon}\right] \\
& \quad +\delta_0\epsilon(\beta_1+\beta_0\phi)\mathcal{G}^2\left(h_0 z\mathcal{G}^{z-1}\dot{a}^3 - a^3\dot{\phi}\right) \\
& \quad +24\alpha_0 a\dot{a}^2\left[h_0 z\mathcal{G}^{z+1}(\beta_1+\beta_0\phi)\dot{\epsilon} + h_0 z(z-1)\epsilon\mathcal{G}^z(\beta_1+\beta_0\phi)\dot{\mathcal{G}}\right. \\
& \quad \left.\left.-\beta_0 f_0 k n\mathcal{G}^{n+1}(\beta_1+\beta_0\phi)^k\dot{\phi}\right]\right\} \\
\\
IV: & \mathcal{X} = (\xi_0 t + \xi_1)\partial_t + \alpha_0 a \partial_a + (\beta_0\phi + \beta_1)\partial_\phi - 4\xi_0 \mathcal{G}\partial_\mathcal{G} + (\delta_0\epsilon + \delta_1)\partial_\epsilon \\
& h(\mathcal{G}) = h_0\mathcal{G}, \quad f(\mathcal{G},\phi) = f_0\mathcal{G}^n(\beta_0\phi+\beta_1)^k = f_0\mathcal{G}^n(\beta_0\Box^{-1}\mathcal{G}+\beta_1)^k \\
& j = 2f_0 n(n-1)\xi_0\mathcal{G}^{n-1}(\beta_1+\beta_0\phi)^k\dot{a}^3 + (\delta_1+\delta_0\epsilon)\left(8h_0\dot{a}^3 - a^3\dot{\phi}\right) \\
& \quad +(\beta_1+\beta_0\phi)\left[-8\beta_0 f_0 k n\mathcal{G}^{n-1}(\beta_1+\beta_0\phi)^{k-1}\dot{a}^3 - a^3\dot{\epsilon}\right] \\
& \quad +24\alpha_0 a\dot{a}^2\left\{h_0\dot{\epsilon} - f_0 n\mathcal{G}^{n-2}(\beta_1\right. \\
& \quad \left.+\beta_0\phi)^{k-1}\left[(n-1)(\beta_1+\beta_0\phi)\dot{\mathcal{G}} + \beta_0 k\mathcal{G}\dot{\phi}\right]\right\} \\
\\
V: & \mathcal{X} = (\xi_0 t + \xi_1)\partial_t + \alpha_0 a \partial_a + \beta_1\partial_\phi - 4\xi_0 \mathcal{G}\partial_\mathcal{G} + \delta_0\epsilon\partial_\epsilon \\
& h(\mathcal{G}) = h_0\sqrt{\mathcal{G}}, \quad f(\mathcal{G},\phi) = f_0\mathcal{G}^n e^{k\phi}, \quad k \equiv \dfrac{\delta_0 + 4n\xi_0}{\beta_1} \\
& j = \mathcal{G}^{-\frac{3}{2}}\left\{-\beta_1 a^3\mathcal{G}^{\frac{3}{2}}\dot{\epsilon} + 12\alpha_0 h_0 a\mathcal{G}\dot{a}^3\dot{\epsilon}\right. \\
& \quad -24\alpha_0 f_0 n(n-1)e^{k\phi}a\mathcal{G}^{-\frac{1}{2}+n}\dot{a}^2\dot{\mathcal{G}} \\
& \quad +8f_0 n e^{k\phi}\mathcal{G}^{\frac{1}{2}+n}\dot{a}^2\left[-(\beta_1 k + 4\xi_0 - 4n\xi_0)\dot{a} - 3\alpha_0 k a\dot{\phi}\right] \\
& \quad \left.+\epsilon\left[4h_0(\delta_0+2\xi_0)\mathcal{G}\dot{a}^3 - 6\alpha_0 h_0 a\dot{a}^2\dot{\mathcal{G}} - \delta_0 a^3\mathcal{G}^{\frac{3}{2}}\dot{\phi}\right]\right\},
\end{cases}
\tag{11.55}
$$

where ξ_0, ξ_1, α_0, β_0, β_1, δ_0, h_0, f_0, f_1, f_2, n, k are integration constants.

It may seem that the theory is overdetermined by the large amount of free parameters. However, after solving the equations of motion, the functions will be further constrained by the cosmological solutions. Specifically, it turns out that not all the functions contained in the system (11.55) admit cosmological solutions for the scale factor. As a matter of fact, while the second and the fourth do not admit any cosmological solution, the first and the third can be analytically solved by setting $\beta_1 = 0$. The fifth admits solutions only after constraining the mutual dependence among the parameters. Let us start by analyzing the Lagrangians corresponding to the functions I and III. They read, respectively, as follows:

$$
\begin{aligned}
\mathcal{L}_I = {} & a^3 \left[f_0(1-n)\mathcal{G}^n - h_0 \left(\frac{n}{k} - \frac{1}{2} \right) \epsilon \mathcal{G}^{\frac{n}{k}+\frac{1}{2}} + \dot{\epsilon}\dot{\phi} \right] \\
& - 8h_0 \left(\frac{n}{k} + \frac{1}{2} \right) \dot{a}^3 \dot{\epsilon} \mathcal{G}^{\frac{n}{k}-\frac{1}{2}} + 8f_0 n(n-1)\dot{a}^3 \dot{\mathcal{G}} \mathcal{G}^{n-2} \\
& - 8h_0 \left(\frac{n^2}{k^2} - \frac{1}{4} \right) \epsilon \dot{a}^3 \dot{\mathcal{G}} \mathcal{G}^{\frac{n}{k}-\frac{3}{2}} + 8f_1 k \dot{a}^3 \dot{\phi} \phi^{k-1}
\end{aligned}
\tag{11.56}
$$

$$
\begin{aligned}
\mathcal{L}_{III} = {} & a^3 \left[f_0(1-n)\mathcal{G}^n \phi^k + h_0(z-1)\epsilon \mathcal{G}^z \right] + 8h_0 z \dot{a}^3 \dot{\epsilon} \mathcal{G}^{z-1} \\
& + 8h_0 z(z-1)\epsilon \dot{a}^3 \dot{\mathcal{G}} \mathcal{G}^{z-2} - 8f_0 n(n-1)\dot{a}^3 \dot{\mathcal{G}} \mathcal{G}^{n-2} \phi^k \\
& - 8f_0 k n \dot{a}^3 \dot{\phi} \mathcal{G}^{n-1} \phi^{k-1} - a^3 \dot{\epsilon}\dot{\phi}.
\end{aligned}
\tag{11.57}
$$

From the former Lagrangian, the EL equations and the energy condition provide a solution given by a power-law of $a(t)$, namely

$$
\begin{aligned}
& a(t) \sim t^{\frac{2}{3}(2n+2kz-k)}, \quad \mathcal{G}(t) \sim t^{-4}, \quad \phi(t) \sim t^{2-4z}, \quad \epsilon(t) \sim t^{2k(1-2z)}, \\
& f(\mathcal{G}, \Box^{-1}h(\mathcal{G})) = f_2 \mathcal{G}^n (\Box^{-1}\mathcal{G}^z)^k .
\end{aligned}
\tag{11.58}
$$

Though exponential solutions do not occur in vacuum, the parameters are not fixed by the equations of motion, so that they might be constrained by observations. For instance, GR can be recovered by setting $n = 1/2$. The other two free parameters can be further constrained by investigating the UV limit of the theory. Furthermore, the Lagrangian (11.56) gives exact de Sitter-like solutions of the form:

$$
\begin{aligned}
& a(t) = a_0 e^{qt}, \quad \mathcal{G}(t) \sim \text{const}, \quad \phi(t) \sim t, \quad \epsilon(t) \sim t, \quad k=1, \ n = \frac{1}{2}, \\
& f(\mathcal{G}, \Box^{-1}h(\mathcal{G})) = f_0\sqrt{\mathcal{G}} + f_1\mathcal{G} + f_2\Box^{-1}\mathcal{G} + f_3 .
\end{aligned}
\tag{11.59}
$$

Since $\mathcal{G}$ is a topological invariant, the linear term in $\mathcal{G}$ does not contribute to dynamics, so the relevant terms are the square root $\sqrt{\mathcal{G}}$ and the nonlocal term $\Box^{-1}\mathcal{G}$. From a cosmological point of view, this action is equivalent to (11.4), so the same considerations in [336] hold. In other words, a dark energy-like behavior, due to nonlocal terms, can be achieved in both R and $\mathcal{G}$ descriptions of cosmological dynamics.

In particular, the dynamics coming from (11.59) is equivalent to

$$
S = \int \sqrt{-g}\,(f_0 R + f_2 \Box^{-1}\mathcal{G})\, d^4x \,,
\tag{11.60}
$$

so that the nonlocal term plays the role of cosmological constant. Similar considerations hold for the case $\mathcal{L}_{III}$.

Finally, the point-like Lagrangian corresponding to the last solution is

$$\mathcal{L}_V = 2\dot{a}^3\left[2\mathcal{G}^{-\frac{1}{2}}\dot{\epsilon} - \epsilon\mathcal{G}^{-\frac{3}{2}}\dot{\mathcal{G}} - 4f_0 n(n-1)\mathcal{G}^{n-2}\dot{\mathcal{G}}e^{k\phi} - 4f_0 kn\mathcal{G}^{n-1}e^{k\phi}\dot{\phi}\right] - \frac{1}{2}\mathcal{G}^2 a^3\left[2\dot{\epsilon}\dot{\phi} + \epsilon\sqrt{\mathcal{G}} + 2f_0(n-1)\mathcal{G}^n e^{k\phi}\right] \tag{11.61}$$

and Eqs (11.52) can be analytically solved, providing two different forms of the scale factor. The first reads as

$$a(t) = a_0 e^{qt}, \quad \phi(t) = \sqrt{\frac{8}{3}}qt, \quad \epsilon(t) \sim e^{\sqrt{\frac{8}{3}}kqt}, \tag{11.62}$$

$$f(\mathcal{G}, \Box^{-1}\sqrt{\mathcal{G}}) = f_0\mathcal{G}^{\frac{12\sqrt{6}}{4k-\sqrt{6}}}e^{k\phi}, \qquad h(\mathcal{G}) = \sqrt{\mathcal{G}}. \tag{11.63}$$

By comparing the function in Eq. (11.63) with the fifth of Eq. (11.55), we notice that a relation between the free parameters n and k occurs, namely

$$n = \frac{12\sqrt{6}}{4k - \sqrt{6}}. \tag{11.64}$$

Furthermore, we also find power-law solutions, namely

$$a(t) = a_0 t^q, \quad \phi(t) = \frac{2\sqrt{6q^3(q-1)}\ln[(1-3q)t]}{3q-1},$$
$$\mathcal{G}(t) = \frac{24q^3(q-1)}{t^4}, \quad \epsilon(t) \sim t^{2-4n+\frac{2k\sqrt{6q^3(q-1)}}{3q-1}}. \tag{11.65}$$

The second solution, coming from the Lagrangian (11.61), introduces a relation among the parameters n, q, and k, enlarging the possibility to compare these cosmological behaviors with observational data. However, no analytic relation for any q, n, k occurs, but it is possible to find the values of such parameters in some particular cases. For example, in what follows we outline the values of q labeling radiation, stiff matter, and dust matter cases:

- Radiation $\rightarrow$ $q = \frac{1}{2}$ $\rightarrow$ $k = \frac{\sqrt{6}}{2}(5 - 8n)i$.

$$a(t) = a_0 t^{\frac{1}{2}}, \quad \mathcal{G}(t) = -\frac{3}{2t^4}, \quad \phi(t) = \sqrt{6}i\ln\left(-\frac{t}{2}\right), \quad \epsilon(t) \rightarrow \infty. \tag{11.66}$$

- Stiff Matter $\rightarrow$ $q = \frac{1}{3}$ $\rightarrow$ $k = 0 \vee n = 0$.

1. $k = 0$:

$$a(t) = a_0 t^{\frac{1}{3}}, \quad \mathcal{G}(t) = -\frac{16}{27t^4}, \quad \phi(t) \rightarrow \infty,$$
$$\epsilon(t) \sim \text{indet}, \quad f(\mathcal{G}, \phi) = f_0\mathcal{G}^n. \tag{11.67}$$

2. $n = 0$:

$$a(t) = a_0 t^{\frac{1}{3}}, \quad \mathcal{G}(t) = -\frac{16}{27t^4}, \quad \phi(t) \rightarrow \infty,$$
$$\epsilon(t) \sim \text{indet}, \quad f(\mathcal{G}, \phi) = f_0 e^{k\phi}. \tag{11.68}$$

- Dust Matter $\rightarrow\ q=\frac{2}{3}\ \rightarrow\ k=\frac{3\sqrt{3}}{8}(3-4n)i.$

$$a(t)=a_0 t^{\frac{2}{3}},\quad \mathcal{G}(t)=-\frac{64}{27t^4},\quad \phi(t)=\frac{8\sqrt{3}}{9}i\ln(-t),\quad \epsilon(t)\rightarrow\infty. \tag{11.69}$$

Finding the values of n that best fit cosmological observations is an important perspective that deserves to be developed in specific studies.

11.5.2 General Relativity with Nonlocal Gauss–Bonnet Corrections

To conclude this section, let us treat the case of GR corrected with nonlocal Gauss–Bonnet terms, considered, for example, in [334, 347]. The action is

$$S=\int\sqrt{-g}\left[\frac{R}{2}+f(\mathcal{G},\Box^{-1}h(\mathcal{G})\right]d^4x\,, \tag{11.70}$$

that is, Eq. (11.44) with the addition of the EH term. Let us make use of the Lagrange multipliers method to find the cosmological Lagrangian and, therefore, to apply the Noether Symmetry Approach. The only difference with respect to the case given by Eq. (11.51) is the presence of the cosmological form of R, so that the Lagrangian reads as follows:

$$\begin{aligned}\mathcal{L}=&\,a^3\left[f(\mathcal{G},\phi)-\mathcal{G}f_{\mathcal{G}}(\mathcal{G},\phi)-\epsilon h(\mathcal{G})+\epsilon\mathcal{G}h_{\mathcal{G}}(\mathcal{G})\right]-8\dot{a}^3\dot{\mathcal{G}}f_{\mathcal{G}\mathcal{G}}(\mathcal{G},\phi)\\&+8\dot{a}^3\dot{\epsilon}h_{\mathcal{G}}(\mathcal{G})-a^3\dot{\phi}\dot{\epsilon}+8\epsilon\dot{a}^3\dot{\mathcal{G}}h_{\mathcal{G}\mathcal{G}}(\mathcal{G})-8\dot{a}^3\dot{\phi}f_{\mathcal{G}\phi}(\mathcal{G},\phi)+3a\dot{a}^2\,,\end{aligned} \tag{11.71}$$

and the only different EL equation is that related to the scale factor, which, in this case, takes the form

$$\begin{aligned}&8\dot{a}\Big[2\ddot{a}\left(-\dot{\mathcal{G}}f_{\mathcal{G}\mathcal{G}}(\mathcal{G},\phi)-\dot{\phi}f_{\mathcal{G}\phi}(\mathcal{G},\phi)+\epsilon\dot{\mathcal{G}}h_{\mathcal{G}\mathcal{G}}(\mathcal{G})\right)\\&+\dot{a}\left(-\ddot{\mathcal{G}}f_{\mathcal{G}\mathcal{G}}(\mathcal{G},\phi)-2\dot{\mathcal{G}}\dot{\phi}f_{\mathcal{G}\mathcal{G}\phi}(\mathcal{G},\phi)-\dot{\mathcal{G}}^2f_{\mathcal{G}\mathcal{G}\mathcal{G}}(\mathcal{G},\phi)-\ddot{\phi}f_{\mathcal{G}\phi}(\mathcal{G},\phi)\right.\\&\left.-\dot{\phi}^2f_{\mathcal{G}\phi\phi}(\mathcal{G},\phi)+\ddot{\epsilon}h_{\mathcal{G}}(\mathcal{G})+\epsilon\ddot{\mathcal{G}}h_{\mathcal{G}\mathcal{G}}(\mathcal{G})+\epsilon\dot{\mathcal{G}}^2h_{\mathcal{G}\mathcal{G}\mathcal{G}}(\mathcal{G})\right)\\&+2\dot{\epsilon}\left(\ddot{a}h_{\mathcal{G}}(\mathcal{G})+\dot{a}\dot{\mathcal{G}}h_{\mathcal{G}\mathcal{G}}(\mathcal{G})\right)\Big]+3(\dot{a}^2+2a\ddot{a})\\&+a^2\left[\mathcal{G}f_{\mathcal{G}}(\mathcal{G},\phi)-f(\mathcal{G},\phi)+\dot{\epsilon}\dot{\phi}+\epsilon\left(h(\mathcal{G})-\mathcal{G}h_{\mathcal{G}}(\mathcal{G})\right)\right]=0\,.\end{aligned} \tag{11.72}$$

The minisuperspace dimension is the same as the previous case, since the scalar curvature does not introduce any new dynamical variable. By replacing $R(t)$ with its cosmological expression, the Noether system turns out to be the same as in the previous section, except for the addition of a further condition on the Noether vector, that is,

$$\alpha-a\partial_t\xi+2a\partial_a\alpha=0\,. \tag{11.73}$$

This new relation between α and ξ, provided by Eq. (11.73), yields an important implication for the solutions of Noether's system, since it uniquely fixes the value of n and the relation between n and k. We obtain five different generators (with corresponding functions) of the form

$$\begin{cases} \mathcal{X} = (3\alpha_0 t + \xi_1)\partial_t + \alpha_0 a\partial_a + (\beta_0\phi + \beta_1)\partial_\phi - 12\alpha_0\mathcal{G}\partial_\mathcal{G} + \delta_0\epsilon\partial_\epsilon \\ h(\mathcal{G}) = h_0\mathcal{G}^{\frac{1}{2}+\frac{1}{2k}}, \quad f(\mathcal{G},\phi) = f_0\mathcal{G}^{\frac{1}{2}} + f_1\mathcal{G} + f_2\left(\beta_0\phi + \beta_1\right)^k \\ \\ \mathcal{X} = (3\alpha_0 t + \xi_1)\partial_t + \alpha_0 a\partial_a + (\beta_0\phi + \beta_1)\partial_\phi - 12\alpha_0\mathcal{G}\partial_\mathcal{G} + (\delta_0\epsilon + \delta_1)\partial_\epsilon \\ h(\mathcal{G}) = h_0\mathcal{G}, \quad f(\mathcal{G},\phi) = f_0\mathcal{G}^{\frac{1}{2}} + f_1\mathcal{G} + f_2(\beta_0\phi + \beta_1) \\ \\ \mathcal{X} = (3\alpha_0 t + \xi_1)\partial_t + \alpha_0 a\partial_a + (\beta_0\phi + \beta_1)\partial_\phi - 12\alpha_0\mathcal{G}\partial_\mathcal{G} + \delta_0\epsilon\partial_\epsilon \\ h(\mathcal{G}) = h_0\mathcal{G}^{\frac{1-2n}{2k}}, \quad f(\mathcal{G},\phi) = f_0\mathcal{G}^n(\beta_0\phi + \beta_1)^k \\ \\ \mathcal{X} = (3\alpha_0 t + \xi_1)\partial_t + \alpha_0 a\partial_a + (\beta_0\phi + \beta_1)\partial_\phi - 12\alpha_0\mathcal{G}\partial_\mathcal{G} + (\delta_0\epsilon + \delta_1)\partial_\epsilon \\ h(\mathcal{G}) = h_0\mathcal{G}, \quad f(\mathcal{G},\phi) = f_0\mathcal{G}^n(\beta_0\phi + \beta_1)^{1-2n} \\ \\ \mathcal{X} = (3\alpha_0 t + \xi_1)\partial_t + \alpha_0 a\partial_a + \beta_1\partial_\phi - 12\alpha_0\mathcal{G}\partial_\mathcal{G} \\ h(\mathcal{G}) = h_0\sqrt{\mathcal{G}}, \quad f(\mathcal{G},\phi) = f_0\mathcal{G}^n e^{k\phi}. \end{cases} \tag{11.74}$$

By comparing Eqs. (11.55) with Eqs. (11.74), we notice that, with regard to the first two solutions, the introduction of R leads to the further constraint $n = 1/2$. Moreover, in the third and in the fourth cases, the further relations $z = \frac{1-2n}{2k}$ and $k = 1 - 2n$ occur, respectively. According to these considerations, it is clear that action (11.70) is fully consistent with (11.44) and then there is no reason to introduce by hand the EH term to recover GR in the context of Nonlocal Gauss–Bonnet Gravity.

11.6 Nonlocality with Torsion

Here we want to take into account nonlocal modification of TEGR action containing nonlocal contributions of torsion T and of boundary term B. In this perspective, Nonlocal Curvature Gravity is recovered by the application of nonlocal operator $\Box^{-1}$ to the relation $R = -T + B$, namely $\Box^{-1}R = -\Box^{-1}T + \Box^{-1}B$. In other words, Nonlocal Curvature Gravity can be exactly recovered as a particular case.

The Noether symmetry condition (6.42) can be applied after introducing the Lagrange multipliers, similarly to Sections 11.4 and 11.5 of this chapter. We start from the action considered in Refs. [344, 348], that is,

$$S = \int \frac{h}{2}\left[-T + (\tau T + \chi B) f(\Box^{-1}T, \Box^{-1}B)\right] d^4x, \tag{11.75}$$

where h is the determinant of tetrad fields and τ, χ real constants. The Ricci scalar can be recovered by the choice $\tau = 1$, $\chi = -1$. On the other hand, when the nonlocal term vanishes, standard TEGR is restored. Notice that the action is not in the most general form: the function f only depends on the nonlocal terms, while the contributions of torsion and boundary term are not included. This ansatz allows one to reduce the dynamics and to restrict the minisuperspace to five dimensions. From the definitions

$$\Box^{-1}T \equiv \phi(t) \;\rightarrow\; T = \Box\phi(t), \qquad \Box^{-1}B \equiv \varphi(t) \;\rightarrow\; B = \Box\varphi(t), \tag{11.76}$$

the action can be localized and a suitable Lagrangian can be obtained. Moreover, in analogy with the two previous sections, two additional scalar fields $\epsilon(t)$ and $\zeta(t)$ must be introduced as Lagrange multipliers. The action therefore reads

$$S = \int \frac{h}{2}\left[-T + (\tau T + \chi B)f(\phi,\varphi) + \epsilon(\Box\phi - T) + \zeta(\Box\varphi - B)\right] d^4x. \tag{11.77}$$

By varying the action with respect to ϵ and ζ we get the Klein–Gordon equations:

$$\Box\epsilon = (\tau T + \chi B)f_\phi(\phi,\varphi), \tag{11.78}$$
$$\Box\zeta = (\tau T + \chi B)f_\varphi(\phi,\varphi). \tag{11.79}$$

In a FLRW space-time with a diagonal set of tetrad fields, second derivatives of the scalar fields and of the scale factor arise into the action. Integrating out the higher-order terms, the canonical point-like Lagrangian turns out to be

$$\begin{aligned}\mathcal{L} = {}& 3a^2\dot a\left[\chi f_\phi(\phi,\varphi)\dot\phi + \chi f_\varphi(\phi,\varphi)\dot\varphi - \dot\zeta\right] \\ & + 3a\dot a^2\left[\epsilon + 1 - \tau f(\phi,\varphi)\right] - \frac{1}{2}a^3\dot\zeta\dot\varphi - \frac{1}{2}a^3\dot\epsilon\dot\phi,\end{aligned} \tag{11.80}$$

whose equations of motion are

$$\begin{cases} 3\dot a^2(\epsilon - \tau f + 1) - \frac{1}{2}\dot\zeta\dot\varphi a^2 - \frac{1}{2}\dot\epsilon\dot\phi a^2 - 3a\dot a\left[\dot\zeta - \chi a^2\left(f_\phi\dot\phi + f_\varphi\dot\varphi\right)\right] = 0 \\ \left(2\frac{\ddot a}{a} + \frac{\dot a^2}{a^2}\right)(\epsilon - \tau f + 1) + \frac{1}{2}\dot\zeta\dot\varphi + \frac{1}{2}\dot\epsilon\dot\phi \\ + \left(f_\phi\dot\phi + f_\varphi\dot\varphi\right)\left[2\frac{\dot a}{a}(\tau + 2\chi) + \chi\right] - 2\frac{\dot a}{a}(2\dot\zeta + \dot\epsilon) - \ddot\zeta = 0 \\ 6\left(\frac{\dot a}{a}\right)^2 + 3\left(\frac{\dot a}{a}\right)\dot\phi + \ddot\phi = 0 \\ 6\left(\frac{\ddot a}{a} + 2\frac{\dot a^2}{a^2}\right) + 3\left(\frac{\dot a}{a}\right)\dot\varphi + \ddot\varphi = 0 \\ -6\left(\frac{\dot a}{a}\right)^2(\tau f_\varphi + 3\chi f_\varphi) - 6\left(\frac{\ddot a}{a} - \frac{\dot a^2}{a^2}\right)\chi f_\varphi + 3\left(\frac{\dot a}{a}\right)\dot\zeta + \ddot\zeta = 0 \\ -6\left(\frac{\dot a}{a}\right)^2(\tau f_\phi + 3\chi f_\phi) - 6\left(\frac{\ddot a}{a} - \frac{\dot a^2}{a^2}\right)\chi f_\phi + 3\left(\frac{\dot a}{a}\right)\dot\epsilon + \ddot\epsilon = 0. \end{cases} \tag{11.81}$$

The generator of the symmetry in the five-dimensional minisuperspace $\mathcal{S} \equiv \{a, \phi, \varphi, \epsilon, \zeta\}$ reads

$$\mathcal{X} = \xi(t)\partial_t + \alpha(a,\phi,\varphi,\epsilon,\zeta,t)\partial_a + \beta(a,\phi,\varphi,\epsilon,\zeta,t)\partial_\phi \\ + \gamma(a,\phi,\varphi,\epsilon,\zeta,t)\partial_\varphi + \delta(a,\phi,\varphi,\epsilon,\zeta,t)\partial_\epsilon + \theta(a,\phi,\varphi,\epsilon,\zeta,t)\partial_\zeta, \tag{11.82}$$

where ξ is supposed to depend on the cosmic time only, according to the result provided in Section 6.3. The application of Noether's identity (6.42) to the preceding Lagrangian yields a system of 43 partial differential equations, but only 18 of them are linearly independent [344]:

$$\begin{cases} a\partial_t\theta - 6\chi f_\varphi \partial_t\alpha = 0 \\ a\partial_t\delta - 6\chi f_\phi \partial_t\alpha = 0 \\ 6\partial_t\alpha + a\partial_t\gamma = 0 \\ a\left(\partial_\varphi\beta + \partial_\epsilon\theta\right) - 6\chi f_\varphi \partial_\epsilon\alpha = 0 \\ a\left(\partial_\zeta\beta + \partial_\epsilon\gamma\right) + 6\partial_\epsilon\alpha = 0 \\ 6\chi f_\phi \partial_\phi\alpha - a\partial_\phi\delta = 0 \\ 6\chi f_\varphi \partial_\varphi\alpha - a\partial_\varphi\theta = 0 \\ 6\partial_\zeta\alpha + a\partial_\zeta\varphi = 0 \\ -6a^2\left(\chi f_\phi \partial_t\beta + \chi f_\varphi \partial_t\gamma - \partial_t\theta\right) + 12a(\tau f - \epsilon - 1)\partial_t\alpha = 0 \\ 6a\left(\chi f_\varphi \partial_\epsilon\gamma - \partial_\epsilon\theta\right) + 12(\epsilon + 1 - \tau f)\partial_\epsilon\alpha - a^2\partial_a\beta = 0 \\ 6\chi\left(f_\varphi\partial_\varphi\alpha + f_\varphi\partial_\phi\alpha\right) - a\left(\partial_\varphi\delta + \partial_\phi\theta\right) = 0 \\ 6\left(\chi f_\varphi \partial_\zeta\alpha - \partial_\phi\alpha\right) - a\left(\partial_\phi\gamma + \partial_\zeta\delta\right) = 0 \\ 6\chi f_\varphi \partial_\epsilon\alpha - a\left(\partial_\phi\beta + \partial_\epsilon\delta - \partial_t\xi\right) - 3\alpha = 0 \\ 6\chi f_\varphi \partial_\zeta\alpha - 6\partial_\varphi\alpha - 3\alpha - a\left(\partial_\varphi\gamma + \partial_\zeta\theta - \partial_t\xi\right) = 0 \\ 6a\left(\chi f_\varphi \partial_\phi\beta + \chi f_{\phi\phi}\beta + \chi f_\phi \partial_a\alpha + \chi f_\varphi \partial_\phi\gamma + \chi f_{\phi\varphi}\gamma - \chi\partial_t\xi f_\phi - \partial_\phi\theta\right) \\ +12\chi f_\phi\alpha + 12(\epsilon + 1 - \tau f)\partial_\phi\alpha - a^2\partial_a\delta = 0 \\ 6a\left(\chi f_\phi \partial_\zeta\beta + \chi f_\varphi \partial_\zeta\gamma - \partial_a\alpha - \partial_\zeta\theta + \partial_t\xi\right) \\ +12(\epsilon + 1 - \tau f)\partial_\zeta\alpha - 12\alpha - a^2\partial_a\gamma = 0 \\ 6a\left[\chi f_\phi \partial_\varphi\beta + \chi f_{\phi\varphi}\beta + \chi f_\varphi\left(\partial_a\alpha + \partial_\varphi\gamma - \partial_t\xi\right) + \chi f_{\varphi\varphi}\gamma - \partial_\varphi\theta\right] \\ +12\chi f_\varphi\alpha + 12(\epsilon + 1 - \tau f)\partial_\varphi\alpha - a^2\partial_a\theta = 0 \\ a\left[\chi a f_\phi \partial_a\beta - \tau f_\phi\beta + \chi a f_\varphi \partial_a\gamma - \tau f_\varphi\gamma - 2\tau f\partial_a\alpha + \tau\partial_t\xi f + 2\epsilon\partial_a\alpha \right. \\ \left. +2\partial_a\alpha + \delta - a\partial_a\theta - (\epsilon + 1)\partial_t\xi\right) + (\epsilon + 1 - \tau f)\alpha = 0 \\ \xi = \xi(t), \quad \beta = \beta(a,\phi,\varphi,\zeta,t). \end{cases} \tag{11.83}$$

Due to the dimension of the minisuperspace, the system turns out to be overdetermined. Therefore, here we outline some solutions of cosmological interest, writing the infinitesimal generators and the corresponding functions containing symmetries. However, we neglect the solutions that do not uniquely determine the form of $f(\phi,\varphi)$ (see [344] for details), namely

$$
\begin{cases}
\mathcal{X}_1 = (\xi_0 t + \xi_1)\,\partial_t + \dfrac{\alpha_0}{3} a\partial_a + [\beta_0 + \beta_1(6\ln a + \phi)]\,\partial_\phi \\
\qquad + [\gamma_0 + \gamma_1(6\ln a + \varphi) + \gamma_2]\,\partial_\varphi + \delta_0 \epsilon\partial_\epsilon + [(\delta_0 - \gamma_1)\,\zeta - \beta_1\epsilon + \theta_0]\,\partial_\zeta \\
f(\phi,\varphi) = \dfrac{1}{\tau} + f_0 \exp\{n(\beta_1\varphi - \gamma_1\phi)\} \\
\\
\mathcal{X}_2 = (\xi_0 t + \xi_1)\,\partial_t + \dfrac{\alpha_0}{3} a\partial_a + [\beta_0 + \beta_1(6\ln a + \varphi)]\,\partial_\phi \\
\qquad + \gamma_0\partial_\varphi + (\delta_0 + \delta_1\epsilon)\,\partial_\epsilon + (\delta_1\zeta - \beta_1\epsilon + \theta_0)\,\partial_\zeta \\
f(\phi,\varphi) = \dfrac{1}{\tau}\left(1 - \dfrac{\delta_0}{\delta_1}\right) + f_0 e^{\frac{\delta_1}{\gamma_0}\varphi} \\
\\
\mathcal{X}_3 = (\xi_0 t + \xi_1)\,\partial_t - \dfrac{\alpha_0}{3} a\partial_a + \beta_0\partial_\phi + (\delta_0 + \delta_1\epsilon)\,\partial_\epsilon \\
f(\phi) = f_0 e^{\frac{\beta_0}{\delta_1}\phi} - \dfrac{\beta_0}{\delta_0} + 1, \quad \text{with} \quad \chi = 0,\, \zeta = 0,\, \tau = 1 \\
\\
\mathcal{X}_4 = (\xi_0 t + \xi_1)\,\partial_t - \dfrac{\alpha_0}{3} a\partial_a + \beta_0\partial_\phi + \delta_0\partial_\epsilon \\
f(\phi) = f_1 + \dfrac{\beta_0}{\delta_0}\phi, \quad \text{with} \quad \chi = 0,\, \zeta = 0,\, \tau = 1,
\end{cases}
\tag{11.84}
$$

where $\xi_0, \alpha_0, \alpha_1, \beta_0, \beta_1, \gamma_0, \gamma_1, \gamma_2, \delta_0, \delta_1, \theta_0, f_0, f_1, n$ are integration constants. The last Noether system solution does not provide any interesting cosmological dynamics; thus in what follows, we investigate the cosmological implications of the first three symmetries. The first symmetry yields the Lagrangian

$$
\begin{aligned}
\mathcal{L} = \frac{a}{2}\Big\{&6(\epsilon - f_0\tau\, e^{n(\beta_1\varphi - \gamma_1\phi)})\dot{a}^2 - a^2\dot{\epsilon}\dot{\phi} \\
&+ 6a\dot{a}\left[\chi f_0 n e^{n(\beta_1\varphi - \gamma_1\phi)}(\beta_1\dot{\varphi} - \gamma_1\dot{\phi}) - \dot{\zeta}\right] - a^2\dot{\varphi}\dot{\zeta}\Big\}
\end{aligned}
\tag{11.85}
$$

whose cosmological solutions are

$$
\begin{aligned}
&a(t) = a_0 e^{H_0 t}, \qquad \phi(t) = -2H_0 t, \qquad \varphi(t) = -6H_0 t, \\
&\epsilon(t) = \epsilon_0 e^{-3H_0 t(1+c_1)} - \epsilon_1 e^{-3H_0 t}, \qquad \zeta(t) = \zeta_0 e^{-3H_0 t(1+c_1)} - \zeta_1 e^{-3H_0 t},
\end{aligned}
$$

and

$$
\begin{aligned}
&a(t) = a_0 t^p, \qquad \phi(t) = \frac{6p^2\ln(t - 3pt)}{1-3p}, \qquad \varphi(t) = -6p\ln t, \\
&\epsilon(t) = \epsilon_0 t^{2-3p} + \epsilon_1 t^{1-3p}, \qquad \zeta(t) = \zeta_0 t^{2-3p} + \zeta_1 t^{1-3p},
\end{aligned}
\tag{11.86}
$$

with H_0, p constants. In the former case, ξ_0 can be related to the other constants by means of the relation

$$
\xi_0 = \alpha_0 - \frac{(3\beta_1 - 2\gamma_1)\tau - 3\gamma_1\chi}{2\chi(2\beta_1 - \gamma_1)^2},
\tag{11.87}
$$

so that n is constrained to be

$$n=-\frac{(3\beta_1-2\gamma_1)\tau-3\gamma_1\chi}{2\chi(2\beta_1-\gamma_1)^2}. \tag{11.88}$$

It is worth noticing that both exponential expansions and power-law solutions for the scale factor are admitted in nonlocal TEGR. As a consequence, this theory can potentially describe all the epochs crossed by the universe, from the early to the late times. Moreover, standard local TEGR is recovered for $\tau=\chi=0$, namely when $n=0$, while modified $f(R)$ gravity is restored as soon as T and B are combined through the relation in Eq. (7.37).

The second solution of Noether's system (11.84) can be related to the first one by appropriate combinations of the free parameters. For this reason, it will not be discussed (see [344] for details).

The Lagrangian corresponding to the third function of (11.84) can also be obtained as a limit of Lagrangian (11.80) and reads as follows:

$$\mathcal{L}=-6a\dot{a}^2\left(f_0 e^{\frac{\beta_0}{\delta_1}\phi}-\epsilon-1\right)-a^3\dot{\epsilon}\dot{\phi}. \tag{11.89}$$

According to the constraints provided by Noether's approach, we set $\chi=0, \zeta=0, \tau=1$. As a consequence, the equations of motion can be solved analytically, providing

$$\begin{aligned}
&a(t)=e^{H_0 t}, \quad \phi(t)=-2H_0 t,\\
&\epsilon(t)=e^{-3H_0 t}\left[f_0\left(3H_0 t+1\right)-\frac{\epsilon_1}{3H_0}\right]-1,\\
&f(\phi,\varphi)=f_0 e^{\frac{\beta_0}{\delta_1}\phi},
\end{aligned} \tag{11.90}$$

and

$$\begin{aligned}
&a(t)=t^p, \quad \phi(t)=\frac{6p^2\ln(t-3pt)}{1-3p},\\
&\epsilon(t)=f_0(1-3p)^{3(1-p)}t^{2-3p}\frac{\epsilon_0 t^{1-3p}}{1-3p}-1,\\
&f(\phi)=f_0 e^{\frac{(9p^2-9p+2)\phi}{6p^2}}.
\end{aligned} \tag{11.91}$$

A further constraint must be imposed in the search for exact solutions, namely $\delta_0=\beta_0$. By means of this imposition, the EL equations and the energy condition can be solved analytically, yielding exponential and time power-law scale factors in Eqs. (11.90) and (11.91).

Finally, it is worth analyzing a further solution, coming from the introduction of a new scalar field ψ, defined as

$$\psi\equiv-\phi+\varphi=-\Box^{-1}T+\Box^{-1}B=\Box^{-1}(-T+B)=\Box^{-1}R. \tag{11.92}$$

In this way, $f(R)$ gravity can be straightforwardly recovered and the minisuperspace can be reduced to $\mathcal{S}\equiv\{a,\psi,\epsilon,\zeta\}$. The Noether system admits two solutions, namely

$$\begin{cases} \mathcal{X}_5 = (\xi_0 t + \xi_1)\,\partial_t - \dfrac{\alpha_0}{3} a\partial_a - 2c_0\partial_\psi + (\delta_0 + \delta_1\epsilon)\,\partial_\epsilon \\ f(\psi) = -1 + \dfrac{\delta_0}{\delta_1} + f_0 e^{-\frac{\delta_1}{\beta_0}\psi} \\ \\ \mathcal{X}_6 = (\xi_0 t + \xi_1)\,\partial_t - \dfrac{\alpha_0}{3} a\partial_a - 2c_0\partial_\psi + \delta_0\partial_\epsilon \\ f(\psi) = f_1 + \dfrac{\delta_0}{2\beta_0}\psi. \end{cases} \tag{11.93}$$

Both functions arise after the redefinition of scalar fields ϕ and φ in Eq. (11.92) and thus they can be obtained from solutions (11.84) with an appropriate change of variables.

The function therefore turns out to be equivalent to $f(\Box^{-1}R)$, so that the action (11.15) can be recovered. As a consequence, also the cosmological solutions are the same as those of Nonlocal Curvature Gravity discussed in Section 11.4.

However, notice that the latter case is not completely equivalent to that in Section 11.4, where three different solutions occurred after the application of Noether's identity. This is due to the different form of the action, since here the torsion scalar and the boundary term appear linearly in the starting Lagrangian. It is also worth mentioning that, although GR and TEGR produce the same field equations, they cannot be distinguished simply by using classical experiments. However, these theories are conceptually different from each other and they could produce different quantum mechanical effects at fundamental level. Even though we do not have a fully developed theory of Quantum Gravity, there are several proposals in this direction. A universal prediction from almost all the approaches is the existence of extended structures of space-time geometry that are expected to give rise to nonlocal deformations, whose effects could be detectable from microscopic scales to cosmology. In principle, cosmology from nonlocal GR predicts different behaviors with respect to nonlocal TEGR. Thus, nonlocal signatures of these cosmological models can be matched with observational data. In [182], for instance, the authors considered cosmological data with the aim of constraining the free parameters of nonlocal TEGR. Indications are that nonlocal TEGR is consistent with present data and then cosmology, besides quantum experiments, could be the ground on which one can discriminate different approaches to gravity. Other tests point out possible violations of the Equivalence Principle to discriminate among concurring theories [111]. In any case, nonlocality could play a fundamental role in this debate.

12

Noether Symmetries in Bianchi Universes

Nothing in life is to be feared, it is only to be understood.
Now is the time to understand more, so that we may fear less.

— *Marie Curie*

(as quoted in *Our Precarious Habitat* [1973])

The Noether symmetries can have an important role also in classifying geometries. Here, we use the Noether Symmetry Approach to study gravitational field equations minimally coupled to a scalar field, in the case of Bianchi universes [349]. When symmetries exist, exact integration is possible, and we discuss, in particular, the cases of Bianchi AI and BV. The main purpose, therefore, is to investigate cosmological anisotropies by the first principle related to the existence of symmetries.

Before developing the topic, it is worth stressing that astronomical observations point out that, at a very large scale, our universe is homogeneous and isotropic. According to this statement, which is substantially the Cosmological Principle, the universe is well described by the FLRW metric. However, such a scenario, according to inflationary prescriptions, is nothing else but the result of a previous evolution where a large amount of primordial anisotropies has been averaged [350]. On the contrary, pre-inflationary anisotropies can play an important role if primordial friction coefficients, responsible for the coupling between the gravitational field and the inflaton, are considered. As a matter of fact, such coefficients are proportional to the volume expansion rate, and their variation can be related to specific inflationary models [350].

Furthermore, anisotropies, in a range between 10^{-7} and 10^{-4}, are detected in the cosmic microwave background. According to this experimental fact, they can affect large and small angular scales [351]. Anyway, observations indicate that anisotropic cosmological models isotropize as time approaches late epochs, so that it is meaningful to study situations where full isotropy is not present in early times. On the other hand, scalar fields are used to obtain inflationary eras and to deal with primordial matter fields according to Grand Unified Theories.

For all these reasons, it is worth studying dynamics of anisotropies and searching for anisotropic cosmological models that are, eventually, solvable.

Let us start from the action

$$S = \int \sqrt{-g}\left(R + \frac{1}{2}g^{\mu\nu}\phi_{;\mu}\phi_{;\nu} - V(\phi)\right) d^4x \tag{12.1}$$

and consider a spatially homogeneous space-time, described by the metric[1]

$$ds^2 = -dt^2 + g_{ij}(t)\omega^i\omega^j\,, \tag{12.2}$$

with ω^i being 1-forms, not necessarily closed ($i, j = 1, 2, 3$), and $V(\phi)$ the potential for ϕ. We search for Noether symmetries of Lagrangian (12.1) to select Bianchi models that are exactly solvable by the Noether Symmetry Approach. Specifically, the Bianchi classification and the Lagrangian description are sketched in Section 12.1. In Section 12.2 we apply the procedure to determine the cases where symmetries are present. Section 12.3 is devoted to the analysis of results, and, in Section 12.4, we give two examples of exact integration of the equations of motion.

12.1 The Bianchi Classification of Space-Times

The Bianchi space-times, in modern geometric language, are G_3-principal fiber bundles over the real numbers $\mathbb{R}$. The bundle, being trivializable, is usually written as $M^4 = \mathbb{R} \times G_3$. G_3 stays for the three-dimensional group of isometries on M^4. If one is not paying attention to the topology[2] of G_3, we can use a parallelization of M^4 with the vector fields $\mathcal{X}_0, \mathcal{X}_1, \mathcal{X}_2, \mathcal{X}_3$, with $\mathcal{X}_0$ the "time"-generator and $\mathcal{X}_1, \mathcal{X}_2, \mathcal{X}_3$ the Killing vector fields of the metric closing on the Lie algebra of G_3. Of course, when $\mathcal{X}_1, \mathcal{X}_2, \mathcal{X}_3$ are known (and complete) we can integrate them to the action of G_3 on M^4.

In the framework of this procedure, in 1897 Bianchi worked out the first classification of three-dimensional Lie algebras. Of course, metrics for which corresponding groups are isometries describe cosmological models with specified symmetry properties. Bianchi started with the adjoint representation, and his approach has been followed since then. Here, we would like to comment briefly on the classification by using the coadjoint representation, as proposed in [352,353].

As it is well known, a Lie group is locally characterized by its structure constants c^i_{jk}. If $\mathcal{X}_i$ is a basis for the vector space underlying the algebra $\mathcal{G}$, we have $[\mathcal{X}_i, \mathcal{X}_j] = c^k_{ij}\mathcal{X}_k$, and the Jacobi identity $[\mathcal{X}_i, [\mathcal{X}_j, \mathcal{X}_k]] + [\mathcal{X}_j, [\mathcal{X}_k, \mathcal{X}_i]] + [\mathcal{X}_k, [\mathcal{X}_i, \mathcal{X}_j]] = 0$ gives a quadratic constraint on the structure constants.

[1] Here, we use the signature $(-,+,+,+)$ in order to deal with a positive definite g_{ij} for the 1-forms.

[2] This means not distinguishing between the simply connected group with respect to the quotient groups obtained by quotienting it with discrete normal subgroups, without discriminating among Lie groups with the same Lie algebra.

By identifying the vector space $\mathcal{G}$ with the linear functions on $\mathcal{G}^*$, $\mathcal{G} = \mathrm{Lin}(\mathcal{G}^*, \mathbf{R})$, we can define a Poisson bracket on $\mathcal{G}^*$ by setting $\{x_i, x_j\} = c^k_{ij} x_k$. The classification problem is reduced now to the problem of classifying linear Poisson brackets on a three-dimensional vector space. By using the transformation properties of a Poisson bracket under a change of coordinates, one can show that, with any bracket, a bivector field Λ is associated. In our case, $\Lambda = c^k_{ij} x_k \frac{\partial}{\partial x_i} \wedge \frac{\partial}{\partial x_j}$. In three dimensions, any bivector field is associated with a one-form α by means of the contraction with a volume form Ω, say $\Omega = dx_1 \wedge dx_2 \wedge dx_3$. Thus, we have to consider and classify quadratic one-forms $\alpha = c^k_{ij} x_k \epsilon^{ijl} dx_l = H^{kl} x_k dx_l$. The Jacobi identity is equivalent to $\alpha \wedge d\alpha = 0$ [352].

The one-form α can be written by decomposing H^{kl} into symmetric and skew-symmetric parts, that is,

$$\alpha = \frac{1}{2} H^{(kl)} (x_l dx_k + x_k dx_l) + \frac{1}{2} H^{[kl]} (x_l dx_k - x_k dx_l) \; . \tag{12.3}$$

By using linear transformations on the three-dimensional vector space, we can bring α to the normal form

$$\alpha = \frac{1}{2} d(N_1 x_1^2 + N_2 x_2^2 + N_3 x_3^2) + \frac{a}{2} (x_2 dx_3 - x_3 dx_2) \; . \tag{12.4}$$

Imposing the Jacobi identity, we find $aN_1 = 0$ and, by means of a suitable change of coordinate basis, we may set $N_i = 0, -1, +1$. Thus, all quadratic one-forms giving rise to three-dimensional Lie algebras can be classified in two classes:

1. Bianchi A , α exact, namely $a = 0$
2. Bianchi B, $d\alpha \neq 0$, namely $a \neq 0$.

In this approach, class A is associated with a foliation of vector space $\mathbf{R}^3$ given by the Casimir function $N_1 x_1^2 + N_2 x_2^2 + N_3 x_3^2$. The level sets of this function are the coadjoint orbits of the Lie group G_3 associated with the Lie algebra $\mathcal{G}$. They give information on the topology of the group. Clearly, if $N_1 N_2 N_3 \neq 0$, we have either $O(3)$ or $O(2,1)$, according to the signs of $N_i N_j$ (the second, if one of them is negative). If at least one of the N is zero, the group is non-compact.

The parameters a, N_i are sufficient to characterize the Bianchi classification, according to Table **12.1**.

A parameter $h \stackrel{\text{def}}{=} a^2/(N_2 N_3)$ is also introduced for subclassifying types VI and VII.

Because any Lie group is parallelizable, we can use left-invariant (or right-invariant) one-forms in the metric (12.2), so that it will be invariant under left (or right) action of G_3 on M^4. By using a diagonal form for g_{ij} and selecting (for obvious reasons) a positive definite one, we can parametrize g_{ij} by factoring $\det \|g_{ij}\|$, namely we obtain [354]

$$g_{ij}(t) = e^{2\lambda(t)} e^{2\beta_{ij}(t)} \; , \tag{12.5}$$

Table 12.1 Classification of Bianchi types

Type	a	N_1	N_2	N_3
I	0	0	0	0
II	0	1	0	0
III	1	0	1	−1
IV	1	0	0	1
V	1	0	0	0
VI_o	0	1	−1	0
VI_h	a	0	1	−1
VII_o	0	1	1	0
VII_h	a	0	1	1
VIII	0	1	1	−1
IX	0	1	1	1

where the matrix β is diagonal and traceless, thus depending on two variables only. A widely used choice is [355]

$$\|\beta_{ij}\| = \mathrm{diag}\left(-\frac{1}{2}\beta_1 + \frac{\sqrt{3}}{2}\beta_2\,,\; -\frac{1}{2}\beta_1 - \frac{\sqrt{3}}{2}\beta_2\,,\; \beta_1\right) . \tag{12.6}$$

With $\sqrt{-g} = e^{3\lambda}$, the expansion of volume is wholly determined by λ. The shear is determined by the β_i. This is typical only for the "orthogonal" case, when there are no effects of rotation and tilt. Spatially homogeneous sections are then orthogonal to the fluid flow vector in the universe [355].

As shown in [355], the second-order Einstein equations of all Bianchi types, except IV and $\mathrm{VII}_{h\neq 0}$, can be derived from a point-like Lagrangian. In Section 12.4, we shall consider also the first-order Einstein equation, which is always a first integral of the second-order ones.

Then the general expression for the Lagrangian is

$$L = e^{3\lambda}\left(R^\star + 6\dot{\lambda}^2 - \frac{3}{2}(\dot{\beta}_1^2 + \dot{\beta}_2^2) - \dot{\phi}^2 + 2\,V(\phi)\right) , \tag{12.7}$$

with the scalar curvature of the space-like hypersurfaces $R^\star$ given by

$$\begin{aligned} R^\star = & -\frac{1}{2}e^{-2\lambda}\Big[N_1^2 e^{4\beta_1} + e^{-2\beta_1}\left(N_2 e^{\sqrt{3}\beta_2} - N_3 e^{-\sqrt{3}\beta_2}\right)^2 \\ & -2N_1 e^{\beta_1}\left(N_2 e^{\sqrt{3}\beta_2} + N_3 e^{-\sqrt{3}\beta_2}\right)\Big] \\ & + \frac{1}{2}N_1 N_2 N_3 (1 + N_1 N_2 N_3), \end{aligned} \tag{12.8}$$

for the Bianchi types belonging to class A, and by

$$R^\star = 2a^2 e^{-2\lambda}\left(3 - \frac{N_2 N_3}{a^2}\right) e^{\beta} , \tag{12.9}$$

with

$$\beta = \frac{2}{3a^2 - N_2N_3}(N_2N_3\beta_1 + \sqrt{-3a^2N_2N_3\beta_2}) \tag{12.10}$$

for Bianchi types in class B, with $n_i^i = 0$.

As just mentioned, for types IV and $\mathrm{VII}_{h\neq 0}$, the Lagrangian is not known, and therefore we shall not consider them here.

12.2 Noether Symmetries in Bianchi Space-Times

In this section, we look for Noether symmetries of the Lagrangian (12.7), using the vanishing Lie derivative condition

$$L_X L = 0 \, . \tag{12.11}$$

As pointed out in the previous applications, conditions imposed by this equation generally puts restrictions on the class of allowed potentials for ϕ.

However, let us first observe that there are a number of trivial cases.

- If $V = \text{const.}$, does not depend on ϕ, which is therefore cyclic. Thus, for all classes and types, we always have the symmetry

$$\mathcal{X} = \frac{\partial}{\partial\phi} \, . \tag{12.12}$$

- In Bianchi AI and BV, we have that $R^\star$ does not depend on β_1, β_2, which are therefore cyclic. We have thus the symmetries

$$\mathcal{X}_1 = \frac{\partial}{\partial\beta_1}, \quad \mathcal{X}_2 = \frac{\partial}{\partial\beta_2} \, . \tag{12.13}$$

  Moreover, since the term $\dot\beta_1^2 + \dot\beta_2^2$ is rotationally invariant, we have the symmetry

$$\mathcal{X}_3 = \beta_1\frac{\partial}{\partial\beta_2} - \beta_2\frac{\partial}{\partial\beta_1} \, . \tag{12.14}$$

- In Bianchi AII, $R^\star$ depends on β_1 only, so that we have the symmetry

$$\mathcal{X} = \frac{\partial}{\partial\beta_2} \, . \tag{12.15}$$

In the following, using (12.11), we investigate the possibility of other symmetries, not so immediately evident.

First of all, we have to say that the configuration space is made of four variables, namely $\mathcal{S} \equiv \{\lambda, \beta_1, \beta_2, \phi\}$. Let us set

$$\mathcal{X} = L\frac{\partial}{\partial\lambda} + B_1\frac{\partial}{\partial\beta_1} + B_2\frac{\partial}{\partial\beta_2} + F\frac{\partial}{\partial\phi} \, , \tag{12.16}$$

where L, B_1, B_2, and F are unknown functions of $\lambda, \beta_1, \beta_2, \phi$. The tangent lift will be

$$X = \mathcal{X} + \frac{dL}{dt}\frac{\partial}{\partial\dot{\lambda}} + \frac{dB_1}{dt}\frac{\partial}{\partial\dot{\beta}_1} + \frac{dB_2}{dt}\frac{\partial}{\partial\dot{\beta}_2} + \frac{dF}{dt}\frac{\partial}{\partial\dot{\phi}} \,. \tag{12.17}$$

Pursuing the same approach as in the previous chapters, we notice that the expression $L_X\mathcal{L}$ is a homogeneous quadratic polynomial in $\dot{\lambda}, \dot{\beta}_1, \dot{\beta}_2, \dot{\phi}$, plus a term of zeroth degree. Therefore, it vanishes if and only if all the coefficients are zero independently. This leads to a system of eleven partial differential equations

$$3L + 2\frac{\partial L}{\partial\lambda} = 0, \;\; 3L + 2\frac{\partial B_1}{\partial\beta_1} = 0, \tag{12.18}$$

$$3L + 2\frac{\partial B_2}{\partial\beta_2} = 0, \;\; 3L + 2\frac{\partial F}{\partial\phi} = 0, \tag{12.19}$$

$$4\frac{\partial L}{\partial\beta_1} - \frac{\partial B_1}{\partial\lambda} = 0, \;\; 4\frac{\partial L}{\partial\beta_2} - \frac{\partial B_2}{\partial\lambda} = 0, \tag{12.20}$$

$$6\frac{\partial L}{\partial\phi} - \frac{\partial F}{\partial\lambda} = 0, \;\; \frac{\partial B_1}{\partial\beta_2} + \frac{\partial B_2}{\partial\beta_1} = 0, \tag{12.21}$$

$$3\frac{\partial B_1}{\partial\phi} + 2\frac{\partial F}{\partial\beta_1} = 0, \;\; 3\frac{\partial B_2}{\partial\phi} + 2\frac{\partial F}{\partial\beta_2} = 0, \tag{12.22}$$

$$L\left(R^\star + 6V(\phi) + N_1N_2N_3(1 + N_1N_2N_3)\right) + B_1\frac{\partial R^\star}{\partial\beta_1} + B_2\frac{\partial R^\star}{\partial\beta_2} + 2FV' = 0 \,, \tag{12.23}$$

which are valid for both classes A and B, with relatively different expressions for $R^\star$.

We first obtain a general solution for equations (12.18–12.22), and then use (12.23) as a constraint to limit the class of solutions and find a condition on the potential.

From Eq. (12.18), we immediately get

$$L = e^{-\frac{3}{2}\lambda}\bar{L}(\beta_1, \beta_2, \phi) \,, \tag{12.24}$$

so that, from (12.20), we have

$$B_{1,2} = -\frac{8}{3}\frac{\partial\bar{L}}{\partial\beta_{1,2}}e^{-\frac{3}{2}\lambda} + \bar{B}_{1,2}(\beta_1, \beta_2, \phi) \,, \tag{12.25}$$

and, from (12.21), it is

$$F = -4\frac{\partial\bar{L}}{\partial\phi}e^{-\frac{3}{2}\lambda} + \bar{F}(\beta_1, \beta_2, \phi) \,. \tag{12.26}$$

Using (12.21), we get then

$$-\frac{8}{3}\frac{\partial^2\bar{L}}{\partial\beta_1\partial\beta_2}e^{-\frac{3}{2}\lambda} + \frac{\partial\bar{B}_1}{\partial\beta_2} = \frac{8}{3}\frac{\partial^2\bar{L}}{\partial\beta_1\partial\beta_2}e^{-\frac{3}{2}\lambda} - \frac{\partial\bar{B}_2}{\partial\beta_1} \,. \tag{12.27}$$

Since $\bar{B}_{1,2}$ do not depend on λ, it must be

$$\frac{\partial^2 \bar{L}}{\partial \beta_1 \partial \beta_2} = 0 \ , \tag{12.28}$$

so that

$$\bar{L} = \bar{L}_1(\beta_1, \phi) + \bar{L}_2(\beta_2, \phi) \ ; \tag{12.29}$$

moreover, (12.21) gives

$$\frac{\partial \bar{B}_1}{\partial \beta_2} = -\frac{\partial \bar{B}_2}{\partial \beta_1} \ . \tag{12.30}$$

Substituting into (12.18), we get

$$3e^{-3/2\lambda}\left(\bar{L}_1 + \bar{L}_2 - \frac{16}{9}\frac{\partial^2 \bar{L}_1}{\partial \beta_1}\right) + \frac{\partial \bar{B}_1}{\partial \beta_1} = 0 \ . \tag{12.31}$$

Now, since only $\bar{L}_2$ and $\bar{B}_1$ depend on β_2, and $\bar{B}_1$ does not depend on λ, it must be

$$\bar{L}_2 = 0, \ \ \frac{\partial \bar{B}_1}{\partial \beta_1} = 0 \ . \tag{12.32}$$

Substituting into (12.19), we get analogously

$$\bar{L}_1 = 0, \ \ \frac{\partial \bar{B}_2}{\partial \beta_2} = 0 \ . \tag{12.33}$$

Thus, we find that $L = 0$ and $B_1 = B_1(\beta_2, \phi)$, $B_2 = B_2(\beta_1, \phi)$. From (12.21) we obtain now

$$B_1 = f(\phi)\beta_2 + g_1(\phi), \ \ B_2 = -f(\phi)\beta_1 + g_2(\phi) \ , \tag{12.34}$$

while, from (12.19) and (712.21), we get $F = F(\beta_1, \beta_2)$, and, from (12.22), it is

$$3f'\beta_2 + 3g_1' = -2\frac{\partial F}{\partial \beta_1} \ . \tag{12.35}$$

The LHS does not depend on β_1, thus

$$F = -\frac{3}{2}(g_1' - f'\beta_2)\beta_1 + \tilde{F}(\beta_2) \ , \tag{12.36}$$

and, analogously, from (12.22)

$$F = -\frac{3}{2}(g_2' - f'\beta_1)\beta_2 + \hat{F}(\beta_1) \ . \tag{12.37}$$

By comparison, we get

$$f' = 0, \ \ g_1'' = g_2'' = 0 \ , \tag{12.38}$$

$$-\frac{3}{2}g_1'\beta_1 - \hat{F}\beta_1 = -F_0, \ \ \frac{3}{2}g_2'\beta_2 + \tilde{F}\beta_2 = F_0 \ , \tag{12.39}$$

where F_0 is an arbitrary constant. The general solution is thus

$$L = 0, \qquad B_1 = c\beta_2 + c_1\phi + c_0, \tag{12.40}$$

$$B_2 = -c\beta_1 + c_2\phi + c_0', \quad F = F_0 - \tfrac{3}{2}(c_1\beta_1 + c_2\beta_2) \ , \tag{12.41}$$

with $c, c_1, c_2, c_0, c_0', F_0$ arbitrary constants.

We have now to plug solutions (12.41) into (12.23) and check for compatibility. With $L = 0$, we get

$$B_1 \frac{\partial R^\star}{\partial \beta_1} + B_2 \frac{\partial R^\star}{\partial \beta_2} = -2FV' \ . \tag{12.42}$$

It is easy to see that the two sides must be zero independently, so that we have two subcases: $V =$ const. and $F = 0$. In both of them, Eq. (12.42) is satisfied for types I and V. Examining the other types, we find that for Class A, Eq. (12.42) is verified only for type II and $B_1 = 0$, while for Class B, Eq. (12.42) is verified for type $\mathrm{VI}_{h\neq 0}$ if and only if

$$c = 0, \quad c_2 = -\frac{1}{a\sqrt{3}} c_1, \quad c_0' = -\frac{1}{a\sqrt{3}} c_0 \ . \tag{12.43}$$

The other types are excluded from the beginning, as said previously.

12.3 Results

Case 1 – $V =$ constant

The situation for $V =$ const. is summarized as follows.

1a) **Bianchi AI**

The most general symmetry is written (on $\mathcal{S}$) as

$$\begin{aligned}\mathcal{X} =&(c\beta_2 + c_1\phi + c_0)\frac{\partial}{\partial \beta_1} + (-c\beta_1 + c_2\phi + c_0')\frac{\partial}{\partial \beta_2} \\ &+ (F_0 - \frac{3}{2}(c_1\beta_1 + c_2\beta_2))\frac{\partial}{\partial \phi} \ ,\end{aligned} \tag{12.44}$$

with obvious lift to $\mathcal{TS}$. A basis of symmetries on $\mathcal{TS}$ is given by

$$X_1 = \frac{\partial}{\partial \beta_1}, \quad X_2 = \frac{\partial}{\partial \beta_2}, \quad X_3 = \frac{\partial}{\partial \phi}, \tag{12.45}$$

$$X_4 = \phi\frac{\partial}{\partial \beta_1} - \frac{3}{2}\beta_1\frac{\partial}{\partial \phi} + \dot\phi\frac{\partial}{\partial \dot\beta_1} - \frac{3}{2}\dot\beta_1\frac{\partial}{\partial \dot\phi}, \tag{12.46}$$

$$X_5 = \phi\frac{\partial}{\partial \beta_2} - \frac{3}{2}\beta_2\frac{\partial}{\partial \phi} + \dot\phi\frac{\partial}{\partial \dot\beta_2} - \frac{3}{2}\dot\beta_2\frac{\partial}{\partial \dot\phi}, \tag{12.47}$$

$$X_6 = \beta_1\frac{\partial}{\partial \beta_2} - \beta_2\frac{\partial}{\partial b} + \dot\beta_1\frac{\partial}{\partial \dot\beta_2} - \dot\beta_2\frac{\partial}{\partial \dot\beta_1}. \tag{12.48}$$

It turns out that only five of them are independent. For instance, we can take the first five. From these symmetries, it is possible to associate the following constants of motion

$$\begin{cases} j_1 = e^{3\lambda}\dot{\beta}_1, \\ j_2 = e^{3\lambda}\dot{\beta}_2, \\ j_3 = e^{3\lambda}\dot{\phi}, \\ j_4 = e^{3\lambda}\left(\phi\dot{\beta}_1 - \frac{3}{2}\beta_1\dot{\phi}\right), \\ j_5 = e^{3\lambda}\left(\phi\dot{\beta}_2 - \frac{3}{2}\beta_2\dot{\phi}\right), \\ j_6 = e^{3\lambda}(\beta_1\dot{\beta}_2 - \dot{\beta}_1\beta_2) \ . \end{cases} \tag{12.49}$$

These symmetries close on the following Lie algebra:

$$[X_i, X_j] = 0, \quad i, j = 1, 2, 3, \quad [X_1, X_4] = -\frac{3}{2}X_3, \quad [X_2, X_4] = 0, \tag{12.50}$$

$$[X_1, X_5] = 0, \quad [X_2, X_5] = -\frac{3}{2}X_3, \quad [X_3, X_4] = X_1, \quad [X_3, X_5] = X_2, \tag{12.51}$$

$$[X_4, X_5] = \frac{3}{2}X_6, \quad [X_1, X_6] = X_2, \quad [X_2, X_6] = -X_1, \tag{12.52}$$

$$[X_3, X_6] = 0, \quad [X_4, X_6] = X_5, \quad [X_5, X_6] = X_4 \ , \tag{12.53}$$

and the j_i close the same algebra in terms of Poisson brackets.

This situation was examined in [356] (with different variables). In that paper, it was shown how to use the symmetries in order to obtain exact integration of the dynamics. We give an example in what follows.

1b) **Bianchi AII**

The general form of $\mathcal{X}$ (on $\mathcal{S}$) is

$$\mathcal{X} = (c_2\phi + c_0)\frac{\partial}{\partial\beta_2} + \left(F_0 - \frac{3}{2}c_2\beta_2\right)\frac{\partial}{\partial\phi} \ , \tag{12.54}$$

from which we get the independent fields (on the tangent space $\mathcal{TS}$)

$$X_1 = \frac{\partial}{\partial\beta_2}, \quad X_2 = \frac{\partial}{\partial\phi}, \quad X_3 = \phi\frac{\partial}{\partial\beta_2} - \frac{3}{2}\beta_2\frac{\partial}{\partial\phi} + \dot{\phi}\frac{\partial}{\partial\dot{\beta}_2} - \frac{3}{2}\dot{\beta}_2\frac{\partial}{\partial\dot{\phi}} \ , \tag{12.55}$$

with Lie algebra

$$[X_1, X_2] = 0, \quad [X_1, X_3] = -\frac{3}{2}X_2, \quad [X_2, X_3] = X_1 \ . \tag{12.56}$$

The associated constants of the motion are a subset of the preceding ones.

In spite of the existence of three symmetries, the exact integration is not straightforward in this case, so that we shall not study the situation here.

1c) **Bianchi BV**

The situation is quite similar to Bianchi AI, with the same symmetries and constants of motion. The equations, of course, are different. An example of exact integration is given in what follows.

1d) **Bianchi BVI$_{h\neq 0}$**

We have

$$X_1 = \frac{\partial}{\partial \phi}, \quad X_2 = \frac{\partial}{\partial \beta_1} - \frac{1}{a\sqrt{3}}\frac{\partial}{\partial \beta_2} \tag{12.57}$$

$$X_3 = \phi\frac{\partial}{\partial \beta_1} - \frac{1}{a\sqrt{3}}\phi\frac{\partial}{\partial \beta_2} - \frac{3}{2}\left(\beta_1 - \frac{3}{2}\beta_2\right)\frac{\partial}{\partial \phi}$$
$$+\dot{\phi}\frac{\partial}{\partial \dot{\beta}_1} - \frac{1}{a\sqrt{3}}\dot{\phi}\frac{\partial}{\partial \dot{\beta}_2} - \frac{3}{2}\left(\dot{\beta}_1 - \frac{3}{2}\dot{\beta}_2\right)\frac{\partial}{\partial \dot{\phi}}, \tag{12.58}$$

with algebra

$$[X_1, X_2] = 0, \quad [X_1, X_3] = X_2, \quad [X_2, X_3] = \left(\frac{1}{3a^2} - \frac{3}{2}\right)X_2 \tag{12.59}$$

and constants of the motion

$$j_1 = e^{3\lambda}\dot{\phi}, \quad j_2 = e^{3\lambda}\left(\dot{\beta}_1 - \frac{3}{2}\dot{\beta}_2\right), \tag{12.60}$$

$$j_3 = e^{3\lambda}\left(-3\phi\beta_1 + \frac{1}{a^2}\phi\dot{\beta}_2 + 3\beta_1\dot{\phi} - \frac{3}{a\sqrt{3}}\right). \tag{12.61}$$

Case 2 – $V \neq$ const.

2a) **Bianchi AI**

We obtain the following independent symmetries:

$$X_1 = \frac{\partial}{\partial \beta_1}, \quad X_2 = \frac{\partial}{\partial \beta_2}, \quad X_3 = \beta_1\frac{\partial}{\partial \beta_2} - \beta_2\frac{\partial}{\partial \beta_1} + \dot{\beta}_1\frac{\partial}{\partial \dot{\beta}_2}\dot{\beta}_2\frac{\partial}{\partial \dot{\beta}_1} \tag{12.62}$$

with algebra

$$[X_1, X_2] = 0, \quad [X_1, X_3] = -X_2, \quad [X_2, X_3] = X_1 \tag{12.63}$$

and constants of the motion

$$j_1 = e^{3\lambda}\dot{\beta}_1, \quad j_2 = e^{3\lambda}\dot{\beta}_2, \quad j_3 = e^{3\lambda}(\beta_2\dot{\beta}_1 - \beta_1\dot{\beta}_2). \tag{12.64}$$

This case was examined already in [356] and, although it was not possible to achieve exact integration, qualitative considerations of the behavior of the solutions are possible.

2b) **Bianchi AII**

Only one symmetry is left, namely

$$X = \frac{\partial}{\partial \phi}. \tag{12.65}$$

2c) **Bianchi BV**

Again the situation is essentially the same as for Bianchi AI.

2d) **Bianchi BVII$_0$**

We have only one symmetry,

$$X = \frac{\partial}{\partial \beta_1} - \frac{3}{2}\frac{\partial}{\partial \beta_2}, \tag{12.66}$$

with the corresponding constant of the motion

$$j = e^{3\lambda}\left(\dot{\beta}_1 - \frac{1}{a\sqrt{3}}\dot{\beta}_2\right). \tag{12.67}$$

12.4 Examples of Exact Integration

In this section, as examples, we obtain exact integration in the Bianchi AI and BV cases, when the potential $V(\phi)$ is zero.

As said earlier, the first-order Einstein equation can be derived from Lagrangian (12.7), and it is always equivalent to the energy condition $E_L = 0$, which is the only equation of motion remaining after reduction. Substituting the constants of the motion found previously, we get

$$6\dot{\lambda}^2 - j^2 e^{-6\lambda} - \alpha^2 e^{-2\lambda} = 0\ , \tag{12.68}$$

where $j^2 = 3(j_1^2 + j_2^2)/2 + j_3^2$ and $\alpha^2 = 6a^2 e^{\frac{2}{3a^2}}$.

In Bianchi AI, namely when $\alpha = 0$, we obtain the immediate integration

$$\lambda = \frac{1}{3}\log\left[\sqrt{\frac{3}{2}}j(t - t_0)\right]\ , \tag{12.69}$$

which coincides with the one found in [356]. In Bianchi BV, setting $x = e^{2\lambda}$, we obtain the equation

$$\frac{3}{2}\dot{x}^2 - \frac{j^2}{x} - \alpha^2 = 0\ , \tag{12.70}$$

which integrates to

$$\sqrt{\frac{2}{3}}t = \sqrt{x(1+x)} - \log(\sqrt{x} + \sqrt{1+x})\ , \tag{12.71}$$

where we have set $j = 1$ (that is $x(0) = 0$), and $\alpha = 1$ (i.e., a rescaling of x and t). This relation cannot be inverted but can be used for a qualitative analysis. Indeed, it is easy to see that, for small t, and hence small x, we have that the mean scale factor e^λ is proportional to $t^{1/2}$, while for large t it is proportional to $t^{1/3}$. It is also possible to see that power law inflation never occurs.

As for the behavior of ϕ, it is immediately derivable from the expression of j_3. In the first case, we get

$$\phi = \sqrt{\frac{2}{3}} \frac{j_3}{j} \log(t - t_0) \, , \tag{12.72}$$

while in the second, we have $\phi \propto t^{-1/2}$ for small t and $\phi \propto \log t$ for large t.

To conclude, we have found all Noether symmetries for the Bianchi universes for which a Lagrangian function is known. This allowed immediate exact integration in simple cases. In the other cases, we have seen that the number of symmetries is often sufficiently high to allow a reduction of the configuration space. These situations are of physical interest and a complete analysis can be pursued.

It is important to observe that Lagrangian (12.7), in the case when we want to add ordinary matter (in form of dust decoupled from the scalar field), changes only by an additive constant. This is not irrelevant, because this constant changes the equation $E_{\mathcal{L}} = 0$; but, of course, the second-order equations and the structure of symmetries are exactly the same. This means that all the preceding discussion is valid for this case, except for the integration, which is now more complicated.

Other types of ordinary matter give only a contribution in the coefficient of L in Eq. (12.23). Since $L = 0$, we see that again the structure of symmetries is the same and the problems for integration may still arise from the energy condition.

To conclude these considerations, it is important to stress the role of symmetries not only for defining the physically viable gravity models but also for classifying the underlying geometries. Clearly, the detected symmetries allow one to reduce dynamics that are completely solvable if the number of symmetries coincides with the number of dimensions of configuration space.

13

The Noether Approach in Spherical Symmetry

> I used to measure the skies, now I measure the shadows of Earth. Although my mind was sky-bound, the shadow of my body lies here.
>
> – *Johannes Kepler*
> (*Epitaph composed for himself*)

So far we have considered the Noether Symmetry Approach in a cosmological context. In this chapter, we apply Noether's identity (13.4) to three representative classes of extended theories, namely $f(R)$, $f(T, B)$, and $f(\mathcal{G})$ gravity assuming spherical symmetry. When the line element is supposed to describe a homogeneous and isotropic universe, field equations are simplified by the assumption that the only dynamical degree of freedom is represented by the scale factor. On the other hand, spherically symmetric backgrounds introduce additional variables into the metric, so that the corresponding field equations become more complicated.

In general, spherically symmetric configurations aim to describe, in a first approximation not considering rotation, the gravitational field induced by self-gravitating systems like planets, stars, or black holes. Though the Schwarzschild solution of GR provides predictions consistent with observations, further issues recently have occurred for compact objects and other self-gravitating systems like galaxies, as pointed out in Chapter 4. In particular, from the mass–radius relations of compact objects like neutron stars, related to the *Tolman–Oppenheimer–Volkoff* (TOV) equation, some inconsistencies emerged with respect to observations. The problem is particularly evident today because several observations cannot be addressed in the framework of GR. Many efforts aim to solve these issues by introducing new equations of state ([357, 358]) or extending the EH action [251, 359]. Furthermore, observations in galaxy structures are generally addressed by invoking the presence of the elusive dark matter [360, 361], but recent research suggests that modified theories of gravity can play an important role in settling the issue from a gravitational point of view [362–365].

As a consequence, searching for spherical solutions beyond the Schwarzschild one is of particular interest in modified theories of gravity from both theoretical and observational points of view. However, due to the high nonlinearity of field equations, in most cases, viable solutions cannot be found analytically. In this context, the search for symmetries can represent an important tool capable of reducing dynamics and providing models mathematically easier to handle.

In order to set the problem, throughout this chapter, unless otherwise indicated, we assume a line element of the form

$$ds^2 = e^{\nu(r,t)}c^2dt^2 - e^{\lambda(r,t)}dr^2 - r^2d\Omega^2, \tag{13.1}$$

where $r^2d\Omega^2$ is the 2-sphere and r the radial coordinate. Notice that Eq. (13.1) is the most general nonrotating spherically symmetric line element. When GR field equations are considered, the metric turns out to be static, and ν and λ satisfy the relation $\nu = -\lambda$. This is the renowned *Birkhoff's Theorem* discovered by George David Birkhoff in 1923. As a consequence, the Schwarzschild solution is the most general spherically symmetric vacuum solution of Einstein's field equations with zero cosmological constant. It is asymptotically flat and then suitable to describe realistic gravitational fields. Specifically, by replacing Eq. (13.1) into the field equations of GR, one gets

$$\begin{cases} e^{-\lambda}\left(\dfrac{\nu'}{r}+\dfrac{1}{r^2}\right)-\dfrac{1}{r^2}=0 \\ e^{-\lambda}\left(\dfrac{\lambda'}{r}-\dfrac{1}{r^2}\right)+\dfrac{1}{r^2}=0, \end{cases} \tag{13.2}$$

where the prime denotes the derivative with respect to the radial coordinate and the solution is given by Eq. (4.19). The Birkhoff theorem is a proper result of GR that, in general, cannot be imposed as a requirement when extensions or modifications of the Einstein theory are considered. However, its validity can be invoked for restricted classes of models. Such a restriction can always be considered in theories like $f(R)$ or $f(T)$ gravity, since the GR limit is recovered for $f(R) = R$ or $f(T) = T$. In other words, it is reasonable that there exists a subclass of solutions satisfying the Birkhoff theorem.

In the search for symmetries, we adopt the above prescriptions. In particular, we consider the first prolongation of a Noether vector of the form

$$X^{[1]} = \xi(r,q^i)\partial_r + \eta^i(r,q^i)\frac{\partial}{\partial q^i} + [\partial_r\eta^i(r,q^i) - \partial_r q^i\partial_r\xi(r,q^i)]\frac{\partial}{\partial(\partial_r q^i)}, \tag{13.3}$$

so that the identity (2.24) can be recast as

$$X^{[1]}\mathcal{L} + \partial_r\xi(r,q^i)\mathcal{L} = \partial_r g(r,q^i). \tag{13.4}$$

When the time coordinate is included in the starting interval, Eqs. (13.3) and (13.4) take the form

$$\begin{cases} X^{[1]} = \xi_r(r,t,q^i)\partial_r + \xi_t(r,t,q^i) + \eta^i(r,t,q^i)\dfrac{\partial}{\partial q^i} \\ + \left\{\partial_r\eta^i(r,t,q^i) - \partial_r q^i\left[\partial_r\xi_r(r,t,q^i) + \partial_t\xi_t(r,t,q^i)\right]\right\}\dfrac{\partial}{\partial(\partial_r q^i)} \\ + \left\{\partial_t\eta^i(r,t,q^i) - \partial_t q^i\left[\partial_r\xi_r(r,t,q^i) + \partial_t\xi_t(r,t,q^i)\right]\right\}\dfrac{\partial}{\partial(\partial_t q^i)} \\ \\ X^{[1]}\mathscr{L} + \partial_r\xi_r(r,t,q^i)\mathscr{L} + \partial_t\xi_t(r,t,q^i)\mathscr{L} = \partial_r g(r,t,q^i) + \partial_t g(r,t,q^i). \end{cases} \tag{13.5}$$

Notice that, in this case, the infinitesimal generator ξ can be split into two terms, namely ξ_r and ξ_t. The former can be addressed to spatial coordinate transformations, and the latter to time transformations. Similarly to the previous chapters, dynamics is provided by the EL equations in the Lagrangian formalism.

13.1 Spherical Symmetry in $f(R)$ Gravity

Let us start by selecting conserved quantities in the form of currents to reduce dynamics of $f(R)$ models compatible with symmetries. To this purpose, we consider a time-dependent spherically symmetric metric of the form (13.1). By this ansatz, the Ricci scalar takes the form[1]

$$\begin{aligned} R(r) = {} & e^{-\lambda}\left[\nu'' + \frac{1}{2}\nu'\left(\nu' - \lambda'\right) + \frac{2}{r}\left(\nu' - \lambda'\right) + \frac{2}{r^2}\right] \\ & + e^{-\nu}\left[\frac{1}{2}\dot{\lambda}\left(\dot{\nu} - \dot{\lambda}\right) - \ddot{\lambda}\right] - \frac{2}{r^2} \end{aligned} \tag{13.6}$$

and therefore the action can be written in terms of Lagrange multipliers as

$$\begin{aligned} S = \int r^2 e^{\frac{\nu+\lambda}{2}} f(R) - \tau\Big\{R - e^{\lambda}\left[\nu'' + \frac{1}{2}\nu'\left(\nu' - \lambda'\right) + \frac{2}{r}\left(\nu' - \lambda'\right) + \frac{2}{r^2}\right] \\ + \frac{2}{r^2} - e^{-\nu}\left[\frac{1}{2}\dot{\lambda}\left(\dot{\nu} - \dot{\lambda}\right) - \ddot{\lambda}\right]\Big\}\, drdt, \end{aligned} \tag{13.7}$$

where τ is the Lagrange multiplier, which allows one to set the Ricci scalar as a constraint for the dynamics. The Lagrange multiplier can be easily found by varying the action with respect to R, and the result is $\tau = r^2 e^{\frac{\nu+\lambda}{2}} f_R(R)$. Using this expression for τ and integrating by parts, the action can be rewritten as

$$\begin{aligned} \mathscr{L}(r,\nu,\lambda,\nu',\lambda',\dot{\nu},\dot{\lambda}) = r^2 e^{\frac{\nu+\lambda}{2}}\Big\{f(R) - Rf_R(R) - 2\frac{f_R(R)}{r^2} \\ + e^{-\lambda}\left[2\frac{f_R(R)}{r^2} - 2\frac{f_R(R)}{r}\lambda' - f_{RR}(R)R'\nu'\right] + e^{-\nu}f_{RR}(R)\dot{R}\dot{\lambda}\Big\}. \end{aligned} \tag{13.8}$$

[1] The prime and the dot denote the derivatives with respect to r and t, respectively.

Dynamics is provided by the EL equations with respect to ν and λ, namely

$$\begin{cases} -\frac{2}{r}f_{RR}(R)R' + \frac{1}{2}f_{RR}(R)\lambda' R' - f_{RRR}(R)R'^2 - f_{RR}(R)R'' \\ +\frac{1}{r}f_R(R)\lambda' - \frac{1}{r^2}f_R(R) + \frac{1}{2}e^{-\nu}f_{RR}(R)\dot{R}\dot{\lambda} \\ +e^{\lambda}\left[\frac{1}{2}Rf_R(R) - \frac{1}{2}f(R) + \frac{1}{r^2}f_R(R)\right] = 0 \\ \\ e^{-\nu}\left[-\frac{1}{2}f_{RR}(R)\dot{\nu}\dot{R} + f_{RRR}(R)\dot{R}^2 + f_{RR}(R)\ddot{R}\right] \\ -\frac{1}{r^2}f_R(R) - \frac{1}{r}f_R(R)\nu' - \frac{2}{r}f_{RR}(R)R' - \frac{1}{2}f_{RR}(R)R'\nu' \\ +e^{\lambda}\left[\frac{1}{2}Rf_R(R) - \frac{1}{2}f(R) + \frac{1}{r^2}f_R(R)\right] = 0. \end{cases} \tag{13.9}$$

The equation for $R(r,t)$ is given by Eq. (13.6), that is, the Euler constraint on the dynamics. Finally, we obtain a system of three partial differential equations in the unknown functions $\nu(r,t)$, $\lambda(r,t)$, and the constraint $R = R(r,t)$, once $f(R)$ is given. As a first check it is easy to see that the preceding system allows us to recover the well-known results of GR by replacing $f(R) = R$. Moreover, it is straightforward to show that when curvature is constant over space and time ($R = R_0$), the following solution can be found:

$$ds^2 = \left[1 - \frac{\Lambda r^2}{3} - \frac{A_1}{2r}\right] A_2(t)dt^2 - \frac{1}{1 - \frac{\Lambda r^2}{3} - \frac{A_1}{2r}}dr^2 - r^2 d\Omega^2, \tag{13.10}$$

where $A_2(t)$ is a generic function of time that cannot be determined. Internal symmetries of Eq. (13.8) can be found by imposing the vanishing Lie derivative condition in the minisuperspace $\mathcal{S} \equiv \{\nu, \lambda, R\}$. The corresponding Noether vector with infinitesimal generators $\eta^i = \{\alpha, \beta, \gamma\}$ reads

$$X = \alpha\frac{\partial}{\partial\nu} + \beta\frac{\partial}{\partial\lambda} + \gamma\frac{\partial}{\partial R} + \alpha'\frac{\partial}{\partial\nu'} + \beta'\frac{\partial}{\partial\lambda'} + \gamma'\frac{\partial}{\partial R'} + \dot{\alpha}\frac{\partial}{\partial\dot{\nu}} + \dot{\beta}\frac{\partial}{\partial\dot{\lambda}} + \dot{\gamma}\frac{\partial}{\partial\dot{R}}. \tag{13.11}$$

The condition $L_X\mathscr{L} = X\mathscr{L} = 0$ provides the following system of differential equations [250]:

$$\begin{cases} \frac{1}{2}r^2\left[f(R) - Rf_R(R)\right](\alpha+\beta) - f_R(R)(\alpha+\beta) - r^2 Rf_{RR}(R)\gamma \\ -2f_{RR}(R)\gamma + e^{-\lambda}\left[f_R(R)(\alpha-\beta)\right. \\ \left.-r\lambda' f_R(R)(\alpha-\beta) - 2r\lambda' f_R(R)\partial_\lambda\beta + 2\gamma f_{RR}(R)\left(1 - r\lambda'\right)\right] = 0 \\ \frac{\alpha}{2} + \partial_\nu\alpha - \frac{1}{2}\beta + \frac{f_{RRR}(R)}{f_{RR}(R)}\gamma + \partial_R\gamma = 0 \\ \frac{1}{2}e^{\nu-\lambda}(\beta-\alpha) + \partial_\lambda\beta + \gamma\frac{f_{RRR}(R)}{f_{RR}(R)} + \partial_R\gamma = 0 \\ \alpha = \alpha(\nu), \qquad \beta = \beta(\lambda), \qquad \gamma = \gamma(R), \end{cases} \tag{13.12}$$

where we neglected all those solutions leading to the function $f(R) = f_0 R + \Lambda$. The system selects the form of the function and the infinitesimal generators, providing a continuity equation of the form $\partial_t j_t + \partial_r j_r = 0$, which allows one to find a two-dimensional conserved quantity $j^\mu = \{j_t, j_r\}$. Specifically, the only solution is [250]

$$\mathcal{X} = \alpha_0 (\partial_\nu + \partial_\lambda) + \gamma_0 R^{\frac{3}{2}} \partial_R,$$

$$f(R) = 2 f_0 \left(2 C_1 e^{-C_2} - 2\right)^{\frac{1}{2}} R^{\frac{1}{2}} + f_1 R + f_2,$$

$$j^\mu = \begin{array}{l} \left\{ \left[\left(r^2 e^{\frac{\nu-\lambda}{2}} \right) \left(\alpha_0 f_{RR}(R) R' + \alpha_0 \frac{2}{r} f_R(R) - \frac{\gamma_0}{2} f_0 \left(2 C_1 e^{-C_2} - 2\right)^{\frac{1}{2}} \nu' \right) \right]; \right. \\ \left. \left[\left(r^2 e^{\frac{\lambda-\nu}{2}} \right) \left(\alpha_0 f_{RR}(R) \dot{R} - \frac{\gamma_0}{2} f_0 \left(2 C_1 e^{-C_2} - 2\right)^{\frac{1}{2}} \dot{\lambda} \right) \right] \right\}, \end{array} \tag{13.13}$$

where C_1, C_2, f_0, f_1, f_2 are integration constants. Here the index μ runs from 0 to 1. Substituting the function into the EL equations (13.9) and assuming that $\nu = -\lambda$, we obtain the line element

$$ds^2 = \left[\left(C_1 - \frac{1}{r} \right) + C_2 \right] dt^2 - \frac{1}{\left[\left(C_1 - \frac{1}{r} \right) + C_2 \right]} dr^2 - r^2 d\Omega^2, \tag{13.14}$$

with

$$R(r) = 2 \frac{\left(C_1 e^{-C_2} - 1 \right)}{r^2}, \tag{13.15}$$

in agreement with the Schwarzschild vacuum solution.

As mentioned previously, under proper coordinate reparameterizations, any diagonal interval can be reduced to that in Eq. (13.1). As a consequence, different line elements share the same solutions, which can be related to each other by a change of variables. This is not true for symmetries, since Noether's approach provides different results depending on the starting metric. For this reason, it is possible to search for symmetries by introducing a new function $M(r)$ in Eq. (13.1):

$$ds^2 = e^{\nu(r,t)} dt^2 - e^{\lambda(r,t)} dr^2 - M(r) d\Omega^2. \tag{13.16}$$

By this ansatz, Lagrangian (13.8) can be recast as[2]

$$\begin{aligned} \mathcal{L} = {} & e^{\frac{\nu-\lambda}{2}} \frac{M'^2}{2M} f_R - e^{\frac{\nu-\lambda}{2}} \nu' M' f_R - e^{\frac{\nu-\lambda}{2}} M \nu' R' f_{RR} \\ & - 2 e^{\frac{\nu-\lambda}{2}} R' M' f_{RR} - e^{\frac{\nu+\lambda}{2}} \left[(2 + MR) f_R - M f \right], \end{aligned} \tag{13.17}$$

[2] In this case the Lagrangian will be denoted by the symbol $\mathcal{L}$, unlike the previous case where it has been denoted by $\mathscr{L}$. This is due to the fact that Eq. (13.17) can be obtained by integrating the two-dimensional surface, so that $\mathscr{L}$ in Eq. (13.17) represents a linear density Lagrangian. On the other hand, the Lagrangian in Eq. (13.8) can be obtained from the integration of the three-dimensional surface.

where we assumed the Birkhoff theorem to hold, namely we set $\dot{\nu} = \dot{\lambda} = \dot{M} = 0$. Notice that the minisuperspace is four-dimensional, while the tangent space contains seven variables:

$$\begin{cases} \mathcal{S} \equiv \{\nu, \lambda, M, R\} \\ \mathcal{TS} \equiv \{\nu, \lambda, M, R, \nu', M', R'\}. \end{cases} \tag{13.18}$$

This is due to the fact that λ' does not explicitly occur in the Lagrangian (13.17). As a consequence, the corresponding EL equation $\partial_\lambda \mathcal{L} = 0$ provides the following relation:

$$e^\lambda = \frac{2M^2 f_{RR}\nu' R' + 2Mf_R \nu' M' + 4Mf_{RR}M'R' + f_R M'^2}{2M[(2+MR)f_R - Mf]}. \tag{13.19}$$

This allows one to reduce the minisuperspace dimension, so that the symmetry generator can be written as

$$\mathcal{X} = \alpha(\nu, M, R)\frac{\partial}{\partial \nu} + \beta(\nu, M, R)\frac{\partial}{\partial M} + \gamma(\nu, M, R)\frac{\partial}{\partial R}. \tag{13.20}$$

Introducing the new variable $M(r)$ yields a not-dissipative Lagrangian, thus dynamics is provided by EL equations together with the energy condition. Nevertheless, the latter turns out to be a linear combination of the former ones. The application of the Noether Symmetry Approach to Lagrangian (13.17) yields the following system:

$$\begin{cases} [(2+MR)f_R - Mf]\left(\partial_\nu \beta f_R + M\partial_\nu \gamma f_{RR}\right) = 0 \\ \dfrac{1}{M}\left[(2+MR)\gamma f_{RR} - \dfrac{2\beta}{M}f_R\right] f_R + [(2+MR)f_R - Mf] \\ \times\left[\left(e^{-\nu}\dfrac{\alpha}{M} + 2e^{-\nu}\partial_M\alpha + \dfrac{2}{M}\partial_M\beta\right) f_R + \left(\dfrac{\gamma}{M} + 4\partial_M\gamma\right) f_{RR}\right] = 0 \\ [(2+MR)f_R - Mf]\left(M\partial_R\alpha + 2e^\nu \partial_R\beta\right) f_{RR} = 0 \\ \beta\,(f - Rf_R)\, f_R - [(2+MR)f_R - Mf] \\ \times\left[(\gamma + M\partial_M\gamma + 2\partial_\nu\gamma)\, f_{RR} + \left(\partial_M\beta + e^{-\nu}\partial_\nu\alpha + \dfrac{1}{M}\partial_\nu\beta\right) f_R\right] = 0 \\ [M(2+MR)\gamma f_{RR} - 2\beta f_R]\, f_{RR} + [(2+MR)f_R - Mf] \\ \times\left[f_R\partial_R\beta + (2\beta + Me^{-\nu}\partial_\nu\alpha + 2\partial_\nu\beta + M\partial_R\gamma)\,\gamma f_{RR} + Mf_{RRR}\right] = 0 \\ 2e^\nu\left[(2+MR)\gamma f_{RR} - (f - Rf_R)\,\beta\right] f_{RR} \\ + [(2+MR)f_R - Mf]\left[\left(\partial_R\alpha + \dfrac{e^\nu}{M}\partial_R\beta\right) f_R\right. \\ \left. + (2\alpha + 2e^\nu\partial_R\gamma + M\partial_M\alpha + 2e^\nu\partial_M\beta)\, f_{RR} + 2e^\nu\gamma f_{RRR}\right] = 0, \end{cases} \tag{13.21}$$

whose solution is [366–368]

$$\begin{cases} \mathcal{X} = \alpha_0(3-2k)e^\nu\partial_\nu - \alpha_0 M\partial_M + \alpha_0 R\partial_R \\ f(R) = f_0 R^k \\ j = 2k\alpha_0 MR^{2k-3}[2k + (k-1)MR]\left[(k-2)R\nu' - (2k^2 - 3k + 1)\,R'\right] e^\nu. \end{cases} \tag{13.22}$$

As before, the Schwarzschild solution is recovered for $R \sim$ const. The integral of motion j can be used to find a relation among ν, M, and R:

$$e^{\nu} = \begin{cases} f_0 R^{\frac{2k^2-3k+1}{k-2}} \left\{ 1 + j \int \frac{R^{\frac{4k^2-9k+5}{2-k}} dr}{2\alpha_0 k(k-2)M[2k+(k-1)MR]} \right\}, & k \neq 2 \\ -\frac{j}{12\alpha_0 r^2 (4 + r^2 R) RR'}, & k = 2. \end{cases} \tag{13.23}$$

The preceding integral cannot be analytically solved for any k, but specific models can be separately analyzed by setting the value of the free parameter. For instance, for $k = \frac{5}{4}$ one gets [369]

$$ds^2 = \frac{1}{\sqrt{5}} (k_2 + k_1 r)\, dt^2 - \frac{1}{2} \left(\frac{1}{1 + \frac{k_2}{k_1 r}} \right) dr^2 - r^2 d\Omega. \tag{13.24}$$

It is worth noticing that the Schwarzschild space-time cannot be recovered as a limit of Eq. (13.24), as well as the asymptotic flatness for large r. Another subcase of (13.23) is discussed in Ref. [370], where the authors consider the first-order approximation of the field equations, around $k = 1$. In particular, by setting $k = 1 + \epsilon$ with $\epsilon \ll 1$, the metric components turn out to be

$$e^{\nu} = r^{\frac{2\epsilon(1+2\epsilon)}{1-\epsilon}} + \frac{C_1}{r^{\frac{1-4\epsilon}{1-\epsilon}}}, \tag{13.25}$$

$$e^{\lambda} = \frac{(1-\epsilon)^2}{(1-2\epsilon+4\epsilon^2)\,[1-2\epsilon(1+\epsilon)]} \left(1 + \frac{C_1}{r^{\frac{1-2\epsilon+4\epsilon^2}{1-\epsilon}}} \right). \tag{13.26}$$

The preceding solution extends the Schwarzschild one, which is clearly recovered for $\epsilon \to 0$. Finally, assuming a constant scalar curvature $R(r) = R_0$, Eq. (13.23) provides

$$e^{\nu} = e^{-\lambda} = \frac{4c_1 - c_2 + 12r + r^3 R_0}{12r}, \tag{13.27}$$

where c_1, c_2 are integration parameters. By means of the definition $c_1 = (c_2 - 24G\mathcal{M})/4$, Eq. (13.27) takes the form

$$e^{\nu} = e^{-\lambda} = 1 - \frac{2G\mathcal{M}}{r} + \frac{R_0}{12} r^2, \tag{13.28}$$

which is a Schwarzschild–de Sitter metric. When physical units are restored, G can be intended as the Newton constant, R_0 as a cosmological constant, and $\mathcal{M}$ as the gravitational mass of the system.

An important remark is in order at this point. As discussed in detail in [366], the conserved quantity j in (13.22) has a fundamental physical meaning. As soon as $k = 1$, we have $j = 2G\mathcal{M}/c^2$, which is the Schwarzschild radius, where $\mathcal{M}$ is the gravitational mass of the system. Clearly, for any k, a generalized mass (and then a gravitational radius) can be recovered. This fact is crucial for addressing the missing matter problem in galaxies. As discussed in [371,372], the gravitational radius for $k = 3/2$ can be given in terms of the MOND acceleration parameter. In other words, the gravitational theory $f(R) = f_0 R^{3/2}$ is capable, in principle, of solving the problem of flat rotation curves of spiral galaxies extending the gravitational sector without dark matter. This argument will be discussed in detail in Chapter 15.

Another interesting property occurs when the extension of Eq. (13.27) to $(d+1)$-dimensions is considered. Specifically, it turns out that $f(R) = f_0 R^k$ field equations can be solved analytically only when the relation $k = (d+1)/2$ holds. This can be directly proven by the $f(R)$ field equations, which, for constant scalar curvature, can be written as

$$f_R(R) R_{\mu\nu} - \frac{1}{2} g_{\mu\nu} f(R) = 0. \tag{13.29}$$

By contracting the indexes and replacing $f(R) = f_0 R^k$, we have the trace

$$R^k \left[\left(\frac{d+1}{2} \right) - k \right] = 0, \tag{13.30}$$

which is satisfied either for $R = 0$ or for $R \neq 0$, $k = (d+1)/2$. The former yields the $(d+1)$-dimensional extension of the Schwarzschild solution, while the latter provides the same solution of GR with cosmological constant. Specifically, the generalization of Eq. (13.27) to any dimension reads

$$\begin{aligned} e^{\nu} &= e^{-\lambda} = 1 - \frac{2(c_2 - c_1 d - c_1)}{d(d+1)(d-1)} \frac{1}{r^{d-2}} + \frac{r^2 R_0}{d(d+1)}, \\ k &= \frac{d+1}{2}, \qquad R(r) = R_0, \qquad f(R) = f_0 R^{\frac{d+1}{2}}. \end{aligned} \tag{13.31}$$

It is interesting to compare Eq. (13.31) with the spherically symmetric solution provided by a $(d+1)$-dimensional Schwarzschild space-time. To this purpose, let us distinguish the cases $R_0 r^2 \ll 1$ and $R_0 r^2 \gg 1$. In the former case, the time-like component can be recast as

$$e^{\nu} = 1 - \frac{\eta_d \mathcal{M}}{r^{d-2}}, \tag{13.32}$$

with η_d and $\mathcal{M}$ being constants depending on c_1, c_2, and d. The constant $\mathcal{M}$ is a generalized mass. In this case, the horizon scales as $r_S \sim \mathcal{M}^{\frac{1}{d-2}}$. The Bekenstein–Hawking entropy $\mathscr{S}$, computed (in natural units) by the well-known formula $\mathscr{S} = A/4$, turns out to scale as

$$\mathscr{S} \sim r_S^{d-1} \sim \mathcal{M}^{\frac{d-1}{d-2}}. \tag{13.33}$$

On the other hand, in the case $\Lambda r^2 \gg 1$, no term can be neglected in the time-like component (13.31) and the solution can be written as

$$1 - \frac{\eta_d \mathcal{M}}{r^{d-2}} + \Lambda r^2 = \Lambda \left(\frac{\frac{r^d}{r^2 \Lambda} - \frac{\eta_d \mathcal{M}}{\Lambda} + r^d}{r^{d-2}} \right), \tag{13.34}$$

where we set $\frac{R_0}{d(d-1)} \equiv \Lambda$. Considering that, for the first term, the condition $\frac{r^d}{r^2\Lambda} \ll 1$ must hold, similar computations show that the horizon is

$$r_S \sim \mathcal{M}^{\frac{1}{d}} \tag{13.35}$$

and thus the entropy scales as

$$\mathscr{S} \sim \mathcal{M}^{\frac{d-1}{d}}. \tag{13.36}$$

Notice that the preceding entropy scales as that of a d-dimensional Conformal Field Theory (CFT), due to the AdS/CFT correspondence.

To conclude this section, it is worth stressing that analytic solutions in modified gravity are very rare, and a selection criterion aimed at finding exact dynamics is often needed. From this point of view, Noether's theorem can provide an integral of motion by means of which it is possible to reduce the order of the field equations.

13.1.1 Axial Symmetry from Spherical Symmetry

It is possible to show that the spherically symmetric interval (13.1) can be used to derive an axially symmetric spacetime. To this purpose, we consider the *Newman–Janis* method [373, 374], based on the so-called *complex null tetrads.* Let us start with the weak field limit solution (13.25), which, under proper redefinitions of the free parameters, can be equivalently written as [375, 376]

$$e^{\nu(r)} = 1 - \frac{r_S}{r}\left[1 + \left(\frac{r}{r_c}\right)^{\beta}\right], \qquad r_S = 2\mathcal{M}, \tag{13.37}$$

with r_c being a constant with dimension of length. Let us first assume the validity of the Birkhoff theorem and start from the following line element:

$$ds^2 = e^{2\phi(r)}dt^2 - e^{2\lambda(r)}dr^2 - r^2 d\Omega^2, \tag{13.38}$$

where we renamed the variables in (13.1). By a suitable change of coordinates $dt = du + F(r)dr$ and $F(r) = \pm e^{\lambda(r)-\phi(r)}$, the interval (13.38) takes the following form:

$$ds^2 = e^{2\phi(r)}du^2 \pm 2e^{\lambda(r)+\phi(r)}du dr - r^2 d\Omega^2. \tag{13.39}$$

A general spherically symmetric metric tensor can be decomposed by means of the complex null tetrads $l^\mu, n^\mu, m^\mu, \overline{m}^\mu$ as $g^{\mu\nu} = l^\mu n^\nu + l^\nu n^\mu - m^\mu \overline{m}^\nu - m^\nu \overline{m}^\mu$, with the conditions

$$l_\mu l^\mu = m_\mu m^\mu = n_\mu n^\mu = 0, \quad l_\mu n^\mu = -m_\mu \overline{m}^\mu = 1, \quad l_\mu m^\mu = n_\mu m^\mu = 0, \tag{13.40}$$

where $\overline{m}^\mu$ is the complex conjugated of m^μ. In the space-time (13.38), with coordinates $x^\mu = (u, r, \theta, \phi)$, the complex null tetrads take the form

$$\begin{cases} l^\mu = \delta_1^\mu \\ n^\mu = -\frac{1}{2} e^{-2\lambda(r,\bar{r})} \delta_1^\mu + e^{-\lambda(r,\bar{r})-\phi(r,\bar{r})} \delta_0^\mu \\ m^\mu = \frac{1}{\sqrt{2}\bar{r}} \left(\delta_2^\mu + \frac{i}{\sin\theta} \delta_3^\mu \right) \\ \overline{m}^\mu = \frac{1}{\sqrt{2}r} \left(\delta_2^\mu - \frac{i}{\sin\theta} \delta_3^\mu \right). \end{cases} \tag{13.41}$$

Moreover, considering the complex coordinate transformation

$$\tilde{x}^\mu = x^\mu + ia\,(\delta_1^\mu - \delta_0^\mu)\cos\theta, \quad i.e. \quad \begin{cases} \tilde{u} = u + ia\cos\theta \\ \tilde{r} = r - ia\cos\theta \\ \tilde{\theta} = \theta \\ \tilde{\phi} = \phi, \end{cases} \tag{13.42}$$

with a a real constant, the complex null tetrads become

$$\begin{cases} \tilde{l}^\mu = \delta_1^\mu \\ \tilde{n}^\mu = -\frac{1}{2} e^{-2\lambda(\tilde{r},\theta)} \delta_1^\mu + e^{-\lambda(\tilde{r},\theta)-\phi(\tilde{r},\theta)} \delta_0^\mu \\ \tilde{m}^\mu = \frac{1}{\sqrt{2}(\tilde{r} - ia\cos\theta)} \left[ia\,(\delta_0^\mu - \delta_1^\mu)\sin\theta + \delta_2^\mu + \frac{i}{\sin\theta}\delta_3^\mu \right] \\ \tilde{\overline{m}}^\mu = \frac{1}{\sqrt{2}(\tilde{r} + ia\cos\theta)} \left[-ia\,(\delta_0^\mu - \delta_1^\mu)\sin\theta + \delta_2^\mu - \frac{i}{\sin\theta}\delta_3^\mu \right]. \end{cases} \tag{13.43}$$

In this way, substituting Eqs. (13.43) into Eq. (13.38), the latter can be written as

$$\begin{aligned} ds^2 =& e^{2\phi(\tilde{r},\theta)} dt^2 + 2a e^{\phi(\tilde{r},\theta)} \left[e^{\lambda(\tilde{r},\theta)} - e^{\phi(\tilde{r},\theta)} \right] \sin^2\theta \, dt d\phi \\ & - \frac{\Sigma^2 d\tilde{r}^2}{\left(\Sigma^2 e^{-2\lambda(\tilde{r},\theta)} + a^2 \sin^2\theta \right)} - \Sigma^2 d\theta^2 \\ & - \left[\Sigma^2 + a^2 \sin^2\theta \, e^{\phi(\tilde{r},\theta)} \left(2 e^{\lambda(\tilde{r},\theta)} - e^{\phi(\tilde{r},\theta)} \right) \right] \sin^2\theta \, d\phi^2, \end{aligned} \tag{13.44}$$

with $\Sigma = \sqrt{\tilde{r}^2 + a^2\cos^2\theta}$. Substituting the solution (13.37) into the preceding line element, we obtain

$$\begin{aligned} ds^2 =& \frac{r(a + \beta r) + a^2 \beta \cos^2\theta}{\Sigma} dt^2 \\ & - 2 \frac{a\left(-2ar - 2\beta\Sigma^2 + \sqrt{2\beta}\Sigma^{3/2} \right) \sin^2\theta}{2\Sigma} dt d\phi \\ & - \frac{\beta\Sigma^2}{2ar + \beta\,(a^2 + r^2 + \Sigma^2)} d\tilde{r}^2 - \Sigma^2 d\theta^2 \\ & - \left[\Sigma^2 - \frac{a^2\left(ar + \beta\Sigma^2 - \sqrt{2\beta}\Sigma^{3/2} \right) \sin^2\theta}{\Sigma} \right] \sin^2\theta \, d\phi^2, \end{aligned} \tag{13.45}$$

which finally shows that a spherically symmetric space-time can be recast as a Kerr-like axially symmetric spacetime. Though we considered the application to solution (13.37), the same result holds for any spherical interval. The dynamics of

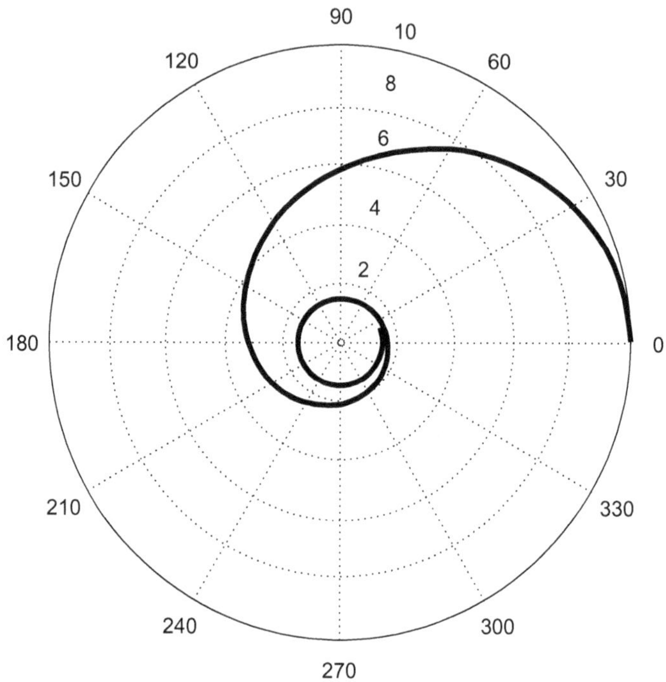

Figure 13.1 Trajectories of a test particle with unit mass in an axially symmetric space-time

a test particle with unit mass can be obtained by considering the particle Lagrangian $\mathcal{L} = g_{\mu\nu}\frac{dx^\mu}{ds}\frac{dx^\nu}{ds}$ and solving the related EL equations. The two-dimensional representation of the dynamics of the test particle moving in the space-time (13.45) is illustrated in Fig. **13.1** (see Ref. [368] for details).

As a final remark, it is important to say that, in general, by the Newman–Janis method it is possible to derive axially symmetric solutions starting from spherical solutions. This statement applies in general as well as to the cases developed in what follows, and not only for $f(R)$ gravity.

13.2 Spherical Symmetry in $f(T, B)$ Gravity

Here we consider $f(T, B)$ gravity in a spherically symmetric space-time, where the search for symmetries can be helpful in providing exact solutions for the field equation. Specifically, we use Noether's symmetry condition (13.4) to select models leading to exact solutions and thus find the corresponding dynamics. Though this theory has been largely studied in literature, especially in the context of GWs [99, 377, 378] or large-scale structure [379–381], few studies rely on exact spherically symmetric solutions (see, e.g., [382–385]). Here we sketch the steps reported in the latter reference and apply the Noether Symmetry Approach to modified $f(T, B)$ gravity in a spherically symmetric background. It is worth pointing out that, unlike $f(R)$ gravity, modified TEGR involves tetrad fields, which cannot be uniquely determined by the starting interval. Indeed, a careful choice of "good" tetrads is often important [100].

Let us start by considering the following ansatz for the line element[3]

$$ds^2 = P(r)^2 dt^2 - Q(r)^2 dr^2 - M(r)^2 d\Omega^2 \,, \tag{13.46}$$

with Ω being the 2-sphere. Before proceeding, an important comment is necessary here. We assume that the metric (13.46) is not dynamical, which means that the Birkhoff theorem should be valid for these models. A "good" set of tetrad fields providing the line element (13.46) is

$$h^a_\mu = \begin{pmatrix} P(r) & 0 & 0 & 0 \\ 0 & Q(r)\sin\theta\cos\phi & M(r)\cos\theta\cos\phi & -M(r)\sin\theta\sin\phi \\ 0 & Q(r)\sin\theta\sin\phi & M(r)\cos\theta\sin\phi & M(r)\sin\theta\cos\phi \\ 0 & Q(r)\cos\theta & -M(r)\sin\theta & 0 \end{pmatrix}, \tag{13.47}$$

which yields the following torsion scalar and boundary term

$$T(r) = \frac{2}{MQ^2}(Q - M')\left[\frac{(Q-M')}{M} - \frac{2P'}{P}\right], \tag{13.48}$$

$$\begin{aligned} B(r) = &-\frac{4}{QM}\left(\frac{P'}{P} + \frac{M'}{M}\right) \\ &+ \frac{2}{Q^2}\left[\frac{P''}{P} - \frac{P'Q'}{PQ} + \frac{2M''}{M} + \frac{2M'}{M}\left(\frac{2P'}{P} - \frac{Q'}{Q} + \frac{M'}{M}\right)\right]. \end{aligned} \tag{13.49}$$

Notice that the scalar curvature (13.6), written in terms of the line element (13.46), is straightforwardly recovered by the combination $R(r) = -T(r) + B(r)$, as expected. The Lagrangian can be obtained by using the constraints (13.48) and (13.49) as Lagrange multipliers, with the starting action given by Eq. (7.38). Setting $\theta = \frac{\pi}{2}$ and integrating out the three-dimensional hypersurface, we obtain[4]

$$\begin{aligned} \mathcal{L} = &-4M(f_T + f_B)\,P' - 4P(f_T + f_B)\,M' + \frac{2}{Q}f_T\,(2MP'M' + PM'^2) \\ &- \frac{2M}{Q}(f_{BT}T' + f_{BB}B')\,(MP' + 2PM') \\ &- PQ\left[M^2\,(Tf_T + Bf_B - f) - 2f_T\right], \end{aligned} \tag{13.50}$$

with f_T and f_B being the first derivatives of $f(T, B)$ with respect to T and B, respectively. The EL equations, obtained by varying the Lagrangian with respect to P, Q, M, read, respectively

[3] For convenience, in this section and in the following one, the interval is not written in terms of exponential functions.

[4] For simplicity, we choose to develop computations in the equatorial plane, but the final result is clearly independent of the value of θ.

$$
\begin{cases}
2f_T\left[\frac{2M''}{M}+\frac{M'^2}{M^2}-\frac{2Q'M'}{QM}-\frac{Q^2}{M^2}\right]+\frac{4f'_T}{M}\left(M'-Q\right) \\
+2f'_B\left(\frac{Q'}{Q}-\frac{2Q}{M}\right)-2f''_B-Q^2\left(f-Tf_T-Bf_B\right)=0 \\
\\
2f_T\left[\frac{M'}{M}\left(\frac{2P'}{P}+\frac{M'}{M}\right)-\frac{Q^2}{M^2}\right]-2f'_B\left(\frac{P'}{P}+\frac{2M'}{M}\right) \\
-Q^2\left(f-Tf_T-Bf_B\right)=0 \\
\\
2f_T\left[\frac{P''}{P}+\frac{M''}{M}+\frac{P'M'}{PM}-\frac{Q'M'}{QM}-\frac{P'Q'}{PQ}\right] \\
+2f'_T\left(\frac{P'}{P}+\frac{M'}{M}-\frac{Q}{M}\right)+2f'_B\left(\frac{Q'}{Q}-\frac{Q}{M}\right) \\
-2f''_B-Q^2\left(f-Tf_T-Bf_B\right)=0,
\end{cases}
\tag{13.51}
$$

where f'_B denotes the total derivative of f_B with respect to r, namely $f'_B=f_{BT}T'+f_{BB}B'$. The other two equations, namely with respect to T and B, provide the spherically symmetric expressions (13.48) and (13.49) by construction.

Notice that the Lagrangian is written in a canonical form, thus the constraint $\xi=\xi(r)$ can be imposed a priori as a requirement. Moreover, the minisuperspace $\mathcal{S}\equiv\{P,Q,M,T,B\}$ is five-dimensional, while the tangent space $\mathcal{TS}\equiv\{P,Q,M,T,B,P',M',T',B'\}$ is only nine-dimensional, due to the absence of Q' in the Lagrangian. This feature preserves the condition of vanishing Hessian determinant. The symmetry generator in the minisuperspace $\mathcal{S}$ is

$$
\begin{aligned}
\mathcal{X}&=\xi(r)\partial_r+\alpha(P,Q,M,T,B)\partial_P+\beta(P,Q,M,T,B)\partial_Q \\
&\quad+\gamma(P,Q,M,T,B)\partial_M+\delta(P,Q,M,T,B)\partial_T+\zeta(P,Q,M,T,B)\partial_B.
\end{aligned}
\tag{13.52}
$$

The application of Eq. (13.4), with generator (13.52), to Lagrangian (13.50) provides a system of 34 differential equations. Neglecting linear combinations and imposing $\xi=\xi(r)$, the system can be solved analytically and selected functions with symmetries naturally arise from the system resolution. In what follows we shall only discuss Noether's system solutions of physical interest, outlining generators of the symmetry, conserved quantities, and corresponding functions.

The first set of solutions comes from considering the function $f(T,B)=f_0T+f_1B$, from which three symmetry generators (and integrals of motion) follow, namely

$$
\begin{cases}
\mathcal{X}_1=-P\partial_P+Q\partial_Q+M\partial_M, \quad j_1=4f_0M^2\frac{P'}{Q} \\
\mathcal{X}_2=\frac{1}{PM}\left(\partial_P-\frac{Q}{P}\partial_Q\right)+\partial_M, \quad j_2=\frac{4_0M'}{PQ} \\
\mathcal{X}_3=\frac{1}{2M}\left(-P\partial_P+Q\partial_Q\right)+\partial_M, \quad j_3=\frac{2f_0M}{PQ}\left(2\frac{P'}{P}+\frac{M'}{M}\right),
\end{cases}
\tag{13.53}
$$

and the associated gauge functions are, respectively,

$$\mathcal{X}_1 \to \forall g, \quad \mathcal{X}_2 \to g_2 = -\frac{4}{P}(f_0 + f_1), \quad \mathcal{X}_3 \to g_3 = -2(f_0 + f_1). \tag{13.54}$$

Exact solutions can be found by either plugging the function $f(T, B) = f_0 T + f_1 B$ into the EL equations (13.51) or making use of the conserved quantities provided by the Noether symmetries. As the boundary term provides a trivial contribution, in both cases we expect to recover a Schwarzschild space-time when $M(r) = r$, due to the equivalence between GR and TEGR. The latter approach takes advantage of Noether symmetries to straightforwardly provide dynamics without solving the equations of motion directly. More precisely, notice that the EL equations can be used to recast the functions $P(r)$ and $Q(r)$ in terms of the conserved quantities j_1, j_2, j_3 as

$$P(r)^2 = \left(\frac{4f_0}{j_2}\right)^2 \left(1 - \frac{j_1}{j_3 M}\right), \quad Q(r)^2 = \frac{M'^2}{1 - \frac{j_1}{j_3 M}}. \tag{13.55}$$

The Schwarzschild solution is thus recovered for $M(r) = r$, as we expect. More generally, another function containing symmetries is an extension of the previously considered $f(T, B) = f_0 T + f_1 B$, that is, $f(T, B) = f_0 T^k + f_1 B$, with $k \in \mathbb{R}$. In this case, Noether's system resolution yields

$$\begin{cases} \mathcal{X}_1 = (2k-3)P\partial_P + Q\partial_Q + M\partial_M - 2T\partial_T \\ j_1 = \dfrac{4f_0 P}{Q} M^2 T^{k-1} \left[\dfrac{P'}{P} + 2(k-1)\left(\dfrac{M'}{M} - \dfrac{Q}{M}\right)\right], \end{cases} \tag{13.56}$$

with associated gauge function of the form $g = -8f_1(k-1)MP$. Also here, the boundary term does not affect the equations of motion, which thus reduce to those of $f(T) = f_0 T^k$ gravity. However, as in the previous case, a unique solution cannot be found as long as (at least) one of the functions in the line element is not constrained by some physical ansatz. To this purpose, we assume the coefficient of the 2-sphere to be $M(r) = r$. In this way, exact solutions can be found for general k. More precisely, we have

$$\begin{aligned} &P(r)^2 = 1 - \frac{c_1}{r}, \quad Q(r) = P(r)^{-1}, \quad k = 1, \\ &P(r)^2 = \left(\frac{r}{r_0}\right)^{\frac{8k(k-1)(2k-3)}{4k^2-8k+5}}, \quad Q(r) = c_2, \quad \forall k \neq \frac{1}{2}, \frac{5}{6}, \frac{5}{4}, \frac{3}{2}, \end{aligned} \tag{13.57}$$

with c_2 being defined as

$$c_2 = \frac{(2k-1)(4k-5)}{4k^2 - 8k + 5}. \tag{13.58}$$

This solution lacks a physical horizon and contains a nontrivial singularity in $r = 0$. However, the function $f(T) = T^k$ can be interesting when it is recast in terms of a parameter $\epsilon \ll 1$ as $f(T) = T^{1+\epsilon}$. It provides corrections to standard TEGR that extend the Schwarzschild solution.

Another class of solutions, corresponding to the function $f(T,B)=f_0T^k+f_1B^\ell$, with ℓ real constant, comes from the solution of Noether's system after imposing $k=\ell$. It reads

$$\mathcal{X}_1=(2k-3)P\partial_P+Q\partial_Q+M\partial_M-2T\partial_T-2B\partial_B, \tag{13.59}$$

with corresponding integral of motion of the form

$$\begin{aligned} j_1=\frac{4f_0kPM^2}{QT^{1-k}}&\left\{\frac{P'}{P}+2(k-1)\left(\frac{M'}{M}-\frac{Q}{M}\right)\right.\\ &\left.+(k-1)\frac{f_1}{2f_0}\left(\frac{B}{T}\right)^{k-1}\left[2\left(\frac{P'}{P}+\frac{2M'}{M}\right)+(1-2k)\frac{B'}{B}-\frac{4Q}{M}\right]\right\}. \end{aligned} \tag{13.60}$$

Once again, after setting $M(r)=r$, the conserved quantity can be used to analytically solve the equations of motion, providing

$$P(r)^2=\left(\frac{r}{r_0}\right)^{c_0},\quad Q(r)^2=c_3, \tag{13.61}$$

with r_0, c_0, and c_3 real constants. This solution yields a vanishing conserved quantity and, moreover, suffers the same problems as that in Eq. (13.57), corresponding to $f(T)=f_0T^k$ gravity. In what follows we outline the last three symmetries and the related functions $f(T,B)$. All of them lead to spherically symmetric solutions of the form (13.61) and therefore suffer the previously mentioned issues of physical interpretation. They read

$$\begin{cases} \mathcal{X}_I=-P\partial_P+Q\partial_Q+M\partial_M-2B\partial_B, \quad g=8f_2PM\\ j_I=\frac{4PM^2}{Q}\left[f_0\frac{P'}{P}+f_2\left(\frac{P'}{P}+\frac{2M'}{M}-\frac{2Q}{M}-\frac{B'}{2B}\right)\right]\\ f_I(T,B)=f_0T+f_1B+f_2B\ln B\\ \\ \mathcal{X}_{II}=-P\partial_P+Q\partial_Q+M\partial_M-2T\partial_T-2B\partial_B, \quad g=8f_2PM\\ j_{II}=\frac{4PM^2}{Q}\left[\left(f_0+f_2\frac{B}{T}\right)\frac{P'}{P}+f_2\left(\frac{P'}{P}+\frac{2M'}{M}-\frac{2Q}{M}-\frac{T'}{2T}\right)\right]\\ f_{II}(T,B)=f_0T+f_2B\ln T\\ \\ \mathcal{X}_{III}=(2k-1)P\partial_P+Q\partial_Q-M\partial_M-2T\partial_T-2B\partial_B, \quad \forall g\\ j_{III}=\frac{4f_1kT^kPM^2}{Q}\left\{\frac{P'}{P}+\frac{2M'}{M}-\frac{2Q}{M}+\frac{B}{T}\left[\frac{P'}{P}+2k\left(\frac{M'}{M}-\frac{Q}{M}\right)\right]-(2k+1)\frac{T'}{2T}\right\}\\ f_{III}(T,B)=f_1BT^k. \end{cases} \tag{13.62}$$

In all cases, analytic solutions, following from the imposition $M(r)=r$, have the same form, that is,

$$P(r)^2=\left(\frac{r}{r_0}\right)^{2q_i},\quad Q(r)=c_i, \tag{13.63}$$

with q_i and c_i constants. In particular, we have

$$
\begin{array}{l} f(T,B)=f_0T \\ +f_1B+f_2B\ln B \end{array} \rightarrow \begin{cases} c_1=\dfrac{f_2\left(f_0^2+8f_0f_2+f_2^2\right)\pm\left(f_0+f_2\right)K_1}{f_0\left(f_0^2+2f_0f_2+5f_2^2\right)} \\ q_1=\dfrac{f_2\left[\left(f_0+f_2\right)\left(2f_2-3f_0\right)\pm 2K_1\right]}{f_0\left(f_0^2+2f_0f_2+5f_2^2\right)}, \quad K_1\in\mathbb{R}, \end{cases} \tag{13.64}
$$

$$
\begin{array}{l} f(T,B)=f_0T \\ +f_2B\ln T \end{array} \rightarrow \begin{cases} c_2=\dfrac{K_2^2+K_2(q_2+3)-11q_2^2+18q_2+21}{6K_2} \\ K_2^3=\left(q_2+3\right)\left(37q_2^2+24q_2-27\right) \\ \quad +\sqrt{75q_2^6+96q_2^5+684q_2^4+702q_2^3 -714q_2^2-864q_2-75}, \end{cases} \tag{13.65}
$$

$$
\begin{array}{l} f(T,B)=f_0T \\ +f_2B\ln T \end{array} \rightarrow \begin{cases} c_3=\dfrac{(q_3+8)(q_3+1)^2\mp q_3C}{2\left(2q_3^2+5q_3+4\right)} \\ k=\dfrac{\left(q_3+1\right)\left(3q_3^3+9q_3^2+20q_3+16\right)\pm q_3^2C}{4(q_3+2)\left(2q_3^2+5q_3+4\right)} \\ C^2=q_3^4+4q_3^3+22q_3^2+48q_3+33, \end{cases} \tag{13.66}
$$

where c_1, q_1 correspond to the first solution of (13.62), c_2, q_2 to the second, and c_3, q_3 to the third.

In this section, we adopted Noether's prescription to $f(T,B)$ gravity in spherical symmetry, outlining solutions of physical interest. All of them yield analytic expressions of the line element in a spherically symmetric background, most of which, however, suffer some shortcomings for physical interpretation (such as the lack of physical horizons, or of Newtonian potentials for large radius). Nonetheless, they are mathematically consistent. This is an indication that they need to be further investigated in the low-energy limit, in order to establish whether Schwarzschild space-time can actually be recovered when small deviations from TEGR are considered.

Summing up, it is still unclear whether such solutions might be astrophysically relevant, but several approaches can be pursued in this direction, such as analyzing the accretion process of these solutions [386] or studying the photon sphere and the perihelion shift [387].

Let us recall that we considered only symmetries leading to exact spherically symmetric analytic solutions. For a complete discussion see [384]. It is worth observing that only standard TEGR yields a Schwarzschild space-time, while the other functions selected by Noether's approach lead to other mathematical solutions, which could not be black holes. All of them formally share the same form, that is $P(r)^2=\left(\dfrac{r}{r_0}\right)^{2q_i}$, $Q(r)=c_i$, with different c_i depending on the considered generator. Also notice that, independently of the starting action, by setting $T(r) = 0$ another solution occurs, that is [388],

$$P(r)^2 = \frac{1}{2}\left(1+\sqrt{1-\frac{r_0^2}{r^2}}\right), \quad Q(r) = 1/P(r). \tag{13.67}$$

While the asymptotic flatness is recovered for $r \to \infty$, no event horizon occurs. However, Eq. (13.67) can be recast in a Schwarzschild-like form after considering the first-order approximation, namely $P(r)^2 \sim 1 - \frac{r_0^2}{4r^2}$.

As a final remark, notice that the interval (13.46) is equivalently provided by the diagonal set of tetrads

$$h^a_\mu = diag(P(r), Q(r), M(r)\sin\theta, M(r)\sin\theta\cos\phi). \tag{13.68}$$

However, after getting the point-like Lagrangian and applying Noether's approach, it turns out that the latter yields a system of three differential equations, which provides only trivial symmetries.

13.3 Spherical Symmetry in $f(\mathcal{G})$ Gravity

Here we adopt the Noether Symmetry Approach for $f(\mathcal{G})$ gravity in a $(d+1)$-dimensional spherically symmetric background, with the aim to find black hole solutions. It is interesting to point out that the only forms of $f(\mathcal{G})$ selected by symmetries are power-law functions, as in Section 8.1.2. As previously pointed out, recent works suggest that the four-dimensional limit of $f(\mathcal{G}) = f_0\mathcal{G}$ gravity provides nontrivial solutions when the gravitational coupling diverges as $d = 3$. In this way, at the level of field equations, the vanishing topological term can be compared to a coupling and gives rise to reliable models.

Let us consider now the higher-dimensional extension of the static and spherically symmetric line element (13.46), with $M(r) = r$, namely

$$ds^2 = P(r)^2 dt^2 - Q(r)^2 dr^2 - r^2 d\Omega^2_{d-1}, \tag{13.69}$$

where

$$d\Omega^2_{d-1} = r^2 \left[d\theta_1^2 + \sum_{i=2}^{d-1}\prod_{j=1}^{i-1} \sin^2\theta_j d\theta_i^2 \right].$$

is the $(d-1)$-sphere. Also here, we proceed by assuming the validity of the Birkhoff theorem. This is not proven and we take it for granted, as we did in the previous sections. However, there are a lot of references in the literature claiming to have found cases where a generalization of the Birkhoff theorem could exist [389–393].

Starting from the metric (13.69), the nonvanishing components of the $(d+1)$-dimensional Riemann tensor, Ricci tensor, and Ricci scalar are

$$
\begin{aligned}
R_{0101} &= P\left(\frac{P'Q'}{Q} - P''\right), \quad R_{0i0j} = -\left.\frac{rP'P}{Q^2}\delta_{ij}\right|_{i,j\geq 2}, \\
R_{1i1j} &= -\left.\frac{rQ'}{Q}\delta_{ij}\right|_{i,j\geq 2}, \quad R_{ikj\ell} = -r^2\left(1-\frac{1}{Q^2}\right)\delta_{ij}\delta k\ell\bigg|_{i,j,k,\ell\geq 2}, \\
R_{00} &= -\frac{P\left[rP'Q' - (d-1)QP' - rQP''\right]}{rQ^3}, \\
R_{11} &= \frac{(d-1)PQ' + r(P'Q' - QP'')}{rPQ}, \\
R_{ij} &= d-2-\left.\frac{(d-2)P+rP'}{PQ^2} + \frac{rQ'}{Q^3}\delta_{ij}\right|_{i,j\geq 2}, \\
R^{d+1} &= \frac{1}{r^2PQ^3}\{2r\{Q[(d-1)P' + rP''] - rP'Q'\} \\
&\quad +(1-d)P[(d-2)Q^3 + (2-d)Q + 2rQ']\},
\end{aligned}
\tag{13.70}
$$

so that the Gauss–Bonnet term in arbitrary $(d+1)$-dimensions takes the form

$$
\begin{aligned}
\mathcal{G}^{(d+1)} = \frac{(d-2)(d-1)}{r^4PQ^5}&\{(d-3)P(Q^2-1)\left[(d-4)Q^3 - (d-4)Q + 4rQ'\right] \\
&-4r\left[(d-3)Q^3P' + rQ^3P'' - (d-3)QP' - rQP''\right. \\
&\left.-rQ^2P'Q' + 3rP'Q'\right]\},
\end{aligned}
\tag{13.71}
$$

where we set for simplicity $\theta_j = \pi/2$. Note that for $d \leq 2$ (i.e. in less than four dimensions), the above scalar vanishes identically, while for $d = 3$, it becomes a topological surface term, as happens in the cosmological case.

In order to calculate the point-like Lagrangian of the theory with respect to the line element (13.69), we introduce a Lagrange multiplier as

$$
S = \int d^{d+1}x\, r^{d-1}PQ\left[f(\mathcal{G}) - \lambda\left(\mathcal{G} - \tilde{\mathcal{G}}\right)\right], \tag{13.72}
$$

with $\tilde{\mathcal{G}}$ being the Gauss–Bonnet term in spherical symmetry (13.71) and λ the Lagrange multiplier given by varying the action with respect to $\mathcal{G}$, that is, $\lambda = \partial_{\mathcal{G}}f$. Substituting $\tilde{\mathcal{G}}$ and integrating out the second derivatives, we obtain

$$
\begin{aligned}
\mathscr{L}(r, P, Q, \mathcal{G}) = r^{d-1}PQ\left[f - \mathcal{G}f_{\mathcal{G}}\right] + \frac{(d-1)(d-2)r^{d-5}(Q^2-1)}{Q^4} \\
\times\{(d-3)Pf_{\mathcal{G}}\left[(d-4)Q(Q^2-1) + 4rQ'\right] + 4r^2QP'\mathcal{G}'f_{\mathcal{G}\mathcal{G}}\}.
\end{aligned}
\tag{13.73}
$$

This is the point-like canonical Lagrangian of our theory in a static and spherically symmetric space-time. Its configuration space is $\mathcal{S} \equiv \{P, Q, \mathcal{G}\}$, and the tangent space $\mathcal{TS} \equiv \{P, P', Q, Q', \mathcal{G}, \mathcal{G}'\}$.

The generator of the point transformations (2.13) is given by

$$\mathcal{X} = \xi(r, \mathcal{G}, P, Q)\partial_r + \alpha(r, \mathcal{G}, P, Q)\partial_P + \beta(r, \mathcal{G}, P, Q)\partial_Q + \gamma(r, \mathcal{G}, P, Q)\partial_{\mathcal{G}} \,. \quad (13.74)$$

By applying the Noether theorem for Eq. (13.4), we obtain a system of twelve equations, which are not all independent. The resolution of the system provides only two different nontrivial solutions for the $f(\mathcal{G})$ function, which we outline below (see [128] for details).

- **Case 1:** In more than three dimensions, we have $f(\mathcal{G}) = f_0\mathcal{G}^k$ with $k \neq 1$. The Noether symmetry of this model is given by

$$\mathcal{X} = \xi_0 r\partial_r + (4k - d)\xi_0 P\partial_P - 4\xi_0\mathcal{G}\partial_{\mathcal{G}} \quad (13.75)$$

and $g = c_2$, with c_1 and c_2 being constants. The invariant quantity (2.11), related to the above symmetry (13.75), is

$$j = -c_1 + \frac{1}{Q^3}\xi_0 f_0 r^{d-4}\mathcal{G}^{k-2}\Big\{(1-k)r^4\mathcal{G}^2 PQ^4 - 4k(k-1)(d-2)(d-1)r\left(Q^2-1\right)\left(rP' + (d-4k)P\right)\mathcal{G}' - (d-2)(d-1)k\mathcal{G}\left(Q^2-1\right)\left[16(k-1)rP' - (d-4)(d-3)P\left(Q^2-1\right)\right]\Big\}, \quad (13.76)$$

where $c_1 \in \mathbb{R}$.

- **Case 2:** In five dimensions ($d = 4$) there is also the possibility of having a linear model of the form $f(\mathcal{G}) = f_0\mathcal{G}$. Its Noether symmetry reads

$$\mathcal{X} = \xi_0 r\partial_r + \alpha_0\partial_P, \quad (13.77)$$

and $g = c_2 - \frac{8f_0\alpha_0(3Q^2-1)}{Q^3}$, with c_2, α_0 being constants. The conserved quantity related to the generator (13.77) is

$$j = -\frac{8f_0}{Q^3}\left(\frac{6r\left(Q^2-3\right)\left(\xi_0 rP' - c_2\right)Q'}{Q} + \alpha_0\left(3Q^2-5\right)\right) - c_2 \,. \quad (13.78)$$

Lagrangian (13.73) for the **Case 1**, that is, $f = f_0\mathcal{G}^k$, becomes

$$\mathscr{L} = \frac{f_0 r^{d-5}\mathcal{G}^{k-2}}{Q^4}\Big\{\mathcal{G}PQ\left[(d-4)(d-3)(d-2)(d-1)k\left(Q^4 - 2Q^2 + 1\right)\right] + 4(d-1)(d-2)k\left(Q^2-1\right)r\left[(d-3)\mathcal{G}PQ' + (k-1)rQ\mathcal{G}'P'\right] - \mathcal{G}^2P(k-1)Q^5r^4\Big\}, \quad (13.79)$$

and the associated EL equations $\dfrac{\partial \mathscr{L}}{\partial q^i} = \dfrac{d}{dr}\dfrac{\partial \mathscr{L}}{\partial q'^i}$ are

$$
\begin{aligned}
&(k-1)r^4 Q^5 + \Big\{4(d-3)(d-2)(d-1)k(k-1)rQ(Q^2-1)\mathcal{G}'\mathcal{G}^{-2}\\
&+4(d-2)(d-1)(k-2)(k-1)kr^2 Q(Q^2-1)\mathcal{G}'^2\mathcal{G}^{-3}\\
&+8(d-2)(d-1)(k-1)kr^2 Q^2 Q'\mathcal{G}'\mathcal{G}^{-2}\\
&-12(d-2)(d-1)(k-1)kr^2(Q^2-1)Q'\mathcal{G}'\mathcal{G}^{-2}\\
&-(d-3)(d-2)(d-1)k(Q^2-1)\Big[(d-4)Q(Q^2-1)+4rQ'\Big]\mathcal{G}^{-1}\\
&+4(d-2)(d-1)(k-1)kr^2 Q(Q^2-1)\mathcal{G}''\mathcal{G}^{-2} = 0,
\end{aligned}
\tag{13.80}
$$

$$
\begin{aligned}
&(k-1)r^4\mathcal{G}^2 PQ^4\\
&-(d-3)(d-2)(d-1)k\mathcal{G}(Q^2-1)\Big[(d-4)P(Q^2-1)-4rP'\Big]\\
&+4(d-2)(d-1)(k-1)kr\mathcal{G}'\Big[(d-5)P(Q^2-1)+r(Q^2-3)P'\Big] = 0\,,
\end{aligned}
\tag{13.81}
$$

$$
\mathcal{G}(r) = \tilde{\mathcal{G}}(r),
\tag{13.82}
$$

for P, Q, and $\mathcal{G}$ respectively.

Solving Eq. (13.82) with respect to $\mathcal{G}(r)$, we recover the spherically symmetric expression of the Gauss–Bonnet scalar $\mathcal{G}(r) = \tilde{\mathcal{G}}$, given by Eq. (13.71). Moreover, Eq. (13.80) and Eq. (13.81) can be further simplified by setting $Q(r) = 1/P(r)$, so that the system of differential equations ends up with an equation of the form

$$
\begin{aligned}
(d-1)(d-2)\tilde{\mathcal{G}}^k \Big\{ &(d-3)(P^2-1)\left[(d-4)(P^2-1)-4(k-2)rPP'\right]\\
&-4(k-1)r^2\left[(P^2-1)PP''+(3P^2-1)P'^2\right]\Big\} = 0\,.
\end{aligned}
\tag{13.83}
$$

Notice that the ansatz $Q(r) = 1/P(r)$ is needed to find the dynamics of $(d+1)$-dimensional $f(\mathcal{G})$ gravity in spherical symmetry. Moreover, it allows to recover the Schwarzschild solution as a particular case.

Obviously, for $d = 1, 2$, Eq. (13.83) is automatically satisfied. The system has the solutions

$$
P(r)^2 = 1 + e^{-2k_2}\sqrt{k_1 - 4r}\, r^{\frac{3}{2}-\frac{d}{2}} \text{ and } \mathcal{G}(r) = 0, \quad k \neq 1, \quad d \geq 3;
\tag{13.84}
$$

$$
P(r)^2 = P_0^2\left(1 - \frac{k_3}{r^{\frac{d}{2}-2}}\right) \text{ and } \mathcal{G}(r) = 0, \quad k = 1, \quad d \geq 4;
\tag{13.85}
$$

$$
\begin{aligned}
P(r)^2 &= 1 \pm r^{2-\frac{d}{2}}\sqrt{\frac{4k_1 d}{120\binom{d+1}{d-4}}} \pm r^2\sqrt{\frac{\mathcal{G}_0(d-3)}{120\binom{d+1}{d-4}}},\\
\mathcal{G}(r) &= \mathcal{G}_0, \quad f(\mathcal{G}) = f_0\mathcal{G}^{\frac{d+1}{4}}, \quad d \geq 4\,,
\end{aligned}
\tag{13.86}
$$

with $k_1, k_2, k_3, P_0, \mathcal{G}_0$ constants. These are general black hole solutions for $f(\mathcal{G})$ gravity, with $f(\mathcal{G}) = f_0\mathcal{G}^k$; in particular, the first one is valid in arbitrary d dimensions with $d \geq 3$, while the others hold in more than four dimensions. Solution (13.86), which is the AdS equivalent of $f(\mathcal{G})$ gravity, holds for any $k = \left(\frac{d+1}{4}\right)$, in agreement with the trace equation (8.44). In any case, the asymptotic flatness is always recovered in more than five dimensions, as well as the presence of horizons. Specifically, solutions (13.85) and (13.84) admit as horizon $r_S \sim \mathcal{M}^{\frac{2}{d-4}}$ where $\mathcal{M}$ is the gravitational mass.

Regarding solution (13.86), it is interesting to compare the Bekenstein–Hawking entropy provided by $f(\mathcal{G}) = f_0\mathcal{G}^k$ gravity, in $(d+1)$-dimensions, with that of CFT. To this purpose, let us rewrite the spherically symmetric solution (13.86) as

$$P^2 = 1 - r^{2-\frac{d}{2}}\mathcal{M}\eta_{1,d} - r^2\eta_{2,d}, \tag{13.87}$$

with $\eta_{1,d}$, $\mathcal{M}$, and $\eta_{2,d}$ being constants depending on the dimension. When the contribution of the cosmological constant is negligible, namely when $r^2\eta_{2,d} \ll 1$, the horizon, and the entropy scale as

$$r_S \sim \mathcal{M}^{\frac{2}{d-4}}, \quad \mathcal{S} \sim \mathcal{M}^{\frac{2d-2}{d-4}}. \tag{13.88}$$

On the other hand, when $r^2\eta_{2,d} \gg 1$, similar computations show that the horizon and the entropy profiles are, respectively,

$$r_S \sim \mathcal{M}^{\frac{2}{d}} \qquad \mathcal{S} = \frac{A}{4} \sim r^{d-1} \sim \mathcal{M}^{\frac{2(d-1)}{d}}. \tag{13.89}$$

Therefore, in both cases the entropy scales differently that that of CFT.

Let us now see some more specific solutions of the system (13.80)–(13.82), analyzing the boundary cases $d = 3$ and $d = 4$. In $d = 3$, we have the following solution for any $P(r)$, that is,

$$Q(r) = \frac{1}{3}\left(A(r) - e^{q_0}P'(r) + \frac{e^{2q_0}}{A(r)}P'(r)^2\right), \tag{13.90}$$

with

$$A(r) = \left(\frac{27e^{q_0}}{2}P'(r) - e^{3q_0}P'(r)^3 + \frac{3e^{q_0}}{2}P'(r)\sqrt{81 - 12e^{2q_0}P'(r)^2}\right)^{1/3}.$$

Here, q_0 is an integration constant, and the Gauss–Bonnet term vanishes in this case. As an example, by introducing the relation $Q(r) = P(r)^n$, we find that the field equations are satisfied by any $P(r)$ that solves the equation

$$k(P^{2n} - 3)P'^2 - P(P^{2n} - 1)P'' = 0. \tag{13.91}$$

The limit $n = -1$ provides back solution (13.84). An interesting analytic solution of Eq. (13.91) occurs for $n = -1/3$, where the components of the interval are

$$P(r)^2 = -2c_1\left[(r + c_2)\left(\frac{6r}{M(r)} + 1\right)\right] + \frac{3}{8}\left[\frac{M(r)^2 + 9}{M(r)} + 3\right],$$

Table 13.1 Exact static and spherically symmetric solutions in $f(G)$ gravity, with $f(\mathcal{G}) = f_0\mathcal{G}^k$, in arbitrary $(d+1)$-dimensions

P(r)^2	Q(r)^2	d	k
$1+e^{-2c_2}\sqrt{c_1-4r}r^{\frac{3}{2}-\frac{d}{2}}$	$1/P(r)^2$	$d\geq 3$	$k>0, \neq 1$
$P_0^2\left(1-\frac{k_3}{r^{\frac{d}{2}-2}}\right)$	$1/P(r)^2$	$d>3$	$k=1$
$1\pm r^{2-\frac{d}{2}}\sqrt{\frac{4k_1 d}{120\binom{d+1}{d-4}}}\pm r^2\sqrt{\frac{\mathcal{G}_0(d-3)}{120\binom{d+1}{d-4}}}$	$1/P(r)^2$	$d>3$	$k=\frac{d+1}{4}$
$\forall P(r)$	$\frac{1}{3}\left(A(r)-e^{q_0}P'(r)+\frac{e^{2q_0}}{A(r)}P'(r)^2\right)$	$d=3$	$k>0$
$-\frac{1}{2}\exp\left[tanh^{-1}\left(\sqrt{\frac{\mathcal{G}_0}{30}}\frac{r^2}{2}\right)\right]\sqrt{4-\frac{\mathcal{G}_0 r^4}{30}}$	$1+\frac{\sqrt{\mathcal{G}_0}r^2}{2\sqrt{30}}$	$d=4$	$k=5/4$
1	1	$d=4$	$\forall k$

with

$$M(r) = \sqrt[3]{\overline{\overline{\overline{128c_1^2r^2+16\left(16c_1c_2-9\right)c_1r}}}} \\ \overline{\overline{+64\sqrt{c_1^3\left(c_2+r\right)^3\left(4c_1r+4c_2c_1-1\right)}}} \\ \overline{+128c_2^2c_1^2-144c_2c_1+27.} \tag{13.92}$$

Moreover, for $d=4$, we get the following solutions for constant $\mathcal{G}$:

$$P(r)^2 = -\frac{1}{2}\exp\left[tanh^{-1}\left(\sqrt{\frac{\mathcal{G}_0}{30}}\frac{r^2}{2}\right)\right]\sqrt{4-\frac{\mathcal{G}_0 r^4}{30}}, \tag{13.93}$$

$$Q(r)^{-2} = 1+\frac{\sqrt{\mathcal{G}_0}r^2}{2\sqrt{30}}, \text{ for } k=5/4; \tag{13.94}$$

and

$$P(r)^2 = Q(r)^2 = 1, \text{ for } \mathcal{G}_0 = 0 \text{ and } \forall k. \tag{13.95}$$

If we Taylor expand $P(r)^2$ in the first solution (13.93), we find that $P(r)^2 = Q(r)^{-2}$, which is an AdS-like solution, where $\frac{\sqrt{\mathcal{G}_0}}{2\sqrt{30}}$ can be considered as the bulk cosmological constant. To conclude, in Table **13.1**, we outline the spherically symmetric solutions provided in this section.

Conclusion of Part II

In this part of the book, we sketched the steps needed to recast the Noether theorem as a method capable of selecting physically reliable theories. To this purpose, we started introducing an unknown potential in a free-particle Lagrangian and selected its form by the Noether Symmetry Approach. It turned out that several potentials are allowed by symmetry considerations. Each of them is invariant under different symmetry transformations, which yield, in turn, different symmetry generators. The approach is worthwhile mainly when dealing with general theories, which must be constrained in order to have predictive power.

Afterwards, we considered gravity actions containing unknown functions of geometric invariants. In particular, we studied the related cosmologies. Selected models turn out to be viable in providing exact solutions of field equations, whose dynamics is often in agreement with the current cosmological observations. However, modified theories of gravity carry out extra geometric terms, which yield new degrees of freedom into the system. Consequently, new free parameters, which must be constrained by observations and experiments, occur. Constraining the functional form of selected actions is the main purpose of the next part of the book, where astrophysical and cosmological applications of modified gravity will be considered in different backgrounds.

Throughout Part II, we dealt with the Lagrangian formalism, so that the dynamics was provided by the EL equations together with the energy condition. The latter represents a consistency relation, which allows to recover the first Friedman equation in modified gravity. Moreover, by means of the Lagrangian formalism, it is possible to apply Noether's identity (2.24) and find the symmetry generator.

In this regard, we mainly considered cosmological point-like Lagrangians, where the minisuperspace variables depend on the cosmic time only. In this way, Eq. (6.42) is satisfied whenever the action admits extra Noether symmetries. Specifically, geometric extensions of the EH action have been firstly treated as $f(R)$, $f(T)$, and $f(T, B)$ gravity. Notice that the former can be recovered from the latter by choosing $f(T, B) \equiv f(-T + B) = f(R)$, because of the relation (4.105). Then, second-order curvature and torsion invariants have been included into the action.

Particular interest is gained by the Gauss–Bonnet scalar, a second-order topological surface term whose integration over the manifold provides the Euler characteristic. Hence, $f(R, \mathcal{G})$ and $f(\mathcal{G})$ models were selected by the Noether Symmetry Approach. In particular, in the latter case, the prescription was generalized to $(d+1)$-dimensions, due to the capability of $f(\mathcal{G})$ gravity to describe higher-dimensional universes. Subsequently, sixth-order gravity was analyzed by introducing the operator $\Box = D_\mu D^\mu$.

In Chapter 10, we selected potential and coupling functions, starting from different scalar-tensor actions. Furthermore, we showed that Noether symmetries are preserved under conformal transformations, so that models with symmetries in the Einstein frame correspond to related models in the Jordan frame. Moreover, we showed the equivalence among different descriptions of scalar-tensor gravity, involving the curvature scalar, the torsion scalar, and the Gauss–Bonnet term, respectively. These three theories admit the same symmetries and lead to equivalent dynamics. On the one hand, the equivalence between R and T is straightforwardly provided by Eq. (4.105). It follows that the teleparallel Lagrangian differs from the EH one only for a boundary term. On the other hand, in some cosmological contexts, the Gauss–Bonnet term can be addressed to R^2, so that $\sqrt{\mathcal{G}}$ turns out to be dynamically equivalent to GR.

Afterward, by means of the propagator $\Box^{-1}$, nonlocal theories have been considered in various representations of gravity.

Finally, in the last two chapters, the Noether Symmetry approach has been applied to Bianchi universes and spherical symmetry. In the former case, we studied the Einstein equations minimally coupled to a scalar field for Bianchi universes of class A and B. In the latter case, we considered three classes of modified theories of gravity in a spherically symmetric background, selecting symmetries by means of Eq. (13.4) where the related Lagrangians are functions of the radial coordinate only.

It is worth pointing out that, in some cases, only internal symmetries have been considered by the vanishing Lie derivative condition, which can be recovered from Eq. (2.24) after setting $\xi = g = 0$. However, the properties of Lie derivatives are of great interest in several contexts. For instance, internal symmetries provide a methodical procedure (summarized in Eq. (2.37)) to change suitably the minisuperspace dimension by introducing a cyclic variable into the system. This will be mainly useful in Chapter 17, where the ADM formalism will be applied to modified cosmological Lagrangians. It will turn out that the corresponding Hamiltonians need to be reduced in order to provide suitable WDW equations with analytic solutions.

Part III

APPLICATIONS

14

Applications to Solar System, Stars, and Our Galaxy

> Two things fill the mind with ever new and increasing admiration
> and awe, the more often and steadily we reflect upon them:
> the starry heavens above me and the moral law within me.
>
> — *Immanuel Kant*
>
> (*Kritik der Reinen Vernunft* [1781])

In this part of the book,[1] we investigate the capability of modified theories of gravity (selected by the Noether Symmetry Approach) to fit several current observations at different scales. Specifically, we consider some of the models selected in the second part and constrain the free parameters by experimental data. To this purpose, we focus on the most promising selected models, such as those containing GR as a limit or not suffering theoretical shortcomings. It is worth stressing that this part is not meant to verify that all the models with symmetries can have a physical meaning; rather it aims to constrain theories with symmetries and check whether some of them might be physically viable.

Specifically, in this chapter, we want to check if models selected by Noether symmetries are of physical interest at astrophysical scales. On the other hand, experiments and observations can be used as a test bed to constrain the parameters coming from the field equations.

As a first step, in order to recover GR as a limit of modified/extended, we can consider a first-order expansion of $f(R) = f_0 R^k$ gravity around $k = 1$, namely $f(R) = f_0 R^{1+\epsilon}$ with $\epsilon \ll 1$. As discussed previously, such a function is selected by the Noether Symmetry Approach (see Section 13.1) and constitute a minimal extension for GR. The power-law function will be constrained in Section 14.1 by Solar System tests. In Section 14.2 we show that little deviations from $\epsilon = 0$ are capable of providing a mass–radius relation for neutron stars, within the range allowed by several new observations that cannot be explained in the context of GR. In Section 14.3, the gravitational lensing in the context of $f(R)$

[1] In some chapters of this part we restore physical units, with the aim of comparing theoretical results with experimental observations.

gravity will be discussed, and finally, in Section 14.4, a specific Nonlocal Gravity model will be constrained by the orbit of the S2 star around Sagittarius A*. In the latter case, the Noether Symmetry Approach will be applied ad hoc to the Deser–Wodard action, with the aim of selecting the unknown function $f(\Box^{-1}R)$ admitting Noether symmetries.

To conclude this introduction, we want to briefly overview the importance of weak-field solutions, when spherically symmetric backgrounds are considered. Often, modified theories of gravity exhibit no predictive power in their fundamental forms, due to the large amount of degrees of freedom. In order to overcome this difficulty and have a term of comparison with GR, it is worth searching for their weak-field limit. In this perspective, it is possible to derive expressions of phenomena, like, for example, the gyroscopic and Lense–Thirring precessions, to compare with experimental data.

The weak-field limit comes from the linearization of metric tensor

$$g_{\mu\nu} = \eta_{\mu\nu} + h_{\mu\nu}, \tag{14.1}$$

with $h_{\mu\nu}$ being a small perturbation of the Minkowski space-time such that $h_{\mu\nu} \ll \eta_{\mu\nu}$. Under this assumption, the first-order Einstein tensor turns out to be [32]

$$G_{\mu\nu} = \frac{1}{2}(\partial_\sigma \partial_\mu h^\sigma_\nu + \partial_\sigma \partial_\nu h^\sigma_\mu - \partial_\mu \partial_\nu h - \Box h_{\mu\nu} - \eta_{\mu\nu}\partial_\rho \partial_\lambda h^{\rho\lambda} + \eta_{\mu\nu}\Box h). \tag{14.2}$$

When higher-order corrections are considered, new scalar and vector potentials arise in the approximation. Moreover, in Kerr-like metrics, the nonvanishing term g_{03} yields a further vector potential $\mathbf{A}$. Up to the fourth order, a generic expansion of the metric tensor can be written as [310]

$$\begin{aligned} g_{\mu\nu} &\sim \begin{pmatrix} 1 + \overset{(2)}{g}_{00} + \overset{(4)}{g}_{00} + \dots & \overset{(3)}{g}_{0i} + \dots \\ \overset{(3)}{g}_{0i} + \dots & -\delta_{ij} + \overset{(2)}{g}_{ij} + \dots \end{pmatrix} \\ &= \begin{pmatrix} 1 + 2\Phi + 2\Xi & 2A_i \\ 2A_i & -\delta_{ij} + 2\Psi\delta_{ij} \end{pmatrix}. \end{aligned} \tag{14.3}$$

It is worth noticing that the second-order expansion of the element g_{00} provides the Newtonian potential Φ. Moreover, two other scalar potentials, Ψ and Ξ, arise: the former comes from the second-order expansion of the metric tensor components g_{ij} (Newtonian limit), while the latter is related to the fourth-order expansion of g_{00} (PN limit). Specifically, $\Phi(r)$, $\Psi(r)$ are scalar potentials proportional to $(v/c)^2$, $\Xi(r)$ is proportional to $(v/c)^4$, while A_i is a vector potential proportional to the power $(v/c)^3$. The latter is the potential linked to the rotation, whose curl operator magnitude (in analogy to the electromagnetic case) provides the Lense–Thirring precession. In GR, the angular precession can be derived analytically from the definition [310]

$$\Omega^{LT}_{(GR)} = \frac{1}{2}(\epsilon^{ijk}\partial_i A_k)(\epsilon_{\ell nk}\partial^\ell A^k) = \frac{G}{r^3}\sqrt{\left(\epsilon_{\ell km}\partial^m \epsilon^{ijk} J_i x_j\right)^2} = \frac{r_S}{4Mr^3}\, J, \tag{14.4}$$

with the obvious meaning of the symbols indicating the curl operator. The form of the vector potential A_i can be found after replacing the potentials defined in the linearized field equations. However, the Kerr metric suffers the lack of perfect fluids matching boundary conditions between external and internal solutions. In this context, modified/extended theories of gravity might solve the problem, since they contain free parameters that can be constrained by experimental observations. As an example, Table **14.1** shows the weak-field solutions of some modified/extended theories of gravity with the corresponding potentials. Yukawa-like corrections are a general feature of several modified/extended theories of gravity. In some sense, further degrees of freedom, related to higher-order terms or scalar fields, give rise to these corrections in the weak field limit. This is a general result, as discussed in [124]. Detailed discussions of these topics are reported in [124, 394]. Notice that GR is recovered when the EH action is considered. It is worth noticing that power-law solution (13.37), selected by Noether symmetry in $f(R)$ gravity, is not included in the preceding scheme, since the effective mass m_R turns out to be indeterminate for $f(R) = R^k$. However, Yukawa-like potentials imply a finite-range gravitational interaction, mediated by massive gravitons.

14.1 Solar System Tests in Modified Gravity

Let us overview now some basic foundations of Solar System tests in modified theories of gravity, considering, in particular, the perihelion precession, the light deflection, and the Shapiro time delay. Then, in Section 14.1.1, as an example we constrain $f(R)$ gravity by the preceding tests. For convenience, here we recast the line element (13.1) in terms of the functions $P(r)$ and $Q(r)$ as

$$ds^2 = P(r)^2 dt^2 - Q(r)^2 dr^2 - r^2 d\Omega^2. \tag{14.5}$$

Let us first consider the perihelion precession of Mercury. To this purpose, notice that in the equatorial plane, Eq. (14.5) can be written as

$$1 = P(r)^2\left(\frac{dt}{d\tau}\right)^2 - Q(r)^2\left(\frac{dr}{d\tau}\right)^2 - r^2\left(\frac{d\phi}{d\tau}\right), \tag{14.6}$$

with τ being the proper time along the worldline considered. By imposing the Birkhoff theorem, the geodesic equations provide

$$P(r)^2\frac{dt}{d\tau} = E, \quad r^2\frac{d\phi}{d\tau} = L, \tag{14.7}$$

Table 14.1 Weak-field solutions of modified theories of gravity [310]

Modified Gravity Model	Corrected potential	Yukawa parameters
$f(R)$	$\Phi(r) = -\frac{GM}{r}\left[1 + \alpha\, e^{-m_R r}\right]$ $\mathbf{A} = \dfrac{G}{r^3}\,\mathbf{r}\times\mathbf{J}$	${m_R}^2 \;=\; -\frac{f_R(0)}{6f_{RR}(0)}$
$f(R, \Box R) = R + a_0R^2 + a_1 R\Box R$	$\Phi(r) = -\frac{GM}{r}\left(1 + c_0 e^{(-r/l_0)} + c_1 e^{(-r/l_1)}\right)$ $\mathbf{A} = \dfrac{G}{r^3}\,\mathbf{r}\times\mathbf{J}$	$c_{0,1} \;=\; \frac{1}{6} \mp \frac{a_0}{2\sqrt{9a_0^2+6a_1}}$ $l_{0,1} \;=\; \sqrt{-3a_0 \pm \sqrt{9a_0^2 + 6a_1}}$
$f(R, \Box R, ..\Box^k R) = R + \Sigma_{k=0}^{p} a_k R\Box^k R$	$\Phi(r) = -\frac{GM}{r}\left(1 + \Sigma_{k=0}^{p} c_i \exp(-r/l_i)\right)$ $\mathbf{A} = \dfrac{G}{r^3}\,\mathbf{r}\times\mathbf{J}$	$c_i,\ l_i$ are functions of a_k. See [124]
$f(R,\ R_{\alpha\beta}R^{\alpha\beta})$	$\Phi(r) = -\frac{GM}{r}\left[1 + \frac{1}{3}\, e^{-m_R r} - \frac{4}{3}\, e^{-m_P r}\right]$ $\mathbf{A} = \dfrac{G}{r^3}\,\mathbf{r}\times\mathbf{J}$	${m_R}^2 \;=\; -\frac{1}{3f_{RR}(0)+2f_P(0)}$ ${m_P}^2 \;=\; \frac{1}{f_P(0)}\,,\ P = R_{\mu\nu}R^{\mu\nu}$
$f(R, \mathcal{G})\,, \quad \mathcal{G} = R^2 - 4R_{\mu\nu}R^{\mu\nu} + R_{\alpha\beta\mu\nu}R^{\alpha\beta\mu\nu}$	$\Phi(r) = -\frac{GM}{r}\left[1 + \frac{1}{3}\, e^{-m_1 r} - \frac{4}{3}\, e^{-m_2 r}\right]$ $\mathbf{A} = \dfrac{G}{r^3}\,\mathbf{r}\times\mathbf{J}$	${m_1}^2 \;=\; -\frac{1}{3f_{RR}(0)+2f_P(0)+2f_Q(0)}$ ${m_2}^2 \;=\; \frac{1}{f_P(0)+4f_Q(0)}\,,\ Q = R_{\alpha\beta\mu\nu}R^{\alpha\beta\mu\nu}$

$f(R,\,\phi)+\omega(\phi)\phi_{;\alpha}\phi^{;\alpha}$	$\Phi(r)=-\frac{GM}{r}\Big[1+g(\xi,\eta)\,e^{-m_R\tilde{k}_R r}+$ $+[1/3-g(\xi,\eta)]\,e^{-m_R\tilde{k}_\phi r}\Big]$ $\mathbf{A}=\dfrac{G}{r^3}\,\mathbf{r}\times\mathbf{J}$	${m_R}^2=-\frac{1}{3f_{RR}(0,\phi^{(0)})}$ ${m_\phi}^2=-\frac{f_{\phi\phi}(0,\phi^{(0)})}{2\omega(\phi^{(0)})}$ $\xi=\frac{3f_{R\phi}(0,\phi^{(0)})^2}{2\omega(\phi^{(0)})}$ $\eta=\frac{m_\phi}{m_R}$ $g(\xi,\,\eta)=\frac{1-\eta^2+\xi+\sqrt{\eta^4+(\xi-1)^2-2\eta^2(\xi+1)}}{6\sqrt{\eta^4+(\xi-1)^2-2\eta^2(\xi+1)}}$ $\tilde{k}^2_{R,\phi}=\frac{1-\xi+\eta^2\pm\sqrt{(1-\xi+\eta^2)^2-4\eta^2}}{2}$
$f(R,\,R_{\alpha\beta}R^{\alpha\beta},\phi)+\omega(\phi)\phi_{;\alpha}\phi^{;\alpha}$	$\Phi(r)=-\frac{GM}{r}\Big[1+g(\xi,\eta)\,e^{-m_R\tilde{k}_R r}+$ $+[1/3-g(\xi,\eta)]\,e^{-m_R\tilde{k}_\phi r}-\frac{4}{3}\,e^{-m_P r}\Big]$ $\mathbf{A}=\dfrac{G}{r^3}\,[1-(1+m_P r)\,e^{-m_P r}]\,\mathbf{r}\times\mathbf{J}$	${m_R}^2=-\frac{1}{3f_{RR}(0,0,\phi^{(0)})+2f_P(0,0,\phi^{(0)})}$ ${m_P}^2=\frac{1}{f_P(0,0,\phi^{(0)})}$ ${m_\phi}^2=-\frac{f_{\phi\phi}(0,0,\phi^{(0)})}{2\omega(\phi^{(0)})}$ $\xi=\frac{3f_{R\phi}(0,0,\phi^{(0)})^2}{2\omega(\phi^{(0)})}$ $\eta=\frac{m_\phi}{m_R}$ $g(\xi,\,\eta)=\frac{1-\eta^2+\xi+\sqrt{\eta^4+(\xi-1)^2-2\eta^2(\xi+1)}}{6\sqrt{\eta^4+(\xi-1)^2-2\eta^2(\xi+1)}}$ $\tilde{k}^2_{R,\phi}=\frac{1-\xi+\eta^2\pm\sqrt{(1-\xi+\eta^2)^2-4\eta^2}}{2}$

where E and L are constants of motion that account for the energy and the angular momentum per mass unit of the test particle. By replacing Eq. (14.7) into Eq. (14.6), we have

$$1 = \frac{E^2}{P(r)^2} - \frac{1}{P(r)^2}\left(\frac{dr}{d\tau}\right)^2 - \frac{L^2}{r^2} \Rightarrow E^2 - \left(\frac{dr}{d\tau}\right)^2 = P(r)^2\left(1 + \frac{L^2}{r^2}\right). \qquad (14.8)$$

To obtain a suitable precession equation, we re-parameterize the radial coordinate in terms of the angle ϕ as $r \equiv r(\phi)$, and define $u(\phi) \equiv r(\phi)^{-1}$. In this way, the derivative of Eq. (14.8) with respect to ϕ can be written as

$$u''(\phi) + u(\phi) = u(\phi)h(u) + \frac{1}{2}\frac{dh(u)}{du}\left[\frac{1}{L^2} + u(\phi)^2\right] \equiv H(u), \qquad (14.9)$$

with $h(u) \equiv 1 - P(u)^2$ and where the prime mark denotes the derivative with respect to ϕ. With the purpose of evaluating the deviations with respect to the circular orbit, we split the function $u(\phi)$ into two contributions, namely $u = u_0 + \epsilon u_1$, with u_0 being the closed trajectory and u_1 the first-order deflection. Therefore, the circular orbit is automatically given by the solution of the equation

$$u_0 = H(u_0), \qquad (14.10)$$

and the first-order displacement can be found by solving

$$u_1'' = -u_1\eta(u_0), \quad \text{where} \quad \eta(u_0) \equiv \left(1 - \left.\frac{dH(u)}{du}\right|_{u=u_0}\right). \qquad (14.11)$$

First, notice that the solution of Eq. (14.11) is

$$u_1 = A\sin\left(\sqrt{\eta(u_0)}\phi + \Phi\right). \qquad (14.12)$$

In order to find the angle displacement, $\Delta\phi$ must be further decomposed as

$$\Delta\phi = 2|\phi(r_+) - \phi(r_-)| - 2\pi, \qquad (14.13)$$

where r_+ and r_- are the radial coordinates of aphelion and perihelion, respectively. By means of the preceding equation, the final angle displacement can be straightforwardly written as a function u:

$$\Delta\phi = \frac{2\pi}{\sqrt{\eta(u_0)}} \sim \pi \left.\frac{dH(u)}{du}\right|_{u=u_0}. \qquad (14.14)$$

Eq. (14.14) generalizes the precession provided by GR, which can be recovered by imposing $h(u) = r_S u$. In this way, Eqs. (14.10) and (14.14) give

$$u_0 = \frac{1}{3r_S}\left(1 - \sqrt{1 - \frac{3r_S^2}{L^2}}\right) \sim \frac{r_S}{2L^2}, \qquad (14.15)$$

$$\Delta\phi = \frac{6\pi M^2}{L^2}. \qquad (14.16)$$

The preceding equation can also be written in terms of the eccentricity of elliptical orbit e and the corresponding semiaxis a, by means of the relation $L^2 = (1 - e^2)aM_\odot$. As a final result, GR agrees with experiments in predicting a precession for Mercury of the order of 43 arcseconds per century.

The next Solar System test here discussed is the *Shapiro time delay*, which is the time delay suffered by a signal passing near a massive body. It is due to space-time modifications provoked by the large mass of a self-gravitating body. This effect was confirmed by several observations throughout the years and was first obtained as an application of GR results. Specifically, referring to the time delay of a signal approaching the Earth due to the presence of the Sun, in 1964 I. Shapiro wrote [395]

Because, according to the general theory, the speed of a light wave depends on the strength of the gravitational potential along its path, these time delays should thereby be increased by almost $2 \cdot 10^{-4}$ sec. when the radar pulses pass near the Sun. Such a change, equivalent to 60 km in distance, could now be measured over the required path length to within about 5 to 10 percent with presently obtainable equipment.

The most significant difference occurs in the path between Earth and Mercury, where experimental results provide a time delay of $\sim 0 : 0002492s$ due to the Sun. In what follows, we discuss the theoretical foundations of such an effect, outlining a general scheme that reduces to GR when Schwarzschild solution is considered. The time taken by a light signal, moving in the spherically symmetric space-time (14.5), is

$$t = \int_{l_1}^{l_2} \frac{dz}{v}, \tag{14.17}$$

where

$$v = \frac{dz'}{dt'} = \frac{Q(r)}{P(r)} \frac{dz}{dt} = c\frac{P(r)}{Q(r)}, \tag{14.18}$$

so that

$$t = \frac{1}{c} \int_{l_1}^{l_2} \frac{Q(r)}{P(r)} dz. \tag{14.19}$$

The difference between Eq. (14.19) and the time evaluated in the Minkowski space-time is

$$\Delta t = \frac{1}{c} \int_{l_1}^{l_2} \left[\frac{Q(r)}{P(r)} - 1 \right] dz. \tag{14.20}$$

The integral can be solved by means of the relation $r \to \sqrt{z^2 + R_\odot^2}$, where $R_\odot$ is the radius of the massive body:

$$\Delta t = \frac{1}{c} \int_{l_1}^{l_2} \left[\frac{Q(\sqrt{z^2 + R_\odot^2})}{P(\sqrt{z^2 + R_\odot^2})} - 1 \right] dz. \tag{14.21}$$

When $P(r) = Q(r) = 1$, it is $\Delta t = 0$ as we expect. On the other hand, when a nontrivial spherically symmetric solution is considered, the integral provides a nonvanishing time delay, in agreement with theoretical considerations. For instance, if one considers a light ray moving from Earth to Mercury, then the Sun generates a time delay of 0.24 ms, if compared to the time taken in absence of the Sun. This result can be achieved by considering a Schwarzschild space-time and replacing the initial position of the two planets l_1, l_2 and the Sun radius $R_\odot$ in Eq. (14.21).

Finally, the effect of light deflection consists of the change of trajectory of a signal crossing a region of curved space-time, due to the gravitational field generated by a massive body. As first confirmed by the solar eclipse in 1919, the deflection angle of a signal moving near the Sun is 1.73 arcseconds. This phenomenon cannot be predicted by Newtonian mechanics, but it is theoretically provided by the geometric description of gravity. Indeed, considering the geodesic equations of a spherically symmetric interval, namely

$$\begin{cases} \dfrac{d^2r}{d\tau^2} + \dfrac{P'(r)}{P^2(r)}\dfrac{dr}{d\tau} - \dfrac{r}{P^2(r)}\left(\dfrac{d\theta}{d\tau}\right)^2 - \dfrac{r\sin^2\theta}{P^2(r)}\left(\dfrac{d\phi}{d\tau}\right)^2 + \dfrac{Q'(r)}{P^2(r)}\left(\dfrac{dt}{d\tau}\right)^2 = 0 \\ \dfrac{d^2\theta}{d\tau^2} + \dfrac{2}{r}\dfrac{d\theta}{d\tau}\dfrac{dr}{d\tau} - \sin\theta\cos\theta\left(\dfrac{d\phi}{d\tau}\right)^2 = 0 \\ \dfrac{d^2\phi}{d\tau^2} + \dfrac{2}{r}\dfrac{d\phi}{d\tau}\dfrac{dr}{d\tau} + 2\cot\theta\dfrac{d\theta}{d\tau}\dfrac{d\phi}{d\tau} = 0 \\ \dfrac{d^2t}{d\tau^2} + \dfrac{2P'(r)}{P^2(r)}\dfrac{dt}{d\tau}\dfrac{dr}{d\tau} = 0 \end{cases} \tag{14.22}$$

and the constants of motion (14.7), Eq. (14.22), in the equatorial plane, becomes

$$\frac{Q^2(r)}{L^2}\left(\frac{dr}{d\tau}\right)^2 + \frac{1}{r^2} - \frac{1}{L^2P^2(r)} + \frac{E}{L^2} = \frac{Q^2(r)}{r^4\left(\frac{d\phi}{d\tau}\right)^2}\left(\frac{dr}{d\tau}\right)^2 + \frac{1}{r^2}$$
$$-\frac{1}{L^2P^2(r)} + \frac{E}{L^2} = \frac{Q^2(r)}{r^4}\left(\frac{dr}{d\phi}\right)^2 + \frac{1}{r^2} - \frac{1}{L^2P^2(r)} + \frac{E}{L^2} = 0. \tag{14.23}$$

By solving Eq. (14.23) with respect to ϕ and assuming that, for a light signal, E/L^2 is negligible with respect to the other terms, one gets

$$\phi = \int \frac{1}{r}\sqrt{\frac{Q^2(r)}{\frac{r^2}{L^2P^2(r)} - 1}}\,dr. \tag{14.24}$$

Moreover, in order to find a suitable expression for L, Eq. (14.23) also gives

$$L^2 = \frac{r^4}{P^2(r)Q^2(r)\left(\frac{dr}{d\phi}\right)^2 + r^2P^2(r)}, \tag{14.25}$$

which, once evaluated at the minimum distance from the Sun, that is, $r_{min} \sim R_\odot$ with $\left.\frac{dr}{d\phi}\right|_{r=r_{\min}} = 0$, provides

$$L^2 = \frac{R_\odot^2}{P^2(R_\odot)}. \tag{14.26}$$

Finally, the light deflection can be obtained from the quantity $\phi(r) - \phi(r \to \infty)$, that is,

$$\Delta\phi = \int_{r'}^{\infty} \frac{1}{r} \sqrt{\frac{Q^2(r)}{\frac{r^2}{R_\odot^2}\frac{P^2(R_\odot)}{P^2(r)} - 1}}\, dr. \tag{14.27}$$

14.1.1 Solar System Constraints in $f(R)$ Gravity

Here we aim to constrain $f(R)$ gravity by Solar System tests. Specifically, we consider solution (13.22) derived from Noether's symmetries, with the purpose of finding the free parameter k that fits observations at the Solar System level. Therefore, we start from action $S = \int \sqrt{-g} f_0 R^k\, d^4x$ and consider the corresponding weak-field solution (13.37), that is,

$$e^{\nu(r)} = 1 - \frac{r_S}{2r}\left[1 + \left(\frac{r}{r_c}\right)^{\beta}\right], \qquad r_S = \frac{2GM}{c^2}, \tag{14.28}$$

where we restore physical units in order to compare theoretical results with observations (see [362] for details on the derivation of the solution). In the PN limit, the potential reads

$$\Phi(r) = -\frac{GM}{2r}\left[1 + \left(\frac{r}{r_c}\right)^{\beta}\right], \tag{14.29}$$

which is a slight generalization of the Newtonian potential. Notice that Eqs. (14.28) and (14.29) must be obtained by setting $f_0 = c^4/(16\pi G)$. Here r_c is a constant with length dimension, which comes from the boundary conditions imposed in the PN limit. The detailed physical meaning will be further discussed in Section 15.1. Here we only consider r_c as a real constant of the order of $1 \div 10^4$ AU. On the other hand, β is a dimensionless constant defined in terms of k as

$$\beta = \frac{12k^2 - 7k - 1 - \sqrt{36k^4 + 12k^3 - 83k^2 + 50k + 1}}{6k^2 - 4k + 2}, \tag{14.30}$$

so that GR is recovered when $\beta = 0$ $(k = 1)$, as shown in Fig. **14.1**.

Though Eq. (14.28) is not an exact solution of the EL equations (13.9), the PN approximation provides a precision of 10^{-6}, which makes it a viable solution

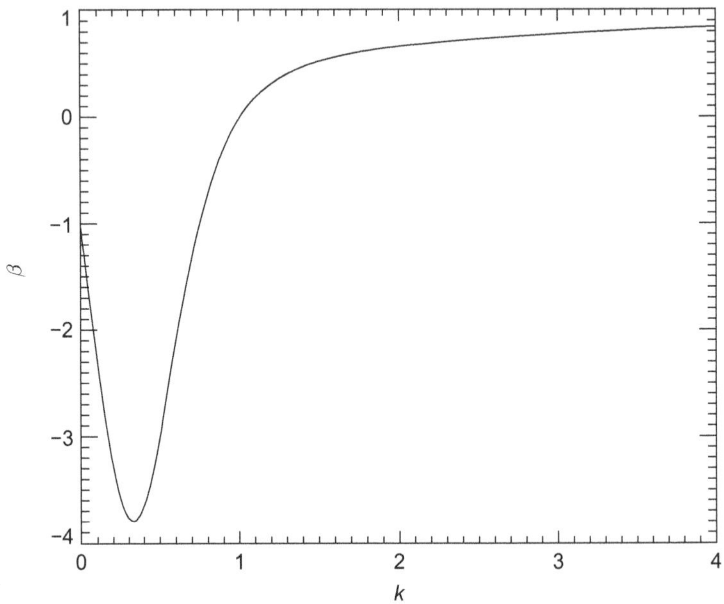

Figure 14.1 Relation between the parameters β and k in the weak-field limit of R^k gravity

to the purpose of constraining β. A first restriction can be obtained by requiring the asymptotic flatness, namely

$$\Phi(r \to \infty) = 0, \tag{14.31}$$

which yields $\beta < 1$. Furthermore, Newtonian potential must be recovered when r is largely greater than r_S, but lower than r_c, namely $r_S \ll r \ll r_c$, which implies $\beta > 0$. Therefore, the validity range of β is further constrained to $0 < \beta < 1$. Let us now consider the planetary orbits around the Sun; the corresponding acceleration is provided by the derivative of Eq. (14.29) with respect to the radius, that is [396],

$$a = -\frac{d\Phi}{dr} = -\frac{GM}{2r^2}\left[1 + \left(\frac{r}{r_c}\right)^\beta - \beta\left(\frac{r}{r_c}\right)^\beta\right], \tag{14.32}$$

with orbital period

$$T = \sqrt{\frac{8\pi^2 r^3}{GM}}\left[1 + \left(\frac{r}{r_c}\right)^\beta - \beta\left(\frac{r}{r_c}\right)^\beta\right]^{-\frac{1}{2}}, \tag{14.33}$$

where $T_K \equiv \sqrt{\dfrac{4\pi^2 r^3}{GM}}$ is the Keplerian period. By defining

$$1 + \left(\frac{r}{r_c}\right)^\beta - \beta\left(\frac{r}{r_c}\right)^\beta \equiv f(\beta, r_c) \tag{14.34}$$

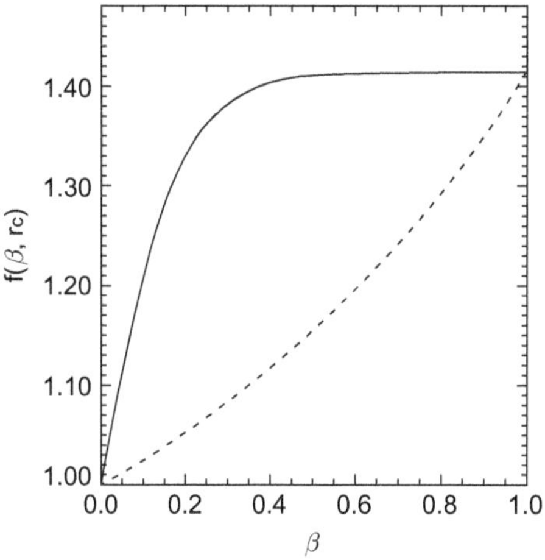

Figure 14.2 $f(\beta, r_c)$ as a function of beta, with r_c fixed to 1AU (dashed line) and 10^4AU (solid line)

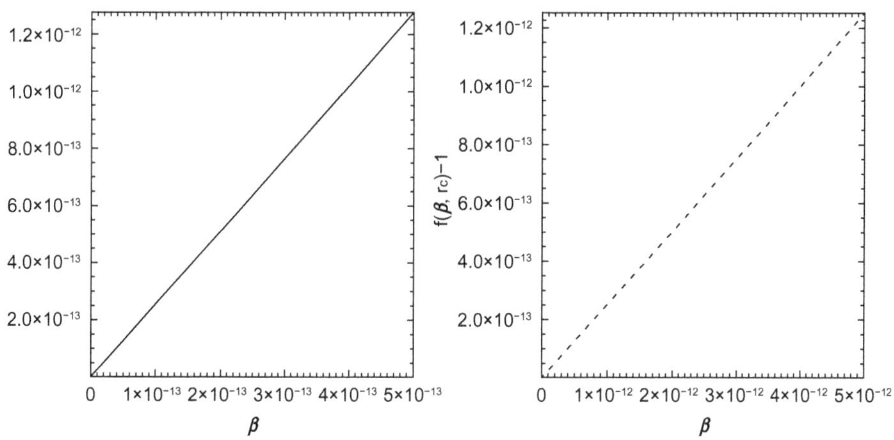

Figure 14.3 Zoomed graph of $f(\beta, r_c) - 1$ as a function of beta, with r_c fixed to 1AU (right) and 10^4AU (left)

and recasting the RHS of Eq. (14.33) in terms of β and r_c, the Keplerian orbit is recovered when

$$f(\beta, r_c) - 1 \sim 0. \tag{14.35}$$

Eq. (14.35) is satisfied for small deviations from $\beta = 0$. The general relation between $f(\beta, r_c)$ and β, with fixed r_c, is reported in Figs. **14.2–14.3** [396].

Figures **14.2–14.3** show the trend of $f(\beta, r_c) - 1$ as a function of β, with r_c fixed. The first plot refers to the validity range of β, namely $\beta \in [0, 1]$, while the second and the third consider β within the interval $[10^{-13}, 5 \cdot 10^{-12}]$. This is due to experimental observations, which suggest that the orbital period of Earth is approximately 365.256363051 *d*, with an error of $\Delta T = 5 \cdot 10^{-10}$ *d*. Therefore, the ratio $\frac{\delta T}{T}$ is of the order of 10^{-12}. Noticing that $\frac{\delta T}{T} = f(\beta, r_c) - 1$, we expect that, in the scale considered, deviations from GR might occur. As a result, β can be constrained by Solar System tests when r_c is fixed, providing that the only allowed parameters must be close to $\beta = 0$ (under the assumption of circular orbit).

The light deflection can also be used as a test in the context of $f(R)$ gravity, with $f(R) = R^k$, to further constrain the free constants r_c and β. In particular, while previously we arbitrarily set the value of r_c by hand, here we use the light deflection test to find a relation between β and r_c that well fits experimental observations. In the Parameterized Post-Newtonian (PPN) formalism, the deflection of a light ray due to the presence of a massive object is given by

$$\phi = \frac{(1+\gamma)GM}{c^2 b}(1+\cos\theta), \tag{14.36}$$

where M is the mass, b the impact parameter, θ the elongation angle, and γ the PN parameter [32, 39]. As pointed out in the introduction of this chapter, GR predicts a light deflection of 1.75 arcseconds when $M = M_{\odot}$, which agrees with experiments within an error of one part in one thousand. The general relation between the deflection angle and the potential is [397]

$$\phi = \frac{2}{c^2}\int \nabla_{\perp}\Phi dl, \tag{14.37}$$

with $\nabla_{\perp}$ defined by means of the spatial vector $\hat{e}$ as

$$\nabla_{\perp} \equiv \nabla - \hat{e}(\hat{e}\cdot\nabla). \tag{14.38}$$

The position of the light ray can be written in terms of the orthogonal vector $\vec{\xi}$, as $\vec{r} = \vec{\xi} + l\hat{e}$, so that the angle (14.37) becomes

$$\phi = \frac{2\xi}{c^2}\int_{-\infty}^{\infty}\left(\frac{1}{r}\frac{d\Phi}{dr}\right) dl. \tag{14.39}$$

Substituting Eq. (14.29) into Eq. (14.39), the deflection angle turns out to be

$$\phi = \frac{2GM}{c^2 r_c}\left(\frac{\xi}{r_c}\right)^{-1}\left[1 + \frac{\sqrt{\pi}(1-\beta)\Gamma(1-\beta/2)}{2\Gamma(3/2-\beta/2)}\left(\frac{\xi}{r_c}\right)^{\beta}\right], \tag{14.40}$$

with Γ being the Euler Gamma function [375]. Notice that, for $\beta = 1$, we get $\phi = \phi_{GR}/2$, so that the net effect of the correction term is to decrease the deflection angle with respect to the classical one. Imposing ϕ to fit observations, the weak field limit solution (14.28) is physically viable when r_c and β are related as shown by Fig. **14.4** [396].

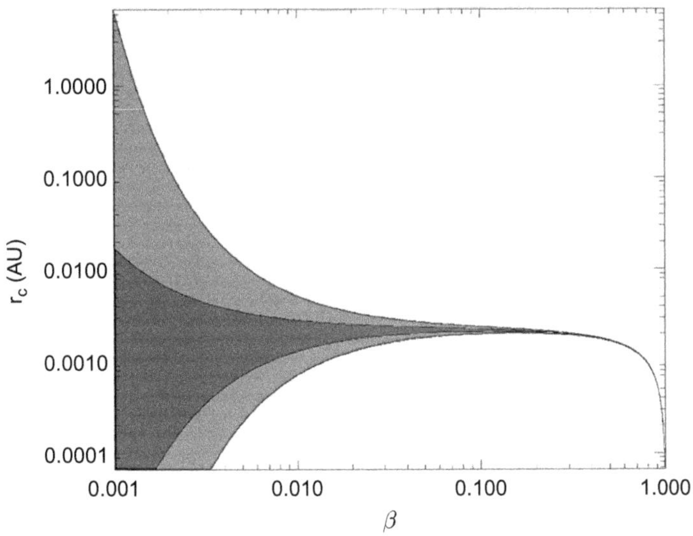

Figure 14.4 Relation between r_c and β provided by the light deflection test, at the confidence level of 2σ (dark grey) and 5σ (light grey) within the allowed values of β

14.2 Mass–Radius Relation of Neutron Stars in $f(R)$ Gravity

In Section 13.1, we proved that $f(R) = f_0 R^k$ gravity has Noether symmetries and integrals of motion. Thus, we considered the corresponding external solutions, showing that Schwarzschild space-time is recovered in the first-order approximation. Taking into account the external solution provided, for example, in Refs. [251,359], it is possible to study the mass–radius diagram for neutron stars with the possibility of explaining anomalous masses recently reported in literature [398]. To this purpose, we consider different equations of state and the internal solution for $f(R) = f_0 R^k$ gravity, under the approximation

$$f(R) = f_0 R^{1+\epsilon} \sim R + \epsilon R \ln R, \tag{14.41}$$

where the RHS is the first-order expansion around $\epsilon = 0$ (see Ref. [399]). In the search for numerical solutions, we assume the Birkhoff theorem to hold, starting from a line element of the form $ds^2 = e^{\nu}(r)dt^2 - e^{\lambda}(r)dr^2 - r^2 d\Omega^2$. Moreover, the energy-momentum tensor of matter fields is supposed to have the diagonal form $\mathcal{T}^{\mu}_{\nu} = diag(\rho(r), -p(r), -p(r), -p(r))$, with $\rho(r)$ and $p(r)$ being the energy density and pressure, respectively. Five equations are needed to find internal spherically symmetric solution. Three of them are the field equations (4.45) in the presence of matter. The remaining two are the continuity equation $D_{\mu}\mathcal{T}^{\mu\nu} = 0$ and the EoS, which, in turn, establishes the relation between matter and energy density. The former set of equations in $f(R)$ gravity, with $f(R) = f_0 R^{1+\epsilon}$, reads

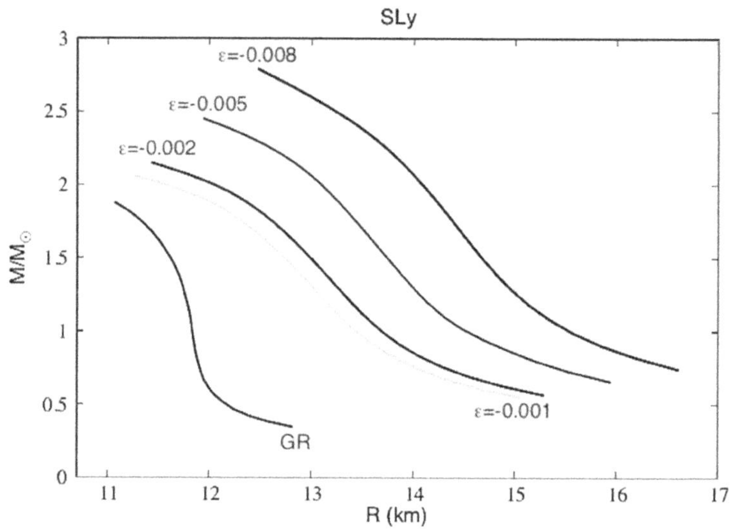

Figure 14.5 Mass–radius diagram in $f(R) = R + \epsilon R \ln R$ gravity with Sly EoS

$$\begin{cases} \dfrac{R^\epsilon}{r^2}\left[r(1-e^{-\lambda})\right] = \rho + \dfrac{1}{2}\epsilon R^{1+\epsilon} + e^{-\lambda}\Big\{ \left(\dfrac{2}{r} - \dfrac{1}{2}\dfrac{d\lambda}{dr}\right)\left[\epsilon(1+\epsilon)R^{\epsilon-1}R'\right] \\ + \left[\epsilon(1+\epsilon)R^{\epsilon-2}\left[RR'' + (\epsilon-1)R'^2\right)\right] \Big\} \\ \\ \dfrac{R^\epsilon}{r}\left[e^{-\lambda}\dfrac{d\nu}{dr} - \dfrac{1}{r}\left(1-e^{-\lambda}\right)\right] = p + \dfrac{1}{2}\epsilon R^{1+\epsilon} \\ + e^{-\lambda}\left(\dfrac{2}{r} + \dfrac{1}{2}\dfrac{d\nu}{dr}\right)\left[\epsilon(1+\epsilon)R^{\epsilon-1}R'\right] \\ \\ 3\epsilon(1+\epsilon)e^{\lambda}\left[(\epsilon-1)R^{\epsilon-2}R'^2 + R^{\epsilon-1}R'\left(\dfrac{2}{r} - \dfrac{1}{2}\dfrac{d\lambda}{dr} + \dfrac{1}{2}\dfrac{d\nu}{dr}\right) + R^{\epsilon-1}R''\right] \\ = -(\rho - 3p) - (\epsilon-1)R^{\epsilon+1}, \end{cases} \tag{14.42}$$

while the continuity equation is

$$\frac{\partial p}{\partial r} + \frac{1}{2}(\rho + p)\frac{\partial \nu}{\partial r} = 0. \tag{14.43}$$

On the other hand, the EoS strictly depends on the fluid considered. Neutron stars are well described by Sly, BSK19, BSK20, and BSK21 equations of state, which provide different numerical relations between density and pressure [400]. In the graphs of Figs. **14.5–14.8**, we report the mass–radius diagram for neutron stars, which can be obtained by numerically solving (14.42) and (14.43) with respect to ρ and p, with the addition of the preceding three equations of state [251,359,398].

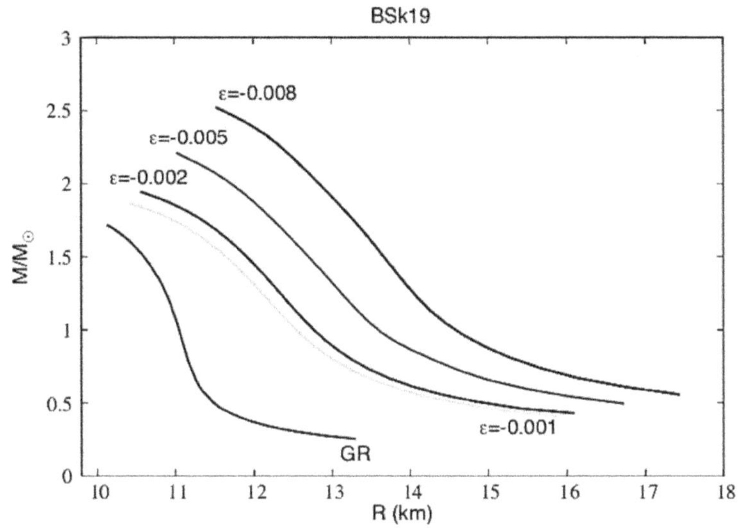

Figure 14.6 Mass–radius diagram in $f(R) = R + \epsilon R \ln R$ gravity with BSK19 EoS

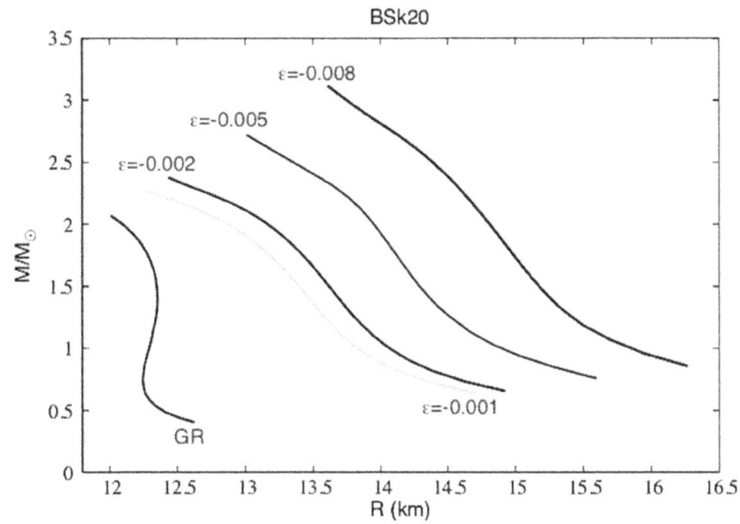

Figure 14.7 Mass–radius diagram in $f(R) = R + \epsilon R \ln R$ gravity with BSK20 EoS

Diagrams compare the mass–radius relation for three different equations of state, namely Sly, BSK19, BSK20, BSK21. Specifically, they are obtained by numerically solving the TOV equation of $f(R)$ gravity, in the first-order approximation. Different relations occur by increasing the value of the free parameter ϵ. Clearly, in the limit $\epsilon = 0$ Einstein GR is recovered, while, for larger values of ϵ, a significant difference occurs with respect to GR. For increasing values of

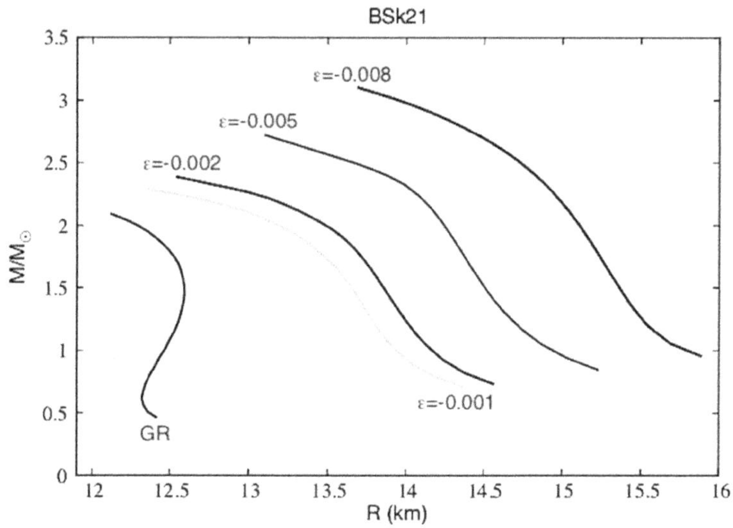

Figure 14.8 Mass–radius diagram in $f(R) = R + \epsilon R \ln R$ gravity with BSK21 EoS

ϵ, the mass–radius diagram assumes a self-similar behavior. As a result, $f(R)$ gravity, with $f(R) \sim R^{1+\epsilon}$, provides a new stability branch for neutron stars, which allows the presence of extreme objects without introducing exotic matter, rather only realistic equations of state, as in the case discussed in [398] related to the GW event GW190814 (see [401]).

However, it must be pointed out that functions like $f(R) \sim R^{1+\epsilon}$, with $\epsilon < 0$, suffer the so-called "Dolgov–Kawasaki instability," occurring when $f_{RR}(R) < 0$. This problem was first addressed in Ref. [402] by A. D. Dolgov and M. Kawasaki. They showed that introducing the function $f(R) \sim 1/R$ into the gravitational action leads to temporal instability in a cosmological background. This result was further generalized to any function $f(R)$ whose second derivative is negative; see Ref. [403]. Therefore, actions considered in this section are not good candidates to address the cosmic expansion of the Universe as a curvature effect. However, these models can be suitable at astrophysical scales in the context of compact objects, though they cannot be adopted in a cosmological framework. A detailed discussion on this topic is contained in Section 18.4.

To conclude this section, it is possible to show that the mass–radius diagram can be also fitted by Starobinsky's model, with function $f(R) = R + \alpha R^2$. Here the Dolgov–Kawasaki instability does not occur, as $f_{RR}(R) \geq 0$ for all $\alpha \geq 0$. Depending on the free parameter α, the theory predicts a better mass–radius correlation with respect to GR. In this case, the numerical solution of the field equations is reported in Fig. **14.9**. Note that the larger the value α is, the larger the neutron star mass becomes.

Therefore, in both cases, modified gravity can address the anomalous mass of compact objects like GW190814 [401] with the hypothesis that it is a supermassive neutron star. Specifically, for $f(R) = R + \alpha R^2$ and for $f(R) = R^{1+\epsilon}$,

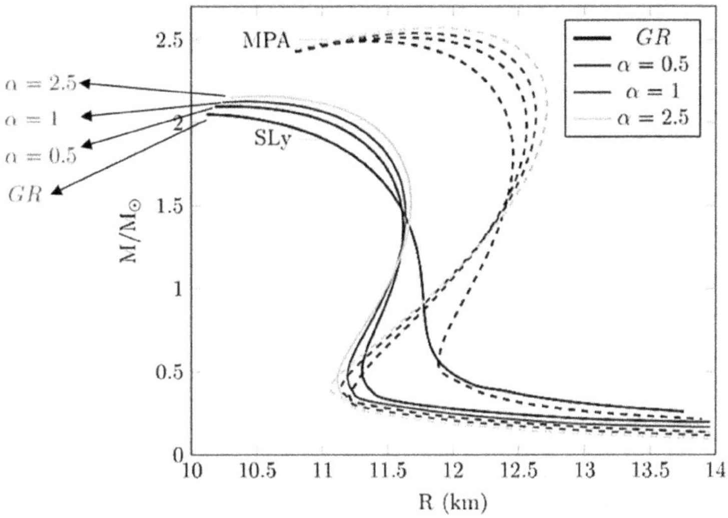

Figure 14.9 Mass–radius diagram for neutron stars in $f(R) = R + \alpha R^2$ compared with GR considering the SLy and MPA1 as equations of state. Parameter α is given in units of $R_\odot = G^2 M_\odot^2/c^4$. Here $R_\odot$ is the gravitational radius of the Sun

it is straightforward to obtain results consistent with LIGO detection without invoking exotic equations of state. The fact that extended theories of gravity are consistent with observations that cannot be explained by standard GR is fundamental not only because we can shed new light on extreme gravity regimes that are realized in compact objects like neutron stars, but also because these observations could validate the theoretical grounding of Extended Theories of Gravity. It is worth noticing that it could be not only an alternative explanation of reported results, but also a sort of *experimentum crucis* for these theories if GR fails to explain this kind of observations. It is also worth noticing again that both forms of $f(R)$ are consistent with the existence of Noether symmetries, which, in this particular case, play an important role in fixing the mass and the radius of neutron stars [251].

14.3 Gravitational Lensing in $f(R)$ Gravity

A gravitational lens is a big distribution of matter capable of curving the space-time to such an extent that it deflects the trajectory of a light signal passing nearby. Considering results provided by light deflection tests in the previous sections, here we consider the gravitational lensing in $f(R)$ gravity, with $f(R) = R^k$. Starting from the potential (14.29), we obtained the deflection angle (14.40) which, in the limit $\beta = 0$, provides back the GR result, that is,

$$\phi(\xi, 0) = \phi_{GR} = \frac{4GM}{c^2\xi}. \tag{14.44}$$

The deflection angle predicted by $f(R)$ gravity can be smaller than that provided by GR. This can be easily proved by considering Eq. (14.39), which establishes a relation between angle and potential. Knowing that potential (14.29) is larger than the GR one only when $r > r_c$, Eq. (14.28) yields

$$I(r) < I(r)_{GR} \iff r < (1-\beta)^{-\beta} r_c, \quad I(r) \equiv \frac{1}{r}\frac{d\Phi}{dr}, \tag{14.45}$$

which is always satisfied when $\beta \in [0,1]$. With the aim of investigating the gravitational lensing in $f(R)$ gravity, let α and α_S be the position of the lens and of the source planes of the image, respectively. Then the lens equation can be written as

$$\alpha - \alpha_s = \frac{D_{ls}}{D_s}\phi, \tag{14.46}$$

where D_l, D_s, and D_{ls} are the angular diameter distances between observer-lens, observer-source, and lens-source. Replacing Eq. (14.46) into Eq. (14.40), the lens equation takes the form

$$\left[1 - \mathcal{N}(\beta, \alpha_c)\left(\frac{\alpha}{\alpha_c}\right)^{\beta-2}\right]\alpha^2 - \alpha_s\alpha - \frac{1}{2} = 0, \tag{14.47}$$

with

$$\mathcal{N}(\beta, \alpha_c) = \frac{\sqrt{\pi}(1-\beta)\Gamma(1-\beta/2)}{4\alpha_c^2\Gamma(3/2-\beta/2)},$$
$$\alpha_c = \frac{r_c}{D_l \alpha_{E,GR}}, \quad \alpha_{E,GR} = \sqrt{\frac{4GMD_{ls}}{c^2 D_l D_s}}, \tag{14.48}$$

which is defined only for $0 \le \beta \le 2$, though we will consider the narrower range $0 \le \beta \le 1$.

Since we are considering the point-like lens approximation, a typical lensing system is represented by a compact object (either visible or not) in the Galaxy halo acting as a lens on the light rays coming from a stellar source in an external galaxy, like the Magellanic Clouds (LMC and SMC) or Andromeda. In such a configuration, the standard Einstein angle $\alpha_{E,GR}$ and the images' angular separation are of the order of a few 10^{-5} arcsec, so that we are in the *microlensing* regime. Microlensing is an astronomical phenomenon where the light is deflected by the gravitational field of a foreground lens to create a distorted or a multiple image. As a fiducial system, we take a stellar lens in the Galactic halo (with $M = 1M_\odot$ and $D_l = 20kpc$) working as a lens on the light rays coming from a source in an external galaxy (with $D_s = 50kpc$). Under these assumptions, the Einstein angle predicted by GR is of the order of 10^{-5} arcseconds. In order to estimate a reasonable range for r_c, we redefine the potential (14.29) as

$$\Phi(\eta) = -\frac{c^2}{2}\left(\frac{v_1}{c}\right)^2 \eta^{-1}\left(1+\eta^\beta\right), \tag{14.49}$$

with

$$\eta \equiv \frac{r}{r_c}, \quad \Phi_c = \frac{GM}{r_c}, \quad v_1 = \sqrt{\Phi_c}. \tag{14.50}$$

The length r_c can also be written in terms of the new variables as

$$r_c = \frac{GM}{c^2 \tilde{v}_1^2}, \tag{14.51}$$

with $\tilde{v}_1 = \frac{v_1}{c}$. Rather than r_c, we will use in the following $\log \tilde{v}_1$ as the parameter, since its range is easier to evaluate. Let us consider, for instance, the Sun as the field source and a planet as a test particle. Since the Earth velocity with respect to the Sun is $\sim 30\,\mathrm{km/s}$, we can indeed take $(-6, -4)$ as a reasonable range for $\log \tilde{v}_1$. Note that Eq. (14.51) clearly shows why r_c depends on both the mass and the scale of the system under examination.

The lens equation (14.47), in the case of perfect alignment among the observer, lens, and source, takes the form

$$\left[1 - \mathcal{N}(\beta, \alpha_c) \left(\frac{\alpha}{\alpha_c}\right)^{\beta-2}\right] \alpha^2 = \frac{1}{2}, \tag{14.52}$$

which for $\beta = 0$ $(k = 1)$ provides back $\alpha = \alpha_{E,GR}$, namely the image is the well-known Einstein ring with angular radius equal to the Einstein angle. In general, Eq. (14.52) can be solved numerically for given values of the model parameters $(\beta, \log \tilde{v}_1)$. As a general result, the Einstein ring still forms, but the corrected Einstein angle α_E deviates from the GR one. It is therefore interesting to study the magnitude of such deviations. To this aim, we can define

$$\epsilon_E = \frac{\Delta\alpha}{\alpha} = \frac{\alpha_E - \alpha_{E,GR}}{\alpha_{E,GR}}, \tag{14.53}$$

by means of which we get the plot of $\log \tilde{v}_1$ as a function of β with different ϵ_E, which quantifies the relative deviation from the GR result [375].

Some interesting remarks may be drawn from the plot. As a general result, ϵ_E could take both positive and negative values, that is, R^k gravity may increase or decrease the Einstein angle with respect to the standard GR result. Actually, over almost the full parameter space, ϵ_E is negative and, in particular, taking $(\beta, \log \tilde{v}_1) \sim (0.58, -5)$ as fiducial values,[2] we find $\epsilon_E \sim -0.25$, which leaves open the possibility of detecting the corrections due to R^k gravity.

Before discussing this issue, it is worth examining how ϵ_E depends on the model parameters. For fixed values of β, the relative deviation is clearly an increasing function of $\log \tilde{v}_1$, a result that can be easily explained qualitatively. Indeed, the

[2] Indeed, fitting the low surface brightness rotation curves, one gets $\beta = 0.58 \pm 0.15$ (see Section 15.1), while $\log \tilde{v}_1 \sim -5$ is suggested by considering a Jupiter-like planet as the test particle in the potential of a Sun-like lens.

larger $\log \tilde{v}_1$ is, the smaller r_c is and hence the earlier the correction term in the modified potential starts contributing. This fact explains why the Einstein angle deviates more and more from the classical one as $\log \tilde{v}_1$ increases.

Modifying the Einstein angle has a deep impact on the microlensing phenomenology. As a first interesting application, let us remember that the duration of a microlensing event is estimated as $t_E = R_E/v_\perp$ with $R_E = D_l \alpha_E$ the Einstein radius and $v_\perp$ the transverse velocity of the lens with respect to the line of sight. Supposing that the distances (D_l, D_s) and the transverse velocity $v_\perp$ were known (for instance, from deviations with respect to the standard lightcurve), a measurement of the Einstein time $t_{E,obs}$ translates into an estimate of the mass of the lens.

This application may represent a useful tool in order to get the mass of the lens considered; in fact, it is possible to translate the duration of the microlensing process into an estimate of the mass of the lens. The true mass μ (in Solar mass units) can be linked to that predicted by GR μ_{GR} as [375]

$$\mu = \frac{\mu_{GR}}{\alpha_E^2 (M = 1 \mathrm{M}_\odot)}. \tag{14.54}$$

As can be inferred from Fig. **14.10**, $\alpha_E < 1$ over almost the full parameter space $(\beta, \log \tilde{v}_1)$ so that the true lens mass is higher than the one estimated by the standard method. However, in most of the applications, one may statistically

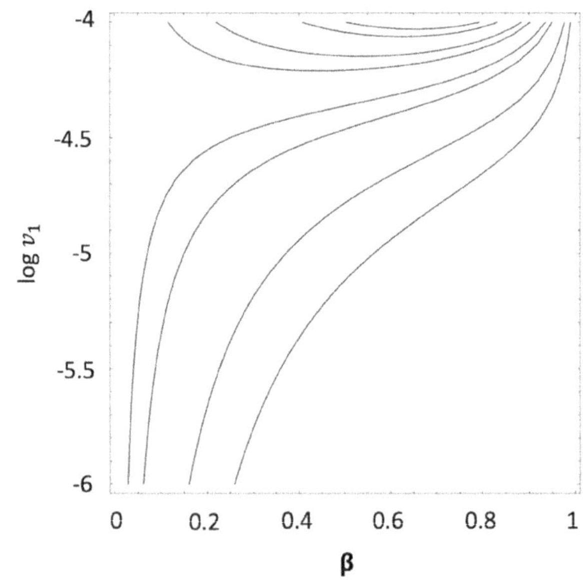

Figure 14.10 Contours of equal ϵ_E in the plane $(\beta, \log \tilde{v}_1)$ for $\epsilon_E = -0.25, -0.20, -0.10, -0.05, 0.05, 0.10, 0.20, 0.25$, from the bottom curve to the top one

infer only the lens transverse velocity, but not its distance, so that what is indeed underestimated by the standard method is the quantity $\mu x(1-x)$ with $x = D_l/D_s$. As such, therefore, taking into account the correction to the Einstein angle may lead to an increase of either the estimated lens mass or the distance ratio x. Disentangling these two effects is quite difficult and requires nonstandard microlensing events where the duration of the event may differ from t_E, so that the preceding qualitative argument does not apply anymore.

As a further constraint to the theory, we can study the amplification of source luminosity $\mathcal{A}$ in our fourth-order theory. In this regard, gravitational lenses work also as a sort of natural telescopes amplifying the source luminosity in such a way as to make visible objects that are otherwise too faint to be detected. As a general rule, the amplification is given by the inverse of the lens-mapping Jacobian. In the case of a lens with cylindrical symmetry (and hence for the point-like lens) and in the plane where $\alpha_s \neq 0$, this reduces to [404]

$$\mathcal{A} = \left| \frac{\alpha_s}{\alpha} \frac{d\alpha_s}{d\alpha} \right|^{-1} . \tag{14.55}$$

Specifically, Eq. (14.55) can be written as

$$\mathcal{A} = \left| 1 - a^2 - [(2-\beta)a + (1-\beta)b + \beta]b \right|^{-1} , \tag{14.56}$$

with

$$\begin{aligned} a &= \left(2\alpha^2\right)^{-1}, \\ b &= \mathcal{N}\left(\beta, \alpha_c\right) \left(\frac{\alpha}{\alpha_c}\right)^{\beta-2} . \end{aligned} \tag{14.57}$$

Note that we have considered the unsigned amplification since we are not interested in the parity of the images, rather only in the luminosity variation. As a check of the theory consistency, it is easy to see that the GR result is recovered for $k = 1$.

Since in a microlensing event the two images are merged together (given the tiny angular separation of order twice the Einstein angle), what we observe is the total luminosity, which is obtained by multiplying the source luminosity by the sum of the amplification of the two images. This procedure does not admit analytic resolution in R^k gravity, since no exact expression of the image positions, in terms of source position, occurs. Assuming that the lens crosses the line of sight moving with a constant velocity, the source position as a function of time t can be written as [405]

$$\alpha_s = \sqrt{\alpha_0^2 + \left(\frac{t-t_0}{t_E}\right)^2} , \tag{14.58}$$

with $\alpha_0 = \alpha_s(t-t_0)$, t_0 the time of closest approach to the lens in the line of sight, and t_E the Einstein time defined previously. As the time goes on, the source position changes according to Eq. (14.58) and hence the images' position

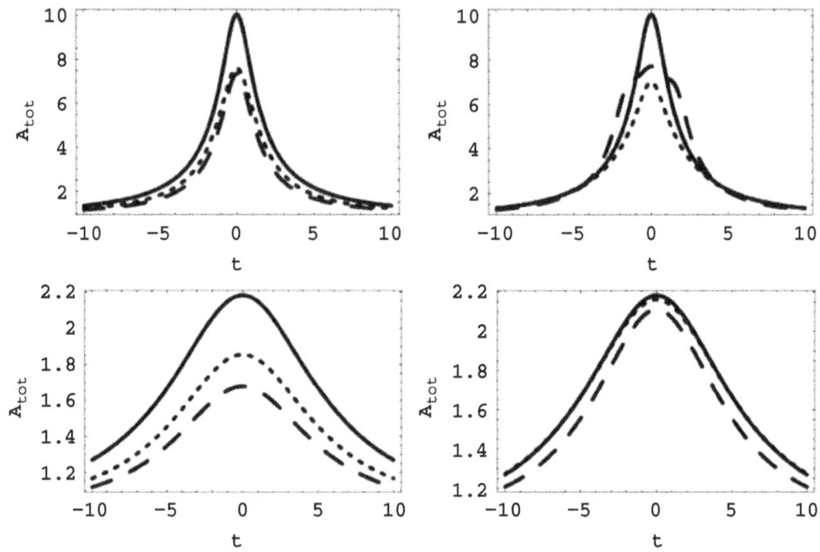

Figure 14.11 Paczynski lightcurves for Einsteinian gravitational lensing (solid line) and R^k gravity with $\beta = 0.25$ (short dashed) and $\beta = 0.75$ (long dashed). Upper (lower) panels refer to $\alpha_0 = 0.1\,(0.5)$, while left (right) panels are for $\log \tilde{v}_1 = -5.5(-4.5)$.

varies. As a consequence, the total amplification $\mathcal{A}_{tot}$ also becomes a function of time. Using the numerical solutions of Eq. (14.52), we can work out a corrected Paczynski lightcurve for R^k models. Some examples are shown in Fig. **14.11** where, without loss of generality, we have set $t_0 = 0$ and $t_E = 10d$ [375].

As it appears from Figs. **14.11**–**14.12**, both the sign and the magnitude of the deviations from the standard lightcurve depend in a complicated way on the theory parameters $(\beta, \log \tilde{v}_1)$ and on the minimum impact parameter α_0. Although a general rule cannot be extracted from the plots, it is interesting to note, however, that, for some combination of the parameters, the microlensing lightcurve may also be distorted with respect to the Paczynski one. Nevertheless, the typical signature of microlensing events (symmetry, uniqueness of the bump, and acromaticity) are preserved. Note that, in the $\beta = 0$ limit, the total amplification is in agreement with that predicted by GR, whose analytic expression exists and turns out to be:

$$\mathcal{A}_{tot,GR} = \left| \frac{\alpha_s^2 + 2}{\alpha_s \sqrt{\alpha_s^2 + 4}} \right| . \qquad (14.59)$$

Considering, for instance, solid lines in the right panels, it is clear that the deviations are of order of 10 percent, which leaves open the possibility for a detection in the highest magnification events, where the measurement errors on the lightcurve points are sufficiently low. A more detailed investigation is,

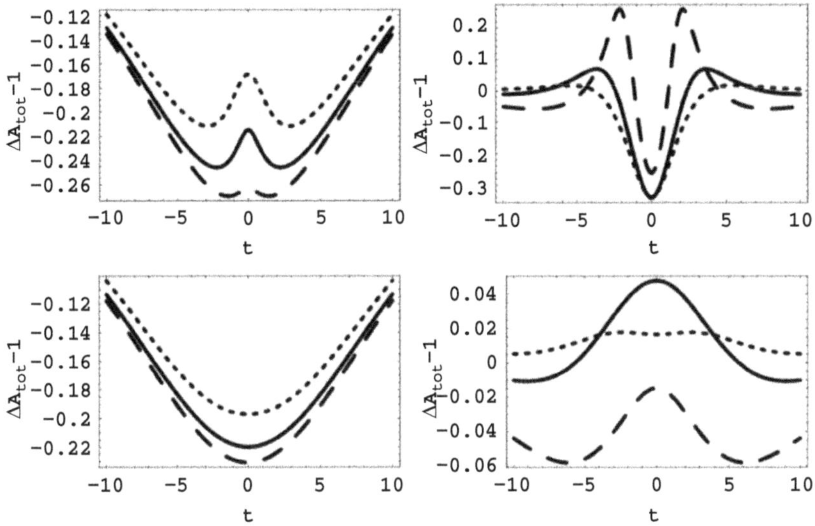

Figure 14.12 Relative deviation with respect to the Paczynski lightcurve of the corrected microlensing lightcurve for R^k gravity with $\beta = 0.43$ (short dashed), $\beta = 0.58$ (solid), and $\beta = 0.73$ (long dashed). α_0 and $\log \tilde{v}_1$ are set as Fig. **14.11**.

however, needed to understand whether such deviations may be obtained by other astrophysical effects.

14.4 Constraining Nonlocal Gravity by S2 Star Orbit

In Chapter 11, we discussed the main features of Nonlocal Gravity, pointing out the advantages provided by nonlocality at low- and high-energy scales. By means of Noether's approach, we selected nonlocal functions containing symmetries and found exact solutions for the corresponding cosmological equations. In this section, we use Noether's theorem to select the nonlocal function in the Deser–Woodard action, assuming a spherically symmetric space-time of the form (13.1). Then, we consider the weak-field limit and constrain the nonlocal gravity parameters by the S2 star, which is a relevant star orbiting around our Galactic Center in Sagittarius A* (Sgr A*).

As previously, we shall start by writing the point-like Lagrangian, by means of which the identity (13.4) can be applied. Introducing the local scalar field $\phi(r,t) = \Box^{-1}R(r,t)$ to localize the theory and the Lagrange multiplier $\varepsilon(r,t)$ (see the above Chapter 11 for details), the Deser–Woodard action turns out to be

$$S = \frac{1}{16\pi G}\int \sqrt{-g}\Big\{R[1+f(\phi)] + \varepsilon(r,t)(\Box\phi - R)\Big\}d^4x. \tag{14.60}$$

In this way, the minisuperspace is four-dimensional, that is, $\mathcal{S} \equiv \{\nu, \lambda, \phi, \varepsilon\}$ because the scalar curvature can be written explicitly in terms of the metric coefficients e^{ν} and e^{λ}. The Klein–Gordon equations, with respect to the two scalar fields ϕ and ε, are respectively

$$\frac{\delta S}{\delta\phi}=0\to\Box\varepsilon=-Rf_{\phi},\qquad\frac{\delta S}{\delta\varepsilon}=0\to\Box\phi=R,\tag{14.61}$$

while the variation of Eq. (14.60) with respect to the metric yields the field equations

$$[1+f(\phi)-\varepsilon]G_{\mu\nu}=\left(D_{\mu}D_{\nu}-g_{\mu\nu}\Box\right)f(\phi)-\frac{1}{2}g_{\mu\nu}D_{\alpha}\varepsilon D^{\alpha}\phi+D_{\mu}\varepsilon D_{\nu}\phi.\tag{14.62}$$

Notice that the position

$$\nu\equiv\nu(r)=-\lambda(r)\tag{14.63}$$

cannot be applied a priori, since the relation between λ and ν must occur from the solution of field equations and the validity of the Birkhoff theorem. Before deriving the point-like Lagrangian, it is worth noticing that the metric depends both on the radial coordinate r and on the cosmic time t. Therefore, the infinitesimal generator ξ^{μ} turns out to be a two-vector with components ξ^{t} and ξ^{r}. Considering the form of the d'Alembert operator in spherical symmetry and integrating out the second derivatives, the canonical Lagrangian reads as

$$\begin{aligned}\mathcal{L}(r,\nu,\lambda)=e^{-\frac{1}{2}(\lambda+\nu)}\big[&-e^{\nu}r^{2}\nu_{r}\phi_{r}f_{\phi}(\phi)+e^{\lambda}r^{2}\lambda_{t}\phi_{t}f_{\phi}(\phi)\\&-2e^{\nu}f(\phi)\left(e^{\lambda}+r\lambda_{r}-1\right)-2e^{\lambda+\nu}+2e^{\nu}+e^{\nu}r^{2}\varepsilon_{r}\phi_{r}+e^{\nu}r^{2}\nu_{r}\varepsilon_{r}\\&-e^{\lambda}r^{2}\varepsilon_{t}\phi_{t}-e^{\lambda}r^{2}\lambda_{t}\varepsilon_{t}+2e^{\nu}\varepsilon\left(e^{\lambda}+r\lambda_{r}-1\right)-2e^{\nu}r\lambda_{r}\big]\,,\end{aligned}\tag{14.64}$$

whose symmetry generator is

$$\mathcal{X}=\xi^{t}\partial_{t}+\xi^{r}\partial_{r}+\alpha\partial_{\nu}+\beta\partial_{\lambda}+\gamma\partial_{\phi}+\delta\partial_{\varepsilon},\tag{14.65}$$

with infinitesimal generators $\xi,\alpha,\beta,\gamma,\delta$ depending on the variables $r,t,\nu,\lambda,\phi,\varepsilon$. The Noether theorem selects the symmetries [406]

$$\begin{cases}\chi=(\xi_0 t+\xi^{t}(r))\partial_{t}-2\xi_0\partial_{\nu}+(\gamma_0+2\xi_0)\partial_{\phi}+\delta_0(\gamma_0+2\xi_0)\partial_{\varepsilon}\\ f(\phi)=\delta_0\phi+f_1\\ \chi=(\xi_0 t+\xi^{r}(r))\partial_{t}-\frac{\xi_1}{2}r\partial_{r}-(2\xi_0+\xi_1)\partial_{\nu}+\gamma_0\partial_{\phi}+\xi_1(\varepsilon-\delta_0-1)\partial_{\varepsilon}\\ f(\phi)=\delta_0+f_1e^{\frac{\gamma_0}{\xi_1}\phi}.\end{cases}\tag{14.66}$$

In order to investigate the weak-field limit, we restrict the space of solutions to a subclass where the Birkhoff theorem holds, namely where both ν and λ are independent of time. This assumption is reasonable as a first approximation, since, in the weak-field limit, a static and spherically symmetric space-time must be a solution of the field equations. Let us therefore consider the line element $ds^2=A(r)dt^2-B(r)\,dr^2-r^2d\Omega^2$. The functions selected by Noether's theorem are still solutions of Noether's identity (13.4). The purpose is to compare the weak-field limit with a set of S2 star orbit observations given by New Technology Telescope/Very Large Telescope (NTT/VLT). The first-order expansion of g_{00} allows to find the expected potential $\Phi(r)$ for time-like particles. To this purpose, let us expand the g_{00} component of the metric up to the sixth order and the g_{11} component up to the fourth:

$$g_{00} \sim \mathcal{O}(6), g_{0i} \sim \mathcal{O}(5) \text{ and } g_{ij} \sim \mathcal{O}(4). \tag{14.67}$$

In the present case, the above expansions provide [406]

$$\begin{cases} A(r) = 1 + \frac{1}{c^2}\Phi(r)^{(2)} + \frac{1}{c^4}\Phi(r)^{(4)} + \frac{1}{c^6}\Phi(r)^{(6)} + \mathcal{O}(8) \\ B(r) = 1 + \frac{1}{c^2}\Psi(r)^{(2)} + \frac{1}{c^4}\Psi(r)^{(4)} + \mathcal{O}(6) \\ \phi(r) = \phi_0 + \frac{1}{c^2}\phi(r)^{(2)} + \frac{1}{c^4}\phi(r)^{(4)} + \frac{1}{c^6}\phi(r)^{(6)} + \mathcal{O}(8) \\ \varepsilon(r) = \varepsilon_0 + \frac{1}{c^2}\varepsilon(r)^{(2)} + \frac{1}{c^4}\varepsilon(r)^{(4)} + \frac{1}{c^6}\varepsilon(r)^{(6)} + \mathcal{O}(8), \end{cases} \tag{14.68}$$

with $\Phi(r)$ and $\Psi(r)$ being the Newtonian potentials coming from the first-order expansions of g_{00} and g_{11}, respectively, while ϕ_0 and ε_0 are constants. Considering the second function of (14.66) and setting $\delta_0 = f_1 = 1$ and $\gamma_0 = \xi_1$, the Klein–Gordon equations together with the field equations yield the following system:

$$\begin{aligned} &2B^2\left(-\varepsilon + e^\phi + 2\right) + rB'\left(-2\varepsilon - r\varepsilon' + re^\phi\phi' + 2e^\phi + 4\right) \\ &- B\Big[-2\varepsilon + 2\left(-r^2\varepsilon'' + r^2e^\phi\phi'' + r^2e^\phi\left(\phi'\right)^2 + 2re^\phi\phi' + e^\phi + 2\right) \\ &+r\varepsilon'\left(r\phi' - 4\right)\Big] = 0, \end{aligned}$$

$$\begin{aligned} &rA'\left(-2\varepsilon - r\varepsilon' + re^\phi\phi' + 2e^\phi + 4\right) \\ &- A\left[2B\left(-\varepsilon + e^\phi + 2\right) + 2\varepsilon + r^2\varepsilon'\phi' + 4r\varepsilon' - 4re^\phi\phi' - 2e^\phi - 4\right] = 0, \end{aligned}$$

$$\begin{aligned} &A^2\left[-4B^2e^\phi + rB'\left(r\varepsilon' - 4e^\phi\right) + B\left(-2r^2\varepsilon'' - 4r\varepsilon' + 4e^\phi\right)\right] \\ &+ Br^2\left(-e^\phi\right)\left(A'\right)^2 + Ar\left\{B\left[2re^\phi A'' + A'\left(4e^\phi - r\varepsilon'\right)\right] - re^\phi A'B'\right\} = 0, \end{aligned}$$

$$\begin{aligned} &A^2\left[-4B^2 - rB'\left(r\phi' + 4\right) + 2B\left(r^2\phi'' + 2r\phi' + 2\right)\right] + B\left(-r^2\right)\left(A'\right)^2 \\ &+ Ar\left\{B\left[2rA'' + A'\left(r\phi' + 4\right)\right] - rA'B'\right\} = 0, \end{aligned} \tag{14.69}$$

where the prime denotes the derivative with respect to r. The solution of the preceding equations can be written in terms of the effective gravitational constant $G_{eff} = G\phi_c$, with ϕ_c a real constant. Replacing the perturbations (14.68) into the preceding system, we get

$$A(r) = 1 - \frac{2GM\phi_c}{c^2 r} + \frac{G^2M^2}{c^4 r^2}\left[\frac{14}{9}\phi_c^2 + \frac{18r_\varepsilon - 11r_\phi}{6r_\varepsilon r_\phi} r\right] - \frac{G^3M^3}{c^6 r^3}\left[\frac{50r_\varepsilon - 7r_\phi}{12r_\varepsilon r_\phi}\phi_c r + \frac{16\phi_c^3}{27} - \frac{r^2\left(2r_\varepsilon^2 - r_\phi^2\right)}{r_\varepsilon^2 r_\phi^2}\right],$$

$$B(r) = 1 + \frac{2GM\phi_c}{3c^2 r} + \frac{G^2M^2}{c^4 r^2}\left[\frac{2\phi_c^2}{9} + \left(\frac{3}{2r_\varepsilon} - \frac{1}{r_\phi}\right) r\right],$$

$$\phi(r) = \frac{4GM\phi_c}{3c^2 r} - \frac{G^2M^2}{c^4 r^2}\left[\left(\frac{11}{6r_\varepsilon} + \frac{1}{r_\phi}\right) r - \frac{2\phi_c^2}{9}\right] - \frac{G^3M^3}{c^6 r^3}\left[\frac{r^2}{r_\phi^2} - \left(\frac{25}{12r_\varepsilon} - \frac{7}{6r_\phi}\right)\phi_c r - \frac{4\phi_c^3}{81}\right],$$

$$\varepsilon(r) = 1 + \frac{G^2M^2}{c^4 r^2}\left[\frac{2\phi_c^2}{3} - \left(\frac{13}{6r_\varepsilon} - \frac{1}{r_\phi}\right) r\right] + \frac{G^3M^3}{c^6 r^3}\left[\frac{20\phi_c^3}{27} - \left(\frac{1}{r_\varepsilon^2} - \frac{1}{r_\phi^2}\right) r^2 - \left(\frac{131}{36r_\varepsilon} + \frac{1}{6r_\phi}\right)\phi_c r\right],$$

where r_ε and r_ϕ are length scales related to the scalar fields ε and ϕ, which can be addressed to the nonlocality. The observations of the S2 star orbit can be used to constrain the free parameters ϕ_c, r_ϕ, and r_ε and then to test possible nonlocal effects. Let us then consider the second-, fourth-, and sixth-order expansions of the potential Φ:

$$\Phi^{(2)}(r) = -\frac{2GM}{r}\phi_c,$$
$$\Phi^{(4)}(r) = \frac{G^2M^2}{r^2}\left[\frac{14}{9}\phi_c^2 + \frac{18r_\varepsilon - 11r_\phi}{6r_\varepsilon r_\phi} r\right],$$
$$\Phi^{(6)}(r) = \frac{G^3M^3}{r^3}\left[\frac{7r_\phi - 50r_\varepsilon}{12r_\varepsilon r_\phi}\phi_c r - \frac{16\phi_c^3}{27} + \frac{2r_\varepsilon^2 - r_\phi^2}{r_\varepsilon^2 r_\phi^2} r^2\right]. \tag{14.70}$$

Notice that the Newtonian potential can be recovered as a particular limit by setting $\phi_c = 1$. The other two free parameters must be constrained by two-body simulation of S2, orbiting around Sgr A*. Let $\dot{\mathbf{r}} = \mathbf{v}$ be the spatial velocity, U_{NL} the potential energy, and $\mu = M \cdot m_s/(M + m_s)$ the reduced mass, with M being the mass of Sgr A* and m_S the mass of S2. According to the data provided in Ref. [407], we vary the values of r_ε and r_ϕ, searching for solutions that provide a lower χ^2 than the Keplerian orbit ($\chi^2 \sim 1.89$). Following Refs. [406, 408], the reduced χ^2 can be written as

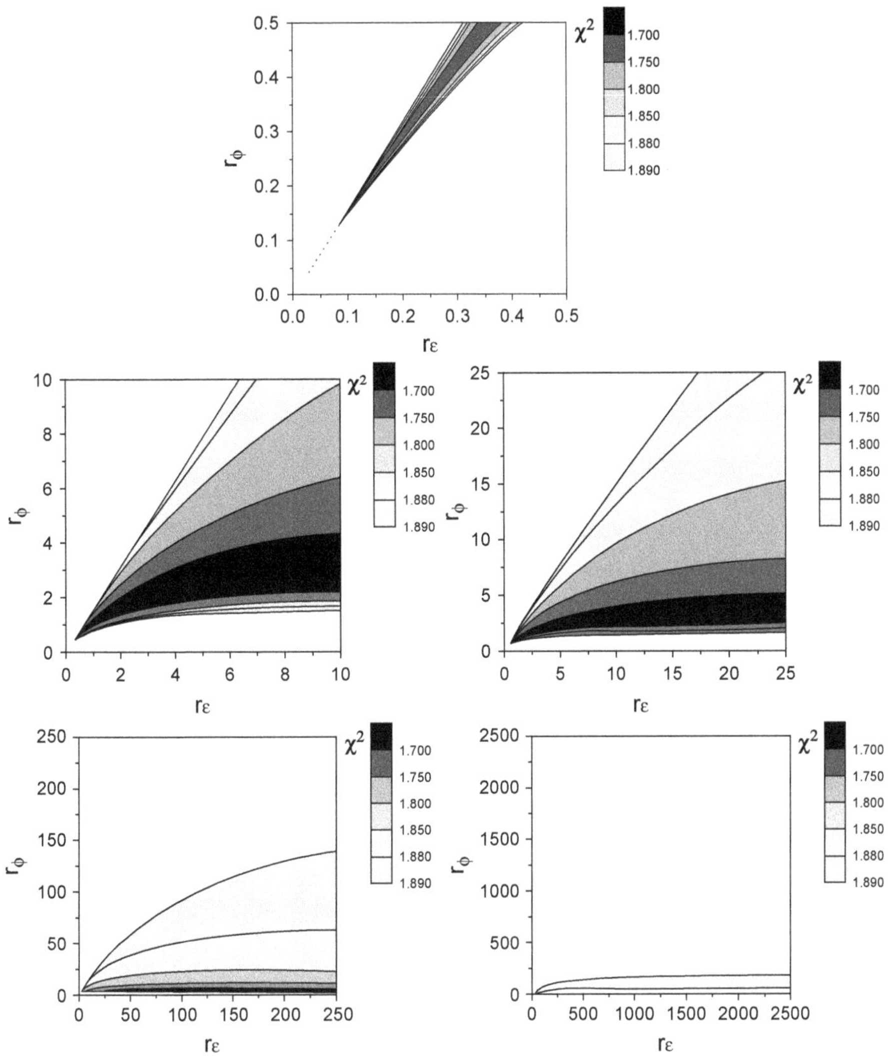

Figure 14.13 Maps of the reduced χ^2 in different regions of the parameter space. Figures show the lengths r_ϕ and r_ε in AU for all those simulated orbits of the S2 star giving a lower χ^2 than the Keplerian orbit. The color is darker when the fit is better, namely when the χ^2 is lower.

$$\chi^2 = \frac{1}{2N-\nu}\sum_{i=1}^{N}\left[\left(\frac{x_i^o - x_i^c}{\sigma_{xi}}\right)^2 + \left(\frac{y_i^o - y_i^c}{\sigma_{yi}}\right)^2\right], \tag{14.71}$$

where (x_i^o, y_i^o) and (x_i^c, y_i^c) are the observed and the theoretical apparent positions, respectively. N is the number of observations, ν the number of initial conditions, and σ_{xi} and σ_{yi} the uncertainties of the observed positions. Fig. **14.13** shows the χ^2 in different regions of the parameter space $r_\phi - r_\varepsilon$ [406].

We notice that the best value for r_ϕ is in the range $r_\phi \sim 0.1 \div 2.5$ AU. However, the analysis of the reduced χ^2 can only constrain the length r_ϕ, since the parameter r_ε is related to the non-dynamical scalar field ε, which is just a mathematical tool aimed at localizing the theory.

The potential energy in the weak-field limit can be calculated by means of Eq. (14.70). It contains nonlocal corrections, related to the function $f(\phi)$, which vanish when standard Newtonian gravity is restored. It reads

$$\begin{aligned} U_{NL} = &-\frac{GM}{r}\phi_c + \frac{G^2M^2}{2c^2r^2}\left[\frac{14}{9}\phi_c^2 + \frac{18r_\varepsilon - 11r_\phi}{6r_\varepsilon r_\phi}r\right] \\ &+ \frac{G^3M^3}{2c^4r^3}\left[\frac{7r_\phi - 50r_\varepsilon}{12r_\varepsilon r_\phi}\phi_c r - \frac{16\phi_c^3}{27} + \frac{2r_\varepsilon^2 - r_\phi^2}{r_\varepsilon^2 r_\phi^2}r^2\right], \end{aligned} \tag{14.72}$$

from which it is possible to compute the precession per orbital period [406] (see Fig. **14.14**).

Notice that the precession can have either the same direction as GR or the opposite one, depending on the parameters range considered. Specifically, for $r_\phi < 0.2$ AU, the precession turns out to be positive; otherwise, it is negative. Finally, considering the values $r_\phi \sim 1.2$ AU and $r_\varepsilon \sim 1.1$ AU, which correspond to $\chi^2 \sim 1.72$, the comparison between the Keplerian orbit of the S2 star and the nonlocal orbit yields the plots in Fig. **14.15** and Fig. **14.16** [406].

It is worth noticing that some regions are in agreement with observations even more than the Keplerian case. To conclude, the parameters r_ϕ and r_ε of Nonlocal Gravity model, selected by the Noether symmetries, perfectly fit astrometric data. This can be a preliminary test bed, at the astronomical level, for this kind of theories. It is important to stress that sensitivities reported by the EHT collaboration are not far from the preceding values [40] so, in principle, nonlocal gravity effects could be detected by investigating stellar orbits in our Galaxy.

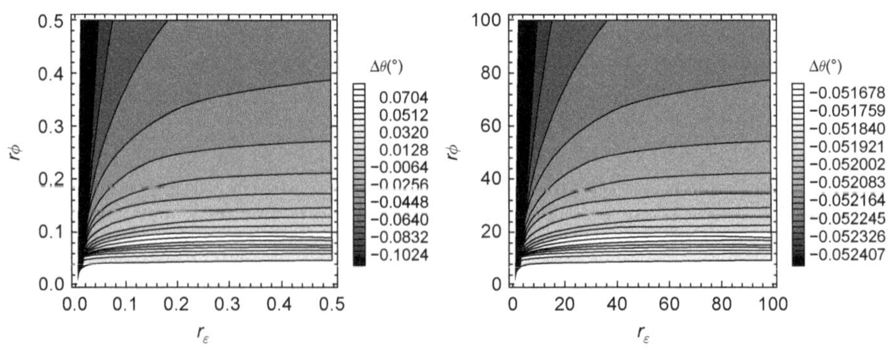

Figure 14.14 Precession per orbital period in two different regions of the parameter space. Darker colors refer to lower values of the precession angles.

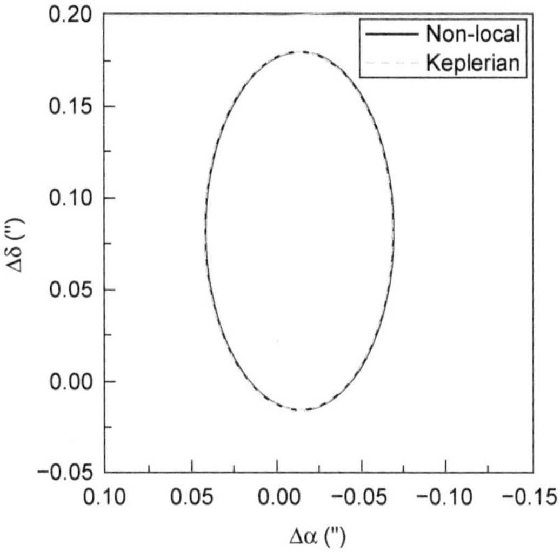

Figure 14.15 Comparison between the Keplerian orbit of the S2 star (dashed line) and the orbit predicted by Nonlocal Gravity (solid line) with parameter values that minimize the χ^2: $r_\phi \sim 1.2$ AU and $r_\varepsilon \sim 1.1$ AU. $\Delta\alpha$ and $\Delta\delta$ are the coordinates of the S2 star, whose experimental values refer to NTT/VLT observations.

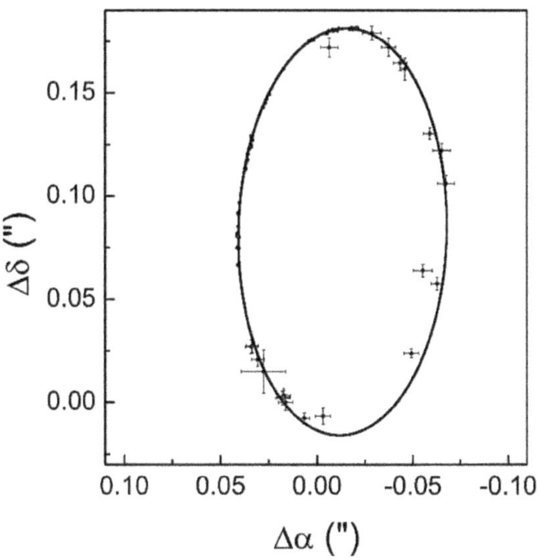

Figure 14.16 Best-fit orbit of S2 star in Nonlocal Gravity, with parameters r_ϕ and r_ε minimizing the χ^2: $r_\phi \sim 1.2$ AU and $r_\varepsilon \sim 1.1$ AU. The fitted orbit is compared with the data provided by NTT/VLT observations. $\Delta\alpha$ and $\Delta\delta$ are the coordinates of S2 star.

15
Applications to Galaxies

The Universe doesn't allow perfection.

— *Stephen Hawking*
(*A Brief History of Time* [1988])

As mentioned in the first part of the book, dark matter (together with dark energy) is one of the main open issues of the current cosmological scenario. It represents a hypothetical kind of matter that does not interact with electromagnetic radiation and thus can be detected by gravitational interaction. It is supposed to account for almost a third of the universe content and was firstly introduced due to inconsistencies between theory and observations of galaxy rotation curves and clusters of galaxies [409, 410].

In particular, dark matter is essential in order to explain formation and dynamics of galaxies and large-scale structure. For example, observations suggest that the velocity of stars and neutral hydrogen clouds, in spiral galaxies, remains unchanged from the edge to the center, unlike what was expected from considering standard Keplerian motions. According to Newtonian Mechanics, stars that are far away from the center have the same velocity as nearer ones only if the mass of the galactic system increases with increasing distances. However, the observed distribution of luminous matter is not linearly dependent on the radial coordinate and this fact gives rise to one of the dark matter issues. Furthermore, galaxies do not result in stable systems if a large amount of nonluminous matter is not supposed to constitute large haloes around luminous components like the bulge and the disc. Finally, clusters of galaxies, considered as thermodynamic systems, evaporate if dark matter is not assumed in their formation and stabilization. From a cosmological point of view, the ΛCDM model, considered the paradigm of modern cosmology, is inconsistent without dark matter. Despite the observational evidence, the dark matter puzzle has no counterpart at the fundamental level. In fact, there is no experimental evidence for new particles capable of addressing dark matter dynamics at the astrophysical and cosmological levels. Several models have been proposed and fine experiments have been realized to detect it in the last decades, but no final result is available so far.

As discussed in Chapter 4, a possible way out could come from theories of gravity involving modification/extensions of GR. From a genuinely astrophysical point of view, the missing matter problem, described by dark matter or by Modified Gravity, has completely equivalent results unless a discriminating evidence, excluding one of the two pictures, is detected. Concretely, discovering some new particle beyond the Standard Model or some anomalous gravitational phenomenon (e.g. new gravitational polarizations) could lead to the solution of this big issue of modern physics.

Here, as examples, we discuss the possibility of addressing the missing matter problem in galaxies by Extended Gravity. In particular, we consider $f(R)$ models selected by Noether symmetries (see Section 7.1). In particular, we constrain the extra degrees of freedom, coming from $f(R)$ gravity, by the rotation curves and the Fundamental Plane of galaxies.

It is worth noticing that several extended and modified theories of gravity can mimic the role of dark matter, see, for example, Ref. [411], where $f(T)$ gravity is considered, or [412, 413], where the clustering of galaxies is discussed in the framework of $f(R)$ gravity. However, this chapter is not meant to investigate the capability of modified theories of gravity in fitting the astrometric data, but it aims to discuss models, selected by the Noether Symmetry Approach, as a realistic approach to the dark matter problem. We restore again physical units in order to compare observations with theoretical predictions.

15.1 The Missing Matter Problem by $f(R)$ Gravity

Let us begin by considering the weak field limit solution (14.28) coming from R^k gravity, namely

$$e^{\nu(r)} = 1 - \frac{r_S}{2r}\left[1 + \left(\frac{r}{r_c}\right)^{\beta}\right], \qquad r_S = \frac{2GM}{c^2}, \tag{15.1}$$

with potential

$$\Phi(r) = -\frac{GM}{2r}\left[1 + \left(\frac{r}{r_c}\right)^{\beta}\right]. \tag{15.2}$$

As before, r_S is the Schwarzschild radius and r_c is a parameter with length dimensions, which labels the deviation from the Newtonian potential. Specifically, r_c can be recast in terms of the Milgrom acceleration constant a_0 [54–56] as $r_c = \sqrt{\frac{GM}{a_0}}$. The real constant β is defined in Eq. (14.30).

As pointed out in [252, 371], Eq. (15.1) can be used to address the missing matter problem. The parameter r_c results as a further gravitational radius, induced by the existence of Noether symmetry, which allows one to reproduce the flat rotation curve of spiral galaxies and the Fundamental Plane of elliptical galaxies. To this purpose, let us search for values of β that best fit the galaxy rotation curves. According to the results provided in Section 14.1.1, imposing the asymptotic flatness for large distances in Eq. (15.1), we obtain $0 < \beta < 1$. If the gravitational field is conservative (and irrotational), then the equation $a = -\nabla\Phi$, where a is the acceleration, must hold for potentials derived in the Newtonian limit. This means that the circular velocity v_c of a point mass moving in the potential (15.2) is given by

$$v_c^2 = r\frac{d\Phi}{dr} = \frac{GM}{2r}\left[1 + (1-\beta)\left(\frac{r}{r_c}\right)^{\beta}\right]. \tag{15.3}$$

In the limit $\beta \to 0$, deviations from GR do not occur and the standard velocity $v_c^2 = \frac{GM}{r}$ is recovered. The relation between mass and asymptotic rotation velocity can be used to study the Tully–Fisher diagram for gas-rich galaxies [371]. Higher-order corrections, arising from R^k gravity, can provide galaxy rotation curves that fit observation better than the Newtonian case. Since we are only interested in the gravitational field produced by a point mass source located at the origin, the energy-momentum tensor is given by [372]

$$\mathcal{T}_{\mu\nu} = diag(\rho c^2, 0, 0, 0), \quad \rho(r) = a\delta(r), \tag{15.4}$$

with $\delta(r)$ being the three-dimensional Dirac delta function. A point mass distribution generates a stationary spherically symmetric spacetime and so, the trace contains all the relevant information relating the field equations. Considering the PN limit

$$\frac{d}{d\chi} \sim \frac{1}{\chi}, \quad \rho \sim \frac{M}{r^3}, \tag{15.5}$$

the trace equation of Eqs. (4.45), in the case of R^k gravity, turns out to be

$$\chi^k(k-2) - 3kL_M^2\frac{\chi^{(k-1)}}{r^2} \sim \frac{8\pi GML_M^2}{c^2r^3}, \tag{15.6}$$

where we defined $\chi \equiv L_M^2 R$. Here L_M defines a length scale fixed by the parameters of the theory. We are using the dimensionless variable χ to avoid the redefinition of coupling for any k.

Note that the second term in the LHS of Eq. (15.6) dominates the first one when the spatial curvature radius is larger than the length r. This occurs in the weak-field regime, where MOND effects are expected. For a metric description of gravity, this limit must correspond to the relativistic regime of MOND. Under this assumption, the trace equation takes the form

$$R^{(k-1)} \approx -\frac{8\pi GM}{3kc^2rL_M^{2(k-1)}}. \tag{15.7}$$

The acceleration can be obtained by comparing Eq. (15.7) with the Ricci curvature scalar at the second order of approximation in the nonrelativistic regime, namely

$$R = -\frac{2}{c^2}\nabla^2\phi \sim \frac{2\phi}{c^2 r^2} \sim \frac{2a}{c^2 r}. \tag{15.8}$$

Therefore, Eqs. (15.7) and (15.8) provide

$$a \sim -\frac{c^2 r}{2L_M^2}\left(\frac{8\pi GM}{3kc^2 r}\right)^{1/(k-1)}, \tag{15.9}$$

which converges to the MOND acceleration for $k = \frac{3}{2}$. Therefore, this shows that R^k gravity can reproduce the MOND theory, recasting dark matter as a geometric effect given by $f(R)$ gravity. In order to exactly recover the MOND regime, we have to notice that, in the nonrelativistic limit, the speed of light should not appear in Eq. (15.9). Thus L_M must depend on a power of c, that is,

$$L_M \propto c^{\frac{k-2}{k-1}}. \tag{15.10}$$

MOND is recovered when $k = 3/2$, namely $L_M \propto c^{-1}$. Moreover, the length L_M can be recovered by parameters describing the theory of gravity, and so, let us assume that

$$L_M = L_0 r_S^{\alpha} r_c^{\zeta}, \quad \text{with} \quad \alpha + \zeta = 1, \tag{15.11}$$

where L_0 is constant and ζ is a dimensionless number of order one. The comparison between Eqs. (15.11) and (15.10) provides

$$L_M \propto \sqrt{r_S r_c}, \qquad \alpha = \beta = 1/2. \tag{15.12}$$

Setting $L_0 = \dfrac{2\sqrt{2}}{9}$, $k = 3/2$ and substituting Eq. (15.12) into Eq. (15.9), the acceleration turns out to be

$$a \sim \frac{\sqrt{a_0 GM}}{r}, \tag{15.13}$$

which is the standard form of MOND in spherical symmetry. The formulation of the theory described so far implies that, as soon as we relax the hypothesis by which the gravitational action is strictly described by standard GR, new characteristic lengths have to be considered.

Assuming that rotation curve is flat within the measurement uncertainties at some finite radius r_f, namely $v_c(r_f) \sim v_f$, Eq. (15.13) provides as a limit

$$r_f = \frac{\sqrt{a_0 GM}}{a_f}. \tag{15.14}$$

Finally, using Eqs. (15.14), (14.10) and the definition of r_c, the baryonic Tullly–Fisher relation of R^k gravity, expressed in terms of MOND accelerations, is

$$M = \frac{4a_0 v_f^4}{G a_f^2\left[1 + (1-\beta)\left(\frac{a_0}{a_f}\right)^{\beta}\right]^2}. \tag{15.15}$$

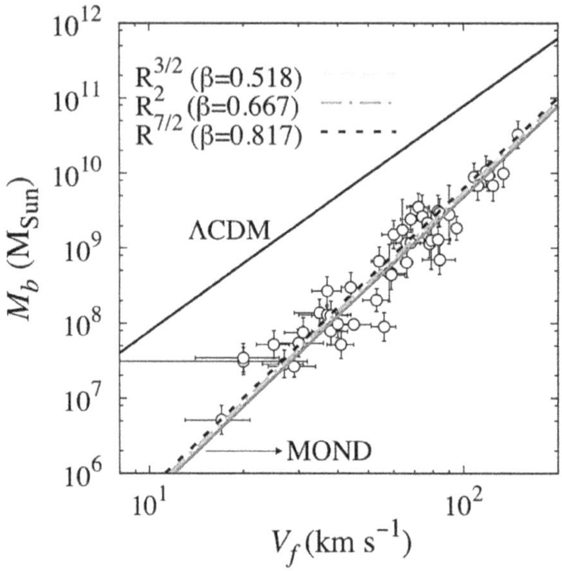

Figure 15.1 Comparison between best-fit baryonic Tully–Fisher relations of gas-rich galaxies (for the sample of galaxies used in [415]), in MOND, R^k gravity for values of k equal to 1.5, 2, and 3.5 (the corresponding values of β are 0.518, 0.667, and 0.817, respectively) and ΛCDM.

According to observations, we set $a_0 = 1.21 \pm 0.24 \cdot 10^{-10}\mathrm{m/s}^2$ [414], so that Eq. (15.15) can be plotted for different values of β (k) [371,415]:
with M_b being the baryonic mass given by the sum of stars and gas. For the preceding values of β, the fits result with the following MOND accelerations:

$$\begin{cases} \beta = 0.518 \to a_f = (2.02 \pm 0.05) \times 10^{-13}\mathrm{km/s}^2 \\ \beta = 0.667 \to a_f = (2.30 \pm 0.06) \times 10^{-13}\mathrm{km/s}^2 \\ \beta = 0.817 \to a_f = (2.56 \pm 0.06) \times 10^{-13}\mathrm{km/s}^2, \end{cases} \tag{15.16}$$

which are within the error interval of the corresponding MOND estimate. Notice that the values of β that best fit observations are in the validity range $0 < \beta < 1$. From Figs. **15.1**–**15.2**, it can be inferred that almost a perfect alignment is obtained between MOND and R^k gravity predictions. On the contrary, ΛCDM fit is in unsatisfactory agreement with observations. Therefore $f(R) = R^k$ gravity can give a theoretical foundation both for rotation curves of galaxies and for the empirical baryonic Tully–Fisher relation. We have to stress that the obtained values for β from the baryonic Tully–Fisher diagram differ from those provided by Solar System scales. A possible explanation for this result, as pointed out in Section 4.3, is that the gravitational action is made of different contributions, whose dominance depends on the considered scale. Mathematically speaking, this can be pursued by multiplying geometric terms by weakly scale-depending coupling constants; thus gravity can be described by a well-defined, yet unknown,

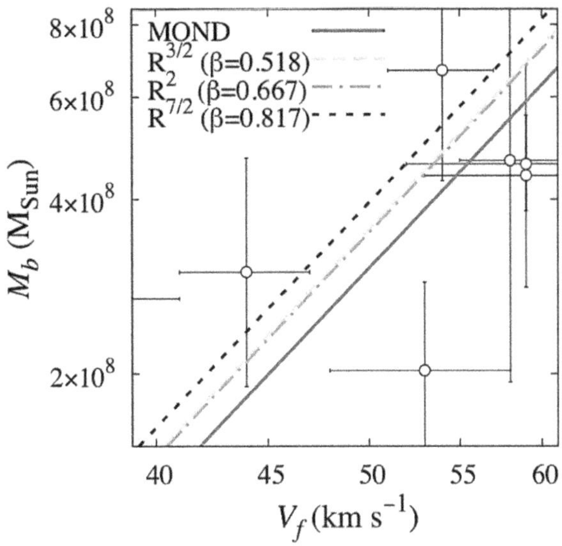

Figure 15.2 A zoomed part of Fig. **15.1** for a small range of parameters

gravitational action. From this point of view, it seems reasonable that constraints working at the Solar System level are not viable at larger scales.

Results obtained in this section for a point-like source can be easily generalized to other gravitational configurations by introducing a given matter density $\rho(r)$ and thus considering the mass as a function of the radial distance $M(r)$. In this way, the Newtonian potential can be written as

$$\Phi_N(r) = \frac{GM(r)}{r} = -\frac{G}{r}\int_0^\infty 4\pi r^2 \rho(r)\, dr. \tag{15.17}$$

With the aim of fitting the rotation curve for a matter distribution $M(r)$, let us recast the total potential Φ' as a sum between the Newtonian potential Φ_N and a correction potential that vanishes when standard Einstein gravity is recovered:

$$\Phi'(r) = \frac{1}{2}\left(\Phi_N + \Phi_c'\right), \tag{15.18}$$

where

$$\Phi_c'(r) = -G\int_0^\infty r'^2 \rho(r')dr' \int_{-\frac{\pi}{2}}^{\frac{\pi}{2}} \sin\theta'\, d\theta' \int_0^{2\pi} \psi_c d\phi', \tag{15.19}$$

with ψ_c being the non-Newtonian part of the modified gravitational potential for the point-like case. In order to generalize the preceding results, we consider a generic modified potential of the following type:

$$\psi \equiv \psi_c + \psi_N = -\frac{GM}{2r}\left[1 + \alpha\left(\frac{r}{r_c}\right)^\beta\right], \tag{15.20}$$

so that

$$\psi_c = -\frac{\alpha GM}{r_c}\left(\frac{r}{r_c}\right)^{\beta-1}, \tag{15.21}$$

with α and β two parameters depending on the particular theory of gravity one is considering. While for R^k gravity $\alpha = 1$, α could also be negative for other theories. However, due to the spherical symmetry, the integral can be evaluated by considering the relation $|\mathbf{r} - \mathbf{r}'| = \left(r^2 + r'^2 - 2rr'\cos\theta'\right)^{1/2}$. Thus, after integrating Eq. (15.19) over the angular variables, we get [362]

$$\Phi_c'(r) = -\frac{\pi G\alpha r_c^2}{3}\left[\mathcal{J}_1(r) + \mathcal{J}_2(r)\right], \tag{15.22}$$

with the integrals

$$\begin{aligned}
\mathcal{J}_1 &= 3\pi\int_0^\infty \left(\xi^2 + \xi'^2\right)^{(\beta-1)/2} \rho(\xi')\,\xi^2\,d\xi' \\
&\times\ {}_2F_1\left[\left\{\frac{1-\beta}{4}, \frac{3-\beta}{4}\right\}, \{2\}, \frac{4\xi^2\xi'^2}{\left(\xi^2+\xi'^2\right)^2}\right], \\
\mathcal{J}_2 &= 4(1-\beta)\xi\int_0^\infty \left(\xi^2 + \xi'^2\right)^{(\beta-3)/2} \rho(\xi')\,\xi'^2\,d\xi' \\
&\times\ {}_3F_2\left[\left\{1, \frac{3-\beta}{4}, \frac{5-\beta}{4}\right\}, \left\{\frac{3}{2}, \frac{5}{2}\right\}, \frac{4\xi^2\xi'^2}{\left(\xi^2+\xi'^2\right)^2}\right],
\end{aligned} \tag{15.23}$$

where we have generically defined $\xi = \dfrac{r}{r_c}$ and used the notation ${}_nF[\{a_1, a_2, \dots, a_n\}, \{b_1, b_2, \dots, b_n\}, x]$ for the hypergeometric functions. Therefore, the total circular velocity turns out to be the sum of the Newtonian velocity $v_{circ,N}^2 = GM(r)/r$ and the one associated to the potential $\Phi_c'(r)$:

$$v_{circ}^2 = r\frac{d\Phi'(r)}{dr} = \frac{1}{2}v_{circ,N}^2 + \frac{1}{2}r\frac{d\Phi_c'(r)}{dr}. \tag{15.24}$$

Since Φ_c has to be evaluated numerically, it is better to differentiate analytically the expressions for $\mathcal{J}_1$ and $\mathcal{J}_2$ and then integrate numerically. It is easy to check that the resulting rotation curve is slowly decreasing so that it vanishes asymptotically as in the Newtonian case. However, the rate of decline is slower than the Keplerian one so that the total rotation curve turns out to be higher than the Newtonian one: this fact allows one to fit galaxy rotation curves without the need of any dark matter halo.

A similar result can also be obtained for galaxies with a thin disk by using the cylindrical symmetry instead of spherical symmetry [362]. In such a case, the potential is a function of the coordinates r, z, ϕ and provides a Newtonian circular velocity of the form:

$$v^2_{N,\,circ} = 2\pi G\Sigma_0 R_d (\eta/2)^2 \left[I_0(\eta/2)K_0(\eta/2) - I_1(\eta/2)K_1(\eta/2)\right], \tag{15.25}$$

with the definitions

- $R_d \rightarrow$ Scale Radius
- $\Sigma(r) \rightarrow$ Surface Mass Density
- $\Sigma_0 = \Sigma(r=0)$
- $\eta = \dfrac{r}{R_d}$
- $I_i \rightarrow$ Bessel function of the first type, of the order i
- $K_i \rightarrow$ Bessel function of the second type, of the order i

With the disk infinitesimally thin, the potential is defined by means of the surfaces mass density Σ as

$$\Phi'(r,z) = \int_0^\infty \Sigma(r')\, r'\, dr' \int_0^{2\pi} \psi\left(|\mathrm{r}-\mathrm{r}'|\right) d\phi'. \tag{15.26}$$

Hereafter, we assume for Σ an exponential profile of the form $\Sigma(r) = \Sigma_0 e^{-\frac{r}{R_d}}$, as proposed in Ref. [416]. Also in this case, the total circular velocity can be split into two contributions, namely the Newtonian one (15.25) and the correction. The latter, for a thin disk, takes the form

$$v^2_{circ,\,corr} = -2^{\beta-5}\eta_c^{-\beta}\pi\alpha(\beta-1)G\Sigma_0 R_d \eta^{\frac{\beta-1}{2}} \mathcal{I}_{disk}(\eta,\beta), \tag{15.27}$$

with $\eta_c = \dfrac{r_c}{R_d}$ and

$$\begin{aligned}
&\mathcal{I}_{disk}(\eta,\beta) \equiv \int_0^\infty \mathcal{F}(\eta,\eta',\beta)\, \ell^{3-\beta} \eta'^{\frac{\beta-1}{2}} \mathrm{e}^{-\eta'} d\eta', \\
&\ell \equiv \frac{4rr'}{(r'+r)^2 + z^2}, \\
&\mathcal{F} \equiv 2(\eta+\eta')\, {}_2F_1\left[\left\{\frac{1}{2}, \frac{1-\beta}{2}\right\}, \{1\}, k^2\right] \\
&+ \left[(k^2-2)\,\eta' + k^2\eta\right] {}_2F_1\left[\left\{\frac{3}{2}, \frac{3-\beta}{2}\right\}, \{2\}, k^2\right].
\end{aligned} \tag{15.28}$$

The function $\mathcal{I}_{disk}(\eta,\beta)$ may not be evaluated analytically, but it is straightforward to estimate it numerically. In order to fit observations without any dark matter halo, our low surface brightness galaxy model is made out of stellar and gaseous components only. We assume stars to be distributed in an infinitely thin and circularly symmetric disk. The surface density $\Sigma(r)$ may be derived from the surface brightness distribution: we consider the galaxy made of gaseous and stellar components, whose total mass depends on the mass density distribution as

$$\mu(r) = -2.5 \log\left(\frac{\Sigma(r)}{\Upsilon_\star}\right), \tag{15.29}$$

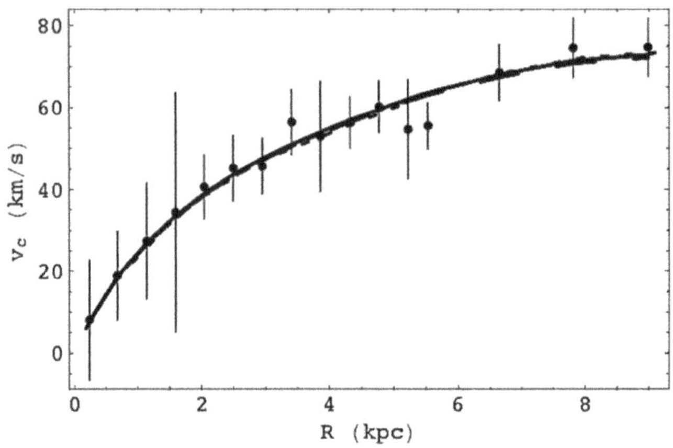

Figure 15.3 Some illustrative examples of simulated rotation curves (smoothing the data for convenience) with the overplotted the input theoretical rotation curve (solid line) and the best-fit one from raw (short dashed line) and smooth (long dashed line) data

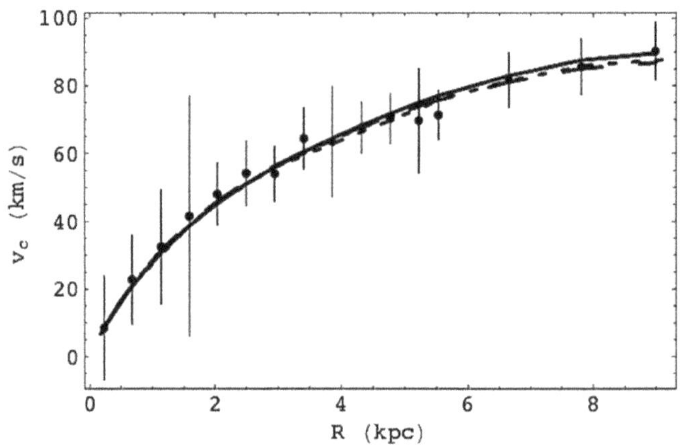

Figure 15.4 Some illustrative examples of simulated rotation curves (smoothing the data for convenience) with the overplotted the input theoretical rotation curve (solid line) and the best-fit one from raw (short dashed line) and smooth (long dashed line) data

with $\Upsilon_\star$ the mass-light ratio. The remaining free parameters can be fixed by minimizing the χ^2 of the theoretical circular velocity with respect to the observed one. However, the procedure provides a degeneracy on the parameters β and r_c, which can be partially alleviated following the steps in Ref. [362].

Figures **15.3**–**15.4** point out that free parameters of R^k gravity can potentially fit experimental observations. However, no definitive conclusion can be drawn from observational data. For instance, in the 2σ confidence range for $\log r_c$, it turns out that β can assume three different values, namely $\beta = 0.74$, 0.817, 0.858. We choose $\beta = 0.74$ to compare the theoretically predicted galaxy rotation curve with data, for different galaxies. In Fig. **15.5**, we show the results provided by a set of twelve galaxies (see [363] for a detailed discussion).

As we can see, modified R^k theory of gravity well fits the galaxy rotation curve for most of galaxies reported. Calculations have been performed both in spherical and in cylindrical symmetry for thin disks. The missing matter problem can be thus overcome without any exotic fluid but only as a gravitational correction at galactic scales. It is worth noticing again that the gravity model has been selected by the Noether symmetries.

15.2 The Fundamental Plane by $f(R)$ Gravity

Another important application of $f(R)$ gravity can come from self-gravitating collisionless spheroidal systems. Let us start from the empirical fact that some global properties of normal elliptical galaxies are correlated. It is well known that there are three main global observables: the central projected velocity dispersion σ_0, the effective radius r_e (containing half of the total luminosity of the galaxy), and the mean effective surface brightness (within r_e) I_e [417,418]. Any of the three parameters may be estimated from the other two, and, varying them, a plane is described within a more general three-dimensional configuration space. This correlated plane is referred to as the Fundamental Plane of galaxies.

Specifically, there are two main properties that can be addressed to the Fundamental Plane: the tilt and the tightness. The former means that the coefficients of the Fundamental Plane equation differ from those predicted by the virial theorem. If written in logarithmic form, the two (fundamental and virial) planes appear to be tilted by an angle of $\sim 15^\circ$. By analyzing the properties of a sample of elliptical galaxies [419], it is possible to show that more than half of the "tilt" of the Fundamental Plane of elliptical galaxies is accounted for by the non-homology in the dynamical structures of the systems [252]. It has also been shown that about 30 percent of the tilt can be explained by stellar population effects and by spatial non-homology. There are also studies pointing out that the tilt is mainly driven by a mass-dependent dark matter fraction, such that more massive galaxies have larger dark matter fractions within effective (half-light) radius. Some authors find that a "hybrid" interpretation of the Fundamental Plane should be preferred, where dynamical non-homology and dark matter effects seem to play a significant role in producing the tilt [420]. The tightness is due to several physical effects working in clusters of galaxies which "tighten" the objects in a given region of the configuration space.

Here we use $f(R) = R^k$ gravity in order to show that the Fundamental Plane can be fully recovered without introducing dark matter but just correcting GR.

The Fundamental Plane can be written as [252]

$$\log(r_e) = a\log(v) + b\log(I_e) + c, \tag{15.30}$$

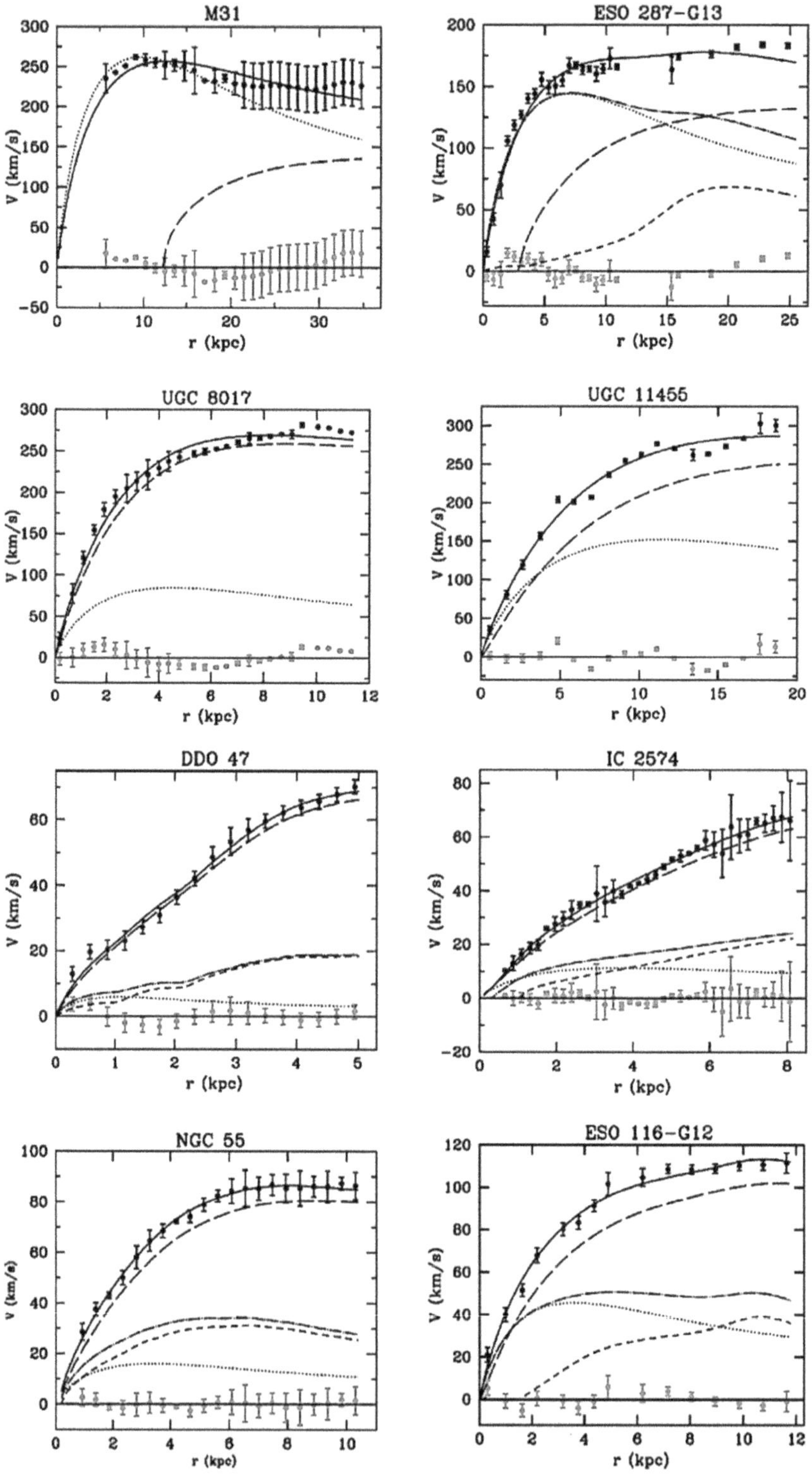
M31
ESO 287-G13
UGC 8017
UGC 11455
DDO 47
IC 2574
NGC 55
ESO 116-G12
V (km/s)
r (kpc)

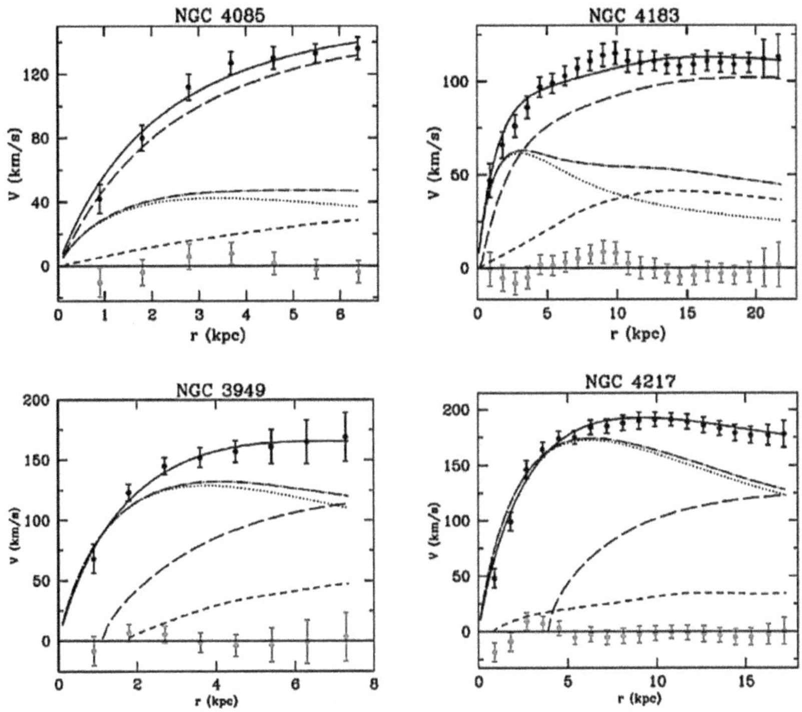

Figure 15.5 Comparison of galaxy rotation curves for twelve different galaxies: the solid line represents the best-fit total circular velocity; the dashed and dotted lines are the Newtonian contributions from the gas and the stars, while the dot-dashed line represents their sum. The long-dashed line is the non-Newtonian contribution of the gas and the stars to the model. Below the rotation curves, the residuals $(v_{obs} - v_{tot})$ are plotted.

with a, b, c constants of integration depending on the properties of the given system; r_e and I_e have been defined previously. The velocity v assumes the meaning of the characteristic velocity of the self-gravitating system considered, which, for elliptical galaxies turns out to be the central velocity dispersion $v = \sigma_0$. In elliptical galaxies, the rotation velocity is negligible with respect to σ_0, so that v represents the only contribution to velocity we will take into account here. This means that kinematics is not dominated by well-defined circular velocities of orbiting stars as in the case of spiral galaxies, but we need a velocity distribution as in thermodynamics. More in detail, most elliptical galaxies exhibit little or no rotation. Their stars have random velocities along the line of sight, whose root mean square dispersion σ_0 can be measured from the Doppler broadening of spectral lines.

To find a consistent relation among a, b, c in Eq. (15.30), we use power-law $f(R)$ gravity, selecting the parameters β (k) best fitting the Fundamental Plane. Let us notice that the circular velocity is equal to the effective velocity when evaluated at the effective radius: $v_c(r_e) = \sigma_0$. This means that it can be calculated by Eq. (15.3), while the radius r_e can be assumed to be proportional (or equal) to the theoretical gravitational radius r_c: $\xi(r = r_e) = r_e/r_c =$ const. In this way, $\mathcal{J}_1$ and $\mathcal{J}_2$ in Eq. (15.23) are also constant, and the only contribution to the circular velocity (evaluated at $r = r_e$) is the Newtonian one:

$$\sigma_0 = \frac{v_{c,N}(r_e)}{\sqrt{2}}, \qquad v_{c,corr} = 0. \tag{15.31}$$

Moreover, the surface brightness I_e can be linked to the effective radius r_e through the empirical relation $r_e \sim I_e v$ [417,418]. Considering the Virgo Cluster elliptical galaxies as a sample, the parameters a and b can be set to 1.4 and -0.85, respectively [417]. Therefore, the empirical relation between the Fundamental Plane (15.30) and the radius r_e is illustrated in Fig. **15.6**[1].

The empirical relation provided by the plot can be compared with the theoretical Fundamental Plane coming from $f(R) = R^k$ gravity. This can be pursued by writing the parameters a, b, and c in terms of β and r_c. To this purpose the three-dimensional fit of the Fundamental Plane equation has to be considered, and thus the corresponding χ^2 must be minimized for each fit (see [252]). As a final result, the relation among r_c/r_e, β, and the three constants a, b, c can be inferred from Figs. **15.7**, **15.8**, and **15.9**.

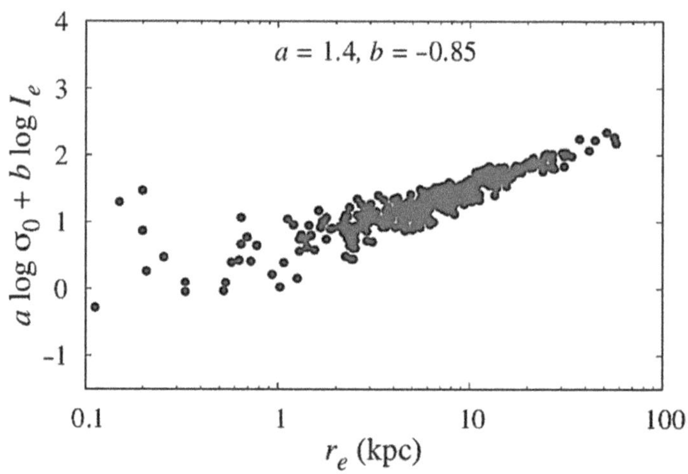

Figure 15.6 Fundamental plane equation for observed values of effective radius r_e, velocity dispersion σ_0, and mean surface brightness within the effective radius I_e, and for values of a and b taken from [417]: $a = 1.4$ and $b = -0.85$

[1] Here the parameter c is assumed $c = 0$.

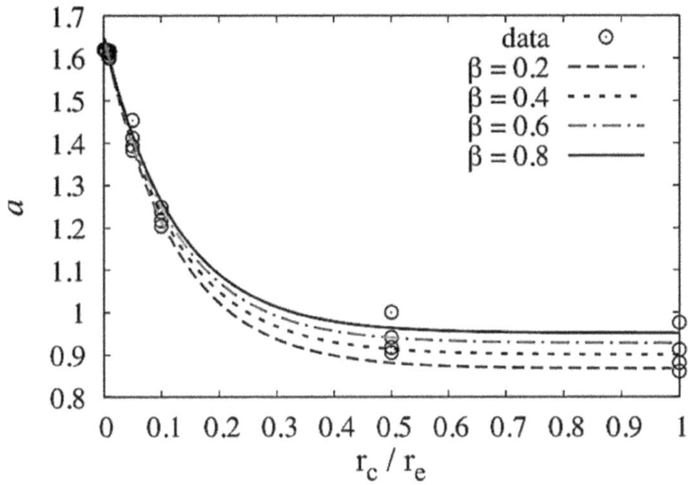

Figure 15.7 Relations between a and r_c/r_e for fixed values of β

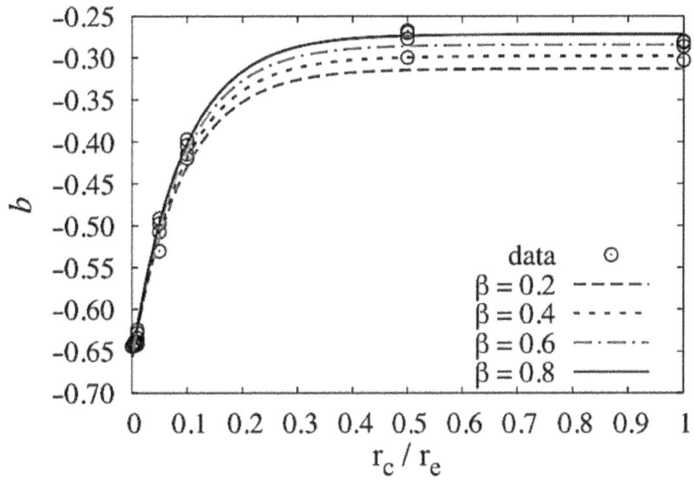

Figure 15.8 Relations between b and r_c/r_e for fixed values of β

The Fundamental Plane relation with respect to the parameters of power-law $f(R)$ gravity is shown in Fig. **15.10**.

As a concluding remark, R^k gravity can fit observational data of the Fundamental Plane. In agreement with the previous section, the parameter β is set to $\beta = 0.8$ and the circular velocity is derived through the weak field potential (15.2). Thus, after fixing the ratio r_c/r_e, the parameters a and b which better fit observations automatically follow. This is a further confirmation that modified gravity behaves like dark matter. Furthermore, the phenomenological r_e parameter can be related to the gravitational radius r_c, which, as discussed in Chapter 14, comes from Noether symmetries.

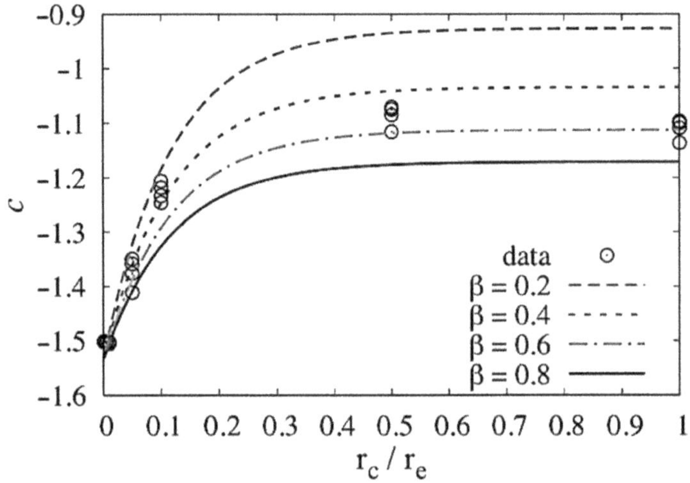

Figure 15.9 Relations between c and r_c/r_e for fixed values of β

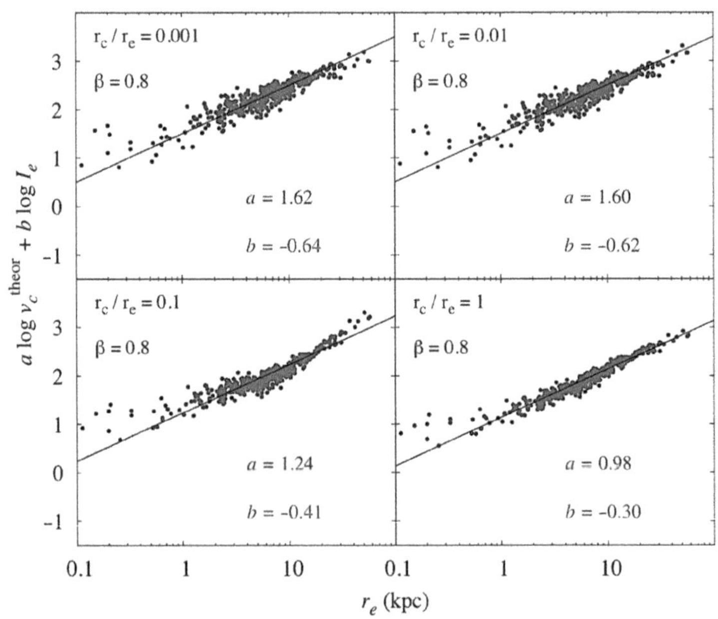

Figure 15.10 Relations between the Fundamental Plane and the radius r_e for the fixed value $\beta = 0.8$

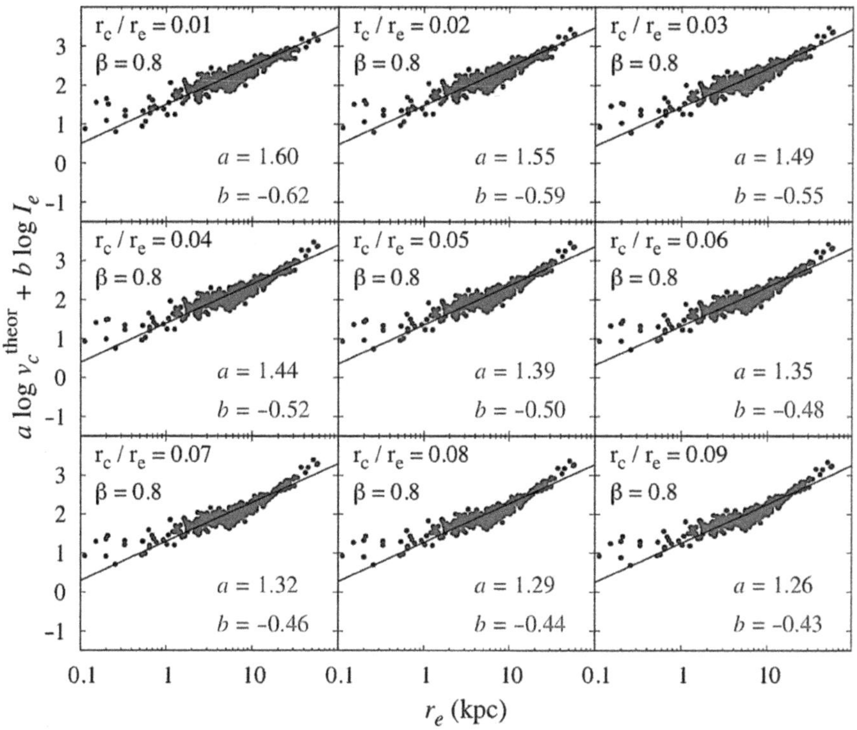

Figure 15.10 (Cont.)

16

Applications to Cosmology

Malum est consilium, quod mutari non potest.

— *Publilius Syrus*
(as quoted in *Sententiae*
by Wilhelm Meyer)

In this chapter, we shall consider some cosmological applications of models selected by the Noether Symmetry Approach. The topic is really vast because most of these models have been considered, with different motivations, elsewhere in literature. However, our aim is only to point out cosmological features that can be directly related to the presence of symmetries. As discussed in the previous two chapters of this part of the book, symmetries have a main role in determining physically viable models, so the search for symmetries can be actually considered a "paradigm" for formulating self-consistent models.

We shall discuss some applications in the early and late universe where the form of gravitational action and related dynamical quantities are determined by symmetries.

After a brief summary of inflation driven by a scalar field (the inflaton), we focus on a scalar-tensor action, where the scalar field is non-minimally coupled to the geometry through the coupling function $f(\phi)$ and the potential $V(\phi)$. The potential and coupling are determined by symmetries. This fact allows to solve dynamics exactly, as done in Chapter 10. Following Ref. [421], we study the cosmological perturbations showing that they strictly depend on the conserved quantities related to the symmetries.

As discussed in Chapter 4, thanks to the Lovelock theorem, several modified/extended theorics can be recast scalar-tensor gravity. From this statement, several other applications follow.

Then, we show how slow-roll inflation can be addressed in the context of models implying Gauss–Bonnet corrections [422]. In this case, we consider energy conditions, with the purpose to show that the late-time acceleration of the universe also can be obtained by modifying the EH action. Clearly other energy regimes have to be taken into account. Specifically, we show that energy conditions violation occur, so that geometric contributions in the effective action can mimic the effects of cosmological constant. In Section 16.4, cosmological quintessence is interpreted as a consequence of geometry extension as first reported in [116].

Clearly applications are countless. As said previously, we want to give here only some remarkable examples.

16.1 Inflation and Cosmological Perturbations in Scalar-Tensor Gravity

Let us recall some main results regarding inflationary models and the reasons why cosmological inflation was first introduced. When the Big Bang model was largely affirmed as the best candidate to describe the initial phase of the universe, it was soon evident that this prescription was not in agreement with standard GR. The spatial flatness of the universe observed nowadays cannot be explained by Einstein's gravity, as well as the reason why photons of the CMB coming from opposite directions have the same temperature, while the size of causally connected regions at the last scattering is at most one degree. When the EH action is considered, these problems can be solved only by imposing ad hoc initial conditions. The inflationary model was introduced by A. Guth and A. Linde [131,132] to address the evidences provided by the cosmological data at the early time. Inflation is usually thought of as generated by a scalar field ϕ, the previously mentioned inflaton, which is supposed to be responsible for the accelerating expansion of the early universe. For this reason, a minimal inflationary model containing a scalar field ϕ can be written as

$$S = \frac{1}{2}\int \sqrt{-g}\left[R + \frac{1}{2}\partial_\mu\phi\partial^\mu\phi - V(\phi)\right] d^4x, \tag{16.1}$$

where $V(\phi)$ is a self-interaction potential. In this way, the field equations together with the Klein–Gordon equation can be written in a cosmological background as

$$\begin{cases} \dfrac{\dot{a}^2}{a^2} = \dfrac{1}{6}\left[\dfrac{\dot{\phi}^2}{2} + V(\phi)\right], \\ \dfrac{\ddot{a}}{a} = -\dfrac{1}{6}\left[\dot{\phi}^2 - V(\phi)\right], \\ \ddot{\phi} + 3\dfrac{\dot{a}}{a}\dot{\phi} + \partial_\phi V(\phi) = 0 . \end{cases} \tag{16.2}$$

Comparing the preceding equations with the two Friedman equations of GR in Eqs. (4.23) and (4.24), it follows that it is possible to define a scalar field effective energy density ρ_ϕ and pressure p_ϕ of the forms

$$\begin{cases} \rho_\phi = \dfrac{1}{2}\left[\dfrac{1}{2}\dot{\phi}^2 + V(\phi)\right], \\ p_\phi = \dfrac{1}{2}\left[\dfrac{1}{2}\dot{\phi}^2 - V(\phi)\right] . \end{cases} \tag{16.3}$$

Nowadays inflation can be realized in several ways [422–426]. Some of them are the so-called *new inflation* or the *slow-roll inflation*, according to which inflation is driven by a scalar field rolling down a potential energy hill. Inflation occurs as soon as the rolling scalar field motion is slow with respect to the universe expansion. Therefore, in the slow-roll approximation, inflation starts when the kinetic energy is negligible with respect to the potential one. In this way, replacing the condition $\dot{\phi}^2 \ll V(\phi)$ into the system (16.3), we obtain the EoS $\rho = -p$, in agreement with an accelerating expansion of the universe (see Section 4.1 for details).

During the inflation period, the magnitude of the slow-roll parameters ε and η must be small, namely

$$|\varepsilon| \equiv \left|-\frac{\dot{H}}{H^2}\right| \ll 1, \quad |\eta| \equiv \left|-\frac{\ddot{H}}{2H\dot{H}}\right| \ll 1, \tag{16.4}$$

where $H = \dot{a}/a$. The first condition of Eq. (16.4) comes from the requirement of an accelerating expansion of the universe through the relation

$$\frac{\ddot{a}}{a} = H^2\left(1 + \frac{\dot{H}}{H^2}\right) = H^2(1-\varepsilon) > 0. \tag{16.5}$$

Regarding the second condition, inflation occurs when the potential energy is predominant with respect to the other terms. Then, the potential should have a minimum at the end of the inflation. In the slow-roll approximation, where the scalar field is expected to roll slowly, the field equations, together with the Klein–Gordon equation, yield the condition

$$-\frac{\ddot{\phi}}{H\dot{\phi}} = -\frac{\ddot{H}}{2H\dot{H}} \ll 1. \tag{16.6}$$

Inflation ends when the potential starts decreasing toward its minimum or, in other words, when the values of the slow-roll parameters start increasing and the velocity of the scalar field cannot be neglected if compared to the universe expansion. Therefore, the scalar field approaches the minimum of the potential and starts oscillating around the equilibrium position. Such an oscillation yields particle creation, which dissipates the energy of the scalar field, raising the temperature of the early universe. This thermalization is called *reheating*, a phenomenon that has dropped during the inflationary expansion. The superluminal expansion of the early universe is able to overcome both the flatness problem and the temperature of photons coming from CMB. All the cosmic structures began to grow after the end of the inflationary epochs. Mathematically speaking, the end of inflation occurs when the slow-roll parameters become of order unity.

Similarly, in modified theories of gravity, the extra geometric terms can play the role of scalar field, naturally providing inflation, as pointed out in Section

16.2. In particular, the Starobinsky inflation [122] is the first model where geometric corrections to the EH action act as an effective scalar field, the scalaron, as discussed in Chapter 4.

An important application, related to inflation, is the study of cosmological perturbations. Here, we want to show that those coming from scalar-tensor gravity can be modeled out from the background solutions derived from Noether's symmetries. Specifically, we consider the second solution of (10.19), whose coupling and potential containing symmetries are $f(\phi) = \ell(s)\phi^2$ and $V(\phi) = V_0\phi^{\frac{6(s+1)}{2s+3}}$, respectively. The action therefore reads

$$S = \frac{1}{2}\int\sqrt{-g}\left[\ell(s)\phi^2 R + \frac{1}{2}\partial_\mu\phi\partial^\mu\phi - V_0\phi^{\frac{6(s+1)}{2s+3}}\right]. \tag{16.7}$$

The corresponding field equations in a spatially flat FLRW background take the form

$$\begin{cases} \rho = \dfrac{1}{2\ell(s)\phi^2}\left[\dfrac{1}{2}\dot{\phi}^2 + V(\phi) - 12\ell(s)\dfrac{\dot{a}}{a}\phi\dot{\phi}\right], \\ p = \dfrac{1}{2\ell(s)\phi^2}\left[\dfrac{1}{2}\dot{\phi}^2 - V(\phi) + 4\ell(s)\left(\phi\ddot{\phi} + 2\dfrac{\dot{a}}{a}\phi\dot{\phi} + \dot{\phi}^2\right)\right], \end{cases} \tag{16.8}$$

where p and ρ can be intended as effective pressure and energy density provided by the scalar field, which, under suitable conformal transformations, can be recast as geometric fields. It is worth stressing again that, in modified theories of gravity, an effective energy-momentum tensor can be derived from the further degrees of freedom of the gravitational field. It contributes as a further source in the field equations. Specifically, any modified theory of gravity field equations can be recast as

$$FG_{\mu\nu} = \mathcal{T}^{(Curv)}_{\mu\nu} + \mathcal{T}^{Matter}_{\mu\nu}, \tag{16.9}$$

where the first term in the RHS can be intended as an effective energy momentum tensor, independent of the matter chosen. In the preceding equations, F is a generic coupling function depending on the theory [427].

In order to study the cosmological perturbation evolution, the variations $\delta\rho$ and δp must be explicitly computed. To this purpose, it is useful to introduce the gauge-invariant variables Φ_H and Φ_A, respectively defined as

$$\begin{cases} \dfrac{\delta\rho}{\rho} = \dfrac{2k^2}{a^2 p}\Phi_H \\ v = \dfrac{2k}{a(\rho+p)}\left[-\dot{\Phi}_H + \dfrac{\dot{a}}{a}\Phi_A\right], \end{cases} \tag{16.10}$$

where k is the comoving wave number of the perturbation. According to Refs. [428–430], we introduce a gauge-invariant frame-independent perturbation ΔF, thanks to which Φ_H and Φ_A can be reparameterized as

$$\tilde{\Phi}_H = \Phi_H + \frac{\Delta F}{2F}, \quad \tilde{\Phi}_A = \Phi_A + \frac{\Delta F}{2F}. \tag{16.11}$$

By means of these definitions, the variation of the scalar field $\Delta\phi$ can be written in terms of Φ_H as

$$\frac{\Delta F}{F} = \frac{2\Delta\phi}{\phi} = -\frac{8\ell(s)}{a[12\ell(s)+1]\phi}\left[a\phi\tilde{\Phi}_H\right], \tag{16.12}$$

so that the evolution equation reads

$$\ddot{\tilde{\Phi}}_H + \left[\frac{\dot{a}}{a} - 2\frac{\ddot{\phi}}{\dot{\phi}} + 4\frac{\dot{\phi}}{\phi}\right]\dot{\tilde{\Phi}}_H + \left\{\frac{k^2}{a^2} - 2\frac{\dot{a}}{a}\frac{\ddot{\phi}}{\dot{\phi}} \right.$$
$$\left. - \frac{1}{2\ell(s)\phi^2}\left[\dot{\phi}^2 + 4\ell(s)\phi\left(\ddot{\phi} - 3\frac{\dot{a}}{a}\dot{\phi}\right)\right] - 2\frac{\dot{\phi}}{\phi}\left[\frac{\dot{a}}{a} + \frac{\dot{\phi}}{\phi}\right]\right\}\tilde{\Phi}_H = 0. \tag{16.13}$$

To simplify the equation and find analytic solutions, we consider the change of variables

$$\begin{cases} u = \left(\dfrac{2\ell(s)}{\sqrt{12\ell(s)+1}}\right) a\dfrac{\phi^2}{\phi'}\tilde{\Phi}_H \\ z = \left(\dfrac{1}{\sqrt{12\ell(s)+1}}\right)\dfrac{(a\phi)'}{a\phi\phi'}, \end{cases} \tag{16.14}$$

where the prime denotes the derivative with respect to the conformal time η. In this way, Eq. (16.13) can be written as

$$u'' + \left[k^2 - \frac{z''}{z}\right]u = 0, \tag{16.15}$$

which admits as solution a linear combination of Bessel functions of order ν. Coming back to the old variable Φ_H, the solution yields

$$\Phi_H = \Phi_0(s)\eta^{m(s)}\left[A_0 J_\nu(k\eta) + B_0 Y_\nu(k\eta)\right], \tag{16.16}$$

with A_0, B_0 constant of integration and with

$$m(s) = 3\left[\frac{s^2+6s+8}{s^2+6s+6}\right], \quad p(s) = \frac{2s^2+9s+6}{s(s+3)}, \quad q(s) = -\frac{2s+3}{s},$$
$$\Phi_0(s) = \frac{12\ell(s)}{\sqrt{12\ell(s)+1}}\left(\frac{1-p(s)}{q(s)}\right)^{3m(s)+1/2} a_0^{\left(\frac{1+q(s)}{1-p(s)}\right)}\phi_0. \tag{16.17}$$

The expression of Φ_H, a function of the conformal time, is particularly interesting. It strictly derives from the Noether symmetry, and its asymptotic behavior determines the large-scale structure of the universe. In the large-scale limit $k^2 \ll z''/z$ [421], the differential equation (16.15) admits another solution of the form

$$\Phi_H = c_d(s)t^{l(s)} + c_g(s), \tag{16.18}$$

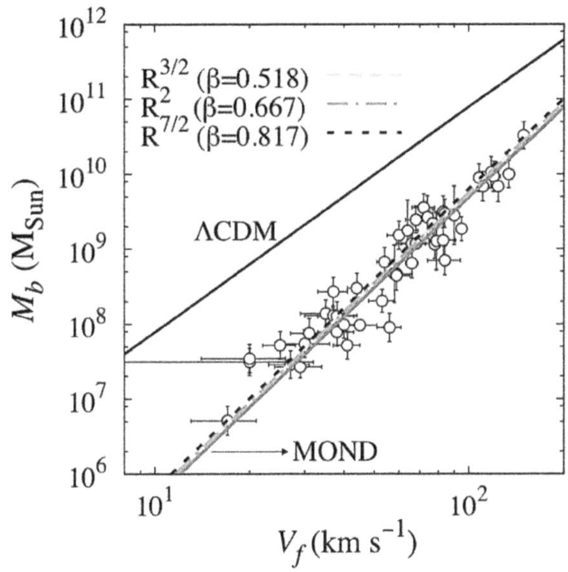

Figure 16.1 Qualitative trend (with arbitrary scale) of the perturbative potential Φ_H. The dotted line describes a pole-like evolution; the continuous line a Friedman evolution; the dashed line is a power-law evolution.

where

$$c_d(s) = \frac{24\,(s^2+3s+2)\,(2s^2+9s+6)}{s(s+3)(2s+3)^2}c_d,$$

$$c_g(s) = \frac{6\,(s^2+5s+6)}{s^2+6s+12}c_g, \quad l(s) = \frac{s^2+6s+12}{2(s+3)}. \tag{16.19}$$

The qualitative trend of Φ_H, as a function of η, is represented in Fig. **16.1**.

Notice that when the limits $s=-3$ and $s=0$ are considered, Eq. (16.18) diverges and the differential equation (16.15), in the large-scale limit, provides other solutions. In the former case, the potential reads

$$\Phi_H = \frac{2\,(e^t+1)\left[8c_d - 3c_g\int \frac{e^{2t/3}\,dt}{(e^t-1)^{1/3}}\right]}{9e^{2t/3}\,(e^t-1)^{2/3}} + 2c_g. \tag{16.20}$$

In the latter case ($s=0$), as shown in Eq. (10.19), the coupling function can be written in terms of a new free parameter f_0 as $f(\phi)=f_0\phi^2$. In the large-scale limit, the potential turns out to be

$$\Phi_H = [e^{2t}-2e^t-1]\left\{c_d - 2c_g k_0\int \frac{\exp\left[\frac{4}{3}\arctan e^t + \frac{1}{3}\ln(2\cosh t)\right]}{2\cosh t}dt\right\}$$
$$\times\left\{3k_0\exp\left[\frac{3t+4\arctan e^t+4\ln(2\cosh t)}{3}\right]\right\}^{-1} + 2c_g. \tag{16.21}$$

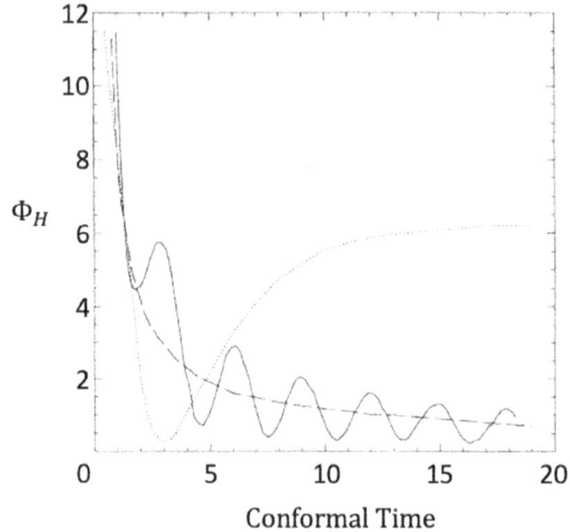

Figure 16.2 Dashed line denotes the evolution of the potential as a function of the conformal time when $s = -3$; dotted line denotes the same evolution when $s = 0$ in the case of non-minimal coupling; continuous line denotes the same evolution when $s = 0$ in the case of minimal coupling.

Finally, in the limit $s = 0$, also the minimal coupling can be considered, so that the potential (under the same assumption $k^2 \ll z''/z$) is

$$\Phi_H = \frac{2(\omega t - \cos\omega t \sin\omega t)\left[c_d - 2c_g \int \left(\omega^2 t^2 - \sin^2\omega t\right)^{1/3} dt\right]}{3\left(\omega^2 t^2 - \sin^2\omega t\right)^{4/3}} + 2c_g. \qquad (16.22)$$

The qualitative trends of these three potentials are compared in Fig. **16.2**.

The potential (16.22) is of particular interest, since it describes the transition from an initial power-law inflation to a matter-dominated era, modulated by a decaying oscillatory term. The free parameter s in the potential can be set, for example, by matching the observed galaxy correlations or the anisotropy of CMB. The evolution of perturbations suggests that the preceding cases can be considered good candidates to describe inflationary models for some value of s. A further analysis involving a quantum treatment is needed to confirm the validity of the theory and to definitely set the values of the free parameter.

16.2 Inflation in $f(R, \mathcal{G})$ Gravity

As a further application, we focus on a slow-roll inflationary model derived from an action containing the Ricci curvature scalar R and the Gauss–Bonnet invariant $\mathcal{G}$. In this specific case, inflation is driven by two scalar fields of geometric origin, as discussed in [422, 431]. The form of $f(R, \mathcal{G})$ gravity is a function selected by the Noether Symmetry Approach in Section 8.1, that is, $f(R, \mathcal{G}) = f_0 R^k \mathcal{G}^{1-k}$. In this case, dynamics is described by the scale factor

$$a(t) = a_0 t^{2k-1}, \qquad f(R,\mathcal{G}) = f_0 R^k \mathcal{G}^{1-k}, \tag{16.23}$$

with k being a real constant. By properly defining a new free parameter $s \equiv 2k-1$, Eq. (16.23) can be written as

$$a(t) = a_0 t^s, \qquad f(R,\mathcal{G}) = f_0 R^{\frac{1+s}{2}} \mathcal{G}^{\frac{1-s}{2}}. \tag{16.24}$$

This definition allows to suitably constrain the value of s admitting inflation in the early time. In order to impose the conditions (16.4) on the slow-roll parameters, we write the first two components of the field equations (8.11) in a cosmological spatially flat background. They read

$$\begin{cases} \dot{H} = -\dfrac{s(s-1)\left[(s+1)(6s-4)-6s(s+1)+8\right]}{4\left[s(s-5)+(s+1)(2s-1)+2\right]t^2} \\ \\ H^2 = -\dfrac{s^2(s-1)(s-1)}{\left[s(s-5)+2(s+1)(2s-1)+2\right]t^2}, \end{cases} \tag{16.25}$$

with H being the Hubble constant $H = \dfrac{\dot{a}}{a}$. Imposing the magnitude of the slow-roll parameters

$$\varepsilon = -\frac{\dot{H}}{H^2}, \qquad \eta = -\frac{\ddot{H}}{2H\dot{H}}, \tag{16.26}$$

to be small, namely

$$\left|\frac{\dot{H}}{H^2}\right| \ll 1, \qquad \left|\frac{\ddot{H}}{2H\dot{H}}\right| \ll 1, \tag{16.27}$$

inflation in the $f(R,\mathcal{G})$ model can be realized only if the following conditions hold:

$$\begin{aligned} &\frac{s(1-s)+2(s-1)-s(s-1)}{2s(s-1)} \ll 1, \\ &\frac{1}{\sqrt{2}}\sqrt{\frac{s^2(s-1)^2}{2s(s-5)+4s(s+1)+4}} \ll 1. \end{aligned} \tag{16.28}$$

The solution of the above inequalities numerically provides

$$s \gg 2.171. \tag{16.29}$$

As a check, we notice that this result is in agreement with the sign of the deceleration parameter q, defined as

$$q = -1 - \frac{d\ln H}{d\ln a}. \tag{16.30}$$

Substituting the scale factor (16.23) in Eq. (16.30), it turns out that the deceleration parameter can be written in terms of s as $q = 1-s$. Accelerating expansion occurs when $s > 1$, corresponding to a negative value of q, in agreement with Eq. (16.29).

16.2.1 Inflation in $f(\mathcal{G})$ Gravity

Inflation can also be achieved also in a purely Gauss–Bonnet gravity. As previously, we impose the magnitude of slow-roll parameters to be small during inflation. To this purpose, we rewrite the first two components of the field equations (8.43) as

$$H^2 = \frac{1}{3(2R-4\Box)f_{\mathcal{G}}(\mathcal{G})}\left\{\frac{f_0}{2}\mathcal{G}^{k-3}\left[-24k(k-1)H^2\left(\mathcal{G}\ddot{\mathcal{G}}+(k-2)\dot{\mathcal{G}}^2\right)\right.\right.$$
$$\left.\left.-72k\mathcal{G}^2H^4-48k\mathcal{G}H^2\left(2\mathcal{G}(\dot{H}+H^2)+(k-1)H\dot{\mathcal{G}}\right)+\mathcal{G}^3\right]\right\},$$

$$\dot{H} = -\frac{3}{2}H^2-\frac{1}{2(2R-4\Box)f_{\mathcal{G}}(\mathcal{G})}$$
$$\times\left\{\frac{f_0}{2}\mathcal{G}^{k-3}\left[16k(k-1)(\dot{H}+H^2)\left(\mathcal{G}\ddot{\mathcal{G}}+(k-2)\dot{\mathcal{G}}^2\right)\right.\right.$$
$$+24k\mathcal{G}^2H^4+16k\mathcal{G}(\dot{H}+H^2)\left(3\mathcal{G}(\dot{H}+H^2)+2(k-1)H\dot{\mathcal{G}}\right)$$
$$\left.\left.+24k\mathcal{G}H^2\left(4\mathcal{G}(\dot{H}+H^2)+(k-1)H\dot{G}\right)-\mathcal{G}^3\right]\right\}.$$

Substituting the solution of $f(\mathcal{G})$ gravity (8.63), obtained by Noether's symmetries, in the preceding equations and considering the explicit expression of the Gauss–Bonnet term

$$\mathcal{G} = 24H^2(H^2+\dot{H}), \tag{16.31}$$

the magnitude of the slow-roll parameters turns out to be

$$|\varepsilon| = |\eta| = \left|\frac{1}{1-4k}\right|. \tag{16.32}$$

The conditions for the inflation

$$|\varepsilon| \ll 1, \qquad |\eta| \ll 1, \tag{16.33}$$

provide the constraint

$$k \ll 0 \vee k \gg \frac{1}{2}. \tag{16.34}$$

This means that cosmological inflation in $f(\mathcal{G})$ gravity occurs only when k is strictly negative or when it is much higher than $1/2$. Interestingly, it is worth noticing again that the value $k = 1/2$ is the limit in which $f(\mathcal{G}) = \mathcal{G}^k$ gravity behaves like Einstein GR in a cosmological spatially flat background. As before, we also notice that the deceleration parameters can be written in terms of k as

$$q = -1+\frac{1}{4k-1}, \tag{16.35}$$

so that when $k \ll 0 \vee k \gg 1/2$, q turns out to be negative, as we expect. The negative value of q corresponds to an accelerating expansion of the universe, in agreement with the result provided by Eq. (16.34). Slow-roll conditions are not sufficient to establish whether the theory is able to fit the cosmological data, and further studies are necessary. Nevertheless, this is a first step toward a theory that can result in a consistent inflationary model.

16.2.2 Inflation in $\{R + f(\mathcal{G})\}$ Gravity

Gauss–Bonnet terms can be considered also as corrections to standard GR. Inflation can be realized also in this case. As pointed out in Section 8.1.1, extra Noether symmetries occur only for an action of the form

$$S = \int \sqrt{-g}\left(\frac{R}{2} + f_0\sqrt{\mathcal{G}} + f_1\mathcal{G}\right) d^4x, \tag{16.36}$$

with f_0, f_1 real constants. The linear contribution of the Gauss–Bonnet term does not play any role in the dynamics, and, at the level of field equations, it identically vanishes. In a cosmological spatially flat background, the field equations can be written as

$$H^2 = \frac{f_0}{6}\mathcal{G}^{-\frac{3}{2}}\left\{12H^2\left[-\frac{1}{2}H\dot{\mathcal{G}} - \mathcal{G}(\dot{H} + H^2)\right] + 1\right\}, \tag{16.37}$$

$$\begin{aligned}\dot{H} = & -\frac{3}{2}H^2 + \frac{f_0}{4}\mathcal{G}^{-\frac{5}{2}}\left[-12H^2\mathcal{G}^2\left(H^2 + \dot{H}\right) + \mathcal{G}^3\right.\\ & \left. -2H\left(2\mathcal{G}\left(H^2 + \dot{H}\right)\dot{\mathcal{G}} + H\left(\mathcal{G}\ddot{\mathcal{G}} - \frac{3}{2}\dot{\mathcal{G}}^2\right)\right)\right].\end{aligned} \tag{16.38}$$

With the aim of constraining the value of the coupling constant f_0 admitting slow-roll inflation, we consider the vacuum solution (8.39), namely

$$a_{\pm}(t) = a_0\, t^{\frac{4f_0^2 + 3 \pm \sqrt{3}\sqrt{16f_0^2 + 3}}{2\left(3 - 2f_0^2\right)}}. \tag{16.39}$$

Replacing a_+[1] in the field equations (16.37) and considering the conditions for the inflation in Eq. (16.4), it turns out that slow-roll inflation is admitted by $R + f_0\sqrt{\mathcal{G}}$ theory of gravity for

$$f_0 \sim \pm\sqrt{\frac{3}{2}}. \tag{16.40}$$

In particular, the more f_0 approaches the values $\pm\sqrt{3/2}$, the faster the scalar field rolls down the potential hill. These values are in agreement with the range provided by the energy conditions in Section 16.3.2.

[1] No coupling constant is selected by $a_-(t)$.

16.2.3 Inflation in $f(R)$ Gravity

The slow-roll inflation can also be studied in the context of $f(R)$ cosmology. Specifically, we choose the function selected by Noether Symmetry Approach in Section 7.1, namely

$$f(R) = f_0 R^k \tag{16.41}$$

and constrain the value of k leading to slow-roll inflation. In a FLRW spatially flat universe, the field equations (4.45) can be written in terms of the Hubble constant as

$$\begin{cases} H^2 = \dfrac{1}{3f_R}\left\{\dfrac{1}{2}(f - Rf_R) - 3H\dot{R}f_{RR}\right\} \\ \dot{H} = -\dfrac{3}{2}H^2 + \dfrac{1}{2f_R}\left\{\dfrac{1}{2}(f - Rf_R) - (\ddot{R} + 2\dot{R}H)f_{RR} - \dot{R}^2 f_{RRR}\right\}. \end{cases} \tag{16.42}$$

The vacuum solution of the preceding field equations for $f(R) = R^k$ is given by Eq. (7.19) and reads as

$$a(t) = a_0 t^{\frac{(2k-1)(k-1)}{2-k}}. \tag{16.43}$$

Substituting Eq. (16.43) in the field equations (16.42), the conditions (16.4) provide the values of k that admit inflation in the early time:

$$k \ll \frac{1}{2}\left(1 - \sqrt{3}\right) \vee k \gg \frac{1}{2}\left(1 + \sqrt{3}\right). \tag{16.44}$$

This value is in agreement with the sign of deceleration parameter that, when k is included in the range (16.44), turns out to be negative, as expected from an accelerating universe.

16.3 Generalized Energy Conditions and Cosmology

Energy conditions [432] play a fundamental role in defining physically viable models. Here, we want to show that their role can be generalized in modified/extended gravity [427] with interesting applications in Cosmology. Specifically, we shall study their role in extended models involving the Ricci curvature scalar and the Gauss–Bonnet scalar. We consider the functions $f(R, \mathcal{G}) = f_0 \mathcal{G}^k$ (in Section 16.3.1), $f(R, \mathcal{G}) = R + f_0 \mathcal{G}^k$ (in Section 16.3.2), and $f(R, \mathcal{G}) = f_0 R^k$ (in Section 16.3.3). All these models have been selected previously by Noether's symmetries.

In general, the energy conditions are a set of inequalities that must be applied to the matter content of the theory, when such a content cannot be specified explicitly. They have been formulated to preserve, for example, the Principle of Causality in GR [432]. They read as

$$
\begin{aligned}
&\text{Null Energy Condition (NEC)} \rightarrow \rho + p \geq 0; \\
&\text{Weak Energy Condition (WEC)} \rightarrow \rho \geq 0,\ ;\ \rho + p \geq 0; \\
&\text{Dominant Energy Condition (DEC)} \rightarrow \rho - |p| \geq 0; \\
&\text{Strong Energy Condition (SEC)} \rightarrow \rho + p \geq 0,\ ;\ \rho + 3p \geq 0
\end{aligned}
\tag{16.45}
$$

and serve to select the states of matter that are allowed in a given space-time. In particular, the NEC validity implies the trace of the energy-momentum tensor to be nonnegative. The WEC assures that in any reference frame a nonnegative energy density is observed. The DEC preserves the Principle of Causality imposing that ordinary matter cannot propagate faster than light. The SEC validity imposes gravitational interaction to be attractive. In GR, ordinary matter identically satisfies all the energy conditions, which however can be violated when exotic fluids are considered. This also implies that traversable wormholes, where a violation of the NEC is needed, are not allowed by Einstein gravity. Here we do not discuss wormhole solutions in modified theories of gravity, but for further readings see [388, 433–439] and references therein.

As we said, energy conditions can also be considered in the context of modified theories of gravity [427] where, according to Eq. (16.9), an extra energy-momentum tensor, related to the gravitational field, occurs in the field equations together with the ordinary matter. From a cosmological point of view, energy conditions can be used along with the cosmographic parameters to test the validity of the theory, as well as to constrain the starting gravitational actions leading to the late-time cosmic expansion. This approach is straightforwardly discussed, for example, in Refs. [180, 427, 440], where different alternatives to GR are considered. By a similar procedure, in Section 16.3.1 we apply the energy conditions to the action (8.42) (with $d = 3$), showing that the Einstein tensor can be recovered even when the GR limit is not assumed as a requirement. Then, in Section 16.3.2, we compare the result with that provided by $R + f(\mathcal{G})$ gravity. In both cases, we consider the function $f(\mathcal{G}) = f_0 \mathcal{G}^k$ and show that the effective energy density and the effective pressure can be written in terms of geometry. Finally, in Section 16.3.3, we apply the same prescription to $f(R)$ gravity with $f(R) \sim R^k$. As a result, in a FLRW space-time, cosmographic parameters allow one select viable values of the real constant k.

16.3.1 Energy Conditions in $f(\mathcal{G})$ Cosmology

Let us derive the energy conditions for $f(\mathcal{G}) \sim \mathcal{G}^k$ cosmology, which is a model containing Noether's symmetries. Then we constrain the function by cosmographic parameters. We start with the action

$$
S = \int \sqrt{-g}\, f_0 \mathcal{G}^k \, d^4x\,, \tag{16.46}
$$

selected in Section 8.1.2. If the standard matter is considered through the action S^m, the field equations can be written as

$$
\begin{aligned}
&-\left(2RR_{\mu\nu} - 4R_{\mu\rho}R^{\rho}{}_{\nu} + 2R_{\mu}{}^{\rho\sigma\tau}R_{\nu\rho\sigma\tau} - 4R^{\alpha\beta}R_{\mu\alpha\nu\beta}\right) f_{\mathcal{G}}(\mathcal{G}) \\
&+\Big(2RD_{\mu}D_{\nu} + 4G_{\mu\nu}\Box - 4R^{\rho}_{\{\nu}D_{\mu\}}D_{\rho} + 4g_{\mu\nu}R^{\rho\sigma}D_{\rho}D_{\sigma} \\
&-4R_{\mu\alpha\nu\beta}D^{\alpha}D^{\beta}\Big) f_{\mathcal{G}}(\mathcal{G}) + \frac{1}{2}g_{\mu\nu}f(\mathcal{G}) + \mathcal{T}_{\mu\nu} = 0\,, \qquad (16.47)
\end{aligned}
$$

with $\mathcal{T}_{\mu\nu}$ being the energy-momentum tensor of matter fields. The energy conditions can be recast in terms of the effective energy-momentum tensor of the gravitational field by isolating the Einstein tensor $G_{\mu\nu}$ in the LHS. To this purpose, the preceding field equations become

$$
\begin{aligned}
G_{\mu\nu}(2R-4\Box)f_{\mathcal{G}}(\mathcal{G}) = \mathcal{T}_{\mu\nu} &-\Big[\left(R^2 - 4R_{\mu\rho}R^{\rho}{}_{\nu} + 2R_{\mu}{}^{\rho\sigma\tau}R_{\nu\rho\sigma\tau} - 4R^{\alpha\beta}R_{\mu\alpha\nu\beta}\right) f_{\mathcal{G}}(\mathcal{G}) \\
&-\Big(2RD_{\mu}D_{\nu} - 4R^{\rho}_{\{\nu}D_{\mu\}}D_{\rho} + 4g_{\mu\nu}R^{\rho\sigma}D_{\rho}D_{\sigma} \\
&-4R_{\mu\alpha\nu\beta}D^{\alpha}D^{\beta}\Big) f_{\mathcal{G}}(\mathcal{G}) - \frac{1}{2}g_{\mu\nu}f(\mathcal{G})\Big]\,. \qquad (16.48)
\end{aligned}
$$

Notice that the term in the square bracket of the RHS can be intended as the effective energy-momentum tensor of the gravitational field, introduced in Eq. (16.9), while $(2R-4\Box)f_{\mathcal{G}}(\mathcal{G})$ accounts for the function F. Therefore, the components of the Einstein tensor can be interpreted as the analogue of energy density and pressure, namely

$$
G^0_0 = \frac{1}{(2R-4\Box)f_{\mathcal{G}}(\mathcal{G})}\left(\rho^{(Curv)} + \rho_0\right), \qquad (16.49)
$$

$$
G^i_j = -\frac{\delta^i_j}{(2R-4\Box)f_{\mathcal{G}}(\mathcal{G})}\left(p^{(Curv)} + p_0\right), \qquad (16.50)
$$

where ρ_0 and p_0 are the matter density and pressure respectively. In a spatially flat cosmological universe of the form

$$
ds^2 = dt^2 - a(t)^2 d\mathbf{x}^2, \qquad (16.51)
$$

the field equations of $f(\mathcal{G}) = f_0\mathcal{G}^k$ gravity can be written as

$$
\begin{aligned}
G^0_0 = &\frac{1}{(2R-4\Box)f_{\mathcal{G}}(\mathcal{G})}\Big\{\rho_0 \\
&+\frac{f_0}{2}\mathcal{G}^{k-3}\Big[-24k(k-1)H^2\left(\mathcal{G}\ddot{\mathcal{G}} + (k-2)\dot{\mathcal{G}}^2\right) \\
&-72k\mathcal{G}^2H^4 - 48k\mathcal{G}H^2\left(2\mathcal{G}(\dot{H}+H^2) + (k-1)H\dot{\mathcal{G}}\right) + \mathcal{G}^3\Big]\Big\}, \qquad (16.52)
\end{aligned}
$$

$$
\begin{aligned}
G^1_1 = &-\frac{1}{(2R-4\Box)f_{\mathcal{G}}(\mathcal{G})}\Big\{p_0 \\
&+\frac{f_0}{2}\mathcal{G}^{k-3}\Big[16k(k-1)(\dot{H}+H^2)\left(\mathcal{G}\ddot{\mathcal{G}} + (k-2)\dot{\mathcal{G}}^2\right) \\
&+24k\mathcal{G}^2H^4 + 16k\mathcal{G}(\dot{H}+H^2)\left(3\mathcal{G}(\dot{H}+H^2) + 2(k-1)H\dot{\mathcal{G}}\right) \\
&+24k\mathcal{G}H^2\left(4\mathcal{G}(\dot{H}+H^2) + (k-1)H\dot{G}\right) - \mathcal{G}^3\Big]\Big\}. \qquad (16.53)
\end{aligned}
$$

The term in the square brackets in Eq. (16.52) is the cosmological density of gravitational field $\rho^{(Curv)}$, as the same term in Eq. (16.53) is the pressure. Specifically:

$$\rho^{(Curv)} = \frac{f_0}{2}\mathcal{G}^{k-3}\Big[-24k(k-1)H^2\left(\mathcal{G}\ddot{\mathcal{G}}+(k-2)\dot{\mathcal{G}}^2\right) \\ -72k\mathcal{G}^2H^4 - 48k\mathcal{G}H^2\left(2\mathcal{G}(\dot{H}+H^2)+(k-1)H\dot{\mathcal{G}}\right)+\mathcal{G}^3\Big], \tag{16.54}$$

$$p^{(Curv)} = \frac{f_0}{2}\mathcal{G}^{k-3}\Big[16k(k-1)(\dot{H}+H^2)\left(\mathcal{G}\ddot{\mathcal{G}}+(k-2)\dot{\mathcal{G}}^2\right)+24k\mathcal{G}^2H^4 \\ +16k\mathcal{G}(\dot{H}+H^2)\left(3\mathcal{G}(\dot{H}+H^2)+2(k-1)H\dot{\mathcal{G}}\right) \\ +24k\mathcal{G}H^2\left(4\mathcal{G}(\dot{H}+H^2)+(k-1)H\dot{\mathcal{G}}\right)-\mathcal{G}^3\Big], \tag{16.55}$$

so that each energy condition can be split into two contributions, namely

$$\begin{aligned} &\text{NEC} \rightarrow \rho_0+\rho^{(Curv)}+p_0+p^{(Curv)} \geq 0; \\ &\text{WEC} \rightarrow \rho_0+\rho^{(Curv)} \geq 0, \quad \rho_0+\rho^{(Curv)}+p_0+p^{(Curv)} \geq 0; \\ &\text{DEC} \rightarrow \rho_0+\rho^{(Curv)}-|p_0|-|p^{(Curv)}| \geq 0; \\ &\text{SEC} \rightarrow \rho_0+\rho^{(Curv)}+p_0+p^{(Curv)} \geq 0, \quad \rho_0+\rho^{(Curv)}+3p_0+3p^{(Curv)} \geq 0 . \end{aligned} \tag{16.56}$$

Starting from the preceding inequalities, it is possible to focus on particular subcases requiring that standard matter and gravitational field respect the energy conditions separately. Assuming that ordinary matter automatically satisfies all the energy conditions, the only constraints capable of providing the validity range of the parameter k are

$$\begin{aligned} &\text{NEC} \rightarrow \rho^{(Curv)}+p^{(Curv)} \geq 0; \\ &\text{WEC} \rightarrow \rho^{(Curv)} \geq 0, \quad \rho^{(Curv)}+p^{(Curv)} \geq 0; \\ &\text{DEC} \rightarrow \rho^{(Curv)}-|p^{(Curv)}| \geq 0; \\ &\text{SEC} \rightarrow \rho^{(Curv)}+p^{(Curv)} \geq 0, \quad \rho^{(Curv)}+3p^{(Curv)} \geq 0 . \end{aligned} \tag{16.57}$$

Substituting Eqs. (16.54) and (16.55) into the system (16.57), the energy conditions yield

$$NEC \rightarrow \; 4f_0k\mathcal{G}^{k-3}\Big[(k-1)\mathcal{G}\left(2\ddot{\mathcal{G}}\dot{H}-H^2\ddot{\mathcal{G}}+4H\dot{\mathcal{G}}\dot{H}+H^3\dot{\mathcal{G}}\right) \\ -(k-1)(k-2)\dot{G}^2\left(H^2-2\dot{H}\right)+6\mathcal{G}^2\dot{H}\left(\dot{H}+2H^2\right)\Big] \geq 0;$$

$$4f_0k\mathcal{G}^{k-3}\Big[(k-1)\mathcal{G}\left(2\ddot{\mathcal{G}}\dot{H}-H^2\ddot{\mathcal{G}}+4H\dot{\mathcal{G}}\dot{H}+H^3\dot{\mathcal{G}}\right) \\ -(k-1)(k-2)\dot{G}^2\left(H^2-2\dot{H}\right)+6\mathcal{G}^2\dot{H}\left(\dot{H}+2H^2\right)\Big] \geq 0,$$

$WEC \; \begin{smallmatrix}\nearrow \\ \searrow\end{smallmatrix}$ and

$$\frac{f_0}{2}\mathcal{G}^{k-3}\left[-24k(k-1)H^2\left(\mathcal{G}\ddot{\mathcal{G}}+(k-2)\dot{\mathcal{G}}^2\right)-72k\mathcal{G}^2H^4\right.$$
$$\left.-48k\mathcal{G}H^2\left(2\mathcal{G}(\dot{H}+H^2)+(k-1)H\dot{\mathcal{G}}\right)+\mathcal{G}^3\right]\geq 0;$$

$$DEC \begin{cases} 4f_0k\mathcal{G}^{k-3}\left[(k-1)\mathcal{G}\left(2\ddot{\mathcal{G}}H-H^2\ddot{\mathcal{G}}+4H\dot{\mathcal{G}}\dot{H}+H^3\dot{\mathcal{G}}\right)\right. \\ \left.-(k-1)(k-2)\dot{G}^2\left(H^2-2\dot{H}\right)+6\mathcal{G}^2\dot{H}\left(\dot{H}+2H^2\right)\right]\geq 0, \quad \text{if } p<0, \\ \text{and} \\ f_0\mathcal{G}^{k-3}\left[-4k(k-1)(k-2)\dot{\mathcal{G}}^2\left(2\dot{H}+5H^2\right)\right. \\ -24k\mathcal{G}^2\left(6H^2\dot{H}+\dot{H}^2+7H^4\right) \\ \left.-4k(k-1)\mathcal{G}\left(2\ddot{\mathcal{G}}H+5H^2\ddot{\mathcal{G}}+4H\dot{\mathcal{G}}\dot{H}+13H^3\dot{\mathcal{G}}\right)+\mathcal{G}^3\right]\geq 0, \quad \text{if } p\geq 0; \end{cases}$$

$$SEC \begin{cases} 4f_0k\mathcal{G}^{k-3}\left[(k-1)\mathcal{G}\left(2\ddot{\mathcal{G}}H-H^2\ddot{\mathcal{G}}+4H\dot{\mathcal{G}}\dot{H}+H^3\dot{\mathcal{G}}\right)\right. \\ \left.-(k-1)(k-2)\dot{G}^2\left(H^2-2\dot{H}\right)+6\mathcal{G}^2\dot{H}\left(\dot{H}+2H^2\right)\right]\geq 0, \\ \text{and} \\ f_0\mathcal{G}^{k-3}\left[12k(k-1)(k-2)\dot{\mathcal{G}}^2\left(2\dot{H}+H^2\right)\right. \\ +24k\mathcal{G}^2\left(10H^2\dot{H}+3\dot{H}^2+7H^4\right) \\ \left.+12k(k-1)\mathcal{G}\left(2\ddot{\mathcal{G}}H+H^2\ddot{\mathcal{G}}+4H\dot{\mathcal{G}}\dot{H}+5H^3\dot{\mathcal{G}}\right)-\mathcal{G}^3\right]\geq 0\,. \end{cases}$$

With the aim of studying the validity conditions, it is possible to introduce the cosmographic parameters [32] j (jerk), q (deceleration), and s (snap), defined as

$$q=-1-\frac{\dot{H}}{H^2},\quad j=1+\frac{\ddot{H}+3\dot{H}H}{H^3},$$
$$s=1+\frac{\dddot{H}+3\ddot{H}H+3\dot{H}^2+6H^2\dot{H}+H\ddot{H}}{H^4}; \tag{16.58}$$

in this way, the energy conditions can be recast in terms of j, q, s, k only. To this purpose, notice that the cosmological expression of the Gauss–Bonnet scalar $\mathcal{G}=24H^2(H^2+\dot{H})$ can be written as [431]

$$\mathcal{G}=-24qH^4,\qquad \dot{\mathcal{G}}=24(3q+2q^2+j)H^5,$$
$$\ddot{\mathcal{G}}=24(s-12q-15q^2-2q^3-6j-6jq)H^6\,. \tag{16.59}$$

Therefore, the pressure and the energy density become

$$\rho^{Curv}=\frac{f_0 2^{3k-1}3^k\left(-H^4q\right)^k}{q^3}\left\{2j(k-1)kq[k(2q+3)-q-4]\right.$$
$$+j^2k(k-1)(k-2)+q[k^3q(2q+3)^2$$
$$\left.-k^2(10q^3+25q^2+21q+s)+k(6q^3+9q^2+15q+s)+q^2]\right\},$$

$$
\begin{aligned}
p^{Curv} = & \frac{f_0 2^{3k-1} 3^{k-1} \left(-H^4 q\right)^k}{q^2} \Big\{ 2j^2 k(k-1)(k-2) \\
& + jk(k-1)[4(2k-1)q^2 + 4(3k-4)q + 3] + 2k^3 q^2 (2q+3)^2 \\
& - k^2 q(20q^3 + 50q^2 + 36q + 2s - 9) \\
& + 2kq(6q^3 + 10q^2 + 15q + s - 6) - 3q^2] \Big\},
\end{aligned}
$$

respectively. Though the term $\left(-qH^4\right)^k$ depends on the model considered, it is a total factor that multiplies all the rest of the inequality. Moreover, the numerical value of the deceleration parameter is negative, thus such total term can be neglected in the computation of the energy conditions. First we choose $f_0 > 0$, in order to recover the GR coupling as soon as $k = 1/2$. Considering Eqs. (16.58) and (16.59), and using the experimental values of j, k, s provided in [441] ($q = -0.81$, $j = 2.16$, $s = -0.22$), the energy conditions are satisfied for

$$
\begin{aligned}
& \text{NEC} \rightarrow k \leq 0 \vee 1.457 \leq k \leq 2.977; \\
& \text{WEC} \rightarrow k \leq 0; \\
& \text{DEC} \rightarrow k \leq 0; \\
& \text{SEC} \rightarrow 1.457 \leq k \leq 2.977.
\end{aligned}
\tag{16.60}
$$

It is worth remarking that the preceding result holds independently of the value of the Hubble constant H, since the latter multiplies all the energy conditions as a total factor with an even power. As we can see, no value of k simultaneously satisfies all the energy conditions when $f_0 > 0$. Considering $f_0 < 0$, the energy conditions provide

$$
\begin{aligned}
& \text{NEC} \rightarrow 0 \leq k \leq 1.457 \vee k \geq 2.977; \\
& \text{WEC} \rightarrow 0.093 \leq k \leq 1.457 \vee k \geq 2.977; \\
& \text{DEC} \rightarrow 0.113 \leq k \leq 1.457 \vee k \geq 2.977; \\
& \text{SEC} \rightarrow 0 \leq k \leq 0.189,
\end{aligned}
\tag{16.61}
$$

and all of them are simultaneously satisfied for k in the range

$$
0.113 < k < 0.189. \tag{16.62}
$$

By considering the vacuum solution of the field equations in $f(\mathcal{G})$ gravity (8.63), we notice that the scale factor also is constrained by the energy conditions. Taking into account the result in Eq. (16.62), the scale factors (8.63) not violating the energy conditions are of the form

$$
a(t) = a_0 t^n, \quad \text{with } 0.24 \leq n \leq 0.55. \tag{16.63}
$$

Nevertheless, the validity of energy conditions for other values of k and n might occur in epochs where the behavior of gravitational interaction is different from the current one. Another possibility is for

$$
f(\mathcal{G}) = f_0 \mathcal{G}^{k_1} + f_1 \mathcal{G}^{k_2}, \tag{16.64}
$$

where the two contributions could depend on the values of the coupling constants at different epochs.

16.3.2 Energy Conditions in $\{R+f(\mathcal{G})\}$ Cosmology

$f(\mathcal{G})$ and $R+f(\mathcal{G})$ gravity can be compared by energy conditions. We consider again the action

$$S=\int\sqrt{-g}\left(\frac{R}{2}+f_0\mathcal{G}^k\right)\,d^4x \tag{16.65}$$

and study the energy conditions by using the numerical values of cosmographic parameters. The preceding action is a particular case of that selected in Section 8.1.1, where it turned out that the general action

$$S=\int\sqrt{-g}\left[\frac{R}{2}+f(\mathcal{G})\right]\,d^4x \tag{16.66}$$

contains Noether symmetry for $f(\mathcal{G})=f_0\sqrt{\mathcal{G}}+f_1\mathcal{G}$. Since the Gauss–Bonnet scalar is a topological surface term, the action (16.65) contains Noether symmetry for $k=1/2$. For general k, the field equations in the presence of matter can be written as

$$\begin{aligned}G_{\mu\nu}&=\mathcal{T}_{\mu\nu}+\frac{f_0}{2}\mathcal{G}^kg_{\mu\nu}-kf_0\Big(2RR_{\mu\nu}-4R_{\mu\rho}R^{\rho}{}_{\nu}+2R_{\mu}{}^{\rho\sigma\tau}R_{\nu\rho\sigma\tau}\\&\quad-4R^{\alpha\beta}R_{\mu\alpha\nu\beta}\Big)\mathcal{G}^{k-1}+kf_0\Big(2RD_{\mu}D_{\nu}+4G_{\mu\nu}\Box-4R^{\rho}_{\{\nu}D_{\mu\}}D_{\rho}\\&\quad+4g_{\mu\nu}R^{\rho\sigma}D_{\rho}D_{\sigma}-4R_{\mu\alpha\nu\beta}D^{\alpha}D^{\beta}\Big)\mathcal{G}^{k-1}\,,\end{aligned} \tag{16.67}$$

so that the RHS can be intended as an effective energy-momentum tensor which vanishes as soon as $f_0=0$. In a cosmological spatially flat space-time, the two nonvanishing components of the field equations read

$$G^0_0=\rho_0+\frac{f_0}{2}\mathcal{G}^{k-2}\left\{24kH^2\left[(k-1)H\dot{\mathcal{G}}-\mathcal{G}(\dot{H}+H^2)\right]+1\right\}, \tag{16.68}$$

$$\begin{aligned}G^1_1=-\Big\{&p_0-\frac{f_0}{2}\mathcal{G}^{k-3}\left[-24H^2k\mathcal{G}^2\left(H^2+\dot{H}\right)+\mathcal{G}^3\right.\\&\left.+\;8H(k-1)k\Big(2\mathcal{G}\left(H^2+\dot{H}\right)\dot{\mathcal{G}}+H\Big(\mathcal{G}\ddot{\mathcal{G}}+(k-2)\dot{\mathcal{G}}^2\Big)\Big)\right]\Big\}.\end{aligned} \tag{16.69}$$

As before, the second terms in the RHS of the preceding equations can be intended as effective energy densities and pressures of the gravitational field. In this way, Eqs. (16.68) and (16.69) can be written in terms of the cosmographic parameters (16.58) as

$$\rho^{Curv}=\frac{2^{3k-1}3^kf_0(-H^4q)^k}{q^2}(k-1)[jk-q^2+kq(3+2q)], \tag{16.70}$$

$$p^{Curv} = \frac{2^{3k-1}3^{k-1}f_0(-H^4 q)^k}{q^3}(k-1)[j^2 k(k-2) + 2jkq(-3+3k+2kq) \\ +3q^3 + k^2 q^2(3+2q)^2 - kq(6q+3q^2+2q^3+s)], \tag{16.71}$$

where we used the form of $\mathcal{G}$ given by Eq. (16.59). As we did in the previous section, we assume that the density and the pressure of matter must satisfy the energy conditions separately, and use the same values of the cosmographic parameters. In this way, for $f_0 > 0$, the energy conditions are satisfied for

$$\begin{aligned} &\text{NEC} \to k \leq 0 \vee 1 \leq k \leq 4.371; \\ &\text{WEC} \to k \leq 0 \vee 1 \leq k \leq 4.371; \\ &\text{DEC} \to -1.866 \leq k \leq 0 \vee 1.573 \leq k \leq 4.371; \\ &\text{SEC} \to k \leq -0.313 \vee 1 \leq k \leq 3.129 . \end{aligned} \tag{16.72}$$

The above inequalities admit a common solution, that is,

$$-1.866 \leq k \leq -0.313 \vee 1.573 \leq k \leq 3.129. \tag{16.73}$$

Moreover, as previously found in Eq. (8.41), vacuum solutions of the field equations yield an exponential scale factor when $k \neq 1/2$, namely

$$a(t) = a_0 \exp\left\{\left[\frac{24^k f_0(1-k)}{3}\right]^{\frac{1}{2-4k}} t\right\}. \tag{16.74}$$

In the first range of Eq. (16.73), namely $-1.866 \leq k \leq -0.313$, the argument of the exponential function is positive and the scale factor describes an exponentially accelerating universe. In the second range, when $1.573 \leq k \leq 3.129$, the argument turns out to be negative, leading to a bouncing cosmological model. When $k = 1/2$, another solution occurs, that is,

$$a(t)_{\pm} = a_0 t^{\frac{4f_0^2+3\pm\sqrt{3}\sqrt{16f_0^2+3}}{2(3-2f_0^2)}}. \tag{16.75}$$

The scale factor $a(t)_+$ describes an accelerating universe when

$$-\sqrt{\frac{3}{2}} < f_0 < \sqrt{\frac{3}{2}}. \tag{16.76}$$

On the contrary, when $a_-(t)$ is considered, the power of t is always negative regardless of the value of f_0. Interestingly, notice that Eq. (16.76) provides the same range as Eq. (16.40). This means that only an energy condition violation may lead to an inflationary universe.

Assuming now a negative coupling constant, namely $f_0 < 0$, the energy conditions yield the constraints

$$
\begin{aligned}
&\text{NEC} \to k \leq 0 \ \vee \ 1 \leq k \leq 4.371;\\
&\text{WEC} \to 0.630 < k < 1;\\
&\text{DEC} \to \nexists k \in \mathbb{R};\\
&\text{SEC} \to 0 \leq k \leq 1 \ \vee \ k \geq 4.371
\end{aligned}
\tag{16.77}
$$

and cannot be simultaneously satisfied for any real value of k. However, it is worth noticing that all k in the range $1 < k < 4.371$ violate the SEC, which means that the geometric contributions in $R+f(\mathcal{G})$ gravity can act as a repulsive source of gravitational field.

16.3.3 Energy Conditions in $f(R)$ Cosmology

Adopting the same prescription as in the two previous sections, here we study the energy conditions for $f(R)$ gravity cosmology. Among the functions selected by Noether's symmetries, we choose $f(R) = f_0 R^k$, because it contains a free parameter to be constrained by the energy conditions. This function is the same chosen to study the previously mentioned slow-roll inflation, galaxy rotation curve, and Solar System tests. Let us now consider the field equations (4.45) in a spatially flat cosmological background, namely

$$
\begin{cases}
G_0^0 = \dfrac{1}{f_R}\left\{T_0^0 + \dfrac{1}{2}(f - Rf_R) - 3H\dot{R}f_{RR}\right\}\\[2ex]
G_1^1 = \dfrac{1}{f_R}\left\{T_1^1 + \dfrac{1}{2}(f - Rf_R) - (\ddot{R} + 2\dot{R}H)f_{RR} - \dot{R}^2 f_{RRR}\right\},
\end{cases}
\tag{16.78}
$$

with T_0^0 and T_1^1 being respectively the time and spatial components of the energy momentum tensor of matter. In this way, the RHS can be intended as an effective energy momentum tensor of the gravitational field, so that energy density and pressure can be written in terms of the geometry as

$$
\begin{cases}
\rho = \rho_0 + \dfrac{1}{2}(f - Rf_R) - 3\dot{R}Hf_{RR}\\[2ex]
p = p_0 - \dfrac{1}{2}(f - Rf_R) + (\ddot{R} + 2\dot{R}H)f_{RR} + \dot{R}^2 f_{RRR}.
\end{cases}
\tag{16.79}
$$

As formerly done, let us assume that the matter components satisfy the energy condition separately. In this way, it is possible to write the energy conditions for the geometry in terms of the cosmographic parameters j, q, s introduced in Eq. (16.58). Considering the cosmographic expressions of the scalar curvature (and its time derivatives)

$$
R = -6H^2(1-q), \quad \dot{R} = -6H^3(j - q - 2), \quad \ddot{R} = -6H^4(s + q^2 + 8q + 6), \tag{16.80}
$$

the energy density and pressure in Eq. (16.79) can be written as [442, 443]

$$\begin{cases} \rho^{Curv} = 18(k-1)H^4\Big[-(1-q)^2 + k(j-q-2)\Big]R^{k-2} \\ p^{Curv} = 36(k-1)H^6\Big[-3(1-q)^3 + k(s+q^2+6q+2+2j)(1-q) \\ \qquad +k(k-2)(j-q-2)^2\Big]R^{k-3}. \end{cases} \tag{16.81}$$

According to the numerical values of the cosmographic parameters provided in Ref. [441], the scalar curvature in Eq. (16.80) turns out to be negative. As a consequence, this case must be treated differently with respect to the previous ones, since only integer values of k are allowed. From the analysis, it turns out that no value of k simultaneously respects the energy conditions. However, any of them can be satisfied separately. For instance, for $f_0 > 0$, the WEC holds for

$$k = 1 \lor k = 3 \lor k < -2. \tag{16.82}$$

If $f_0 < 0$, the values of k not violating the WEC are

$$k = 2 \lor k < -1. \tag{16.83}$$

Therefore, Eqs. (16.82) and (16.83) can be used to put physical constraints on $f(R)$ gravity. Moreover, considering, for example, $k = 3/2$, the function becomes

$$f(R) = f_0 R^{3/2} \tag{16.84}$$

and the SEC can be written as

$$\frac{2}{f_0\sqrt{R}}\left\{-\frac{3}{8}f_0 R^{-\frac{3}{2}}\dot{R}^2 + \frac{3}{4}f_0 H R^{-\frac{1}{2}}\dot{R} + \frac{3}{4}f_0 R^{-\frac{1}{2}}\ddot{R} + \frac{1}{6}f_0 R^{\frac{3}{2}}\right\} \geq 0. \tag{16.85}$$

Substituting the cosmological solution of $f(R) = f_0 R^{3/2}$ gravity, namely

$$a(t) = a_0\left[c_4 t^4 + c_3 t^3 + c_2 t^2 + c_1 t + c_0\right]^{1/2}, \tag{16.86}$$

in Eq. (16.85), it turns out that SEC is identically violated and the deceleration parameter is positive, as expected for an accelerating universe.

16.4 Geometric Quintessence

Observations of late universe manifest a result going against natural intuition, according to which cosmic expansion should decelerate, due to the attractive interaction among structures. The generally accepted answer to this puzzle is addressed to dark energy, an ingredient that should cover between 60 and 70 percent of the total amount of mass-energy. Despite this dynamical evidence, no new fundamental particle has been detected, until now, capable of explaining coherently the phenomenology, so the conceptual situation is frustrating, like in the case of dark matter. The straightforward explanation for dark energy is considering a generic cosmological constant Λ, which arises in the Einstein field equations as a vacuum energy density but, as widely discussed in literature, this ingredient presents severe problems of fine-tuning and orders of magnitude to match the present value with those of previous epochs [58].

As a consequence, in the last two decades, nonstandard forms of energy-momentum tensors have been taken into account, and many models have been pursued in order to give an explanation to the present universe behavior. As largely discussed previously and first reported in [116], a possibility of addressing the late-time acceleration comes by considering geometric invariants (see, for example, [91, 97, 444]). In general, as discussed in detail in [445], it is possible to define "perfect scalars," derived from geometric ingredients, by which the field equations naturally contain perfect fluid terms. In this case, it is possible to construct effective EoS giving rise to accelerated behaviors. Specifically, the observed accelerated behavior of the Hubble flow, usually related to the existence of mysterious unknown fluids, can be framed in natural extensions of GR that work at IR scales. This approach gives rise to a geometric view of the phenomenon that does not require any new particle to be addressed but only extensions of GR at different scales.

However, besides the first models related to $f(R)$ gravity, the approach has been enlarged in view to encompass also other geometric invariants of curvature and torsion.

In this section, without claiming completeness, we will give some examples of such an approach by considering models like $f(R)$ gravity, $f(R, \mathcal{G})$ gravity, and $f(T, B)$ gravity. We show that these theories provide an equivalent paradigm where the observed cosmological acceleration can be addressed by extending the gravitational sector. The aim is to show that the EoS for the models provide a sort of *geometric adiabatic index*, which can naturally explain the accelerated phase of the today universe.

It is interesting to stress again that such a feature is often related to the existence of Noether's symmetries [116], so the approach reported in this book is useful also to deal with late-time acceleration.

16.4.1 Quintessence in $f(R)$ Gravity

As pointed out in Eq. (16.79), in a spatially flat FLRW background, it is possible to recast the cosmological field equations in terms of effective energy density and pressure provided by curvature. Specifically, we have

$$\begin{cases} \rho^{Curv} = \dfrac{1}{2}\left(f - Rf_R\right) - 3\dot{R}Hf_{RR} \\ p^{Curv} = -\dfrac{1}{2}\left(f - Rf_R\right) + (\ddot{R} + 2\dot{R}H)f_{RR} + \dot{R}^2 f_{RRR}. \end{cases} \tag{16.87}$$

Clearly, it is assumed that $f_R > 0$ to keep the effective gravitational coupling positive, and that $f_{RR} > 0$ to avoid local instabilities (see Section 18.4).

The effective density ρ^{Curv} is nonnegative in a spatially flat FLRW universe in the limit $\rho \to 0$. One can define the effective EoS parameter w^{Curv}, showing that

$f(R)$ gravity produces accelerated expansion without any material dark energy counterpart. In particular, we obtain

$$w^{Curv} \equiv \frac{p^{Curv}}{\rho^{Curv}} = \frac{-(f - Rf_R) + 2(\ddot{R} + 2\dot{R}H)f_{RR} + 2\dot{R}^2 f_{RRR}}{(f - Rf_R) - 6\dot{R}Hf_{RR}}. \tag{16.88}$$

Plugging the power-law function coming from Noether's approach, that is, $f(R) = f_0 R^k$, into Eq. (16.88), we obtain immediately a de Sitter EoS with $w^{Curv} = -1$. Specifically, computations yield

$$w^{Curv} = -\frac{4kR\dot{a}\dot{R} + a\left[2kR\ddot{R} + 2k(k-2)\dot{R}^2 + R^3\right]}{6kR\dot{a}\dot{R} + aR^3}. \tag{16.89}$$

Replacing the exponential solution of Eqs. (7.20), we obtain $w^{Curv} = -1$ exactly like at early inflationary epochs but with different energy scales. The form of w^{Curv} in R^k gravity, if we substitute the solution (7.19) in Eq. (16.88), is

$$w^{Curv} = \frac{-6k^2 + 7k + 1}{6k^2 - 9k + 3}. \tag{16.90}$$

The above mechanism can, in principle, not only address the dark energy behavior but also the cosmological constant problem related to the fact that Λ does not evolve [446]. In this context, $f(R)$ gravity can achieve cosmic acceleration through the effective EoS parameter $w^{Curv} \simeq -1$, as is well known from R^2-inflation. Being the mechanism possible in the early and late universe, it has been used in the attempt to unify early inflation and late-time acceleration in the context of a unified gravity model [333, 447, 448]. However, modeling the late-time cosmic acceleration should not spoil the successes of the standard cosmological model, which requires early inflation, a radiation era allowing the Big Bang nucleosynthesis, a matter era during which over-densities grow and form structures, and the present accelerated epoch. The transitions between consecutive eras must be smooth. Smoothness may not be guaranteed in all $f(R)$ models, and the radiation–matter transition, in particular, can originate problems in specific $f(R)$ models as reported in [449–451]. However, the problem can be overcome. In Ref. [452], the authors showed that $f(R)$ gravity can give rise to cosmological viable models compatible with a matter-dominated epoch evolving into a late accelerated phase. Additionally, it is shown that large classes of $f(R)$ gravity models, including matter and accelerated phases, can be phenomenologically reconstructed by means of observational data, thus several quintessence models can be framed as $f(R)$ models [453].

16.4.2 Quintessence in $f(R, \mathcal{G})$ Gravity

Another extension of the EH action, as discussed in Chapter 9, includes the Gauss–Bonnet invariant $\mathcal{G}$. The field equations of $f(R, \mathcal{G})$ gravity, in a cosmological spatially flat space-time, can be recast as

$$3f_R H^2 = \rho_0 - \frac{1}{2}(f_R R - f + 6H\dot{f}_R - \mathcal{G}f_{\mathcal{G}} + 24H^3\dot{f}_{\mathcal{G}}), \tag{16.91}$$

$$2f_R \dot{H} = -(\rho_0 + p_0) + H\dot{f}_R - \ddot{f}_R + 4H^3\dot{f}_{\mathcal{G}} - 8H\dot{H}\dot{f}_{\mathcal{G}} - 4H^2\ddot{f}_{\mathcal{G}}. \tag{16.92}$$

Therefore, the effective energy density and pressure are

$$\rho^{(\mathcal{GB})} = \frac{1}{2}\left(f - Rf_R - 6H\dot{f}_R + \mathcal{G}f_{\mathcal{G}} - 24H^3\dot{f}_{\mathcal{G}}\right), \tag{16.93}$$

$$p^{(\mathcal{GB})} = 2H\dot{f}_R + \ddot{f}_R + 8H^3\dot{f}_{\mathcal{G}} + 8H\dot{H}\dot{f}_{\mathcal{G}} + 4H^2\ddot{f}_{\mathcal{G}} + \frac{1}{2}\left(Rf_R - \mathcal{G}f_{\mathcal{G}} - f\right). \tag{16.94}$$

For cosmic acceleration, $\rho + 3p < 0$, assuming that matter components have nonnegative pressure, we can write the EoS, $w_{(\mathcal{GB})} = p_{(\mathcal{GB})}/\rho_{(\mathcal{GB})}$, as

$$w_{(\mathcal{GB})} = \frac{\mathcal{G}f_{\mathcal{G}} - Rf_R + f - 4H\left[2H\ddot{f}_{\mathcal{G}} + 4\dot{f}_{\mathcal{G}}\left(H^2 + \dot{H}\right) + \dot{f}_R + \ddot{f}_R/(2H)\right]}{-f + 24H^3\dot{f}_{\mathcal{G}} - \mathcal{G}f_{\mathcal{G}} + 6H\dot{f}_R + Rf_R},$$

where the quintessence behaviour is obtained for $-1 \leq w_{(\mathcal{GB})} < 0$ while the phantom behavior is achieved for $w_{(\mathcal{GB})} < -1$.

Specifically, the Gauss–Bonnet term plays the role of dark energy, like in the case of $f(R)$ gravity, then it contributes to the effective cosmological density as

$$H(z) = H_0\left[\Omega_m a^{-3} + \Omega_r a^{-4} + \Omega_{(\mathcal{GB})} a^{-3(1+w_{(\mathcal{GB})})}\right]. \tag{16.95}$$

This means that the effective value of Gauss–Bonnet contribution can be indicated by evaluating the standard matter and radiation contributions at the various epochs, that is,

$$\Omega_{(\mathcal{GB})} = 1 - \Omega_m - \Omega_r. \tag{16.96}$$

However, the form of $f(R, \mathcal{G})$ determines the evolution of $w_{(\mathcal{GB})}$. Choosing the function provided by the Noether Symmetry Approach in Chapter 9, namely $f(R, \mathcal{G}) = f_0 R^k \mathcal{G}^{1-k}$, and the related solution in Eq. (8.31), we obtain the relation

$$w_{(\mathcal{GB})} = \frac{5-6k}{6k-3}. \tag{16.97}$$

Asymptotically $w_{(\mathcal{GB})}$ behaves like the ΛCDM model with $w_{(\mathcal{GB})} \to -1$. By the cosmographic analysis reported in Ref. [431], it is possible to trace the whole cosmic history in terms of $f(R, \mathcal{G})$ gravity. It emerges that the Gauss–Bonnet term is dominant at early epochs, while the Ricci scalar term dominates in the late universe. It is worth noticing, that, in the case $f(R, \mathcal{G}) = f_0\mathcal{G}^k$, which is the function selected in Section (9.1.2), the EoS parameter (16.95) becomes

$$w_{(\mathcal{GB})} = \frac{8k(k-1)\left\{(k-2)H^2\dot{\mathcal{G}}^2 + \mathcal{G}H\left[H\ddot{\mathcal{G}} + 2\dot{\mathcal{G}}(\dot{H} + H^2)\right]\right\} - (k+1)\mathcal{G}^3}{(k+1)\mathcal{G}^3 - 24k(k-1)\mathcal{G}H^3\dot{\mathcal{G}}}. \tag{16.98}$$

Plugging Eq. (8.63) into Eq. (16.98) and considering the cosmological expression of the Gauss–Bonnet invariant, we obtain $w_{(\mathcal{GB})} = -1$ independently of the value of k, as in $f(R) \sim R^k$ gravity. Therefore, $f(\mathcal{G})$ gravity also can address the cosmic expansion without dark energy.

16.4.3 Quintessence in Extended Teleparallel Gravity

Let us finally consider $f(T, B)$ cosmology in a flat FLRW universe, where the Friedman equations read

$$-3H^2(3f_B + 2f_T) + 3H\dot{f}_B - 3\dot{H}f_B + \frac{1}{2}f(T, B) = \kappa\rho_0\,, \tag{16.99}$$

$$-(3H^2 + \dot{H})(3f_B + 2f_T) - 2H\dot{f}_T + \ddot{f}_B + \frac{1}{2}f(T, B) = -p_0\,. \tag{16.100}$$

Immediately, from the preceding equations, we get

$$\rho^{(T,B)} = -3H\dot{f}_B + 3\dot{H}f_B - \frac{1}{2}f, \tag{16.101}$$

$$p^{(T,B)} = -2H\dot{f}_T + \ddot{f}_B + \frac{1}{2}f, \tag{16.102}$$

so that the EoS takes the form

$$w^{(T,B)} = \frac{-2H\dot{f}_T + \ddot{f}_B + \frac{1}{2}f}{-3H\dot{f}_B + 3\dot{H}f_B - \frac{1}{2}f}, \tag{16.103}$$

from which it is easy to verify that $f(R)$ gravity can be recovered by the choice $f(T, B) = f(-T + B) = f(R)$. Let us now evaluate the EoS when the function $f(T, B)$ contains Noether symmetries. To this purpose, we consider the cases $f(T, B) = B^k T^{\frac{1}{2}(1-k)}$ and $f(T, B) = f_0 T^k$, selected in Chapter 7. In the former case, using Eq. (7.54), the EoS reads

$$w^{(T,B)} = \frac{2(2k+1)}{1-k^2}, \tag{16.104}$$

while in the latter case, we have

$$w^{(T,B)} = -1 + \frac{4k(k-1)H\dot{T}}{T^2}. \tag{16.105}$$

Therefore, when the second term in the RHS of the preceding equation goes to zero, the teleparallel theory is able to mimic dark energy.

To conclude, the approach used in this section is based on the fact that the further degrees of freedom, related to geometric invariants, behave as scalar fields capable of leading the expansion in the various cosmic epochs. Taking into account viability criteria to avoid inconsistencies and instabilities, large classes of theories, involving geometric invariants, can be considered. The underlying philosophy is that geometry can give a comprehensive picture of the universe ranging from inflation up today. Clearly, the final cosmological model will be derived once a consistent amount of data, tracing back the evolution at any epoch, is matched in a comprehensive precision cosmology picture.

17

Applications to Quantum Cosmology

The higher we soar, the smaller
we appear to those who cannot fly.

— *Friedrich Wilhelm Nietzsche*
(Also Sprach Zarathustra [1885])

We consider now some models selected by the Noether Symmetry Approach in the second part of the book, with the aim of developing the Hamiltonian formalism, the Quantum Cosmology, and finally find the Wave Function of the Universe according to prescriptions developed in Chapter 5. As we will show, in most cases, the related WDW equation can be solved analytically only after performing a suitable change of variables suggested by symmetries. This procedure can be adopted only for internal symmetries, as pointed out in Eq. (2.37). As mentioned in Chapter 5, the presence of Noether symmetry permits one to restrict the Superspace to a suitable minisuperspace where the Hamiltonian can be quantized and classical trajectories can be found. We showed that, according to the canonical quantization, conjugate momenta can be promoted to operators by means of the relation $\pi_a = -i\dfrac{\partial}{\partial q^a}$. Moreover, in the WKB approximation, the Wave Function of the Universe can be recast as $\psi(h_{ij}, \phi) \sim e^{iS}$ so that the Hamilton–Jacobi equations $\dfrac{\partial S}{\partial q^a} = \pi_a$ can be applied to the action in the exponent, providing the dynamics for the generic variable q^a. According to the considerations in Chapter 5, the wave function can be also used to recover the dynamics that describes observable universes. This confirms that the presence of symmetries implies the validity of the Hartle criterion, according to which the wave function is oscillating in the classically permitted region, as discussed in [220].

To be more precise, the ADM formalism can be used to show that the Hartle criterion can be straightforwardly linked to the Noether theorem and then the presence of symmetries allows the birth of classically observable universes. To this purpose, by means of Eq. (2.37), the set of conserved quantities j_i, associated to symmetry transformations, can be linked to the conjugate momenta of the new cyclic variables Q^i, according to the relation

$$j_i \equiv \pi^i_{Q^i} = \frac{\partial \mathcal{L}}{\partial \dot{Q}^i}. \tag{17.1}$$

The quantum counterpart of Eq. (17.1), together with the primary constraints on the quantized Hamiltonian, provide the following set of $m+1$ differential equations:

$$\begin{cases} \hat{\mathscr{H}} H|\psi>=0 \\ -i\partial_1|\psi>=j_1|\psi> \\ -i\partial_2|\psi>=j_2|\psi> \\ \vdots \\ -i\partial_m|\psi>=j_m|\psi>, \end{cases} \tag{17.2}$$

whose common solution, in general, is

$$|\psi>\sim e^{ij_k Q^k}, \tag{17.3}$$

with m being the number of Noether symmetries. The oscillating wave function confirms that the Hartle criterion can always be addressed to symmetries [220].

In this chapter, we adopt the above prescriptions to modified/extended theories of gravity-containing symmetries in a cosmological background. In Section 17.1, Quantum Cosmology is considered in the framework of $f(R)$ gravity and the initial action is chosen according to symmetry considerations. In Section 17.2, the same approach is applied to teleparallel equivalent gravity. In Section 17.3, the Hamiltonian formalism is applied to Gauss–Bonnet gravity. In Section 17.4, Quantum Cosmology is adopted for higher-order gravity. Finally, in Section 17.5, we use the ADM formalism to compare three different scalar-tensor actions containing symmetries and show their equivalence.

17.1 Quantum Cosmology in $f(R)$ Gravity

Let us discuss now Quantum Cosmology for $f(R)$ gravity, by considering the Lagrangian containing symmetries, selected in Section 7.1. Therefore, we deal with the Hamiltonian formalism and solve the WDW equation to find the Wave Function of the Universe. Specifically, we focus on the function $f(R) = f_0 R^{\frac{3}{2}}$, and recover classical trajectories in the semiclassical limit. Before considering the ADM formalism and find the related Hamiltonian, we use the features of Noether's approach to reduce dynamics, by introducing a cyclic variable in the minisuperspace. To this purpose, Eq. (2.37) allows one to pass from the minisuperspace $\mathcal{S} \equiv \{a, R\}$ to $\mathcal{S}' \equiv \{z, w\}$, by means of the following system of differential equations:

$$\begin{cases} \alpha\partial_a z(a,R) + \beta\partial_R z(a,R) = 1 \\ \alpha\partial_a w(a,R) + \beta\partial_R w(a,R) = 0, \end{cases} \tag{17.4}$$

with z being the new cyclic variable. A possible solution is

$$\begin{cases} w = w_0(a\sqrt{R})^\ell, & z = \dfrac{a^2}{2\alpha_0} \\ a = \sqrt{2\alpha_0 z}, & R = \dfrac{1}{2\alpha_0} z \left(\dfrac{w}{w_0}\right)^{\frac{2}{\ell}}, \end{cases} \tag{17.5}$$

being ℓ and w_0 integration constants. Substituting Eq. (17.5) into Eq. (7.5), the Lagrangian turns out to be

$$\mathcal{L} = \frac{f_0}{\ell w}\left(\frac{w}{w_0}\right)^{\frac{1}{\ell}}\left[18\alpha_0 \dot{w}\dot{z} - w\ell\left(\frac{w}{w_0}\right)^{\frac{2}{\ell}}\right]. \tag{17.6}$$

The new equations of motion are

$$\begin{cases} (\ell-1)\dot{w}^2 - \ell w\ddot{w} = 0 \\ \left(\dfrac{w}{w_0}\right)^{\frac{2}{\ell}} + 6\alpha_0 \ddot{z} = 0\,, \end{cases} \tag{17.7}$$

where z is clearly the cyclic variable. The Legendre transformation $\mathcal{H} = \pi_i\dot{q}^i - \mathcal{L}$ provides the Hamiltonian

$$\mathcal{H} = \frac{\ell}{18\alpha_0 f_0} w \left(\frac{w}{w_0}\right)^{-\frac{1}{\ell}} \pi_w\pi_z + f_0\left(\frac{w}{w_0}\right)^{\frac{3}{\ell}}. \tag{17.8}$$

By promoting the momenta to operators, that is, $\pi_i \to -i\partial_i$, the primary and secondary constraints in Eqs. (5.13) and (5.22) read as

$$\begin{cases} \hat{\mathcal{H}}\psi = \left[\dfrac{\ell}{18\alpha_0 f_0} w\left(\dfrac{w}{w_0}\right)^{-\frac{1}{\ell}} \partial_w\partial_z + f_0\left(\dfrac{w}{w_0}\right)^{\frac{3}{\ell}}\right]\psi = 0 \\ \hat{\pi}_z\psi = -i\partial_z\psi = j_0\psi, \end{cases} \tag{17.9}$$

where the former is the WDW equation. The solution of the above system yields the following wave function:

$$\psi(z,w) = \psi_0 \exp\left\{ i\left[j_0 z - \frac{9\alpha_0 f_0^2}{2j_0}\left(\frac{w}{w_0}\right)^{\frac{4}{\ell}}\right]\right\}. \tag{17.10}$$

In the semiclassical limit, according to the WKB approximation, the wave function can be recast in terms of the action S as

$$\psi \sim e^{iS}, \tag{17.11}$$

so that, comparing Eq. (17.11) with Eq. (17.10), the action turns out to be

$$S = j_0 z - \frac{9\alpha_0 f_0^2}{2j_0}\left(\frac{w}{w_0}\right)^{\frac{4}{\ell}}. \tag{17.12}$$

It can be easily proven that the Hamilton–Jacobi equations $\dfrac{\partial S}{\partial q^i} = \pi_i$ exactly provide the same system as Eq. (17.7) and then classical solutions. Also notice that, due to the oscillatory behavior of the wave function, the Hartle criterion is recovered by the Noether Symmetry Approach (see also Chapter 5). In this section, we considered only $f(R) = f_0 R^{3/2}$ gravity, but the same procedure can be applied to the other functions provided in Eqs. (7.17).

In general, finding symmetries allows one to reduce dynamics and introduce cyclic variables. In this way, the Hamiltonian can be straightforwardly quantized and the WDW equation can be analytically solved. Other applications of $f(R)$ gravity to Quantum Cosmology can be found, for example, in [89] or [454]. In the latter reference, the authors use the reduced dynamics provided by Noether symmetries in order to obtain a class of tachyonic potentials.

17.2 Quantum Cosmology in $f(T)$ Gravity

Considering Noether's system solutions of $f(T)$ gravity (7.33), internal symmetries can be recovered by setting $\xi(t) = 0$. Equivalently, it is possible to impose the Lie derivative of Lagrangian (7.29) to vanish identically. See, for example, [254]. The latter condition yields a reduced Noether system of differential equations of the form

$$\begin{cases} \alpha f_T(T) + 2af_T(T)(\partial_a \alpha) + \beta a f_{TT}(T) = 0 \\ \partial_T \alpha = 0 \\ 3\alpha a^2 f(T) - 3\alpha a^2 T f_T(T) - \beta a^3 T f_{TT}(T) = 0, \end{cases} \tag{17.13}$$

whose only solution is

$$\alpha = \alpha_0 a^{1-\frac{3}{2k}}, \qquad \beta = \frac{-3\alpha_0}{k} T a^{-\frac{3}{2k}}. \tag{17.14}$$

The preceding solution is equivalent to the first one of Eq. (7.33), with $\xi_0 = \beta_1 = \beta_2 = 0$, and can be used to find a suitable change of variables that can solve the WDW equation. Thus, relations (2.37) allow one to introduce a cyclic variable into the system, passing from the minisuperspace $\mathcal{S} \equiv \{a, T\}$ to $\mathcal{S}' \equiv \{z, w\}$ with z being a cyclic variable, that is:

$$\begin{cases} \alpha \partial_a z + \beta \partial_T z = 1 \\ \alpha \partial_a w + \beta \partial_T w = 0. \end{cases} \tag{17.15}$$

By replacing the expressions of α and β in Eq. (17.15), a possible solution of the preceding system is

$$\begin{cases} z = \dfrac{2k}{3\alpha_0} a^{\frac{3}{2k}}, & \rightarrow \quad a = \left(\dfrac{3\alpha_0 z}{2k}\right)^{\frac{2k}{3}} \\ w = a^3 T^k, & \rightarrow \quad T = w^{\frac{1}{k}} \left(\dfrac{3\alpha_0 z}{2k}\right)^{-2}, \end{cases} \tag{17.16}$$

so that the Lagrangian written in terms of the new variables reads

$$\mathcal{L} = w(1-k) - 6k\alpha_0^2 \dot{z}^2 w^{\frac{k-1}{k}} , \tag{17.17}$$

where z is the cyclic variable. It is worth noticing that, once the minisuperspace coordinates are changed, the equations of motion coming from Lagrangian (17.17) are simpler than those in Eq. (7.30). They take the form

$$z \rightarrow kw\ddot{z} + \dot{w}\dot{z}(k-1) = 0, \tag{17.18}$$

$$w \rightarrow w = -6^k \alpha_0^{2k} \dot{z}^{2k}. \tag{17.19}$$

Following the same steps as in $f(R)$ gravity, from Lagrangian (17.17), with a straightforward Legendre transformation, we get the Hamiltonian:

$$\mathcal{H} = w(k-1) + \frac{3\pi_z^2}{24k\alpha_0^2 w^{\frac{k-1}{k}}} . \tag{17.20}$$

The canonical quantization procedure can be pursued by replacing the momentum π_z with $i\partial_z$. Thus, by applying the relation (5.22) to the Hamiltonian (17.20), we get the system

$$\begin{cases} \mathcal{H}\psi = 0 \\ i\partial_z\psi = j_0\psi, \end{cases} \tag{17.21}$$

where the first is the WDW equation and the second is the momentum conservation equation. Explicitly, we have

$$\begin{cases} \left[w(k-1) - \dfrac{3\partial_z^2}{24k\alpha_0^2 w^{\frac{k-1}{k}}} \right] \psi = 0 \\ i\partial_z\psi = j_0\psi . \end{cases} \tag{17.22}$$

The solution of the second equation is

$$\psi = \psi_0 e^{ij_0 z} , \tag{17.23}$$

which, combined with the first, yields the Wave Function of the Universe, that is:

$$\psi \sim \exp\left\{ i \left[2\alpha_0 w^{\frac{2k-1}{2k}} \sqrt{2k(k-1)} \right] z \right\} . \tag{17.24}$$

The oscillating wave function, even this time, suggests that the Hartle criterion is recovered and classical trajectories can be selected. To this purpose, we consider the semiclassical approach and recast ψ in terms of the action S as

$$\psi \sim \exp\{iS\} . \tag{17.25}$$

Identifying the action with the quantity

$$S = \left[2\alpha_0 w^{\frac{2k-1}{2k}} \sqrt{2k(k-1)} \right] z,$$

the Hamilton–Jacobi equations give

$$\begin{cases} \dfrac{\partial S}{\partial z} = \pi_z = j_0 \\ \dfrac{\partial S}{\partial w} = \pi_w = 0 \end{cases} \tag{17.26}$$

Note that the preceding equations are nothing else but the EL equations for the new variables defined in Eqs. (17.18) and (17.19). The general solution is

$$z(t) = z_0 t, \quad w = -6^k \alpha_0^{2k} z_0^{2k}. \tag{17.27}$$

Coming back to the old variables, the scale factor and the torsion scalar can be written as functions of time for any $f(T) = f_0 T^k$:

$$a(t) = a_0 t^{\frac{2k}{3}}, \qquad T(t) = -\frac{8k^2}{3}\frac{1}{t^2}. \tag{17.28}$$

Once merged with the energy condition $(\dot{z}\partial_{\dot{z}} + \dot{w}\partial_{\dot{w}} - 1)\mathcal{L} = 0$, the EL equations are further constrained to the subcase $k = 1/2$, which turns out to be one of the cases discussed in Section 7.2. In summary, the existence of the Noether symmetry allows one to integrate exactly the dynamical system and to find classical trajectories describing observable universes.

17.3 Quantum Cosmology in $f(\mathcal{G})$ Gravity

The Hamiltonian formalism can be easily developed also for $f(\mathcal{G})$ gravity by considering the $d = 3$ limit of Lagrangian (8.52), namely

$$\mathcal{L} = a^3[f(\mathcal{G}) - \mathcal{G}f_{\mathcal{G}}(\mathcal{G})] - 8\dot{a}^3\dot{\mathcal{G}}f_{\mathcal{G}\mathcal{G}}(\mathcal{G}). \tag{17.29}$$

Taking into account the function selected by symmetry, here we use the WDW equation to find the Wave Function of the Universe and the corresponding classical trajectories. Substituting $f(\mathcal{G}) = f_0\mathcal{G}^k$ into Eq. (17.29), we get

$$\mathcal{L} = -\frac{1}{3}\mathcal{G}^{k-2}\left[3(k-1)a^3\mathcal{G}^2 + 24k(k-1)\dot{a}^3\dot{\mathcal{G}}\right]. \tag{17.30}$$

Starting from the preceding Lagrangian, we can calculate the related Hamiltonian as a function of momenta:

$$\mathcal{H} = \frac{f_0}{k}\mathcal{G}^k a^3 + \pi_a\left(-\frac{\pi_{\mathcal{G}}}{8f_0}\mathcal{G}^{2-k}\right)^{\frac{1}{3}}, \tag{17.31}$$

where $\pi_a = \dfrac{\partial\mathcal{L}}{\partial\dot{a}}$ and $\pi_{\mathcal{G}} = \dfrac{\partial\mathcal{L}}{\partial\dot{\mathcal{G}}}$, according to the Legendre transformations.

In this form, the Hamiltonian cannot be quantized due to the presence of the fractional exponent. Nonetheless, thanks to the Noether symmetries, we can insert a cyclic variable into Eq. (17.31), which allows one to fully quantize the theory. Specifically, from Eq. (8.63), it is easy to see that the quantity $\frac{\dot{a}}{\mathcal{G}^k}$ is a constant of motion. Therefore, by defining

$$\frac{\dot{a}^3}{\mathcal{G}^{3k}} = \Sigma_0, \tag{17.32}$$

the momentum $\pi_{\mathcal{G}}$ can be rewritten in terms of Σ_0 as

$$\pi_{\mathcal{G}} = -8f_0\Sigma_0\mathcal{G}^{4k-2}. \tag{17.33}$$

When this result is substituted into Eq. (17.31), the new Hamiltonian written in the tangent minisuperspace $\mathcal{TS} \equiv \{a, \pi_a, \mathcal{G}\}$ reads

$$\mathcal{H} = \frac{f_0}{k}\mathcal{G}^k a^3 + \pi_a \left(\Sigma_0\mathcal{G}^{3k}\right)^{\frac{1}{3}} . \tag{17.34}$$

Now, thanks to the quantization rules coming from the ADM formalism, it is possible to recast the conjugate momenta as operators:

$$\begin{cases} \pi_{\mathcal{G}} = -i\dfrac{\partial}{\partial\mathcal{G}} \\ \pi_a = -i\dfrac{\partial}{\partial a} \\ \mathcal{H}\psi = 0 \,. \end{cases} \tag{17.35}$$

The third equation is the WDW equation, and ψ is the Wave Function of the Universe. Being $\pi_{\mathcal{G}} = -8f_0\Sigma_0\mathcal{G}^{4k-2}$, it follows that the quantity $\pi_{\mathcal{G}}\mathcal{G}^{2-4k}$ is a constant of motion. More precisely, the quantized equation of momentum can be written as

$$i\frac{\partial}{\partial\mathcal{G}}\psi(a,\mathcal{G}) = 8f_0\Sigma_0\mathcal{G}^{4k-2}\psi(a,\mathcal{G}), \tag{17.36}$$

so that we get the system

$$\begin{cases} \pi_{\mathcal{G}}\psi = -i\dfrac{\partial}{\partial\mathcal{G}}\psi \quad \rightarrow \quad \psi(a,\mathcal{G}) = A(a)\, \exp\left\{i\,\dfrac{8f_0\Sigma_0\mathcal{G}^{4k-1}}{1-4k}\right\} \\ \mathcal{H}\psi = 0 \quad \rightarrow \quad \dfrac{f_0}{k}\left(\Sigma_0\right)^{-\frac{1}{3}} a^3 A(a) - i\dfrac{\partial A(a)}{\partial a} = 0\,. \end{cases} \tag{17.37}$$

The latter equation can be solved with respect to $A(a)$, and the result can be substituted into the former one. After some computations, the wave function turns out to be

$$\psi(a,\mathcal{G}) = \psi_0 \exp\left\{i\left[-\frac{f_0}{4k}\left(\Sigma_0\right)^{-\frac{1}{3}} a^4 + \frac{8f_0\Sigma_0\mathcal{G}^{4k-1}}{1-4k}\right]\right\} . \tag{17.38}$$

According to the Hartle criterion, an oscillating wave function means correlations among variables and then the possibility of finding classical trajectories (i.e. observable universes). In fact, considering the WKB approximation, the wave function can be recast as $\psi(a, \mathcal{G}) \sim e^{iS}$ (where S is the action), so that, in the $f(\mathcal{G})$ model, the action can be identified with the quantity

$$S = -\frac{f_0}{4k}(\Sigma_0)^{-\frac{1}{3}} a^4 + \frac{8f_0\Sigma_0}{1-4k}\mathcal{G}^{4k-1}. \tag{17.39}$$

The Hamilton–Jacobi equation with respect to the scale factor provides the $d = 3$ limit of the third equation of (8.54), namely:

$$\frac{\partial S}{\partial a} = \pi_a \quad \to \quad \mathcal{G}^k a^3 = 24\mathcal{G}^{k-2}\dot{a}^3\dot{\mathcal{G}}\,. \tag{17.40}$$

The second Hamilton–Jacobi equation $\left(\frac{\partial S}{\partial \mathcal{G}} = \pi_{\mathcal{G}}\right)$, instead, is nothing else but the identity $\pi_{\mathcal{G}} = \Sigma_0$. The energy condition is trivially recovered as a linear combination of the two, so that observable universes are selected by Noether's approach.

Notice, however, that this approach is quite different from those of the previous sections. In order to get a suitable Hamiltonian, capable of providing an analytic expression for the wave function, we considered the cosmological solutions (8.62). Then, we replaced a conserved quantity into the Hamiltonian without using the methodical procedure (2.37). This is due to the fact that we did not develop the analogue of Eq. (2.37) for the first prolongation of Noether vector, so that a system that allows one to introduce a cyclic variable has not been written in this case.

17.4 Quantum Cosmology in Higher-Derivatives Gravity

Here we apply the ADM formalism to the last solution (IV) of Eq. (9.16), introducing a cyclic variable and finding the Wave Function of the Universe. Considering the notation adopted in Chapter 9, where we set $\Box R \equiv \chi$, the starting action is

$$S = \int \sqrt{-g}\,[f_0\sqrt{R\chi} + f_1 R]\,d^4x. \tag{17.41}$$

The action (17.41), after applying the Lagrange multipliers method and integrating out four divergences, yields the Lagrangian

$$\mathcal{L} = 6f_1 a\dot{a}^2 + 3f_0 a\dot{a}^2\sqrt{\frac{\chi}{R}} + \frac{3}{2}f_0 a^2\dot{a}\dot{\chi}\left(\frac{1}{\sqrt{R\chi}}\right) - \frac{3}{2}f_0 a^2\dot{a}\dot{R}\sqrt{\frac{\chi}{R^3}}$$
$$+ \frac{1}{4}f_0 a^3\dot{R}\dot{\chi}\sqrt{\frac{R}{(\chi)^3}} - \frac{1}{4}f_0 a^3\dot{R}^2\left(\frac{1}{\sqrt{R\chi}}\right). \tag{17.42}$$

In order to introduce a cyclic variable z into Lagrangian (17.42), we use Eq. (2.37), by means of which the minisuperspace $\mathcal{S} \equiv \{a, R, \chi\}$ can be transformed to $\mathcal{S}' \equiv \{z, w, y\}$. Therefore, we substitute the third generator of (9.12) into Eq. (2.37), obtaining

$$\begin{cases} \partial_R z + \dfrac{\chi}{R}\partial_\chi z = \dfrac{1}{\beta_0} \\ \partial_R w + \dfrac{\chi}{R}\partial_\chi w = 0 \\ \partial_R y + \dfrac{\chi}{R}\partial_\chi y = 0. \end{cases} \tag{17.43}$$

A possible solution is

$$z = R, \quad y = \frac{\chi}{R}, \quad w = a. \tag{17.44}$$

By means of the preceding change of variables, the Lagrangian can be recast as

$$\mathcal{L} = 3f_0 w^2 \dot{y}\dot{w} + 3f_0 y w \dot{w}^2 + 6f_1 w \dot{w}^2 + \frac{1}{2} f_0 \frac{w^3}{y} \dot{y}\dot{z}, \tag{17.45}$$

and z is cyclic, as expected. The equations of motion are considerably reduced with respect to those in Eq. (9.10). They read

$$\begin{cases} 3y\dot{y}\dot{w} + w(y\ddot{y} - 2\dot{y}^2) = 0 \\ 2(2f_1 + f_0 y)\dot{w}^2 + \dfrac{w^2}{y^2}(2f_0 y^2 \ddot{y} - f_0 \dot{y}\dot{z}) + 4wf_0\dot{y}\dot{w} + 4w(2f_1 + f_0 y)\ddot{w} \\ 6y^2(\dot{w}^2 + w\ddot{w}) + w(3\dot{w}\dot{z} + w\ddot{z}). \end{cases} \tag{17.46}$$

Adopting the ADM formalism, we quantize the cosmological Hamiltonian

$$\mathcal{H} = \left(\frac{24}{f_0} y^2 \pi_y \pi_z\right) - \frac{(6\pi_z y^2 - w\pi_w)^2}{2f_1 + f_0 y} \tag{17.47}$$

and get the WDW equation. For simplicity, we choose $f_1 = 0$, and then the Hamiltonian operator can be written as

$$\hat{\mathcal{H}} = -24y^3\partial_y\partial_z - 36y^4\partial_z^2 + 12y^2 w\partial_w\partial_z - w^2\partial_w^2. \tag{17.48}$$

Imposing the conditions (5.13) and (5.22), we obtain a system of differential equations that must be solved with respect to the wave function ψ, that is:

$$\begin{cases} \left(-24y^3\partial_y\partial_z - 36y^4\partial_z^2 + 12y^2 w\partial_w\partial_z - w^2\partial_w^2\right)\psi = 0 \\ \\ -i\partial_z\psi = j_0\psi \quad \rightarrow \quad \psi \equiv \psi(w, y)e^{-ij_0 z}. \end{cases} \tag{17.49}$$

The only solution of the preceding system is [89]

$$\psi(z, w, y) = e^{-ij_0 z}\chi(w)\Phi(y), \tag{17.50}$$

with $\chi(w)$ and $\Phi(y)$ being linear combinations of Bessel functions. Due to their asymptotic oscillating behavior, the Hartle criterion is recovered and the wave function describes classical universes in the late-time.

17.4.1 Higher-Order Teleparallel Equivalent

Here we focus on the first solution of the Noether system (9.29), discussing the Hamiltonian dynamics and solving the WDW equation in the minisuperspace of the three variables $a, T, \Box T$. According to the notation used in Chapter 9, we set $\Box T \equiv Y$. Therefore, the function considered is $f(T, Y) = f_0 T^{1-k}(Y)^k$. In the limit $k = \frac{1}{2}$, such a function takes the form $f(T, Y) = f_0\sqrt{TY}$ and it is formally equivalent to the third solution of Eq. (9.16).

To find the Hamiltonian and the corresponding wave function, we use the condition in Eq. (2.37), by means of which the minisuperspace $\mathcal{S} \equiv \{a, T, Y\}$ can be transformed to $\mathcal{S}' \equiv \{z, w, y\}$, where z is a cyclic variable. The change of variables yields the system

$$\begin{cases} \alpha\dfrac{\partial z}{\partial a} + \beta\dfrac{\partial z}{\partial T} + \gamma\dfrac{\partial z}{\partial Y} = 1 \\ \alpha\dfrac{\partial w}{\partial a} + \beta\dfrac{\partial w}{\partial T} + \gamma\dfrac{\partial w}{\partial Y} = 0 \\ \alpha\dfrac{\partial y}{\partial a} + \beta\dfrac{\partial y}{\partial T} + \gamma\dfrac{\partial y}{\partial Y} = 0\,, \end{cases} \tag{17.51}$$

whose possible solution is

$$z = z(T) = \frac{T}{\beta_0}, \qquad w = \frac{Y}{T}, \qquad y = a, \tag{17.52}$$

from which the variables T and Y can be written in terms of the new variables w and z as

$$T = \beta_0 z, \qquad Y = \beta_0 wz, \quad a = y. \tag{17.53}$$

The Lagrangian $\mathcal{L}(z, w, y)$ turns out to be[1]

$$\mathcal{L}_{new} = f_0(k-1)\left\{6a\dot{a}^2 w^k - k\beta_0 a^3 w^{k-2}\dot{z}\dot{w}\right\}. \tag{17.54}$$

As expected, the form of the Lagrangian suggests that z is the cyclic variable, so that the constant of motion j_0 is the conjugate momentum related to z:

$$j_0 = \frac{\partial \mathcal{L}}{\partial \dot{z}} = -f_0\beta_0 k(k-1)a^3 w^{k-2}\dot{w}\,. \tag{17.55}$$

For the sake of simplicity, we define $f_0 k(k-1)\beta_0 = p_0$ and $12 f_0(k-1) = p_1$, so that the conjugate momenta can be written as

$$\begin{cases} \pi_z = \dfrac{\partial \mathcal{L}}{\partial \dot{z}} = -p_0 a^3 w^{k-2}\dot{w} \\ \pi_w = -p_0 a^3 w^{k-2}\dot{z} \\ \pi_a = p_1 a w^k \dot{a} \end{cases} \tag{17.56}$$

[1] Since $y = a$, we write the Lagrangian in terms of a.

and, finally, the Hamiltonian reads

$$\mathcal{H} = -\frac{1}{p_0 a^3 w^{k-2}}\left(\pi_z^2 + \pi_w^2 - \pi_z \pi_w\right) - \frac{1}{2}\frac{\pi_a^2}{p_1 a w^k} \,. \tag{17.57}$$

The constraints imposed by Eq. (5.22) provide the system of differential equations

$$\begin{cases} \left(\partial_z^2 + \partial_w^2 + \partial_z \partial_w\right)\psi + \frac{1}{2}\frac{p_0}{p_1}\frac{a^2}{w^2}\partial_a^2\psi = 0 \\ -i\partial_z \psi = j_0 \psi \,. \end{cases} \tag{17.58}$$

The general solution is a linear combination of Bessel functions, whose asymptotic limit provides an oscillating wave function of the form

$$\begin{aligned} \psi \sim \quad & \exp\left\{ i \left[\frac{1}{2}\ln w - \frac{\sqrt{3}}{2} j_0 w - \frac{\pi}{4}\left(1 + \sqrt{4c_0 + 1}\right) \right.\right. \\ & \left.\left. - \; j_0 z + \sqrt{c_0}\ln a \right] \right\} . \end{aligned} \tag{17.59}$$

Notice that the Hartle criterion is then recovered for large-scale factors, where the wave function is peaked in the minisuperspace $\mathcal{S}'$. Moreover, classical trajectories can be recovered by identifying the exponential factor of Eq. (17.59) with S, namely:

$$S = \frac{1}{2}\ln w - \frac{\sqrt{3}}{2} j_0 w - \frac{\pi}{4}\left(1 + \sqrt{4c_0 + 1}\right) - j_0 z + \sqrt{c_0}\ln a, \tag{17.60}$$

and the wave function can be recast as $\psi \sim e^{iS}$. The Hamilton–Jacobi equations coming from the action (17.60) read as

$$\frac{\partial S}{\partial a} = \pi_a \,, \quad \frac{\partial S}{\partial z} = \pi_z \,, \quad \frac{\partial S}{\partial w} = \pi_w \,, \tag{17.61}$$

with

$$\pi_a = \frac{\sqrt{c_0}}{a}, \qquad \pi_w = \frac{1}{2w} - \frac{\sqrt{3}}{2} j_0, \qquad \pi_z = j_0 \,, \tag{17.62}$$

from which a system of three differential equations follows:

$$\begin{cases} p_1 a w^k \dot{a} = \frac{\sqrt{c_0}}{a} \\ p_0 a^3 w^{k-2} \dot{z} = \frac{\sqrt{3}}{2} j_0 - \frac{1}{2w} \\ p_0 a^3 w^{k-2} \dot{w} = j_0 \,. \end{cases} \tag{17.63}$$

The solutions of the preceding system are the same as those for Eq. (9.33), so that classical trajectories, and then observable universes, are immediately recovered [184].

17.5 Quantum Cosmology in Scalar-Tensor Gravity

Noether's symmetries provide transformations allowing one to introduce cyclic variables in point-like scalar-tensor Lagrangians. In order to get internal symmetries, we set the infinitesimal generator related to the time variation equal to zero. In this way, by the change of variable (2.37), we select minisuperspaces where the Hamiltonian formalism can be dealt with. Therefore, we use the WDW equation to find the Wave Function of the Universe and recover classical trajectories. Specifically, in the last subsection, three different representations of scalar-tensor cosmology are considered, namely the couplings between scalar field and curvature, torsion, and Gauss–Bonnet terms.

17.5.1 Curvature Scalar-Tensor Gravity

We consider the second solution of (10.19), that is,

$$\begin{cases} \mathcal{X} = -\dfrac{2(s+1)}{2s+3}\beta_0 a^{s+1}\phi^{\frac{2s^2+4s}{2s+3}}\partial_a + \beta_0 a^s \phi^{\frac{2s^2+6s+3}{2s+3}}\partial_\phi, \qquad g=0 \\ f(\phi)=\ell(s)\phi^2, \quad V(\phi)=V_0\phi^{\frac{6(s+1)}{2s+3}}, \qquad \ell(s)=\dfrac{(2s+3)^2}{48(s+1)(s+2)} \\ j = a^{s+2}\phi^{\frac{2s^2+6s+3}{2s+3}}\left[(2s+3)\phi\dot{a}+2(s+1)a\dot{\phi}\right], \quad s\neq 0, \end{cases} \tag{17.64}$$

discussed in [455], to develop the Hamiltonian formalism in scalar-tensor gravity, where the scalar field is non-minimally coupled to the Ricci curvature scalar as in Eq. (10.14). In Section 17.5.3, the limit $s=0$ of the preceding symmetry is considered and the comparison with other scalar-tensor theories is discussed. For further readings about the Hamiltonian formalism of scalar-tensor theories with Noether symmetries, see [316,456–458], where the authors start from Eq. (10.15) and use relations (2.37) to properly change variables and reduce dynamics.

Here we apply Eq. (2.37) to solution (17.64) with the aim of introducing a cyclic variable z into the Lagrangian. The procedure yields the system

$$\begin{cases} -\dfrac{2(s+1)}{2s+3}\beta_0 a^{s+1}\phi^{\frac{2s^2+4s}{2s+3}}\partial_a z + \beta_0 a^s\phi^{\frac{2s^2+6s+3}{2s+3}}\partial_\phi z = 1 \\ -\dfrac{2(s+1)}{2s+3}\beta_0 a^{s+1}\phi^{\frac{2s^2+4s}{2s+3}}\partial_a w + \beta_0 a^s\phi^{\frac{2s^2+6s+3}{2s+3}}\partial_\phi w = 0, \end{cases} \tag{17.65}$$

whose possible solution is

$$z = -\frac{1}{2\beta_0}\frac{2s+3}{s}a^{-s}\phi^{-\left(\frac{2s^2+4s}{2s+3}\right)}, \qquad w = w_0 a^3 \phi^{\frac{6(s+1)}{2s+3}}. \tag{17.66}$$

Considering Eqs. (17.66) and the definition $\gamma(s) = \dfrac{1}{w_0^s}\dfrac{2s+3}{12(s+2)(s+1)}$, the Lagrangian can be written in terms of the new variables as

$$\mathcal{L} = \gamma(s) w^{\frac{s}{3}}\dot{z}\dot{w} - V_0 w. \tag{17.67}$$

Dynamics of the system is now reduced and thus the Hamiltonian can be quantized, providing a solution for the WDW equation. The Legendre transformation of (17.67) gives

$$\mathcal{H} = \frac{\pi_z \pi_w}{\gamma(s) w^{\frac{s}{3}}} + V_0 w, \tag{17.68}$$

where the first term of Eq. (17.68) can be intended as a kinetic term and the second as a potential. Imposing the canonical quantization rules (5.22), we get a system of two differential equations:

$$\begin{cases} \left[(i\partial_z)(i\partial_w) + V_0 \gamma(s) w^{(s+3)/3}\right]\psi = 0 \\ -i\partial_z \psi = j_0 \psi, \end{cases} \tag{17.69}$$

whose solution is

$$\psi(z, w) = \psi_0 \exp\left\{-i\left[j_0 z + \gamma(s) V_0 w^{\frac{s+6}{3}}\right]\right\}. \tag{17.70}$$

According to Hartle's criterion, the oscillating behavior of the wave function can be addressed to observable universes, so that the Hamilton–Jacobi equations can provide classical cosmological solutions. By means of the WKB approximation, the exponent in Eq. (17.70) can be intended as the scalar-tensor action, namely:

$$S = j_0 z + \gamma(s) V_0 w^{\frac{s+6}{3}}. \tag{17.71}$$

Therefore, the Hamilton–Jacobi equations provide

$$z(t) = [k_1 t + k_2]^{\frac{s+6}{s+3}} + z_0, \qquad w(t) = [k_1 t + k_2]^{\frac{3}{s+3}}. \tag{17.72}$$

Finally, coming back to the old variables, we obtain

$$a(t) = a_0 t^{\frac{5}{3} + \frac{2}{s}}, \qquad \phi = \phi_0 t^{-(2 + \frac{3}{s})},$$

which can be directly addressed to Eq. (10.22) by means of the definition $k = \dfrac{6(s+1)}{2s+3}$.

17.5.2 Torsion Scalar-Tensor Gravity

The Wave Function of the Universe can be derived also when torsion scalar is non-minimally coupled to a scalar field. We focus on the last symmetry of Eq. (10.72), that is,

$$\begin{cases} \mathcal{X} = (-\frac{\xi_0}{3} t + \xi_1)\partial_t - \frac{2}{3}\beta_0\sqrt{V_0}\, a\partial_a + \beta_0\sqrt{V_0}\,\phi\partial_\phi \\ f(\phi) = f_0\phi^2, \qquad V(\phi) = V_0\phi^2 \\ j_0 = \beta_0\sqrt{V_0}\, a^2\phi(8f_0\phi\dot{a} + a\dot{\phi}). \end{cases} \tag{17.73}$$

Here we show that (10.75) and (10.76) can be recovered from the Hamilton–Jacobi equations in the semiclassical limit. However, we have to change the variables by using the inner derivative in Eq. (2.37). Now, by replacing the generator (17.73) in Eq. (2.37) and setting $\xi_0 = \xi_1 = 0$, we obtain

$$\begin{cases} -\frac{2}{3}a\partial_a z + \phi\partial_\phi z = 1 \\ -\frac{2}{3}a\partial_a w + \phi\partial_\phi w = 0\,. \end{cases} \tag{17.74}$$

Assuming $\partial_a z = 0$, one of the possible solutions is

$$\begin{cases} z = \ln\phi \\ w = a^3\phi^2 \end{cases} \qquad \begin{cases} \phi = e^z \\ a = w^{\frac{1}{3}}e^{-\frac{2}{3}z} \end{cases} \tag{17.75}$$

and the new Lagrangian is then

$$\mathcal{L}_{new} = -\frac{2}{3}f_0\frac{\dot{w}^2}{w} + \frac{8}{3}f_0\dot{w}\dot{z} + F_0 w\dot{z}^2 - V_0 w\,, \tag{17.76}$$

where we set $F_0 = 1 - 8/3\,f_0$. Here z is the cyclic variable, and the conserved quantity can be written in terms of the momentum π_z as

$$\partial_{\dot{z}}\mathcal{L} = \pi_z = \frac{8}{3}f_0\dot{w} + 2f_0 w\dot{z} = j_0. \tag{17.77}$$

The ADM formalism can be pursued by performing a Legendre transformation. The procedure yields the Hamiltonian:

$$\hat{\mathcal{H}} = -\frac{\partial_z^2}{2w} - 2\partial_w\partial_z + \frac{1}{2}\frac{f_0}{F_0}w\partial_w^2 + V_0 w. \tag{17.78}$$

The Wave Function of the Universe is the solution of the system

$$\begin{cases} -\frac{1}{2w}\partial_z^2\psi(z,w) - 2\partial_w\partial_z\psi(z,w) \\ +\frac{1}{2}\frac{f_0}{F_0}w\partial_w^2\psi(z,w) + V_0 w\psi(z,w) = 0 \\ i\partial_z\psi(z,w) = j_0\psi(z,w)\,. \end{cases} \tag{17.79}$$

The first equation is the WDW equation, coming from the condition $\hat{\mathcal{H}}\psi = 0$. The second equation is the conservation relation provided by the Noether symmetry, namely $\pi_z = j_0$. The solution of (17.79) is a linear combination of Bessel's functions, which, for large arguments, are peaked in the minisuperspace variables z, w. Therefore the wave function has an asymptotic behavior of the form

$$\psi(z,w) \sim e^{i(j_0 z - \ln w + Ww - p\frac{\pi}{2} - \frac{\pi}{4})}\,, \tag{17.80}$$

with the definitions

$$p = \frac{1}{2}\sqrt{-A^2 - 2iA - 4B + 1}$$

$$-4j_0\frac{f_0}{F_0} = A; \qquad \frac{2F_0}{f_0}V_0 = W^2; \qquad \frac{F_0}{f_0}j_0^2 = B\,. \tag{17.81}$$

Notice that Hartle's criterion is recovered after imposing the change of variables suggested by the Noether symmetry. Furthermore, in the semiclassical limit where the wave function can be recast as $\psi \sim e^{iS}$, the Hamilton–Jacobi equations yield a system of two differential equations, that is,

$$\pi_z = \partial_z S = j_0; \qquad \pi_w = \partial_w S = W - \frac{1}{w}, \tag{17.82}$$

whose solution is

$$a(t) \sim e^{kt}, \quad \phi(t) \sim e^{\ell t}. \tag{17.83}$$

Eqs. (17.82) provide the same exponential solution as the EL equations, as expected by construction. The power-law solution (10.76) can be recovered for vanishing potential $V(\phi) = 0$. Notice that even in this case Noether symmetry allows one to derive the solutions of equations of motion by means of the semiclassical limit of the ADM formalism.

17.5.3 Equivalence of R, T, and $\mathcal{G}$ Scalar-Tensor Gravity via Hamiltonian Dynamics

As pointed out in Section 10.5, the second generator of (10.19), the first of (10.72), and that of (10.88), under appropriate conditions, are equivalent. In light of this result, we investigate the Hamiltonian dynamics provided by the following generators:

$$R :\rightarrow \begin{cases} \mathcal{X} = -\frac{2}{3}\beta_0 a\partial_a + \beta_0\phi\partial_\phi \\ f(\phi) = \frac{3}{32}\phi^2, \quad V(\phi) = V_0\phi^2, \end{cases} \tag{17.84}$$

$$T :\rightarrow \begin{cases} \mathcal{X} = -\frac{2\beta_0}{3} a\partial_a + \beta_0\phi\partial_\phi \\ f(\phi) = \frac{3}{16}\phi^2, \quad V(\phi) = V_0\phi^2, \end{cases} \tag{17.85}$$

$$\mathcal{G} :\rightarrow \begin{cases} \mathcal{X} = \frac{2\ell}{3} a\partial_a - \ell\phi\partial_\phi \\ f(\phi) = f_0\phi^2, \quad V = V_0\phi^2, \end{cases} \tag{17.86}$$

which can be obtained from (10.95), (10.96), and (10.97) by setting $s = 0$ and $k = 2$. Under these conditions, all the preceding generators describe internal symmetries. Let us first consider the generator (17.84). Thanks to Eq. (2.37), we can perform the change of variables induced by the Noether symmetry, which allows one to introduce a cyclic variable into the Lagrangian. The system (2.37) takes the form

$$\begin{cases} \mathcal{X}z = \frac{2\beta_0}{3} a\partial_a z - \beta_0\phi\partial_\phi z = 1 \\ \mathcal{X}w = \frac{2}{3} a\partial_a w - \phi\partial_\phi w = 0, \end{cases} \tag{17.87}$$

where z represents the cyclic variable and the minisuperspace of configurations is transformed from $\mathcal{S} \equiv \{a, \phi\}$ to $\mathcal{S}' \equiv \{z, w\}$. A possible solution of the above system is

$$\begin{cases} z = -\dfrac{1}{\beta_0} \ln \phi \\ w = a^{\frac{3}{2}} \phi \end{cases} \qquad \rightarrow \qquad \begin{cases} \phi = e^{-\beta_0 z} \\ a = w^{\frac{2}{3}} e^{\frac{2\beta_0}{3} z}. \end{cases} \tag{17.88}$$

Replacing the new variables w, z into the Lagrangian (10.5), we get

$$\mathcal{L}_R = -V_0 w^2 + \frac{1}{2}\ell^2 w^2 \dot{z}^2 - \frac{1}{4}\ell w \dot{w} \dot{z} + \frac{1}{4}\dot{w}^2, \tag{17.89}$$

where we set $\ell \equiv \beta_0$ in order to conform the notation to the other examples. Clearly this form of $\mathcal{L}_R$ is cyclic with respect to z. After finding the time-derivatives of the variables as functions of the conjugate momenta, we can easily get the Hamiltonian:

$$\mathcal{H}_R = \pi_w^2 + \frac{4}{7\ell^2}\frac{\pi_z^2}{w^2} + V_0 w^2. \tag{17.90}$$

Classical trajectories (10.11) can be recovered by means of the Hamilton–Jacobi equations, after going back to the old variables (17.88). It is worth noticing that, in order to provide a comparison among the three equivalent classes of theories, the Hamiltonian dynamics has been studied for the solution (17.84) only. Clearly, suitable changes of variables can provide other solutions for the Noether system.

The next case is the torsion non-minimally coupled to the scalar field, whose Noether's solution is Eq. (17.85). The system that allows one to change the variables is

$$\begin{cases} -\dfrac{2\beta_0}{3} a\partial_a z + \beta_0 \phi \partial_\phi z = 1 \\ -\dfrac{2}{3} a\partial_a w + \phi \partial_\phi w = 0 \,, \end{cases} \tag{17.91}$$

and a possible solution is the same as before, namely

$$\begin{cases} z = -\dfrac{1}{\beta_0} \ln \phi \\ w = a^{\frac{3}{2}} \phi \end{cases} \qquad \rightarrow \qquad \begin{cases} \phi = e^{-\beta_0 z} \\ a = w^{\frac{2}{3}} e^{\frac{2\beta_0}{3} z} \end{cases} \tag{17.92}$$

Setting $\beta_0 \equiv \ell$, the new Lagrangian reads

$$\mathcal{L}_T = -\frac{1}{8}\frac{\dot{w}^2}{w} + \frac{1}{2}\ell \dot{w}\dot{z} + \frac{1}{2}\ell^2 w \dot{z}^2 - V_0 w \,. \tag{17.93}$$

Also, here z is the cyclic variable that permits one to write the Hamiltonian as

$$\mathcal{H}_T = \frac{\pi_z^2}{2w} + 2\pi_z \pi_w + \frac{3}{16} w \pi_w^2 + V_0 w, \tag{17.94}$$

where π_z, π_w are the conjugate momenta.

Finally, in the Gauss–Bonnet coupling, we consider the symmetry generator in Eq. (10.88) to find the Hamiltonian dynamics when the geometry is coupled to the scalar field through the function $F(\mathcal{G}) = \sqrt{\mathcal{G}}$. For $k = 2$, the Lagrangian can be written as

$$\mathcal{L} = -V_0 a^3 \phi^2 + \frac{1}{2} F_0 a^3 \sqrt{\mathcal{G}} \phi^2 + 2F_0 \phi^2 \dot{a}^3 \dot{\mathcal{G}}) \mathcal{G}^{-\frac{3}{2}} - 8F_0 \phi \dot{a}^3 \dot{\phi} \mathcal{G}^{-\frac{1}{2}} + \frac{1}{\ell^2} a^3 \dot{\phi}^2. \quad (17.95)$$

The condition (2.37) permits one to change the minisuperspace variables from $\mathcal{S} \equiv \{a, \phi, \mathcal{G}\}$ to $\mathcal{S}' \equiv \{z, w, \mathcal{G}\}$ and gives rise to the same system of differential equations as Eq. (17.87) and Eq. (17.91), that is,

$$\begin{cases} \mathcal{X} z = \frac{2\ell}{3} a \partial_a z - \ell \partial_\phi z = 1 \\ \mathcal{X} w = \frac{2}{3} a \partial_a w - \partial_\phi w = 0, \end{cases} \quad (17.96)$$

with z being the cyclic variable. One possible solution is

$$\begin{cases} z = -\frac{1}{\ell} \ln \phi \\ w = a^{\frac{3}{2}} \phi \end{cases} \quad \rightarrow \quad \begin{cases} \phi = e^{-\ell z} \\ a = w^{\frac{2}{3}} e^{\frac{2\ell}{3} z}. \end{cases} \quad (17.97)$$

Substituting the new variables w, z into the Lagrangian (17.95), we get

$$\mathcal{L}_{\mathcal{G}} = \frac{\mathcal{G}^{-\frac{3}{2}}}{54w} \left[27F_0 \mathcal{G}^2 w^3 - 54 \mathcal{G}^{\frac{3}{2}} w^3 (V_0 - \ell^2 \dot{z}^2) + 32F_0 \dot{\mathcal{G}} (\ell w \dot{z} + \dot{w})^3 \right.$$
$$\left. + 128 F_0 \ell \mathcal{G} \dot{z} (\ell w \dot{z} + \dot{w})^3 \right], \quad (17.98)$$

where, as expected, z is cyclic. By a straightforward Legendre transformation, we find the Hamiltonian

$$\mathcal{H}_{\mathcal{G}} = \frac{1}{4w^2} \left[16 \mathcal{G}^2 \pi_{\mathcal{G}}^2 + 8w \mathcal{G} \pi_{\mathcal{G}} \pi_w - 2F_0 \sqrt{\mathcal{G}} w^4 + w^2 \pi_w^2 \right.$$
$$\left. + 4V_0 w^4 + 3 \cdot 2^{\frac{2}{3}} w \left(\frac{\mathcal{G}^{\frac{3}{2}} w \pi_{\mathcal{G}}}{F_0 \ell^3} \right)^{\frac{1}{3}} (\pi_z + \ell \pi_w) \right]. \quad (17.99)$$

Therefore, dynamical equivalence is exhibited both at the level of EL equations (see Section 10.5) and of Hamiltonian dynamics. The important point to stress is that different representations of gravity coincide when the same solutions are derived.

To conclude this chapter, it is important to stress again the relevance of the Noether symmetries as a selection rule to derive observable universes. In this perspective, the Hartle criterion can be formulated as a Theorem:

In the semiclassical limit of Quantum Cosmology and in the framework of minisuperspace approximation, the reduction procedure of dynamics, due to the existence of Noether symmetries, allows one to select a subset of the solution of WDW equation where oscillatory behaviors are found. As consequence, correlations between coordinates and canonical conjugate momenta emerge so that classical cosmological solutions can be recovered. Vice versa, if a subset of the solution of WDW equation has an oscillatory behavior, conserved momenta have to exist and Noether symmetries are present. In other words, Noether symmetries select classical universes.

18

Strings, Swampland, Renormalizability, and Viability

There is geometry in the humming of the strings.
There is music in the spacings of the spheres.

— *Pythagoras*
(as quoted in *Music of the Spheres* [1961])

Up to now, we have considered modified/extended gravity actions by showing that, in most cases, symmetries can be used to find exact solutions. In this way, many issues carried by GR can be overcome at cosmological and astrophysical scales. In particular, we pointed out that inflation, dark energy, and dark matter can be framed into the alternative gravity paradigm and, eventually, answered if symmetries are detected in dynamics.

To conclude this book, we want to discuss some fundamental issues that, in principle, could be addressed, or at least clarified, with the help of Noether symmetries.

In particular, we shall discuss Strings, Swampland Conjecture, renormalizability, and viability conditions in the perspective of Noether symmetries. Section 18.1 is devoted to the String–Dilaton Cosmology and scale factor duality. We will show that scalar-tensor Lagrangians coming from Noether symmetries can be recast as string–dilaton effective Lagrangians. In particular, we will show that the same duality can be reduced under the standard of Noether symmetries. Exact solutions will be derived from String–Dilaton Cosmology. In this perspective, we will consider also the Swampland Conjecture, which consists of a set of criteria for the realizability of physical models in the String Theory landscape. In Section 18.2, we shall discuss some of these criteria in the light of Noether symmetries and derive cosmological models emerging as a consequence.

Furthermore, higher-order corrections to GR can be useful for fixing UV divergences. In this context, the advantages provided by modified theories toward the quantization of gravity are clear. Specifically, extensions of GR can yield couplings necessary for regularization and renormalizability of effective theories. In this regard, we will focus on naive power-counting renormalizability. Specifically, modified actions coming from Noether symmetries will be considered in Section 18.3 with the aim of discussing renormalizability of GR extensions.

Finally, Section 18.4 is a general discussion about the criteria that a modified theory of gravity must satisfy to be viable. Most of them are satisfied by models presenting Noether's symmetries.

18.1 String–Dilaton Cosmology and Scale Factor Duality

We use Noether symmetry prescription to recover the low-energy limit of String Theory, namely the *String–Dilaton Cosmology*, from generic scalar-tensor actions. As is well known, String Theory represents one of the most promising attempts toward the unification of fundamental interactions. Nonetheless, two main obstacles forbid the realization of this purpose, namely the false vacua and the lack of predictive power at low energies. In addition, the high-energy results cannot be easily tested, since Planck scales are far from current experimental capabilities. However, as stated by its theoretical foundations, String Theory is potentially able to provide explanations for the initial singularity by assuming the universe emerging with a finite size, related to the string scale, instead of a null size.

According to String Theory, the separation between large and small scales is not fixed; rather it depends on the probes used to measure distances and how the probe states are counted. The symmetry that deals with large and small scales under the same standard is the T-*duality*: it represents a global property of the universe, explicitly exhibited by the cosmological scale factor $a(t)$. Duality symmetry is characterized by the invariance of the action integral; therefore, the corresponding EL equations, under this transformation, remain the same for small and large scales. This feature occurs when the "radius" $\mathcal{R}$ of the geometry is transformed as $\mathcal{R} \to \mathcal{R}^{-1}$. Similarly, scale factor duality symmetry can occur when a FLRW space-time is considered. This is due to the fact that $a(t)$ can be understood as the time-depending radius of the expanding universe. More precisely, when the scale factor changes as $a \to a^{-1}$ and the dilaton field is opportunely transformed, the gravitational action remains invariant.

GR is not invariant under duality transformations, hence a modified theory of gravity is needed to map the scale factor in its dual. In particular, the dilaton field, occurring in scalar-tensor Lagrangians, can determine the expansion of spatial dimensions, while the other dual field is due to the geometry, namely $a(t)$. In this regard, String Theory can be recast as a scalar-tensor theory of the form (10.1). In what follows we analyze the string–dilaton effective action, showing that it is invariant under duality transformations of the scale factor and the dilaton field.

Let us start from a D-dimensional action

$$S = \frac{1}{2}\int \sqrt{-g}e^{-2\psi}\left[R + 4\partial_\mu\psi\partial^\mu\psi + \Lambda\right] d^D x, \tag{18.1}$$

which depends on the scalar field ψ through the coupling $e^{-2\psi}$, the kinetic term coefficient $4e^{-2\psi}$, and the potential $e^{-2\psi}\Lambda$. Notice that the kinetic term coefficient $4e^{-2\psi}$ is chosen such that the usual scalar-tensor action can be suitably recovered, but the duality of the resulting Lagrangian holds for any term of the form $\sim\gamma e^{-2\psi}$, with γ constant. By varying the action with respect to the metric we have [459, 460]

$$G_{\mu\nu} = \frac{1}{2}\Lambda g_{\mu\nu} + 2g_{\mu\nu}\Box\psi - 2g_{\mu\nu}D_\mu\psi D^\mu\psi - 2D_\mu D_\nu\psi. \tag{18.2}$$

The Klein–Gordon equation is provided by the variation with respect to the dilaton field, that is,

$$\Box\psi = D_\mu\psi D^\mu\psi - \frac{1}{4}(R+\Lambda). \tag{18.3}$$

Combining Eqs. (18.2) and (18.3), two useful relations occur, namely

$$\Box\psi = -\frac{1}{2}R, \qquad D_\mu\psi D^\mu\psi = \frac{1}{4}(\Lambda - R), \tag{18.4}$$

and will be used mostly in Section 18.2. Before recasting Eq. (18.1) in terms of a canonical scalar-tensor action, let us show its duality symmetry in a FLRW background, generalizing the constant Λ to any potential $W(\psi)$. Substituting the cosmological expression of the Ricci scalar in Eq. (18.1) and integrating out second derivatives, the D-dimensional point-like Lagrangian can be written as

$$\mathcal{L} = a^{D-1}\, e^{-2\psi}\left[(D-1)(D-2)\left(\frac{\dot a}{a}\right)^2 - 4(D-1)\left(\frac{\dot a}{a}\right)\dot\psi + 4\dot\psi^2 - W(\psi)\right]. \tag{18.5}$$

At the one-loop level, the transformation of the dilaton field also involves geometric contributions and in D dimensions can be written as [461–463]

$$\psi(t) \to \psi + \frac{D-1}{2}\ln a. \tag{18.6}$$

Under the transformation (18.6), Lagrangian (18.5) becomes

$$\mathcal{L} = e^{-\psi}\left[(D-1)\left(\frac{\dot a}{a}\right)^2 - 4\dot\psi^2 + W(\psi)\right], \tag{18.7}$$

By setting $W(\psi) = \Lambda$, Eq. (18.7) is invariant under duality transformation of the form $a(t) \to \dfrac{1}{a(t)}$. Also notice that by means of the transformations

$$\begin{cases} \phi(\psi) = e^{-\psi} \\ f(\phi) = \dfrac{1}{8}e^{-2\psi} \\ V(\phi) = e^{-2\psi}\, W(\psi), \end{cases} \tag{18.8}$$

the action (18.1) is equivalent to the D-dimensional scalar-tensor action (10.26). More precisely, the coupling and the potential function-containing symmetries, namely[1] $f(\phi) = \frac{1}{8}\phi^2$ and $V(\phi) = \Lambda\phi^2$, are recovered by the choices $\phi \sim e^{-\psi}$ and $W(\psi) = \Lambda$. When the four-dimensional limit is considered, no further transformation is needed to recover Noether symmetry solutions. Indeed, the action (18.1), with generic kinetic term coefficient γ, reads

$$S = \frac{1}{2}\int \sqrt{-g}\left[e^{-2\psi}R + \gamma e^{-2\psi}\partial_\mu\psi\partial^\mu\psi - e^{-2\psi}W(\psi)\right] d^4x. \tag{18.9}$$

Comparing the coupling, potential, and kinetic terms in Eq. (18.9) with the last solution of Eq. (10.9), that is,

$$f(\phi) = f_0 e^{k\phi}, \quad u(\phi) = u_0 e^{k\phi}, \quad V(\phi) = V_0 e^{k\phi}, \quad k = \pm\sqrt{\frac{2u_0}{3f_0}}, \tag{18.10}$$

it turns out that the D-dimensional scalar-tensor action can be related to the string–dilaton action (18.9). The exact equivalence is provided by

$$f_0 = 1, \quad u_0 = \gamma, \quad V_0 = W(\psi) = \Lambda, \quad k = -2. \tag{18.11}$$

However, it must be considered that γ is not a free parameter, but it is constrained by the relation

$$k = \pm\sqrt{\frac{2u_0}{3f_0}}. \tag{18.12}$$

Therefore, substituting Eq. (18.12) into Eq. (18.11), we obtain $\gamma = 6$. More generally, duality symmetry can be also obtained from general scalar-tensor action of the form

$$S = \int \sqrt{-g}\left[F(\phi)R + \frac{\gamma}{2}\partial_\mu\phi\partial^\mu\phi - V(\phi)\right] d^4x, \tag{18.13}$$

where we focus on the four-dimensional case for simplicity. Our purpose is to show that the duality property is independent of the coupling constants. By generalizing the transformations (18.8) to

$$\begin{cases} \phi(\psi) = e^{-\psi} \\ F(\phi) = e^{-2\psi}f(\psi) \\ V(\phi) = e^{-2\psi}\Lambda \end{cases} \tag{18.14}$$

and integrating second derivatives that arise in the FLRW background, the cosmological Lagrangian takes the form

$$\mathcal{L} = a^3 e^{-2\psi}\left\{f(\psi)\left(\frac{\dot{a}}{a}\right)^2 + \left[-2f(\psi) + f_\psi(\psi)\right]\dot{\psi}\left(\frac{\dot{a}}{a}\right) + \frac{\gamma}{2}\dot{\psi}^2 - \Lambda\right\}. \tag{18.15}$$

[1] $f(\phi)$ and $V(\phi)$ can be obtained from the second solution of (10.32), or from the $s = 0$ limit of the first.

Dilaton field transformation (18.6) can be further generalized by introducing the function $g(w)$, with w being a new variable. Therefore, Eq. (18.6) can be recast as [464]

$$\psi \to g(w) + \frac{3}{2}\ln a, \tag{18.16}$$

whose corresponding Lagrangian, invariant under duality transformations of the form $a(t) \to a(t)^{-1}$, reads

$$\begin{aligned}\mathcal{L} = e^{-2g(w)} \Bigg[&\left(-2f + \frac{3}{2}f_\psi + \frac{9}{8}\gamma\right)\left(\frac{\dot a}{a}\right)^2 + \left(-2f - f_\psi + \frac{3}{2}\gamma\right) g_w \dot w \left(\frac{\dot a}{a}\right) \\ &+ \frac{\gamma}{2} g_w(w)^2 \dot w^2 - \Lambda \Bigg].\end{aligned} \tag{18.17}$$

The equivalence between Eq. (18.17) and Eq. (18.7) can be shown by setting

$$\begin{cases} -2f + \dfrac{3}{2}f_\psi + \dfrac{9}{8}\gamma = \text{const.} \\ -2f - f_\psi + \dfrac{3}{2}\gamma = 0, \end{cases} \tag{18.18}$$

whose only solution is

$$f(\psi) = \frac{3}{4}\gamma. \tag{18.19}$$

In this way, the resulting Lagrangian,

$$\mathcal{L} = e^{-2g(w)} \left[-\frac{3}{8}\gamma \left(\frac{\dot a}{a}\right)^2 + \frac{\gamma}{2} g_w(w)^2 \dot w^2 - \Lambda \right], \tag{18.20}$$

is invariant with respect to duality transformations of the scale factor. The equivalence between scalar-tensor gravity and the string–dilaton effective Lagrangian is straightforward. Let us introduce now the function $G(w)$, defined as

$$G(w) = 2g(w), \qquad \lambda = \frac{8\Lambda}{\gamma}, \tag{18.21}$$

and consider the reparameterization

$$z = \ln\left(\frac{a}{\Sigma}\right) - \mu(w), \tag{18.22}$$

with $\mu(w)$ being a generic function of w and Σ a constant parameter. Thus, up to constant factors, Lagrangian (18.20) becomes

$$\mathcal{L} = e^{-G(w)} \left[-3\dot z^2 - \left(3\mu_w^2 - G_w^2\right)\dot w^2 - 6\mu_w \dot z \dot w - \lambda \right]. \tag{18.23}$$

With $G(w)$ and $\mu(w)$ being arbitrary functions, they can be assumed to be even and odd functions, respectively. As a consequence, Lagrangian (18.23) is invariant under the parity transformation $\mathcal{L}(w, z) \to \mathcal{L}(-w, -z)$. From the definitions of z, it follows that parity transformations of the Lagrangian are equivalent to duality transformations of the scale factor. This result can be considered as a further confirmation that duality symmetry is independent of the kinetic term coefficient γ. Moreover, as largely discussed in the previous chapters, introducing a cyclic variable z into the minisuperspace permits one to find exact dynamics. Indeed, by setting $\mu \sim$ const. and $G(w) = -\ln(w^2)$, Eq. (18.23) takes the form

$$\mathcal{L} = 3\dot{z}^2 w^2 - 4\dot{w}^2 + \lambda w^2. \tag{18.24}$$

The corresponding equations of motion are

$$\begin{cases} 4\ddot{w} + 3\dot{z}^2 w + \lambda w = 0 \\ \dot{z} w^2 = j_0 \\ 3\dot{z}^2 w^2 - 4\dot{w}^2 - \lambda w^2 = 0, \end{cases} \tag{18.25}$$

whose solution is

$$w(t) = \pm \left(c_1 e^{c_2 t} - c_3 e^{-c_2 t}\right)^{\frac{1}{2}}, \qquad z(t) = \pm c_5 \ln \left| \frac{e^{c_2 t} - c_4}{e^{c_2 t} + c_4} \right|, \tag{18.26}$$

with c_1, c_2, c_3, c_4, c_5 constants. Coming back to the previous minisuperspace $\mathcal{S} \equiv \{a(t), \psi(t)\}$, Eq. (18.26) can be written in terms of the old variables as

$$\begin{aligned} a(t) &= a_0 \left(\frac{e^{c_2 t} - c_4}{e^{c_2 t} + c_4} \right)^{c_5}, \\ \psi(t) &= \frac{1}{2} \ln \left| c_1 e^{c_2 t} - c_3 e^{-c_2 t} \right| + \frac{3}{2} c_5 \ln \left| \frac{e^{c_2 t} - c_4}{e^{c_2 t} + c_4} \right| + \psi_0. \end{aligned} \tag{18.27}$$

For simplicity, computations are performed under the assumption of constant potential $W(\psi) = \Lambda$. In the next subsection, viable forms of the potential will be provided by the Noether Symmetry Approach.

18.1.1 Noether Symmetry for String–Dilaton Lagrangian

Let us apply the Noether Symmetry Approach to Lagrangian (18.5), with the aim of finding models with symmetries and the corresponding potentials. Let us start by considering the four-dimensional case of Eq. (18.5), that is,

$$\mathcal{L} = 2e^{-2\psi} \left[3a\dot{a}^2 - 6a^2 \dot{a}\dot{\psi} + 2a^3 \dot{\psi}^2 - a^3 W(\psi) \right], \tag{18.28}$$

where the factor $1/2$ has been included in $W(\psi)$. Thus, the minisuperspace is two-dimensional and yields the symmetry generator

$$\mathcal{X} = \alpha \partial_a + \beta \partial_\psi. \tag{18.29}$$

The search for internal symmetry can be pursued by applying the vanishing Lie derivative condition to Eq. (18.28). It leads to the following Noether system of four partial differential equations

$$\begin{cases} \alpha - 2a\beta + 2a\partial_a\alpha - 2a^2\partial_a\beta = 0 \\ 3\alpha - 2a\beta - 3\partial_\psi\alpha + 2a\partial_\psi\beta = 0 \\ 6\alpha - 6a\beta + 3a\partial_a\alpha - 3\partial_\psi\alpha - 2a^2\partial_a\beta + 3a\partial_\psi\beta = 0 \\ 3\alpha\, W(\psi) + a\beta\left[W_\psi(\psi) - 2\,W(\psi)\right] = 0, \end{cases} \tag{18.30}$$

whose solutions are

$$\begin{cases} \mathcal{X} = \beta_0\zeta a^{s+1}e^{\psi}\partial_a + \beta_0 a^s e^{\psi}\partial_\psi, & W(\psi) = \lambda e^{\xi\psi} \\ \mathcal{X} = \dfrac{2}{3}\beta_0 a\partial_a + \beta_0\partial_\psi, & W(\psi) = \Lambda, \end{cases} \tag{18.31}$$

with

$$s = -\frac{3}{2}\left(\frac{\xi}{\xi+1}\right), \quad \zeta = \frac{1}{3}(2-\xi), \quad \xi = -1 \pm \sqrt{3} \tag{18.32}$$

and $\beta_0, \xi, \Lambda, \lambda$ constants. A constant potential can be recovered also from the first solution by imposing, for example, $\xi = 0$. Dynamics can be further reduced by the suitable change of variables (2.37), suggested by symmetries. To this purpose, we analyze the first solution of Eq. (18.31), which, once replaced in Eq. (2.37), provides

$$\begin{cases} \beta_0 a^s e^{\psi}\left(a\zeta\partial_a + \partial_\psi\right) z = 1 \\ \left(a\zeta\partial_a + \partial_\psi\right) w = 0, \end{cases} \tag{18.33}$$

whose possible solution is

$$w = \frac{\ln a}{\zeta} - \psi, \qquad z = \eta_0 a^{-s} e^{-\psi}, \tag{18.34}$$

with $\eta_0 = -\dfrac{1}{\beta_0}\dfrac{2(\xi+1)}{\xi^2+2}$. In the minisuperspace $\mathcal{S}' \equiv \{z, w\}$ the Lagrangian takes the form

$$\mathcal{L} = e^{w}(\chi\dot{z}\dot{w} - \lambda e^{wv}), \quad \chi = 2\left(1 \pm \frac{\sqrt{3}}{3}\right), \quad v = 2 \mp \sqrt{3}, \tag{18.35}$$

and the equations of motion read

$$\begin{cases} \chi\ddot{z} + \lambda(\xi+1)e^{vw} = 0 \\ \chi\dot{w} = j_0 e^{-w} \\ \chi\dot{z}\dot{w} + \lambda e^{vw} = 0. \end{cases} \tag{18.36}$$

The solution is

$$z(t) = -\frac{c_1\left[c_2 t + c\right]^{v+2}}{c_2^2(v+1)(v+2)} + z_1 t + z_0, \qquad w(t) = \ln\left[\left(\frac{j_0}{\chi}\right)t + c\right], \tag{18.37}$$

where we defined

$$c_1 = \left(\frac{v-1}{\chi}\right)\lambda, \quad c_2 = \frac{j_0}{\chi}, \tag{18.38}$$

with c being an integration constant and j_0 the conserved quantity related to the symmetry transformation. Moreover, a further constraint occurs, that is $j_0 z_1 = 0$, from which three different conditions can be distinguished: $z_1 = 0$, $j_0 \neq 0$; $z_1 \neq 0$, $j_0 = 0$; $z_1 = j_0 = 0$. Coming back to the old variables, the first case gives

$$a(t) = a_0 \left[\frac{c_2 t + c}{(c_2 t + c)}^{v+2} + z_0 \right]^{\gamma_1},$$
$$\psi(t) = \psi_0 - \gamma_2 \ln \left[c_3 (c_2 t + c)^{v+2} + z_0 \right] - \gamma_3 \ln (c_2 t + c), \tag{18.39}$$

with

$$\gamma_1 = \pm \frac{\sqrt{3}}{3}, \quad \gamma_2 = -\frac{1}{1 \mp \sqrt{3}}, \quad \gamma_3 = \frac{1 \mp \sqrt{3}}{2}. \tag{18.40}$$

Similarly, setting $j_0 = 0$ and $z_1 \neq 0$, we have

$$a(t) = a_0 \left[c_4 t^2 + \dot{z}_0 t + z_0 \right]^{-\gamma_1},$$
$$\psi(t) = \psi_0 - \gamma_2 \ln \left[c_4 t^2 + z_1 t + z_0 \right]. \tag{18.41}$$

Finally, when $z_1 = j_0 = 0$, the cosmological solution is

$$a(t) = a_0, \quad \psi(t) = \psi_0. \tag{18.42}$$

However, the cases where j_0 vanishes are not of physical interest, since they yield a vanishing potential and constant $w(t)$. Nevertheless, in the asymptotic regime, Eq. (18.41) provides a dust matter–dominated era. On the other hand, from Eq. (18.39), all cosmological eras, led by the string fluid, can be recovered according to the values of $\gamma_1, \gamma_2, \gamma_3$. This means that, from String–Dilaton Cosmology, one can recover, in principle, all the universe epochs, from the early to the late time.

18.1.2 The Wave Function of the Universe for String–Dilaton Cosmology

Here, considering the second solution of (18.31) and adopting the ADM formalism, we solve the related WDW equation and find the Wave Function of the Universe. Though a constant potential can also be recovered by setting $\xi = 0$ in the first solution of Eq. (18.31), the generators and the conserved quantities of the two cases are different. Therefore, starting from the second generator, we apply the condition (2.37) in order to change the minisuperspace and recover classical universes from the Hamilton–Jacobi equations. Following Ref. [312], the Noether Symmetry Approach yields the change of variables:

$$z = \ln a, \quad w = \psi - \frac{3}{2} \ln a, \tag{18.43}$$

by means of which Lagrangian (18.28) can be recast as

$$\mathcal{L} = -e^{-2w} \left(3\dot{z}^2 - 4\dot{w}^2 + 2\Lambda \right), \tag{18.44}$$

where z is cyclic. The Hamiltonian is then provided by a straightforward Legendre transformation of the preceding Lagrangian. Thus, considering the canonical quantization rules $\pi_z = -i\partial_z$, $\pi_w = -i\partial_w$ and the WDW equation

$$\left[\frac{1}{12}\partial_z^2 - \frac{1}{16}\partial_w^2 + 2\Lambda e^{-4w}\right]\Psi(z,w) = 0, \tag{18.45}$$

we obtain three solutions for the wave function:[2]

$$\begin{aligned}
&\Psi(z,w) \sim \exp\left(\pm\sqrt{\frac{3}{2}}z\right)\left[c_0 \mp \frac{1}{8\sqrt{2\Lambda}}e^{\pm 4\sqrt{2\Lambda}e^{-2w}}\right] e^{w\mp 2\sqrt{2\Lambda}e^{-2\Phi}},\\
&\Psi(z,w) \sim \pm\sin(mz)K_{\frac{im}{2\Lambda}}\left(\sqrt{2\Lambda}e^{-2w}\right),\\
&\Psi(z,w) \sim zK_0\left(\sqrt{2\Lambda}e^{-2w}\right),
\end{aligned} \tag{18.46}$$

with K_0 being the modified zeroth-order Bessel function and c_0, m constants of integration. Hamilton–Jacobi equations permit one to recover classical trajectories in the minisuperspace $\mathcal{S}' \equiv \{z, w\}$. After doing so, Eq. (18.43) can be used to restore the old variables. From this prescription, two sets of solutions occur, namely

$$\begin{aligned}
a(t) &= a_0 \exp\left\{\mp\frac{1}{\sqrt{6}}\arctan\left[\frac{1-2e^{\pm 4\sqrt{\Lambda/2}t}}{2e^{\pm 2\sqrt{\Lambda/2}t}\sqrt{1-e^{\pm 4\sqrt{\Lambda/2}t}}}\right]\right\},\\
\psi(t) &= \frac{1}{4}\ln\left[\frac{\Lambda e^{\pm 4\sqrt{\Lambda/2}t}}{k\left(1-e^{\pm 4\sqrt{\Lambda/2}t}\right)}\right]\\
&\quad \mp\frac{1}{\sqrt{6}}\arctan\left[\frac{1-2e^{\pm 4\sqrt{\Lambda/2}t}}{2e^{\pm 2\sqrt{\Lambda/2}t}\sqrt{1-e^{\pm 4\sqrt{\Lambda/2}t}}}\right] + \psi_0
\end{aligned} \tag{18.47}$$

and

$$\begin{aligned}
a(t) &= a_0\left[\frac{\cos\sqrt{\Lambda/2}t \pm \sin\sqrt{\Lambda/2}t}{\cos\sqrt{\Lambda/2}t \mp \sin\sqrt{\Lambda/2}t}\right]^{\pm\sqrt{3}/3},\\
\psi(t) &= \frac{1}{4}\ln\left[\frac{\Lambda/2}{k\cos^2 2\sqrt{\Lambda/2}t}\right] \pm \frac{\sqrt{3}}{2}\ln\left|\frac{\cos\sqrt{\Lambda/2}t \pm \sin\sqrt{\Lambda/2}t}{\cos\sqrt{\Lambda/2}t \mp \sin\sqrt{\Lambda/2}t}\right| + \psi_0,
\end{aligned} \tag{18.48}$$

where k is a constant of integration. As Eq. (18.47) holds for any $\Lambda > 0$, in Eq. (18.48) Λ must be negative. As a final remark, we can say that String–Dilaton Cosmology is in agreement with prescriptions of Quantum Cosmology and, in particular, Noether's symmetries guarantee the validity of the Hartle criterion and then the emergence of observable universes.

[2] In order to distinguish the wave function from the dilaton field ψ, here the former is denoted by Ψ.

18.2 String–Dilaton Cosmology and Swampland Conjecture

In Section 10.1.2 we showed the equivalence between fourth-order gravity and scalar-tensor gravity. Specifically, they can be related to each other by means of conformal transformations. This is also the case with the string–dilaton Lagrangian, which can be recovered from the $f(R)$ theory of gravity. The purpose of this section is to show that duality symmetry is present in several $f(R)$ models after performing a conformal transformation of the metric tensor $\tilde{g}_{\mu\nu} = e^{-2\psi} g_{\mu\nu}$. As a consequence $f(R)$ gravity can be seen under the standard of String–Dilaton Cosmology and then compared with the Swampland Conjecture.

Thus, we start from the string–dilaton action in the Jordan frame in order to recover $f(R)$ cosmology. In this way, the two actions are equivalent if the following condition holds:

$$\sqrt{-g}e^{-2\psi}\left(R + 4\partial_\mu\psi\partial^\mu\psi + \Lambda\right) = \sqrt{-\tilde{g}}f(\tilde{R}). \tag{18.49}$$

Using Eqs. (18.4), the preceding equation can be written as

$$f(\tilde{R}) = 2\Lambda e^{2\psi}. \tag{18.50}$$

Interestingly, duality symmetry in $f(\tilde{R})$ gravity is naturally provided by Noether Symmetry Approach. As pointed out in Chapter 7, for a general action containing Noether symmetry, the following condition must be satisfied [460]:

$$f_{\tilde{R}\tilde{R}}(\tilde{R})\dot{\tilde{R}} = f_{\tilde{R}}(\tilde{R})\left(\frac{\dot{a}}{a}\right), \tag{18.51}$$

so that any $f(\tilde{R})$ Lagrangian with symmetries can be written as

$$\mathcal{L} = a^3\left[f(\tilde{R}) - f_{\tilde{R}}(\tilde{R})\tilde{R}\right] + 12a\dot{a}^2 f_{\tilde{R}}(\tilde{R}). \tag{18.52}$$

By means of Eq. (18.50), the preceding Lagrangian becomes

$$\mathcal{L} = a^3\left\{\left[2\Lambda e^{2\psi} - 4\Lambda R e^{2\psi}\phi_{\tilde{R}}(\tilde{R})\right] + 48\left(\frac{\dot{a}}{a}\right)^2 \Lambda e^{2\psi}\psi_{\tilde{R}}(\tilde{R})\right\} \tag{18.53}$$

and, performing the dilaton shift (18.6) in four dimensions, Eq. (18.53) takes the following form:

$$\mathcal{L} = \left[2\Lambda e^{2\psi} - 4\Lambda\tilde{R}e^{2\psi}\psi_{\tilde{R}}(\tilde{R})\right] + 48\left(\frac{\dot{a}}{a}\right)^2 \Lambda e^{2\psi}\psi_{\tilde{R}}(\tilde{R}). \tag{18.54}$$

Finally, we have

$$\mathcal{L} = [f(R) - f_R(R)R] + 12\left(\frac{\dot{a}}{a}\right)^2 f_R(R). \tag{18.55}$$

Notice that Eq. (18.55) is manifestly invariant under duality transformations of the scale factor. Moreover, in order to obtain Eq. (18.55), the condition (18.51) must be considered, thus the preceding Lagrangian is directly related to the Noether symmetry. This means that duality symmetry is preserved under conformal transformation only if both Lagrangians contain Noether Symmetries, otherwise Eq. (7.5) and Eq. (18.55) do not yield the same dynamics. See also [249].

18.2.1 The Swampland Conjecture in $f(R)$ Gravity

The *Swampland* can be defined as the set of (apparently) consistent effective field theories that cannot be completed in any Quantum Gravity in the high-energy regime.

Since there can be nothing manifestly wrong with the effective theory, inconsistencies would manifest if one were to try to complete it in the UV regime. The Swampland Conjecture recently has been proposed to connect early-time cosmological models with the string landscape, and then to understand if related scalar fields and potentials can come from some fundamental theory in the high-energy regime. In some sense, this conjecture aims to select physically viable models.

Here, we discuss swampland criteria for $f(R)$ gravity, considering models where duality symmetry is present and it is related to the presence of Noether symmetries. In this perspective, specific $f(R)$ models can naturally belong to the string landscape. The selected $f(R)$ models, satisfying the Swampland Conjecture, are consistent, in principle, with both early- and late-time cosmological behaviors.

Consistently, embedding effective field theories into a general quantum theory involving gravity, in particular, in the context of String Theory, requires distinguishing low-energy effective field theories coupled to gravity from inconsistent counterparts. Establishing the correct criteria to identify the boundary between the landscape and the swampland leads to a series of conjectures known as weak gravity [465] and swampland conjectures, motivated by black hole physics [466] and string compactification [467].

Recently, it was proposed that an effective field theory, to be consistently embedded into Quantum Gravity, must satisfy two specific criteria. Based on them, it was argued that if String Theory should be the ultimate Quantum Gravity theory, there should be evidence that exact de Sitter solutions, with a positive cosmological constant, cannot describe the fate of the late-time universe [468–470]. On the other hand, some models with varying EoS can still be consistent with the conjecture [469]. Models with curvaton-like mechanisms [471] or with electroweak axion potential energy [472] were studied, showing that other mechanisms are still at stake (see also [473–476] for recent works on the topic). In addition, the two swampland criteria practically rule out simple single-field slow-roll models of inflation [477], while more complex models like multi-field inflation [478] and warm inflation [479,480] are still allowed. The possibility that String Theory does not allow for de Sitter vacua is not new [481]; however, it has recently gained impetus from new evidence on the instability of a de Sitter phase, and especially from the recent quantitative proposal for a more detailed constraint on potentials that are not in the swampland region.

Swampland criteria can be investigated also for alternative theories of gravity [482,483]. In our analysis, the important aspect is that some of these models can be framed into fundamental theories. As previously pointed out, one of the main characteristics of string–dilaton theory, the scale factor duality, holds also for some classes of $f(R)$ models presenting Noether symmetries. We demonstrated that string duality is related to Noether symmetries, so that it is possible to ask for the validity of swampland criteria for $f(R)$ gravity as a natural class of theories emerging from the string landscape [484]. On the other hand, since $f(R)$ gravity is also working at late epochs to address the accelerating expansion, it is realistic to search for models addressing swampland criteria, inflation, and dark energy issues under the same standard [119].

If the criteria hold, it follows that the universe would cross a phase transition within some Hubble times, so that such criteria can be used to select effective potentials. In order to be compatible with String Theory and provide the expected current cosmological behavior, an effective field theory must satisfy the following constraints. They impose an upper bound on the range traversed by scalar fields and a lower bound on the ratio V_ϕ/V (with $V(\phi) > 0$). In particular they set the order of the scalar field and the potential to [485]

$$\begin{cases} |\Delta\phi| < \Delta \sim \mathcal{O}(1) \\ \left|\dfrac{V_\phi}{V}\right| > c \sim \mathcal{O}(1), \end{cases} \tag{18.56}$$

with c constant. In particular, the first criterion affirms that any effective Lagrangian has a proper field range for $|\Delta\phi| < \Delta$, where the expectation is that $\Delta \sim \mathcal{O}(1)$. The second criterion imposes a lower bound to any effective theory of gravity.

Eq. (18.56) can be recast in terms of $f(R)$ gravity by considering the equivalence between Einstein and Jordan frames. More precisely, as shown in Section 4.4, the $f(R)$ action $S = \int \sqrt{-g} f(R)\, d^4x$ is dynamically equivalent, in the Einstein frame, to

$$S = \frac{1}{2}\int \sqrt{-g}\left[R + \partial_\mu\phi\partial^\mu\phi - 2V(\phi)\right] d^4x, \tag{18.57}$$

with the definition

$$V(\phi) = \frac{f(R) - Rf_R(R)}{f_R(R)^2}. \tag{18.58}$$

The two frames are related through the general conformal transformation $\tilde{g}_{\mu\nu} = e^{-2k\phi} g_{\mu\nu}$, with k real constant. Substituting Eq. (18.58) in Eq. (18.56), the second criterion provides [119]

$$\left|\frac{V_\phi}{V}\right| = \left(\frac{4k}{f - Rf_R}\right)\left[-f + \frac{Rf_R}{2} + \frac{f_R}{2}\left(\frac{\partial f}{\partial f_R}\right) - \frac{f_R^2}{2}\frac{\partial R}{\partial f_R}\right] > c. \tag{18.59}$$

We consider for $f(R)$ the power-law model selected by Noether Symmetry Approach. In particular, we choose $f(R) = R^{1+\epsilon}$ with $\epsilon \ll 1$, in order to analyze little deviations from GR. As formerly discussed, power-law functions guarantee duality invariance and can be understood as effective models related to the string landscape. By this choice, the function can be expanded around $\epsilon = 0$, providing

$$R^{1+\epsilon} \sim R + \epsilon R \ln R, \tag{18.60}$$

so that the potential (18.58) turns out to be

$$V(\phi) = \frac{\epsilon}{2(1+\epsilon)^2}\left[\frac{e^{2k\phi}}{1+\epsilon}\right]^{\frac{1-\epsilon}{\epsilon}}. \tag{18.61}$$

It is worth stressing that the form of potential in Eq. (18.61) does not rely on the perturbative expansion (18.60), thus it is valid for any ϵ. According to this model, it is possible to recover the cosmological constant as soon as $\epsilon \to 1$, while for $\epsilon \to \infty$, Eq. (18.61) goes to an exponentially suppressed plateau. If one considers the early universe, it has been argued that models of this form, with a noncanonical inflaton kinetic term, may be naturally obtained even if the original potential is not particularly flat. Specifically, such a potential can satisfy the swampland conditions, taking into account that mechanisms in this scenario invalidate the slow-roll condition.

On the other hand, at late epochs, such $f(R)$ theory gives rise to cosmological models that well fit SNeIa and CMB data for $\epsilon \sim -0.6$ and $\epsilon \sim 0.4$ [486]. However, they have to be carefully considered to address the accelerating/decelerating transition necessary for structure formation. In order to discuss if some values of ϵ satisfy the swampland criteria, let us recast Eq. (18.59) for power-law models. This implies

$$\left|\frac{V_{,\phi}}{V}\right| = \left|\left(\frac{2(1-\epsilon)}{\epsilon}\right)\right| > \mathcal{O}(1), \tag{18.62}$$

defined for $\epsilon \neq 0$. The $\mathcal{O}(1)$ for the modulus in Eq. (18.62) is satisfied in the regions $\epsilon < 2/3$ and $\epsilon > 2$, where the first is compatible with values required to describe an accelerating late expansion of the universe. We have to stress that Eq. (18.62) is valid in general for any $f(R)$ power-law models, while Eq. (18.60) represents the expansion of $f(R)$ around $\epsilon \ll 1$. This last case can be taken into account in order to describe small deviations from GR but, in the strong field regime, other power-law models can be physically interesting. Note that the preceding swampland criteria are not defined for $\epsilon \to 0$, which is required in the Solar System. In some sense, this condition is obvious because standard GR cannot be recovered in the string landscape if not equipped with further fields as dilaton, eventually emerging for $f(R) \neq R$, according to the preceding discussion. It is worth noticing also that the value $\epsilon \sim 0.4$ gives rise to $f(R) \sim R^{3/2}$, which is selected by the Noether Symmetry Approach in Chapter 7.

To conclude, Noether's symmetries selects power-law models where criteria (18.56) are in agreement with the Swampland Conjecture and with late-time expansion of the Universe. In this regard, the approach could be understood as a first step toward linking effective models with observations. As discussed in [119], besides power-law ones, other $f(R)$ gravity models can be discussed according to the Swampland Conjecture, giving reliable conditions to connect early and late epochs.

18.3 Renormalizability

The preceding considerations can be related to the wide problem of regularization and renormalization of theories of gravity. In this section we introduce a pedagogical analysis based on the coupling constant mass dimension of gravitational actions showing that models exhibiting Noether symmetries can be, in principle, regularized and renormalized.

We will start by showing that the EH action is not renormalizable. To this purpose, we use a perturbative approach relying on the coupling constant mass dimension. Specifically, physical theories can be split into three main categories, depending on the mass dimension of a given coupling constant χ, namely

- $[\chi] > 0 \rightarrow$ Super-Renormalizable Theories
- $[\chi] = 0 \rightarrow$ Renormalizable Theories
- $[\chi] < 0 \rightarrow$ Not-Renormalizable Theories.

Power-counting is a useful tool to study the renormalizability of modified theories of gravity, where a renormalization group flow equation cannot be easily achieved. However, the latter is needed to have precise information on whether UV divergences occur in the renormalized effective action considered. Nonetheless, power-counting argument provides reliable results for renormalizable (or super-renormalizable) field theories, while it may fail when the effective coupling constant turns out to have a negative mass dimension. The latter case is usually referred to as the Asymptotic Safety (see [487, 488] for reviews on the subject). This belief relies on the possibility that there may be another nontrivial UV fixed point in the renormalization group flow of the gravitational field with respect to which GR is a perturbation by a relevant operator. In this case, the non-renormalizability of gravity would just mean that we are expanding in the wrong variables, and power counting fails for this reason.

Moreover, checking the coupling constant mass dimension to investigate the renormalizability of field theories can be motivated by simple considerations. Indeed, when the mass dimension of the coupling constant χ is positive, then the perturbative expansion yields a cutoff Λ, which occurs in the denominator. Thus, as soon as $\Lambda \rightarrow \infty$, the amplitude converges. On the contrary, if $[\chi] < 0$, in the limit $\Lambda \rightarrow \infty$ the amplitude diverges.

In modified theories of gravity, the analysis of the coupling constant mass dimension can provide only an indication about the renormalizability. Indeed, most of them turn out to be not-renormalizable when more detailed aspects are considered, as well the propagator homogeneity in the momenta space [489]. Detailed discussions on the renormalizability of $f(R)$ theories of gravity can be found, for example, in [490–493]. One of the first attempts toward a renormalization scheme was pursued by Stelle in [82]. He showed that the gravitational action can be renormalized only when all possible combinations of second-order curvature invariants are considered along with the EH action. Adding terms with more than four derivatives, indeed, makes the theory finite after a certain order in the loop expansion, but does not remove the divergences at the one-loop order.

In order to check the coupling constant mass dimension of power-law models, the actions must be rewritten in terms of a sum between a free term, acting as a Gaussian fixed point, and other higher perturbative contributions. In this way, it is possible to check the mass dimension of the coupling constant of each term. The first step is to recast the metric tensor as a perturbation of the Minkowski background, namely

$$g_{\mu\nu} = \eta_{\mu\nu} + h_{\mu\nu}, \tag{18.63}$$

so that the Ricci scalar can be written in terms of $\eta_{\mu\nu}$ and $h_{\mu\nu}$. Consider the Levi–Civita connection

$$\Gamma^{\rho}_{\mu\nu} = \frac{1}{2}(\eta^{\rho\lambda} - h^{\rho\lambda})(\partial_\nu h_{\lambda\mu} + \partial_\mu h_{\nu\lambda} - \partial_\lambda h_{\mu\nu}) \tag{18.64}$$

and the form of the scalar curvature

$$R = (\eta^{\mu\nu} - h^{\mu\nu})(D_\rho \Gamma^{\rho}_{\mu\nu} - D_\nu \Gamma^{\rho}_{\mu\rho}). \tag{18.65}$$

It turns out that in the de Donder gauge, the divergent part of the one-loop effective EH action can be written as [494, 495]

$$\Gamma^{EH}_{\infty} = \int \sqrt{-g}\left(c_1 R^2 + c_2 R^{\mu\nu} R_{\mu\nu} + c_3 R^{\mu\nu\rho\sigma} R_{\mu\nu\rho\sigma}\right) d^4x, \tag{18.66}$$

with c_1, c_2, c_3 constants. Eq. (18.66) can be recast in terms of the Gauss–Bonnet term as

$$\Gamma_{\infty} = \int \sqrt{-g}\left(c_5 R^2 + c_5 R^{\mu\nu} R_{\mu\nu} + c_6 \mathcal{G}\right) d^4x, \tag{18.67}$$

where the first two terms vanish due to Einstein's field equations, while the third is a pure boundary. This means that no incurable divergence occurs at the one-loop level of the EH action. However, when two-loop expansion is considered, the divergent part of the action reads

$$\Gamma^{EH}_{\infty} = \int \sqrt{-g}(R^{\mu\nu}_{\rho\sigma} R^{\rho\sigma}_{\alpha\beta} R^{\alpha\beta}_{\mu\nu}) d^4x. \tag{18.68}$$

Eq. (18.68) shows that at the two-loop level nonvanishing terms arises. As a consequence, the S-matrix of Einstein's theory leads to non-renormalizable UV divergences.

The same result can be obtained through a power-counting evaluation. Considering Eq. (18.65), we see that the perturbed Ricci scalar depends on second derivatives of $h_{\mu\nu}$ multiplied by $h_{\mu\nu}$ itself. Since we are not interested in writing the exact form of R, rather to check the mass dimension of each term, a naive procedure permits to recast EH action as

$$S = \frac{1}{\chi^2}\int [(\partial\partial h) + (\partial h\partial h) + h(\partial h\partial h) + h^2(\partial h\partial h) + ...]\, d^4x. \tag{18.69}$$

Notice that the notation does not take into account constant factors, while it only considers the amount of h and ∂h occurring in each term. For instance, the quantity $\partial h\partial h$ includes all terms made by the product of first derivatives of the tensor $h_{\mu\nu}$. By an appropriate gauge choice, it is always possible to delete the first term. Thus, by means of the reparameterization $h = \chi\tilde{h}$, the action can be written as

$$S = \int \left[(\partial\tilde{h}\partial\tilde{h}) + \chi\tilde{h}(\partial\tilde{h}\partial\tilde{h}) + \chi^2\tilde{h}^2(\partial\tilde{h}\partial\tilde{h}) + ...\right] d^4x. \tag{18.70}$$

In this form, the action accounts for a perturbation of the Gaussian fixed point $\int(\partial\tilde{h})^2\, d^Dx$, which is the expression needed to check the coupling constant mass dimension of any terms. To this aim, there is no need to study the mass dimension of higher-order terms, since it can be inferred from the mass dimension of the first one. Because of the negative mass dimension of χ, the preceding action shows the reason why GR is not a renormalizable field theory.

Considering the linearized form of R in Eq. (18.65), the prescription can be suitably extended to more general actions, though the renormalizability of modified theories of gravity cannot be addressed so easily. Let us consider the action

$$S = \frac{1}{\chi^2}\int \sqrt{|g|}R^k\, d^{d+1}x, \tag{18.71}$$

which, as shown in Section 4.1, contains Noether's symmetry. By the expansion (18.69), the action (18.71) (up to first order) reads as [496]

$$\begin{aligned}
S &= \frac{1}{\chi^2}\int [(\partial h\partial h) + h(\partial h\partial h)]^k\, d^{d+1}x \\
&= \frac{1}{\chi^2}\int \sum_{i=0}^{k}\binom{k}{i}(\partial h\partial h)^{k-i}h^i(\partial h\partial h)^i\, d^{d+1}x \\
&= \frac{1}{\chi^2}\int \sum_{i=0}^{k}\binom{k}{i}(\partial h\partial h)^k h^i\, d^{d+1}x \\
&= \frac{1}{\chi^2}\int \left[(\partial h\partial h)^k + kh(\partial h\partial h)^k + \frac{k(k-1)}{2}h^2(\partial h\partial h)^k + ...\right] d^{d+1}x
\end{aligned}$$

$$= \frac{1}{\chi^2} \int \left[(\partial h \partial h)^k \left(1 + kh + \frac{k(k-1)}{2} h^2 + \ldots + h^k \right) \right] d^{d+1}x$$
$$\sim \frac{1}{\chi^2} \int \left[(\partial h \partial h)^k + kh(\partial h \partial h)^k \right] d^{d+1}x. \tag{18.72}$$

Similar to the previous case, to check the coupling constant dimension we introduce the new variable $h = (\chi)^{\frac{1}{k}} \tilde{h}$, so that the action takes the form

$$S = \int \left[(\partial \tilde{h} \partial \tilde{h})^k + k \chi^{\frac{1}{k}} \tilde{h} (\partial \tilde{h} \partial \tilde{h})^k \right] d^{d+1}x. \tag{18.73}$$

From the initial action (18.71), it is possible to infer the mass dimension of χ in $(d+1)$-dimensional R^k gravity, that is $\chi \to [M^{k-\frac{d+1}{2}}]$. Therefore, the $f(R)$ theory of gravity, with $f(R) \sim R^k$, must satisfy at least one of the following conditions,

$$1 - \frac{d+1}{2k} \geq 0 \quad \to \quad k < 0 \vee k \geq \frac{d+1}{2}, \tag{18.74}$$

to be a possible candidate for renormalizability. Thus, according to the power-counting argument, only $f(R) = R^2$ gravity is selected in four dimensions.

By similar arguments, it is straightforward to show that higher-order theories with action $S = \int \sqrt{|g|} \Box^\ell R^k \, d^{d+1}x$ must satisfy the conditions

$$\begin{cases} k \geq \dfrac{d+1}{2} - \ell \vee k \leq 0, & \text{if } d > 2\ell - 1 \\ \\ k \leq \dfrac{d+1}{2} - \ell \vee k \geq 0, & \text{if } d < 2\ell - 1. \end{cases} \tag{18.75}$$

Notice that this result holds for any $\ell \in \mathbb{Z}$, which means that even negative powers of the operator $\Box$ can occur in the action. In this case, the corresponding theory turns out to be nonlocal. The most general higher-order action containing a single term that depends on powers of the d'Alembert operator and on the scalar curvature is

$$S = \frac{1}{\chi^2} \int \sqrt{|g|} R^k \Box^\ell R^p \, d^{d+1}x. \tag{18.76}$$

In this case the mass dimension of the coupling constant is $[\chi] = M^{k+\ell+p-\frac{d+1}{2}}$, and the first-order linearization of the action (18.76) reads as

$$S = \frac{1}{\chi^2} \int \left[(\partial h \partial h)^k + h(\partial h \partial h)^k \right] \left\{ \Box^\ell (\partial h \partial h)^p + \Box^\ell \left[h(\partial h \partial h)^p \right] \right\} d^{d+1}x. \tag{18.77}$$

The reparametrization of h, which allows one to write the action in the form required by power-counting analysis, is $h = \tilde{h} \chi^{\frac{1}{p+k}}$. Therefore, Eq. (18.77) becomes

$$S = \int (\partial \tilde{h} \partial \tilde{h})^k \Box^\ell (\partial \tilde{h} \partial \tilde{h})^p \, d^{d+1}x \tag{18.78}$$
$$+ \int \left\{ (\partial \tilde{h} \partial \tilde{h})^k \Box^\ell \left[\tilde{h} (\partial \tilde{h} \partial \tilde{h})^p \right] + \tilde{h} (\partial \tilde{h} \partial \tilde{h})^k \Box^\ell (\partial \tilde{h} \partial \tilde{h})^p \right\} \chi^{\frac{1}{p+k}} \, d^{d+1}x,$$

from which we finally get the relation that the free parameters must satisfy to be considered in a renormalization scheme:

$$\begin{cases} k \geq \dfrac{d+1}{2} - p - \ell \ \vee k \leq -p, & \text{if } d > 2\ell - 1 \\ \\ k \leq \dfrac{d+1}{2} - p - \ell \ \vee k \geq -p, & \text{if } d < 2\ell - 1. \end{cases} \tag{18.79}$$

Results so far achieved can be extended to actions containing higher-order curvature invariants. To this purpose, we consider as equivalent all those scalar quantities with the same mass dimension.

It is worth pointing out the analogy with the teleparallel case, where the linearization of the metric is replaced by the expansion of the tetrad fields around the unity matrix, namely

$$e^a_\mu = \delta^a_\mu + h^a_\mu. \tag{18.80}$$

By means of Eqs. (18.80) and (4.102), the first-order teleparallel action can be written as

$$S = \frac{1}{\chi^2} \int (\partial h \partial h + h \partial h \partial h)\, d^{d+1}x, \tag{18.81}$$

which is formally equivalent to the first-order EH action. This implies that the same results, provided in Eq. (18.79), also hold in the teleparallel case and a separate treatment is not needed. This is mainly due to the mass dimension of the torsion scalar T, which turns out to be the same as that of the curvature scalar.

To conclude, from the preceding considerations, it is clear that models selected by the Noether symmetries can have an important role in the renormalizability of theories of gravity.

18.4 Viability Conditions

The final section of this part is devoted to a general discussion about the criteria that modified/extended gravity theories must satisfy to be viable. Throughout the third part of the book, we compared selected models with observational data with the purpose of fitting dynamics at all scales. We want to conclude by pointing out that comparison with experiments is a necessary condition for a theory to be viable, but it is not sufficient. For instance, in modifying gravity, it may occur that there could be violations of experimental limits on the PPN parameters at terrestrial and Solar System scales. Instabilities and ghost fields can emerge and the Cauchy problem could be ill-posed. All these problems are studied in literature, and we will briefly focus on them. Thus far, no theory is capable of describing the gravitational interaction in all its aspects, but in this debate criteria like Noether Symmetries can be extremely useful in selecting physically consistent models.

Taking into account $f(R)$ gravity as the leading example, in what follows we summarize some viability criteria that any modification/extension of GR must satisfy to be a self-consistent formulation for a relativistic theory of gravitation. In summary they are

- correct Newtonian, PN and PPN limits;
- reproducing cosmological dynamics;
- not instabilities and ghosts;
- cosmological perturbations compatible with data from CMB and large-scale structure;
- having a well-posed and well-formulated Cauchy problem.

Let us analyze in some detail each of the listed points.

18.4.1 Weak Field Limit and Solar System Tests

In order to assess whether the limits set on the PPN parameter γ by the available Solar System experiments hold or not, one needs to find the weak-field solution of the field equations and compute this parameter [39]. It is possible to consider a static, spherically symmetric, non-compact body that constitutes a perturbation of a background (de Sitter) solution. The line element is written as

$$ds^2 = \left[1 + 2\Psi(r) - H_0^2 r^2\right] dt^2 - \left[1 + 2\Phi(r) + H_0^2 r^2\right] dr^2 - r^2 d\Omega^2 \tag{18.82}$$

in Schwarzschild coordinates, where $d\Omega^2$ is the line element on the unit 2-sphere and Ψ and Φ are gravitational potentials. These are of small amplitude, $|\Psi(r)|, |\Phi(r)| << 1$, and one considers small (non-cosmological) scales so that $H_0 r << 1$ while expanding the Ricci scalar around the constant curvature of the background de Sitter space, that is $R(r) = R_0 + R_1$. The PPN parameter γ is given by $\gamma = -\Phi(r)/\Psi(r)$ [39]. Three assumptions can be made [497]:

1. $f(R)$ is analytical in R_0;
2. $mr << 1$, where m is the effective mass of the scalar degree of freedom of the theory, that is, it is assumed that this scalar field is light and has a range larger than the size of the Solar System. (We remind the reader that, at the moment, there is no experimental constraint of scalars in the range $m^{-1} < 0.2$ mm.)
3. For matter composing the spherical body, the pressure is negligible, $p \simeq 0$, so that $T = T_0 + T_1 \simeq \rho$.

The first and the last assumptions are not stringent, but the second one is, as will be clear in what follows. The trace of the $f(R)$ gravity field equations in Eq. (4.47), that is,

$$3\Box f_R(R) + R f_R(R) - 2f(R) = \mathcal{T}\ , \tag{18.83}$$

yields the equation for the Ricci scalar perturbation in the weak-field approximation, that is,

$$\nabla^2 R_1 - m^2 R_1 = \frac{\rho}{3f''_0} , \tag{18.84}$$

where

$$m^2 = \frac{(f'_0)^2 - 2f_0 f''_0}{3f'_0 f''_0} \tag{18.85}$$

is the effective mass squared of the scalar, where ∇^2 is the Laplace operator. Here f_0 is defined as $f_0 \equiv f(R_0)$ and f'_0 its first derivative evaluated in $R = R_0$. Eq. (18.85) coincides with the expression obtained in the gauge-invariant stability analysis of de Sitter space and in propagator calculations.

If $mr << 1$, the solution of the linearized field equations is

$$\Psi(r) = -\frac{M}{6\pi f'_0}\frac{1}{r} , \tag{18.86}$$

$$\Phi(r) = \frac{M}{12\pi f'_0}\frac{1}{r} . \tag{18.87}$$

Therefore, the PPN parameter is

$$\gamma = -\frac{\Phi(r)}{\Psi(r)} = \frac{1}{2} , \tag{18.88}$$

in gross violation of the experimental limits [498]

$$|\gamma - 1| < 2.3 \cdot 10^{-5} . \tag{18.89}$$

This result would be the end of metric $f(R)$ gravity if the assumptions made in the calculation were satisfied. However, this is not the case for assumption 2; mr is not always less than unity due to the *Chameleon Effect* [499]. This consists of the effective mass m depending on the curvature or, alternatively, the matter density of the environment. The scalar degree of freedom can be short-ranged (say $m > 10^{-3}$ eV, corresponding to a range $\lambda < 0.2$ mm) at Solar System densities and can evade the experimental constraints, while being long-ranged at cosmological densities, and thus it is able to affect the cosmological dynamics [500, 501].

18.4.2 Cosmological Dynamics

In order to be acceptable and compatible with observations, a cosmological model has to exhibit some minimal requirements, which are as follows: *i*) an early inflationary epoch or an alternative mechanism to solve horizon, flatness, and monopole problems together with a mechanism to generate density perturbations; *ii*) a following radiation-dominated era; *iii*) a matter-dominated era, and, finally, *iv*) the model has to be compatible with the present acceleration. The future eras are usually found to be either an eternal de Sitter attractor phase or a Big Rip singularity truncating the history of the universe at a finite time.

In particular, smooth transitions between different eras are required. It has been pointed out that the exit from the radiation era, in particular, may have problems in many models, a warning that has to be considered in building up $f(R)$ cosmologies. In general, exit from the radiation, or any era can be achieved for given $f(R)$ models. Consider, for example, the so-called "designer $f(R)$ gravity": one can prescribe a desired expansion history of the universe by choosing the scale factor $a(t)$ and then integrating a differential equation determining the function $f(R)$ producing this $a(t)$ [452, 453]. As reported in [247], Noether's symmetries also play a major role in producing models with smooth phase transitions between the various cosmological epochs.

18.4.3 Instabilities and Ghosts

The prototype model for discussing instabilities is $f(R) = R - \mu^4/R$ with $\mu \sim H_0 \sim 10^{-33}$ eV [446]. Shortly after it was proposed, this model was found to suffer a catastrophic instability [402]. The stability analysis was later generalized to any metric $f(R)$ model [403] and the extension to even more general gravitational theories has been pursued in [502, 503]. The approach consists of parameterizing the deviations from GR as

$$f(R) = R + \epsilon\varphi(R), \tag{18.90}$$

where ϵ is a small positive constant with the dimensions of a squared mass and the function φ is dimensionless. The trace equation for the Ricci scalar R takes the form

$$\Box R + \frac{\varphi_{RRR}}{\varphi_{RR}}\nabla^\alpha R\nabla_\alpha R + \left(\frac{\epsilon\varphi_R - 1}{3\epsilon\varphi_{RR}}\right) R = \frac{\mathcal{T}}{3\epsilon\varphi_{RR}} + \frac{2\varphi}{3\varphi_{RR}} . \tag{18.91}$$

Next, one expands around a de Sitter background and writes the metric *locally* as

$$g_{\alpha\beta} = \eta_{\alpha\beta} + h_{\alpha\beta} \,, \tag{18.92}$$

while the scalar degree of freedom R is expanded as

$$R = -\mathcal{T} + R_1 \,, \tag{18.93}$$

with R_1 being a perturbation. To first order, the trace equation yields the dynamical equation for R_1:

$$\begin{aligned} &\ddot{R}_1 - \nabla^2 R_1 - \frac{2\varphi_{RRR}}{\varphi_{RR}}\dot{\mathcal{T}}\dot{R}_1 + \frac{2\varphi_{RRR}}{\varphi_{RR}}\vec{\nabla}\mathcal{T}\cdot\vec{\nabla}R_1 + \frac{1}{3\varphi_{RR}}\left(\frac{1}{\epsilon} - \varphi_R\right)R_1 \\ &= \ddot{\mathcal{T}} - \nabla^2\mathcal{T} - \frac{(\mathcal{T}\varphi^2 + 2\varphi)}{3\varphi_{RR}} . \end{aligned} \tag{18.94}$$

The last term on the LHS is dominated by the term in ϵ^{-1} and gives the effective mass squared of R_1,

$$m^2 \simeq \frac{1}{3\epsilon\varphi''}\,, \tag{18.95}$$

from which one deduces that the theory is

- *stable* if $f_{RR}(R) > 0$
- *unstable* if $f_{RR}(R) < 0$.

The case of GR is excluded by the assumption $f_{RR} \neq 0$, but one can extend the stability criterion for metric $f(R)$ gravity to be $f_{RR} \geq 0$, including GR.

As an example, the prototype model $f(R) = R - \mu^4/R$, which has $f_{RR} < 0$ is unstable. The time scale for the onset of this instability is dictated by the smallness of the scale μ, which corresponds to $\sim 10^{-26}$ s, making this an explosive instability.

One can give a physical interpretation of this result as follows [504]: remembering that the effective gravitational coupling is $G_{eff} = G_N/f_R(R)$, with G_N the Newton constant, if $dG_{eff}/dR = -f_{RR}G_N/(f_R)^2 > 0$ (which corresponds to $f_{RR} < 0$), then G_{eff} increases with R and a large curvature causes gravity to become stronger, which, in turn, causes a larger R, in a positive feedback mechanism driving the system away. If instead $dG_{eff}/dR < 0$, then a negative feedback damps the increase in the gravitational coupling strength.

Palatini $f(R)$ gravity, by contrast, is described by second-order field equations. The corresponding trace equation $f_{\tilde{R}}(\tilde{R})\tilde{R} - 2f(\tilde{R}) = \mathcal{T}$ is not a differential equation but rather a non-dynamical algebraic one and, therefore, there is no instability [505].

From a more conceptual point of view, instabilities can be related to ghosts that are massive states of negative norm, which cause a lack of unitarity and are common for generalizations of Einstein's gravity. The good news is that $f(R)$ gravity is ghost-free. More general theories like $f\big(R, R_{\alpha\beta}R^{\alpha\beta}, R_{\alpha\beta\gamma\delta}R^{\alpha\beta\gamma\delta}, ...\big)$, in general, contain ghost fields. A possible exception is the case of Gauss–Bonnet gravity: if the field equations are of second order, no ghost occurs [506].

18.4.4 Cosmological Perturbations

The expansion history of the universe alone is not sufficient to discriminate between various models, but the growth of structures depends on the theory of gravity and has the potential to achieve this goal. For example, [501, 507–510] assumed an expansion history $a(t)$ typical of the ΛCDM model and found that vector and tensor modes are not affected by $f(R)$ at lowest order, while perturbation scalar modes depend on the specific $f(R)$ model. They also found the condition $f_{RR}(R) > 0$ for the stability of scalar perturbations to be in agreement with the arguments just discussed. The most interesting results are that $f(R)$

corrections lower the large angle anisotropies of the CMB and produce correlations between CMB and galaxy surveys different from those of dark energy models.

Overall, the study of structure formation in modified gravity is still work in progress, and often it is performed within the context of specific models, some of which are already in trouble because they do not pass the weak-field limit or the stability constraints. A similar situation holds for Palatini $f(R)$ models.

18.4.5 The Cauchy Problem

A physical theory must have predictive power and, therefore, a well-posed and well-formulated initial value problem. GR satisfies this requirement for "reasonable" forms of matter. The well-posedness of the Cauchy problem for vacuum $f(R)$ gravity was briefly discussed for special metric models in earlier research [511]. Thanks to the equivalence between $f(R)$ gravity and scalar-tensor theory when $f_{RR}(R) \neq 0$, the Cauchy problem can be reduced to the analogous one for Brans–Dicke gravity with $\omega = 0, -3/2$.

A general discussion of the Cauchy problem in the case of Palatini $f(R)$ gravity is reported in [512–514], where the well-posedness and the well-formulation are discussed according to the fluids sourcing the field equations. It is worth stressing that the well-position and the well-formulation strictly depend on the form of the fluid.

To conclude, we summarized the main viability conditions to achieve physically reliable $f(R)$ models. Clearly, this analysis is necessary in any modified/extended theory of gravity whose aim is to give a self-consistent description of cosmic history. In this debate, Noether's symmetries have a main role because they allow integration of dynamics by the identification of conserved quantities.

Conclusion of Part III

Part III of this book has been devoted to the physical application of results coming from the Noether Symmetry Approach to some observational evidences. The leading philosophy has been to fix GR shortcomings related to the mismatch between observations and theory, taking into account possible extensions and modifications of the Einstein theory. In particular, the existence of symmetries has been considered as a selection principle aimed to discriminate among competing models. In other words, the paradigm consists of assuming that models with symmetries can be physically meaningful and comparable with observations.

Starting from this statement, we investigated some signatures at astrophysical scales. We considered Solar System tests as the standard approach capable of probing theories of gravity. The main result is that in the weak-field limit also, signatures of modified gravity can be detected depending on the precision of experiments.

Further astrophysical evidence comes, for example, from the mass-radius diagram of neutron stars. Recent observations, in particular the gravitational wave event GW190814, point out that mass limits imposed by GR do not match the reported data. Specifically, the resulting self-gravitating object is too massive to be a neutron star and too light to be a black hole. Extending GR as a power-law $f(R)$ gravity selected by the Noether symmetries allows one to fit observations indicating more massive neutron stars as the solution of the puzzle. This fact could indicate further evolutionary branches within the Hertzsprung–Russell diagram [359].

Gravitational lensing also is a useful tool to test theories of gravity. In particular, features coming from corrections of Newton gravitational potential have effects both in modeling the gravitational lens and in the trajectories of light rays. We discussed in detail this issue, which could be a test bed for models at various scales.

Finally, interesting probes come from the study of stellar orbits around massive sources. We considered an the star S2 orbiting around the Galactic Center (Sgr A^*) in order to compare observational data with nonlocal models emerging from symmetries. The best fit procedure between data and theoretical curves puts in evidence that nonlocal gravity features, like characteristic lengths, can be tested by the sensitivities of today experiments like EHT collaboration [40].

Applications involve also galaxies. We showed that the missing matter problem for both spiral and elliptical galaxies can be addressed in the framework of Extended Gravity. In particular, we showed that MOND features are reproduced by considering corrections to the Newtonian potential coming from $f(R)$ gravity. The interesting point is that MOND is not only a phenomenological approach to galactic dynamics but it can be recovered as the weak field limit of a field theory. Besides this result, the Fundamental Plane of elliptical galaxies can be gravitationally reconstructed starting from $f(R)$ gravity. It is worth noticing that observational parameters like the effective radius r_e have a specific gravitational meaning and are related to the conserved quantities selected by the Noether symmetries.

Applications to Cosmology like inflation, perturbations, and quintessence are reviewed in the light of Noether's symmetries. The emerging feature is that models are physically workable as soon as symmetries are present. For example, the further degrees of freedom related to GR extensions can always be seen as scalar fields leading dynamics. The existence of symmetries determines couplings and potentials as well as the possibility of solving the related dynamical systems. This fact allows one to develop coherent models, which can be matched with observations.

Finally, we discussed special applications like the String–Dilaton Cosmology, the Swampland Conjecture, and the renormalizability of theories of gravity. In all the topics, the leading role of symmetries clearly emerges.

We concluded this part of the book with a critical discussion on the viability conditions. Matching them is extremely important both from conceptual and experimental points of view in the perspective of deriving reliable models.

Epilogue

God used beautiful Mathematics in creating the World.
— *Paul Adrien Maurice Dirac*
(as quoted in *The Cosmic Code: Quantum Physics as the Language of Nature* [1982])

In this book, we discussed the role of symmetries in Physics, with particular attention to theories of gravity. After summarizing the general aspects concerning symmetries and conserved quantities, we introduced the two Noether Theorems, which provide a fundamental criterion to seek for symmetries in any physical system. We mainly focused on Noether's First Theorem, as an approach to select physically viable models. Specifically, we developed a method for theories of gravity where the existence of symmetries and conserved laws allows one to select the forms of couplings, potentials and geometric invariants in the effective Lagrangians. The related dynamics results are solvable thanks to the presence of first integrals, namely conserved quantities.

Throughout the text, we mainly focused on cosmological and astrophysical applications; however, the Noether Symmetry Approach can be successfully applied to several other cases than those presented here. For instance, in Refs. [515–517] symmetries are used to find wormhole solutions. In [518–521] authors select symmetries in Kasner anisotropic universes. Kantowski-Sachs spacetimes are dealt in Refs. [522–524].

Further readings on the foundation of the Noether Symmetry Approach can be found in [525–533] while other applications are in [534–558].

Some considerations are in order at this point on the role of Geometry, in general, and symmetries, in particular.

As discussed in [2], Geometry is the science of space that, eventually, is also the science of phenomena substantiating space, namely Physics. This means that Geometry and Physics are intimately linked and, according to this point of view, they coincide. In other words, since Physics studies the correlation between phenomena and tends to describe them through a unified vision, resulting in self-consistent models eventually testable by experiments and observations, Geometry, from a more general point of view, seeks for formal relations among objects like measurements and properties of space. Symmetry, in the sense used here, unifies the visions of Physics and Geometry, referring to invariance under transformations like space-time translations, reflections, rotations, and scaling.

In this perspective, Noether's First Theorem acquires a fundamental role because it unifies *operationally* Physics and Geometry. If the theorem works for a given then symmetries allow one to solve the dynamics and, as we tried to demonstrate throughout this book, models can be physically viable. Clearly this is only a philosophical point of view because, often, it is really difficult to find out workable models with a physical meaning, even if symmetries exist.

However, an *aesthetic* approach can be an important criterion to understand reality, and the Noether Theorems are two cornerstones in this perspective.

Appendix A

Variational Principles

In this appendix, we sketch the procedure that allows one to obtain the field equations of modified/extended theories of gravity from a variational principle. To this purpose, $f(R)$ gravity is firstly considered. Then, we consider the non-minimal coupling between geometry and scalar fields, also providing the related Klein–Gordon equations. Then we introduce higher-derivative terms and finally we show how to recast the variational principle in terms of tetrad fields by considering TEGR.

Unlike GR, modified/extended gravity field equations can be of fourth (or higher) order. As shown in detail here, this is due to $g^{\alpha\beta}\delta R_{\alpha\beta}$, which is a boundary term in GR but carries extra derivatives of the metric in modified gravity.

Therefore, with the aim of finding a suitable form for the quantity $g^{\alpha\beta}\delta R_{\alpha\beta}$, let us set $\delta g^{\alpha\beta} \equiv h^{\alpha\beta}$ so that the following relations hold:

$$\begin{cases} \delta g^{\alpha\beta} = h^{\alpha\beta} \\ \delta g_{\alpha\beta} = -h_{\alpha\beta}, \end{cases} \tag{A.1}$$

where the second is a straightforward consequence of the identity $\delta(g^{\alpha\beta}g_{\alpha\beta}) = 0$. The form of the Levi–Civita connection as a function of the metric can be found by imposing the metricity principle. Specifically, from the identity

$$D_\alpha g_{\mu\nu} + D_\nu g_{\alpha\mu} - D_\mu g_{\nu\alpha} = 0, \tag{A.2}$$

the Levi–Civita connection reads:

$$\Gamma^\alpha_{\mu\nu} = \frac{1}{2} g^{\alpha\gamma} \left(\partial_\mu g_{\gamma\nu} + \partial_\nu g_{\mu\gamma} - \partial_\gamma g_{\mu\nu} \right). \tag{A.3}$$

Therefore, by means of (A.1), the variation of Christoffel symbols reads

$$\begin{aligned}\delta\Gamma^{\alpha}_{\beta\gamma} &= \frac{1}{2}(\partial_\gamma g_{\sigma\beta} + \partial_\beta g_{\sigma\gamma} - \partial_\sigma g_{\beta\gamma})\delta g^{\alpha\sigma} + \frac{1}{2}g^{\alpha\sigma}(\partial_\gamma h_{\sigma\beta} + \partial_\beta h_{\sigma\gamma} - \partial_\sigma h_{\beta\gamma}) \\ &= \frac{1}{2}g^{\alpha\sigma}(\partial_\gamma h_{\sigma\beta} + \partial_\beta h_{\sigma\gamma} - \partial_\sigma h_{\beta\gamma}) - h^{\alpha\sigma}\{\sigma,\beta\gamma\} \\ &= \frac{1}{2}g^{\alpha\sigma}(\partial_\gamma h_{\sigma\beta} + \partial_\beta h_{\sigma\gamma} - \partial_\sigma h_{\beta\gamma}) - g^{\alpha\tau}h_{\sigma\tau}\Gamma^{\sigma}_{\beta\gamma}.\end{aligned} \tag{A.4}$$

Also notice that the sum

$$D_\gamma h_{\sigma\beta} + D_\beta h_{\sigma\gamma} - D_\sigma h_{\beta\gamma} \tag{A.5}$$

can be recast in terms of Christoffel symbols as

$$\begin{aligned}&D_\gamma h_{\sigma\beta} + D_\beta h_{\sigma\gamma} - D_\sigma h_{\beta\gamma} \\ &= \partial_\gamma h_{\sigma\beta} + \partial_\beta h_{\sigma\gamma} - \partial_\sigma h_{\beta\gamma} - \Gamma^{\rho}_{\sigma\gamma}h_{\rho\beta} - \Gamma^{\rho}_{\gamma\beta}h_{\rho\sigma} \\ &\quad - \Gamma^{\rho}_{\beta\sigma}h_{\rho\gamma} - \Gamma^{\rho}_{\gamma\beta}h_{\rho\sigma} + \Gamma^{\rho}_{\beta\sigma}h_{\rho\gamma} + \Gamma^{\rho}_{\gamma\sigma}h_{\rho\beta} \\ &= \partial_\gamma h_{\sigma\beta} + \partial_\beta h_{\sigma\gamma} - \partial_\sigma h_{\beta\gamma} - 2\Gamma^{\rho}_{\gamma\beta}h_{\rho\sigma}.\end{aligned} \tag{A.6}$$

Substituting Eq. (A.6) into Eq. (A.4), the latter becomes

$$\delta\Gamma^{\alpha}_{\beta\gamma} = \frac{1}{2}g^{\alpha\sigma}(D_\gamma h_{\sigma\beta} + D_\beta h_{\sigma\gamma} - D_\sigma h_{\beta\gamma}). \tag{A.7}$$

Finally, by means of Eq. (A.7), the Ricci tensor variation

$$\delta R_{\alpha\beta} = D_\sigma(\delta\Gamma^{\sigma}_{\alpha\beta}) - D_\beta(\delta\Gamma^{\sigma}_{\alpha\sigma})$$

takes the form

$$\begin{aligned}g^{\alpha\beta}\delta R_{\alpha\beta} &= -\frac{1}{2}(D_\sigma D_\beta h^{\beta\sigma} + D_\sigma D_\alpha h^{\alpha\sigma} - g^{\sigma\rho}D_\rho D_\sigma h) + \frac{1}{2}(g^{\alpha\beta}D_\beta D_\alpha h) \\ &= -D_\beta D_\alpha h^{\alpha\beta} + g^{\alpha\beta}D_\beta D_\alpha h = \Box h - D_\beta D_\alpha h^{\alpha\beta},\end{aligned} \tag{A.8}$$

where $g^{\alpha\beta}h_{\alpha\beta} \equiv h$ and $g^{\alpha\beta}D_\beta D_\alpha h = \Box h$.

Before applying the variational principle to modified theories of gravity, let us get some relations that will be useful for the next computations. Let us first show that the definition of the covariant derivative $D_\nu A_\mu = \partial_\nu A_\mu - \Gamma^{\alpha}_{\nu\mu}A_\alpha$, automatically implies $D_\nu A^\mu = \partial_\nu A^\mu + \Gamma^{\mu}_{\nu\tau}A^\tau$. To this purpose, let us evaluate the quantity

$$D_\nu A^\mu = D_\nu(g^{\mu\gamma}A_\gamma) = g^{\mu\gamma}\left(\partial_\nu A_\gamma - \Gamma^{\sigma}_{\nu\gamma}A_\sigma\right), \tag{A.9}$$

where we used the metricity principle. Therefore we have

$$\begin{aligned}D_\nu A^\mu &= g^{\mu\gamma}\left[\partial_\nu(g_{\gamma\tau}A^\tau) - \Gamma^{\sigma}_{\nu\gamma}g_{\alpha\sigma}A^\alpha\right] \\ &= g^{\mu\gamma}\left(A^\tau\partial_\nu g_{\gamma\tau} - A^\tau\Gamma^{\sigma}_{\nu\gamma}g_{\tau\sigma} + g_{\gamma\tau}\partial_\nu A^\tau\right) \\ &= g^{\mu\gamma}\left[\partial_\nu g_{\gamma\tau} - \frac{1}{2}\left(\partial_\nu g_{\tau\gamma} + \partial_\gamma g_{\nu\tau} - \partial_\tau g_{\gamma\nu}\right) + g_{\gamma\tau}\partial_\nu\right]A^\tau \\ &= \frac{1}{2}g^{\mu\gamma}\left(\partial_\nu g_{\gamma\tau} + \partial_\tau g_{\nu\gamma} - \partial_\gamma g_{\tau\nu}\right)A^\tau\partial_\nu A^\mu \\ &= \partial_\nu A^\mu + \Gamma^{\mu}_{\nu\tau}A^\tau.\end{aligned} \tag{A.10}$$

Finally, let us demonstrate the relation

$$\delta\sqrt{-g} = -\sqrt{-g}g_{\mu\nu}\,\delta g^{\mu\nu}. \tag{A.11}$$

The preceding identity can be proved by considering a diagonal metric tensor. (If the starting metric tensor is not diagonal, it can be diagonalized and the resulting diagonal matrix will have the same determinant as the initial one.) In such a case, the determinant g can be written as

$$g = \prod_{\alpha=0}^{D} g_{\alpha\alpha}, \tag{A.12}$$

with D being the number of space-time dimensions. Therefore, using the product differentiation, for the variation δg we have

$$\delta g = \sum_{\beta=0}^{D}\prod_{\alpha\neq\beta}^{D} g_{\alpha\alpha}\delta g_{\beta\beta}. \tag{A.13}$$

Considering Eq. (A.12), we have that

$$\prod_{\alpha\neq\beta}^{D} g_{\alpha\alpha} = \frac{\prod_{\alpha=0}^{D} g_{\alpha\alpha}}{g_{\beta\beta}} = \prod_{\alpha=0}^{D} g_{\alpha\alpha}g^{\beta\beta} = gg^{\beta\beta}. \tag{A.14}$$

Therefore, replacing Eq. (A.14) into Eq. (A.13), we get

$$\delta g = \sum_{\beta=0}^{D} gg^{\beta\beta}\delta g_{\beta\beta} \tag{A.15}$$

and, restoring the Einstein notation, the preceding equation can be written as

$$\delta g = gg^{\mu\nu}\delta g_{\mu\nu} \tag{A.16}$$

or, equivalently, as

$$\delta g = -gg_{\mu\nu}\delta g^{\mu\nu} \tag{A.17}$$

where we used the relations in (A.1). Once we find the transformation rule for δg, it is straightforward to obtain the following:

$$\delta(\sqrt{-g}) = -\frac{1}{2\sqrt{-g}}\delta g = -\sqrt{-g}g_{\mu\nu}\delta g^{\mu\nu}. \tag{A.18}$$

f(*R*) Gravity in Metric Formalism

Using the result provided in Eq. (A.8), here we apply the variational principle to the $f(R)$ action. First, the connection is supposed to be metric compatible, then the Palatini $f(R)$ action is considered. In the latter case, the connection is independent of the metric tensor. Let us start from the action

$$S = \int \sqrt{-g}f(R)\,d^4x, \tag{A.19}$$

whose variation with respect to the metric provides

$$\delta S = \delta \int f(R)\sqrt{-g}\, d^4x = \int [f(R)\, \delta\sqrt{-g} + \sqrt{-g}\, \delta f(R)]\, d^4x$$
$$= -\frac{1}{2}\int \sqrt{-g}\, f(R) g_{\mu\nu}\, \delta g^{\mu\nu}\, d^4x + \int \sqrt{-g} f_R(R)\, R_{\mu\nu} \delta g^{\mu\nu}\, d^4x$$
$$+ \int \sqrt{-g}\, f_R(R)\, g^{\mu\nu} \delta R_{\mu\nu}\, d^4x. \tag{A.20}$$

Using Eq. (A.8), the third integral can be written as

$$\int \sqrt{-g}\, f_R(R)\, R_{\mu\nu}\, \delta g^{\mu\nu}\, d^4x + \int \sqrt{-g}\, f_R(R)\, (\Box h - h^{\mu\nu}{}_{;\mu;\nu})\, d^4x, \tag{A.21}$$

where we made use of the relation

$$\delta\sqrt{-g} = -\sqrt{-g} g_{\mu\nu}\, \delta g^{\mu\nu}, \tag{A.22}$$

just demonstrated. Integrating by parts and neglecting the boundaries, Eq. (A.21) yields

$$\int \sqrt{-g}\, f_R(R)\, \delta R\, d^4x$$
$$= \int \sqrt{-g}\, \delta g^{\mu\nu} \left[R_{\mu\nu} f_R(R) - f_R(R)_{;\mu;\nu} + g_{\mu\nu} \Box f_R(R)\right] d^4x. \tag{A.23}$$

Substituting Eq. (A.23) into Eq. (A.20), we finally obtain

$$-\frac{1}{2} f(R) g_{\mu\nu} + R_{\mu\nu} f_R(R) - f_R(R)_{;\mu;\nu} + g_{\mu\nu} \Box f_R(R) = 0, \tag{A.24}$$

which can be recast in the following equivalent form:

$$G_{\mu\nu} = \frac{1}{f_R(R)} \left\{ \frac{1}{2} g_{\mu\nu} [f(R) - R f_R(R)] + f_R(R)_{;\mu;\nu} - g_{\mu\nu} \Box f_R(R) \right\}. \tag{A.25}$$

f(*R*) Gravity in Palatini Formalism

In the Palatini formalism, the connection is disentangled from the metric tensor, so that two sets of field equations occur. The variation with respect to the metric provides

$$\delta_g S = \int \delta_g \sqrt{-g}\, f(\mathcal{R})\, d^4x = \int [\delta_g \sqrt{-g}\, f(\mathcal{R}) + \sqrt{-g} \delta_g f(\mathcal{R})]\, d^4x$$
$$= \int \left[-\frac{1}{2} f(\mathcal{R}) \sqrt{-g}\, g_{\mu\nu} \delta_g g^{\mu\nu} + \sqrt{-g}\, f_{\mathcal{R}}(\mathcal{R})\, \delta_g \mathcal{R}\right]$$
$$= \int \left[-\frac{1}{2} f(\mathcal{R}) \sqrt{-g}\, g_{\mu\nu} \delta_g g^{\mu\nu} + \sqrt{-g}\, f_{\mathcal{R}}(\mathcal{R}) (\delta_g g^{\mu\nu}) \mathcal{R}_{\mu\nu}\right] d^4x = 0, \tag{A.26}$$

so that the first set of field equations is

$$f_{\mathcal{R}}(\mathcal{R})\mathcal{R}_{\mu\nu} - \frac{1}{2}f(\mathcal{R})g_{\mu\nu} = 0, \tag{A.27}$$

where $\mathcal{R}$ is the Palatini curvature scalar $\mathcal{R} = \mathcal{R}(\Gamma)$. The only term that must be considered in the variation with respect to the connection is the Palatini–Ricci tensor $\mathcal{R}_{\mu\nu}(\Gamma)$. The variation provides

$$\begin{aligned}\delta_\Gamma S &= \int \delta_\Gamma[\sqrt{-g}\, f(\mathcal{R})]\, d^4x = \int \delta_\Gamma[\sqrt{-g}\, f(\mathcal{R})]\, d^4x = \int [\sqrt{-g}\, \delta_\Gamma f(\mathcal{R})]\, d^4x \\ &= \int \sqrt{-g}\delta_\Gamma f(g^{\mu\nu}\mathcal{R}_{\mu\nu})\ d^4x = \int \sqrt{-g} f_{\mathcal{R}}(\mathcal{R}) g^{\mu\nu}\ \delta_\Gamma \mathcal{R}_{\mu\nu}\ d^4x \\ &= \int \sqrt{-g} f_{\mathcal{R}}(\mathcal{R}) g^{\mu\nu}\ (D_\sigma\ \delta\Gamma^\sigma_{\mu\nu} - D_\nu\ \delta\Gamma^\sigma_{\mu\sigma})\, d^4x.\end{aligned} \tag{A.28}$$

Integrating out higher-order terms and neglecting the boundaries, we obtain

$$D_\sigma[\sqrt{-g} f_{\mathcal{R}}(\mathcal{R}) g^{\mu\nu}] = 0, \tag{A.29}$$

so that the Palatini field equations are

$$\begin{cases} f_{\mathcal{R}}(\mathcal{R})\mathcal{R}_{\mu\nu} - \frac{1}{2}g_{\mu\nu}f(\mathcal{R}) = 0 \\ D_\sigma[\sqrt{-g} f_{\mathcal{R}}(\mathcal{R}) g^{\mu\nu}] = 0. \end{cases} \tag{A.30}$$

Scalar-Tensor Gravity

Let us now consider the following scalar-tensor action with coupling $f(\phi)$ and potential $V(\phi)$:

$$S = \int \sqrt{-g}\left\{f(\phi)\ R + \frac{1}{2}g^{\mu\nu}\phi_{,\mu}\phi_{,\nu} - V(\phi) + \mathscr{L}_m\right\} d^4x, \tag{A.31}$$

where for simplicity the kinetic term coefficient is set to 1/2. In this way, Brans–Dicke gravity can be recovered from Eq. (A.31) by properly reparameterizing the scalar field. By varying the action (A.31), one gets

$$\begin{aligned}\delta S &= \int \delta\sqrt{-g}\left[f(\phi)\ R + \frac{1}{2}g^{\mu\nu}\phi_{,\mu}\phi_{,\nu} - V(\phi) + \mathscr{L}_m\right] d^4x \\ &\quad + \int \sqrt{-g}\left[f(\phi)R_{\mu\nu}\ \delta g^{\mu\nu} + f(\phi)g^{\mu\nu}\ \delta R_{\mu\nu} + \frac{1}{2}\phi_{,\mu}\phi_{,\nu}\ \delta g^{\mu\nu}\right] d^4x \\ &= \int \sqrt{-g}\left[-\frac{1}{2}f(\phi)R\ g_{\mu\nu}\ \delta g^{\mu\nu} + f(\phi)g^{\mu\nu}\ \delta R_{\mu\nu} + f(\phi)R_{\mu\nu}\ \delta g^{\mu\nu}\right. \\ &\quad \left. - \frac{1}{4}\phi^{,\alpha}\phi_{,\alpha}\ g_{\mu\nu}\ \delta g^{\mu\nu} + \frac{1}{2}\phi_{,\mu}\phi_{,\nu}\ \delta g^{\mu\nu} + \frac{1}{2}\ V(\phi)g_{\mu\nu}\delta g^{\mu\nu}\right] d^4x = 0.\end{aligned} \tag{A.32}$$

The term $f(\phi)g^{\mu\nu}\,\delta R_{\mu\nu}$ can be easily computed by means of Eq. (A.8). Thus, after integrating by parts, it yields

$$\int\sqrt{-g}f(\phi)g^{\mu\nu}\delta R_{\mu\nu}\,d^4x=\int\sqrt{-g}\left(g_{\mu\nu}\Box f(\phi)-f(\phi)_{;\mu;\nu}\right)\delta g^{\mu\nu}\,d^4x. \tag{A.33}$$

Finally, substituting Eq. (A.33) into Eq. (A.32), the field equations read as follows:

$$f(\phi)\,G_{\mu\nu}=\frac{1}{4}g_{\mu\nu}\,\phi^{,\alpha}\phi_{,\alpha}-\frac{1}{2}\phi_{,\mu}\phi_{,\nu}-\frac{1}{2}V(\phi)g_{\mu\nu}-g_{\mu\nu}\Box f(\phi)+f(\phi)_{;\mu;\nu}. \tag{A.34}$$

Higher Derivatives Gravity

In Eq. (4.55) we considered $(2k+4)$th-order field equations coming from the general action (4.54). Here we show how to obtain field equations at the lowest order, where the action is

$$S=\int\sqrt{-g}f(\Box R)\,d^4x. \tag{A.35}$$

We do not include the scalar curvature in the function, since its contribution in the field equations can be inferred by generalizing Eq. (A.24). Therefore, by varying Eq. (A.35) one gets

$$\begin{aligned}\delta S&=\int\sqrt{-g}\,\delta f\,d^4x+\int f\delta\sqrt{-g}\,d^4x\\&=\int\sqrt{-g}\left(\frac{\partial f}{\partial\Box R}\right)\delta\Box R\,d^4x+\int f\delta\sqrt{-g}\,d^4x.\end{aligned} \tag{A.36}$$

It is worth noticing that the operators δ and $\Box$ do not commute, therefore the terms $\delta\Box R$ and $\Box\delta R$ must be evaluated separately. Let us start from the former:

$$\begin{aligned}\delta\Box R&=\delta(g^{\alpha\beta}D_\alpha D_\beta R)=D_\alpha D_\beta R\,\delta g^{\alpha\beta}+g^{\alpha\beta}\delta(D_\alpha D_\beta\,R)\\&=D_\alpha D_\beta R\,\delta g^{\alpha\beta}+g^{\alpha\beta}\delta(\partial_\alpha\partial_\beta R-\Gamma^\gamma_{\alpha\beta}\partial_\gamma R)\\&=D_\alpha D_\beta R\,\delta g^{\alpha\beta}+g^{\alpha\beta}(\partial_\alpha\partial_\beta\,\delta R-\Gamma^\gamma_{\alpha\beta}\partial_\gamma\,\delta R-\partial_\gamma R\,\delta\Gamma^\gamma_{\alpha\beta})\\&=D_\alpha D_\beta R\,\delta g^{\alpha\beta}+\Box\,\delta R-g^{\alpha\beta}\partial_\gamma R\,\delta\Gamma^\gamma_{\alpha\beta}.\end{aligned} \tag{A.37}$$

From Eq. (A.37) the following commutation relation occurs:

$$\delta\Box=\delta g^{\alpha\beta}D_\alpha D_\beta+\Box\,\delta-g^{\alpha\beta}\,\delta\Gamma^\gamma_{\alpha\beta}\partial_\gamma. \tag{A.38}$$

By using Eq. (A.38), the first integral in Eq. (A.36) can be decomposed as

$$\int \sqrt{-g}\frac{\partial f}{\partial \Box R}\,\delta\Box R\, d^4x$$
$$= \int \sqrt{-g}\left(\frac{\partial f}{\partial \Box R}\right)\left(D_\alpha D_\beta R\,\delta g^{\alpha\beta} + \Box\,\delta R - g^{\alpha\beta}\partial_\gamma R\,\delta\Gamma^\gamma_{\alpha\beta}\right)\, d^4x. \tag{A.39}$$

The second term in Eq. (A.39), that is, $\Box\delta R$, can be recast in the required form by means of Eq. (A.8) (or, equivalently, Eq. (A.38)):

$$\int \sqrt{-g}\left(\frac{\partial f}{\partial \Box R}\right)\delta\Box R\, d^4x = \int \sqrt{-g}\left(\frac{\partial f}{\partial \Box R}\right)\Box\delta R\, d^4x$$
$$= \int \sqrt{-g}\left(\frac{\partial f}{\partial \Box R}\right)\Box(R_{\alpha\beta}\delta g^{\alpha\beta} + g^{\alpha\beta}\delta R_{\alpha\beta})\, d^4x$$
$$= \int \sqrt{-g}\left(\frac{\partial f}{\partial \Box R}\right)\Box(R_{\alpha\beta}\delta g^{\alpha\beta} + \Box h - D_\beta D_\alpha h^{\alpha\beta})\, d^4x$$
$$= \int \sqrt{-g}\,\Box\left(\frac{\partial f}{\partial \Box R}\right)(R_{\alpha\beta}\delta g^{\alpha\beta} + \Box h - D_\beta D_\alpha h^{\alpha\beta})\, d^4x$$
$$= \int \sqrt{-g}\left[-D_\alpha D_\beta \Box\left(\frac{\partial f}{\partial \Box R}\right) + g_{\alpha\beta}\Box^2\left(\frac{\partial f}{\partial \Box R}\right) + R_{\alpha\beta}\Box\left(\frac{\partial f}{\partial \Box R}\right)\right]\delta g^{\alpha\beta}\, d^4x. \tag{A.40}$$

The third term in Eq. (A.39) can be suitably reduced by using the Christoffel connection variation (A.4), which provides

$$\int -\sqrt{-g}\left(\frac{\partial f}{\partial \Box R}\right) g^{\alpha\beta}\partial_\gamma R\,\delta\Gamma^\gamma_{\alpha\beta}\, d^4x$$
$$= \int -\frac{1}{2}\sqrt{-g}\left(\frac{\partial f}{\partial \Box R}\right) g^{\alpha\beta}\partial_\gamma R\, g^{\gamma\sigma}(D_\beta h_{\sigma\alpha} + D_\alpha h_{\sigma\beta} - D_\sigma h_{\alpha\beta})d^4x$$
$$= \int \frac{1}{2}\sqrt{-g}\left(\frac{\partial f}{\partial \Box R}\right) g^{\alpha\beta}\partial_\sigma R(D^\beta h^{\sigma\alpha} + D^\alpha h^{\sigma\beta} - D^\sigma h^{\alpha\beta})d^4x, \tag{A.41}$$

where, in the last equality, we raised the indexes α, β, σ and used the convention $\delta g_{\alpha\beta} = -h_{\alpha\beta}$. Integrating by parts and neglecting the boundaries, we obtain

$$\int -\frac{1}{2}\sqrt{-g}\left[D_\alpha\left(\left(\frac{\partial f}{\partial \Box R}\right)D_\sigma R\right)\delta g^{\sigma\alpha} + D_\beta\left(\left(\frac{\partial f}{\partial \Box R}\right)D_\sigma R\right)\delta g^{\sigma\beta} - g_{\alpha\beta}D^\sigma\left(\left(\frac{\partial f}{\partial \Box R}\right)D_\sigma R\right)\delta g^{\alpha\beta}\right] d^4x$$

$$= \int -\frac{1}{2}\sqrt{-g}\left[2\left(\frac{\partial f}{\partial \Box R}\right)D_\alpha D_\beta R + D_\alpha\left(\frac{\partial f}{\partial \Box R}\right)D_\beta R + D_\beta\left(\frac{\partial f}{\partial \Box R}\right)D_\alpha R - g_{\alpha\beta}\left(\frac{\partial f}{\partial \Box R}\right)\Box R - g_{\alpha\beta}g^{\mu\nu}D_\mu\left(\frac{\partial f}{\partial \Box R}\right)D_\nu R\right]\delta g^{\alpha\beta}\, d^4x. \tag{A.42}$$

Substituting finally Eqs. (A.40) and (A.42) into the action (A.36), we get the variation

$$\int \sqrt{-g}\left(\frac{\partial f}{\partial \Box R}\right)\delta\Box R\, d^4x = \int \sqrt{-g}\Big[-D_\alpha D_\beta \Box\left(\frac{\partial f}{\partial \Box R}\right)$$
$$+g_{\alpha\beta}\Box^2\left(\frac{\partial f}{\partial \Box R}\right) + R_{\alpha\beta}\Box\left(\frac{\partial f}{\partial \Box R}\right) - \frac{1}{2}D_{\{\alpha}\left(\frac{\partial f}{\partial \Box R}\right)D_{\beta\}}R$$
$$+\frac{1}{2}g_{\alpha\beta}\left(\left(\frac{\partial f}{\partial \Box R}\right)\Box R + g^{\mu\nu}D_\mu\left(\frac{\partial f}{\partial \Box R}\right)D_\nu R\right)\Big]\,\delta g^{\alpha\beta}\, d^4x, \tag{A.43}$$

from which the field equations read as

$$-D_\alpha D_\beta\Box\left(\frac{\partial f}{\partial \Box R}\right) + g_{\alpha\beta}\Box^2\left(\frac{\partial f}{\partial \Box R}\right) + R_{\alpha\beta}\Box\left(\frac{\partial f}{\partial \Box R}\right)$$
$$+\frac{1}{2}g_{\alpha\beta}\left(\left(\frac{\partial f}{\partial \Box R}\right)\Box R + g^{\mu\nu}D_\mu\left(\frac{\partial f}{\partial \Box R}\right)D_\nu R\right)$$
$$-\frac{1}{2}D_{\{\alpha}\left(\frac{\partial f}{\partial \Box R}\right)D_{\beta\}}R - \frac{1}{2}g_{\alpha\beta}f = 0. \tag{A.44}$$

Note that by summing the quantity $1/2\, g_{\alpha\beta}R\Box\left(\frac{\partial f}{\partial \Box R}\right)$ to both sides of Eq. (A.44), the latter becomes

$$G_{\alpha\beta}\Box\left(\frac{\partial f}{\partial \Box R}\right) = \frac{1}{2}g_{\alpha\beta}\left(f - R\Box\left(\frac{\partial f}{\partial \Box R}\right)\right)$$
$$+D_\alpha D_\beta\Box\left(\frac{\partial f}{\partial \Box R}\right) - g_{\alpha\beta}\Box^2\left(\frac{\partial f}{\partial \Box R}\right) + \frac{1}{2}D_{\{\alpha}\left(\frac{\partial f}{\partial \Box R}\right)D_{\beta\}}R$$
$$-\frac{1}{2}g_{\alpha\beta}\left(\left(\frac{\partial f}{\partial \Box R}\right)\Box R + g^{\mu\nu}D_\mu\left(\frac{\partial f}{\partial \Box R}\right)D_\nu R\right), \tag{A.45}$$

so that the RHS can be intended as the energy-momentum tensor of the gravitational field (up to the factor $\Box\left(\frac{\partial f}{\partial \Box R}\right)$). Taking the limit $k = 1$ and $f_R = 0$ of Eq. (4.55), Eq. (A.45) is recovered. Also notice that the scalar curvature can be easily included in the starting action by merging the results provided in Eq. (A.45) and Eq. (A.24), which yield

$$G_{\alpha\beta}\left(f_R + \Box\left(\frac{\partial f}{\partial \Box R}\right)\right) = \frac{1}{2}g_{\alpha\beta}\left(f - Rf_R - R\Box\left(\frac{\partial f}{\partial \Box R}\right)\right)$$
$$+D_\alpha D_\beta\left(\Box\left(\frac{\partial f}{\partial \Box R}\right) + f_R\right) - g_{\alpha\beta}\Box^2\left(\frac{\partial f}{\partial \Box R}\right) - g_{\alpha\beta}\Box f_R$$
$$-\frac{1}{2}g_{\alpha\beta}\left(\left(\frac{\partial f}{\partial \Box R}\right)\Box R + g^{\mu\nu}D_\mu\left(\frac{\partial f}{\partial \Box R}\right)D_\nu R\right)$$
$$+\frac{1}{2}D_{\{\alpha}\left(\frac{\partial f}{\partial \Box R}\right)D_{\beta\}}R. \tag{A.46}$$

Teleparallel Gravity

To conclude this appendix, we apply the variational principle to TEGR and recover the field equations previously written in Eq. (4.108). The main difference with respect to the metric-compatible theories is that the variation must be performed with respect to tetrad fields. Therefore, the following conditions must be considered instead of Eq. (A.1):

$$\begin{aligned} &\bullet\, \delta g^{\mu\nu} = \left(h_b^{\nu} g^{\rho\mu} + h_b^{\mu} g^{\rho\nu}\right) \delta h^b_{\rho}, \\ &\bullet\, \delta h_c^{\mu} = -h_c^{\nu} h_a^{\mu} \delta h^a_{\nu}, \end{aligned} \tag{A.47}$$

where h^a_{μ} are the nontrivial tetrad fields. The first identity can be demonstrated from the definition of tetrad fields $g^{\mu\nu} = h_a^{\mu} h_b^{\nu} \eta^{ab}$, whose variation yields

$$\begin{aligned} \delta g^{\mu\nu} &= \eta^{ab} \left(h_b^{\nu} \delta h_a^{\mu} + h_a^{\mu} \delta h_b^{\nu}\right) = \eta^{ba} h_a^{\nu} \delta h_b^{\mu} + \eta^{ab} h_a^{\mu} \delta h_b^{\nu} \\ &= \eta^{ba} h_a^{\nu} \delta^{\mu}_{\rho} \delta h_b^{\rho} + \eta^{ab} h_a^{\mu} \delta^{\nu}_{\rho} \delta h_b^{\rho} = \left(\delta^a_b h_a^{\nu} g^{\rho\mu} + \delta^a_b h_a^{\mu} g^{\rho\nu}\right) \delta h^b_{\rho} \\ &= \left(h_b^{\nu} g^{\rho\mu} + h_b^{\mu} g^{\rho\nu}\right) \delta h^b_{\rho} = h_b^{\{\nu} g^{|\rho|\mu\}} \delta h^b_{\rho}. \end{aligned} \tag{A.48}$$

On the other hand, the orthogonality condition $h_a^{\mu} h^a_{\nu} = \delta^{\mu}_{\nu}$ can be used to prove the second identity of Eq. (A.47). Straightforwardly, considering the variation of the orthogonality condition

$$\delta(h_a^{\mu} h^a_{\nu}) = h^a_{\nu} \delta h_a^{\mu} + h_a^{\mu} \delta h^a_{\nu} = 0 \tag{A.49}$$

and multiplying both sides by h_c^{ν}, one gets

$$h_c^{\nu} h^a_{\nu} \delta h_a^{\mu} + h_c^{\nu} h_a^{\mu} \delta h^a_{\nu} = \delta^a_c \delta h_a^{\mu} + h_c^{\nu} h_a^{\mu} \delta h^a_{\mu} = \delta h_c^{\mu} + h_c^{\nu} h_a^{\mu} \delta h^a_{\mu} = 0. \tag{A.50}$$

Now, from the teleparallel equivalent of Eq. (A.22), that is, $\delta h = h h_a^{\mu} \delta h^a_{\mu}$, the variation of TEGR action reads

$$\delta S = \delta \int h\, T\, d^4x = \int (h \delta T + T \delta h)\, d^4x = \int h \left(\delta T + T h_a^{\mu} \delta h^a_{\mu}\right) d^4x. \tag{A.51}$$

Taking into account the definition of the torsion scalar (4.84) and of the superpotential (4.83), we have

$$\delta T = \left[\frac{1}{4} \delta \left(T^{\mu\nu\beta} T_{\mu\nu\beta}\right) + \frac{1}{2} \delta \left(T^{\mu\nu\beta} T_{\nu\mu\beta}\right) - \delta \left(T_{\nu}^{\ \nu\mu} T^{\nu}_{\ \nu\mu}\right)\right]. \tag{A.52}$$

The variation of each term in the RHS of Eq. (A.52) can be computed explicitly with the help of Eq. (A.47) and provides

$$\begin{aligned} \delta \left(T^{\mu\nu\beta} T_{\mu\nu\beta}\right) &= -4 T^{\mu\nu\beta} T_{\mu\nu\sigma} h_a^{\sigma} \delta h^a_{\beta} + 4 T_{\mu}^{\ \nu\beta} h_a^{\mu} \partial_{\nu} \delta h^a_{\beta}, \\ \delta \left(T^{\mu\nu\beta} T_{\nu\mu\beta}\right) &= 2 \left(T^{\lambda\nu\mu} - T^{\mu\nu\lambda}\right) T_{\nu\mu\beta} h_a^{\beta} \delta h^a_{\lambda} + \left(T^{\mu}{}_{\nu}{}^{\lambda} - T^{\lambda}{}_{\nu}{}^{\mu}\right) h_a^{\nu} \partial_{\mu} \delta h^a_{\lambda}, \\ \delta \left(T_{\nu}^{\ \nu\mu} T^{\nu}_{\ \nu\mu}\right) &= -2 \left(T_{\rho}^{\ \rho\lambda} T^{\beta}_{\ \lambda\mu} + T_{\gamma}^{\ \gamma\beta} T^{\rho}_{\ \rho\mu}\right) h_a^{\mu} \delta h^a_{\beta} \\ &\quad + 2 \left(T_{\gamma}^{\ \gamma\beta} h_a^{\mu} - T_{\alpha}^{\ \alpha\mu} h_a^{\beta}\right) \partial_{\beta} \delta h^a_{\mu}, \end{aligned} \tag{A.53}$$

which, together with the definition of superpotential (4.83), leads to

$$h\delta T = 4\left(-\partial_\mu\left(hS_a{}^{\mu\beta}\right) + hT^\sigma{}_{\mu a}S_\sigma{}^{\beta\mu}\right)\delta h^a_\beta. \tag{A.54}$$

Therefore, the field equations in vacuum finally read as follows:

$$\frac{4}{h}\partial_\mu\left(hS_a{}^{\mu\beta}\right) - 4T^\sigma{}_{\mu a}S_\sigma{}^{\beta\mu} - Th^\beta_a = 0. \tag{A.55}$$

Appendix B

Differential Forms and Variations of Gauge Actions

We introduce the formalism of differential forms, used in Section 5.2, to obtain Cartan's structure equations. Then, we find the equations of motion of the $SU(n)$-invariant Chern–Simons actions. To this purpose, let $\mathcal{P}$ and $\mathcal{Q}$ be differential forms of orders n and m, respectively. The *exterior product* T is defined as the $(m+n)$-form

$$T = \mathcal{P} \wedge \mathcal{Q} = \mathcal{P}_{[\alpha_1,\alpha_2,\ldots\alpha_n}\mathcal{Q}_{\alpha_{n+1},\alpha_{n+2},\ldots\alpha_{n+m}]}\ dx^{\alpha_1} \wedge dx^{\alpha_2} \wedge .. \wedge dx^{\alpha_{m+n}}\ .$$

Consequently, the *exterior derivative* $d\mathcal{P}$ consists of the exterior product between the one-form derivative d and the n-form $\mathcal{P}$:

$$d\mathcal{P} = \partial_{[\alpha_1}\mathcal{P}_{\alpha_2\alpha_3\ldots\alpha_n]}\ dx^{\alpha_1} \wedge dx^{\alpha_2} \wedge \ldots \wedge dx^{\alpha_n}\ . \tag{B.1}$$

We define the Hodge dual $\star\mathcal{P}$ of a n-form $\mathcal{P}$ as the $(m-n)$-form

$$\star\,\mathcal{P} = \frac{1}{(m-n)!}\mathcal{P}^{\alpha_1,\alpha_2,\ldots\alpha_n}\sqrt{|g|}\epsilon_{\alpha_1,\alpha_2,\ldots\alpha_m}\ dx^{\alpha_{n+1}} \wedge dx^{\alpha_{n+2}} \wedge \ldots \wedge dx^{\alpha_m}\ , \tag{B.2}$$

with ϵ being the antisymmetric Levi–Civita tensor. From Eq. (B.2), the general identity

$$\star\, 1 = (-1)^{D-1}\sqrt{|g|}\, d^D x \tag{B.3}$$

follows. The exterior product can be equivalently written as

$$dx^{\alpha_1} \wedge dx^{\alpha_2} \wedge \ldots \wedge dx^{\alpha_D} = \epsilon^{\alpha_1,\alpha_2\ldots\alpha_D} d^D x\ . \tag{B.4}$$

Finally, by means of the definition of the vielbein field, that is,

$$g_{\mu\nu} = \partial_\mu x^a \partial_\nu x^b \eta_{ab} \equiv e^a_\mu e^b_\nu \eta_{ab}\ , \tag{B.5}$$

the generic n-form $\mathcal{P}$ can be recast as

$$\begin{aligned}\mathcal{P} &= \mathcal{P}_{[\alpha_1,\alpha_2\ldots\alpha_n]} dx^{\alpha_1} \wedge dx^{\alpha_2} \wedge \ldots \wedge dx^{\alpha_n} \\ &= \mathcal{P}_{[a_1,a_2\ldots a_n]} e^{a_1} \wedge e^{a_2} \wedge \ldots \wedge e^{a_n}\ ,\end{aligned} \tag{B.6}$$

with e^{a_j} being the set of basis $e^a = e^a_\mu dx^\mu$.

Field Equations and Gauge Invariance of Three-Dimensional Chern–Simons Action on $SU(n)$

Here we show how to perform the variation of the $SU(n)$ invariant Chern–Simons action. It turns out that only a two-dimensional boundary term survives after varying the action with respect to the gauge connection, thus proving the gauge-invariance of the theory. The Chern–Simons $SU(n)$-invariant three-dimensional action

$$S = \int Tr\left[\mathbf{A}d\mathbf{A} + \frac{2}{3}\mathbf{AAA}\right] \tag{B.7}$$

can be equivalently written in coordinates representation as

$$S = Tr\left[\int \epsilon^{ijk}\left(A_i\partial_j A_k + \frac{2}{3}A_iA_jA_k\right) d^3x\right]. \tag{B.8}$$

The second term is the so called *Wess–Zumino–Witten* term, whose variation is

$$\delta\left(\epsilon^{ijk}A_iA_jA_k\right) = 3\epsilon^{ijk}\delta A_iA_jA_k. \tag{B.9}$$

Action (B.7) is holographically dual to the two-dimensional Wess–Zumino–Witten model, which describes propagating strings on the given group. Furthermore, considering the relation

$$\epsilon^{ijk}A_jA_k = -\epsilon^{ijk}\partial_jA_k \tag{B.10}$$

and substituting Eq. (B.10) into Eq. (B.9), the latter yields

$$\delta\left(\epsilon^{ijk}A_iA_jA_k\right) = 3\epsilon^{ijk}\delta A_i\partial_jA_k. \tag{B.11}$$

On the other hand, the variation of the first term of Eq. (B.7) can be recast as

$$\delta S_{AdA} = Tr\left[\int \left(\epsilon^{ijk}\delta A_i\partial_jA_k + \epsilon^{ijk}A_i\partial_j\delta A_k\right) d^3x\right]. \tag{B.12}$$

Integrating the quantity $\epsilon^{ijk}A_i\partial_j\delta A_k$, two contributions arise, which can be understood as a boundary and a bulk term. Therefore, Eq. (B.12) takes the form

$$\delta S_{AdA} = Tr\left[\int \epsilon^{ijk}\delta A_i\partial_jA_k\, d^3x + \int_{\mathcal{B}} \epsilon^{ik}A_i\delta A_k\, d^2x - \int \epsilon^{ijk}\partial_jA_i\delta A_k d^3x\right]. \tag{B.13}$$

By using the antisymmetric property of the Levi–Civita tensor, the action becomes

$$\delta S_{AdA} = Tr\left[2\int \epsilon^{ijk}\delta A_i\partial_jA_k\, d^3x + \int_{\mathcal{B}} \epsilon^{ik}A_i\delta A_k\, d^2x\right]. \tag{B.14}$$

Merging Eq. (B.11) and Eq. (B.14), it turns out that the final variation only yields the term on the boundary:

$$\delta S = Tr\left[\int_{\mathcal{B}} \epsilon^{ik}A_i\delta A_k\, d^2x\right], \tag{B.15}$$

which makes the action *quasi*-gauge invariant. In order to find the field equations on $SU(n)$, the action must be varied with respect to the gauge connection **A**. First, notice that the latter can be generally written in terms of the group element g as

$$A_i = g^{-1}\partial_i g. \tag{B.16}$$

Thus, the variation of the connection with respect to g reads

$$\delta A_i = -g^{-1}\delta g\, A_i + g^{-1}\partial_i \delta g. \tag{B.17}$$

The variation of the Chern–Simons action can be split in two contributions, namely

$$\delta S = \delta S_{AdA} + \delta S_{AAA}, \tag{B.18}$$

that will be separately evaluated in what follows.

1) Term $\boldsymbol{A}\ d\boldsymbol{A}$ In coordinates representation, the first term of Eq. (B.18) can be written as

$$\delta S_{AdA} = Tr\left[\int \epsilon^{ijk}\delta A_i \partial_j A_k\, d^3x + \int \epsilon^{ijk} A_i \partial_j \delta A_k\, d^3x\right]. \tag{B.19}$$

It can be easily shown that the first term of the RHS in Eq. (B.19) vanishes identically. To this purpose, notice that the identity (B.17) yields

$$\int \epsilon^{ijk}\delta A_i \partial_j A_k\, d^3x = \int \epsilon^{ijk}\left(-g^{-1}A_i\partial_j A_k\,\delta g + g^{-1}\partial_j A_k \partial_i \delta g\right) d^3x. \tag{B.20}$$

Integrating out the pure boundary $\partial_i \delta g$ and using again the relation (B.16), the preceding variation provides

$$\begin{aligned}\int \epsilon^{ijk}\delta A_i \partial_j A_k\, d^3x &= \int \epsilon^{ijk}\left(-g^{-1}A_i\partial_j A_k + g^{-1}\partial_i g g^{-1}\partial_j A_k\right)\delta g\, d^3x \\ &= \int \epsilon^{ijk}\left(-g^{-1}A_i\partial_j A_k + g^{-1}A_i\partial_j A_k\right)\delta g\, d^3x = 0.\end{aligned} \tag{B.21}$$

Therefore, the only term surviving in Eq. (B.19) is

$$\delta S_{AdA} = Tr\left[\int \epsilon^{ijk} A_i \partial_j \delta A_k\, d^3x\right], \tag{B.22}$$

which can be made explicit by means of the relation (B.22), providing

$$\begin{aligned}&\int \epsilon^{ijk} A_i \partial_j \delta A_k\, d^3x \\ &= \int \epsilon^{ijk} A_i g^{-1}\left(\partial_j g\, g^{-1} A_k \delta g - \partial_j A_k \delta g - A_k \partial_j \delta g - \partial_j g\, g^{-1}\partial_k \delta g\right) d^3x.\end{aligned} \tag{B.23}$$

Using Eqs. (B.16) and (B.10), it turns out that the last two terms in the preceding equation cancel out, so that we finally have

$$\delta S_{AdA} = Tr\left[\int \epsilon^{ijk} A_i \partial_j \delta A_k \, d^3x\right] = Tr\left[-2\int \epsilon^{ijk} g^{-1} A_k \partial_i A_j \, \delta g \, d^3x\right]. \qquad \text{(B.24)}$$

2) Term ***AAA*** In order to show that the variation of the Wess–Zumino–Witten term yields a boundary term, let us consider Eq. (B.9), by means of which the quantity δS_{AAA} takes the form

$$\delta S_{AAA} = Tr\left[2\int \epsilon^{ijk} \delta A_i A_j A_k \, d^3x\right]. \qquad \text{(B.25)}$$

The preceding equation can be equivalently written in terms of the element g as

$$\delta S_{AAA} = Tr\left[2\int \epsilon^{ijk} g^{-1} \left(-A_i \delta g + \partial_i \delta g\right) A_j A_k \, d^3x\right]. \qquad \text{(B.26)}$$

Integrating by parts and neglecting the boundary term, the variation of the action yields

$$\begin{aligned} \delta S_{AAA} = \; & Tr\left[2\int \epsilon^{ijk} \left(-g^{-1} A_i A_j A_k - \partial_i (g^{-1} A_j A_k)\right) \delta g \, d^3x\right] \\ = \; & Tr\left[2\int \epsilon^{ijk} g^{-1} \left(-A_i A_j A_k + A_i A_j A_k - \partial_i A_j A_k - \partial_i A_k A_j\right) \delta g \, d^3x\right], \end{aligned} \qquad \text{(B.27)}$$

which vanishes due to the antisymmetric property of the Levi–Civita symbol. Merging Eqs. (B.24) and (B.27), the variation of the total Chern–Simons three-dimensional action finally reads

$$\delta S_{CS} = Tr\left[-2\int \epsilon^{ijk} g^{-1} A_k \partial_i A_j \, \delta g \, d^3x\right] = 0. \qquad \text{(B.28)}$$

Application of $U(1)$ Invariant Chern–Simons Action to Electromagnetic Theory

We show that the three-dimensional $U(1)$ invariant Chern–Simons action can be used to describe the electromagnetic theory, providing a massive wave equation. To this purpose, we consider

$$S = \int \mathbf{A} \mathrm{d} \mathbf{A}, \qquad \text{(B.29)}$$

whose exterior derivative provides the four-dimensional Pontryagin density $P_4 = \mathbf{F} \wedge \mathbf{F}$, with $\mathbf{F}$ being the curvature two-form.

By computing the exterior derivative, the action can be written as

$$\int \mathbf{A} \mathrm{d} \mathbf{A} = \int (\partial_\mu A_\nu - \partial_\nu A_\mu) A_\rho \, dx^\mu \wedge dx^\nu \wedge dx^\rho = \int \epsilon^{\mu\nu\rho} F_{\mu\nu} A_\rho \, d^3x. \qquad \text{(B.30)}$$

Introducing the Chern–Simons form (B.30) in the free electromagnetic Lagrangian, the latter turns out to be

$$\mathcal{L} = -\frac{1}{4} F^{\mu\nu} F_{\mu\nu} + \frac{1}{2} m\, \epsilon^{\mu\nu\rho} F_{\mu\nu} A_\rho, \tag{B.31}$$

with m being a constant term with mass dimension. Notice that under gauge transformations, the Lagrangian only changes by a total derivative. In fact, under the transformation

$$\delta A_\mu \to \partial_\mu \Theta, \tag{B.32}$$

the variation of Lagrangian is

$$\begin{aligned} \delta\mathcal{L} &= -\frac{1}{2} F_{\mu\nu} (\partial^\mu \partial^\nu - \partial^\nu \partial^\mu)\Theta + \frac{1}{2} m\, \epsilon^{\mu\nu\rho} [F_{\mu\nu} \partial_\rho \Theta + A_\rho (\partial_\mu \partial_\nu - \partial_\nu \partial_\mu)\Theta] \\ &= \partial_\rho \left(\frac{1}{2} m\, \epsilon^{\mu\nu\rho} F_{\mu\nu} \Theta \right). \end{aligned} \tag{B.33}$$

The last equality is motivated by the identity $\partial_\rho (\epsilon^{\mu\nu\rho} F_{\mu\nu}) = 0$, coming from the field equations of the theory. The field equations can be obtained by varying the action with respect to the gauge connection, namely

$$\begin{aligned} \delta_A S &= \int \left\{ -\frac{1}{4} \delta_A \left(F^{\mu\nu} F_{\mu\nu} \right) + \frac{1}{2} m\, \epsilon^{\mu\nu\rho} \delta_A (F_{\mu\nu} A_\rho) \right\} d^3x \\ &= \int \left\{ \partial_\mu F^{\mu\nu} + \frac{1}{2} m\, \epsilon^{\mu\rho\nu} F_{\mu\rho} \right\} \delta A_\nu \, d^3x, \end{aligned} \tag{B.34}$$

so that they read

$$\partial_\mu F^{\mu\nu} + \frac{1}{2} m\, \epsilon^{\mu\rho\nu} F_{\mu\rho} = 0. \tag{B.35}$$

Using the identity $\partial_\mu \partial_\nu F^{\mu\nu} = 0$ and taking the three-dimensional divergence of Eq. (B.31), we get

$$\partial_\nu \left(m\, \epsilon^{\mu\rho\nu} F_{\mu\rho} \right) = 0, \tag{B.36}$$

which finally provides the result used to obtain Eq. (B.33). After demonstrating the $U(1)$ invariance of the Chern–Simons Lagrangian, by means of Eq. (B.35), it is possible to show that the field equations can be manifestly recast as a Klein–Gordon equation for massive fields. Multiplying Eq. (B.35) by the Levi–Civita symbol, we have

$$\begin{aligned} \epsilon_{\sigma\tau\nu} \partial_\mu F^{\mu\nu} + \frac{1}{2} m\, \epsilon_{\sigma\tau\nu} \epsilon^{\mu\rho\nu} F_{\mu\rho} &= \epsilon_{\sigma\tau\nu} \partial_\mu F^{\mu\nu} + \frac{1}{2} m (\delta^\mu_\sigma \delta^\rho_\tau - \delta^\rho_\sigma \delta^\mu_\tau) F_{\mu\rho} \\ &= \epsilon_{\sigma\tau\nu} \partial_\mu F^{\mu\nu} + m F_{\sigma\tau} = 0. \end{aligned} \tag{B.37}$$

Let us evaluate the first term ($\epsilon_{\sigma\tau\nu}\partial_\mu F^{\mu\nu}$):

$$\begin{aligned}\epsilon_{\sigma\tau\nu}\partial_\mu F^{\mu\nu} &= \frac{1}{2}\epsilon_{\sigma\tau\nu}\partial_\mu \left[F^{\mu\nu} - F^{\nu\mu}\right] \\ = \frac{1}{2}\epsilon_{\sigma\tau\nu}\partial_\mu \left[\delta^\mu_\alpha \delta^\nu_\beta F^{\alpha\beta} - \delta^\mu_\beta \delta^\nu_\alpha F^{\alpha\beta}\right] &= \frac{1}{2}\epsilon_{\sigma\tau\nu}\epsilon^{\mu\nu\lambda}\epsilon_{\alpha\beta\lambda}\partial_\mu F^{\alpha\beta} \\ = -\frac{1}{2}\left[\delta^\mu_\sigma \delta^\lambda_\tau - \delta^\mu_\tau \delta^\lambda_\sigma\right]\epsilon_{\alpha\beta\lambda}\partial_\mu F^{\alpha\beta} &= -\frac{1}{2}\epsilon_{\alpha\beta[\tau}\partial_{\sigma]} F^{\alpha\beta}. \end{aligned} \tag{B.38}$$

Substituting Eq. (B.38) into Eq. (B.37), the field equations take the following form:

$$-\frac{1}{2}\epsilon_{\alpha\beta[\tau}\partial_{\sigma]} F^{\alpha\beta} + mF_{\sigma\tau}. \tag{B.39}$$

Taking the divergence and using Eqs. (B.35) – (B.36), Eq. (B.39) finally reads

$$\left(\Box + m^2\right)\left(\epsilon_{\alpha\beta\tau}F^{\alpha\beta}\right) = 0. \tag{B.40}$$

Eq. (B.40) is a Klein–Gordon equation for the vector field $\epsilon_{\alpha\beta\tau}F^{\alpha\beta}$, which finally demonstrates the capability of the Chern–Simons three-dimensional theory to describe massive particles without breaking the gauge symmetry.

Appendix C

Noether–Bessel-Hagen Symmetries in Scalar-Tensor Cosmology

The purpose of this appendix is to show that symmetries in modified theories of gravity can also be selected starting from the related equations of motion. This prescription can be pursued by applying the Noether–Bessel-Hagen Theorem, introduced in Section 2.4. However, this approach is not a fundamental part of the book, since modified theories can always rely on a Lagrangian description. Here we consider a cosmological model derived from a scalar-tensor theory of gravity with the aim to finding potentials containing symmetries. The two-dimensional configuration space $S = \{a, \phi\}$ forms a global chart on $M \subset \mathbb{R}^2$. We start from the cosmological point-like Lagrangian

$$\mathcal{L}(a, \phi, \dot{a}, \dot{\phi}) = 3a\dot{a}^2 - a^3 \left(\frac{1}{2}\dot{\phi}^2 - V(\phi)\right), \tag{C.1}$$

studied, for example, in Ref. [456]. The approach sketched in Section 2.4, is developed for Lagrangians of the form

$$\mathcal{L} = \mathcal{T} - \mathcal{U}, \tag{C.2}$$

with $\mathcal{T}$ being the kinetic term and $\mathcal{U}$ the potential. With the aim of recasting Eq. (C.1) in terms of Lagrangian (C.2), we define $\mathcal{T}(\dot{a}, \dot{\phi}) = \frac{1}{2}g_{11}\dot{a}^2 + g_{12}\dot{a}\dot{\phi} + \frac{1}{2}g_{22}\dot{\phi}^2$ and $\mathscr{U}(a, \phi) = -a^3 V(\phi)$, where

$$g_{ij} = \frac{\partial \mathcal{L}}{\partial \dot{q}^i \partial \dot{q}^j}. \tag{C.3}$$

The associated EL equations and the energy condition read

$$a\colon 6a\ddot{a} + 3\dot{a}^2 + \frac{3}{2}a^2\dot{\phi}^2 - 3a^2 V(\phi) = 0, \tag{C.4}$$

$$\phi\colon -a^3\ddot{\phi} - 3a^2\dot{a}\dot{\phi} - a^3\frac{dV}{d\phi} = 0, \tag{C.5}$$

$$E_{\mathcal{L}} = 0\, 3a\dot{a}^2 - a^3\left(\frac{1}{2}\dot{\phi}^2 + V(\phi)\right) = 0. \tag{C.6}$$

Under the influence of an external force $(\varphi_a, \varphi_\phi)$, Eqs. (C.4) and (C.5) take the form

$$6a\ddot{a} + 3\dot{a}^2 + \frac{3}{2}a^2\dot{\phi}^2 - 3a^2 V(\phi) = \varphi_a, \tag{C.7}$$

$$-a^3\ddot{\phi} - 3a^2\dot{a}\dot{\phi} - a^3\frac{dV}{d\phi} = \varphi_\phi, \tag{C.8}$$

where $\varphi_a = \varphi_a(a, \phi, \dot{a}, \dot{\phi})$, $\varphi_\phi = \varphi_\phi(a, \phi, \dot{a}, \dot{\phi})$ are smooth real-valued functions on TM. The source form ε, associated with Lagrangian $\lambda = \mathcal{L}dt$ and force $\varphi = \varphi_a da + \varphi_\phi d\phi$, reads

$$\varepsilon = \left(\varepsilon_a \omega^a + \varepsilon_\phi \omega^\phi\right) \wedge dt, \tag{C.9}$$

where its coefficients ε_a, ε_ϕ are real-valued functions on T^2M, given by

$$\varepsilon_a = 6a\ddot{a} + 3\dot{a}^2 + \frac{3}{2}a^2\dot{\phi}^2 - 3a^2 V(\phi) - \varphi_a,$$

$$\varepsilon_\phi = -a^3\ddot{\phi} - 3a^2\dot{a}\dot{\phi} - a^3\frac{dV}{d\phi} - \varphi_\phi,$$

and $\omega^a = da - \dot{a}dt$, $\omega^\phi = d\phi - \dot{\phi}dt$ are contact 1-forms on $\mathbb{R} \times TM$.

The vector field on M is locally expressed by

$$X = \alpha(a, \phi)\frac{\partial}{\partial a} + \beta(a, \phi)\frac{\partial}{\partial \phi}, \tag{C.10}$$

and its second jet prolongation J^2X is

$$J^2X = \alpha\frac{\partial}{\partial a} + \beta\frac{\partial}{\partial \phi} + \dot{\alpha}\frac{\partial}{\partial \dot{a}} + \dot{\beta}\frac{\partial}{\partial \dot{\phi}} + \ddot{\alpha}\frac{\partial}{\partial \ddot{a}} + \ddot{\beta}\frac{\partial}{\partial \ddot{\phi}}. \tag{C.11}$$

Computing the Lie derivative of Eq. (C.9) with respect to J^2X, we obtain that the Noether–Bessel-Hagen equation is equivalent to the following system of conditions:

$$\alpha + 2a\frac{\partial \alpha}{\partial a} = 0, \quad \frac{\partial \beta}{\partial a}a^2 - 6\frac{\partial \alpha}{\partial \phi} = 0, \quad 3\alpha + 2\frac{\partial \beta}{\partial \phi}a = 0. \tag{C.12}$$

Eqs. (C.12) for the components α, β of vector field X can be integrated. From the first equation of (C.12), we get

$$\alpha = c\frac{1}{\sqrt{a}}e^{h(\phi)},$$

for a generic function $h = h(\phi)$ and $c \in \mathbb{R}$. Differentiating α, we have

$$\frac{\partial \alpha}{\partial \phi} = \alpha\frac{dh}{d\phi}, \quad \frac{\partial^2 \alpha}{\partial \phi^2} = \alpha\left(\frac{dh}{d\phi}\frac{dh}{d\phi} + \frac{d^2h}{d\phi^2}\right),$$

$$\frac{\partial \alpha}{\partial a} = -\frac{1}{2a}\alpha, \quad \frac{\partial^2 \alpha}{\partial a \partial \phi} = -\frac{1}{2a}\alpha\frac{dh}{d\phi}$$

and, from the second and third Eqs. (C.12), we get

$$\frac{\partial^2 \beta}{\partial a \partial \phi} = \frac{6}{a^2}\alpha\left(\frac{dh}{d\phi}\frac{dh}{d\phi} + \frac{d^2 h}{d\phi^2}\right) = \frac{9}{4a^2}\alpha = \frac{\partial^2 \beta}{\partial \phi \partial a}.$$

Therefore, a second-order ordinary equation for $h = h(\phi)$ occurs, namely

$$\frac{d^2 h}{d\phi^2} + \frac{dh}{d\phi}\frac{dh}{d\phi} = \frac{3}{8},$$

which is directly integrable with the solution

$$h(\phi) = \ln\left(\exp\left(\sqrt{\frac{3}{2}}\phi\right) + C\right) - \sqrt{\frac{3}{8}}\phi + \text{const.}, \tag{C.13}$$

for real numbers $C > 0$, $c_0 \in \mathbb{R}$. Thus, α becomes of the form (C.14), where $A = c\exp(c_0)$. The third equation of (C.12) now reads

$$\frac{\partial \beta}{\partial \phi} = -\frac{3}{2a}A\frac{1}{\sqrt{a}}\left(\exp\left(\sqrt{\frac{3}{8}}\phi\right) + C\exp\left(-\sqrt{\frac{3}{8}}\phi\right)\right),$$

which can be directly integrated, providing

$$\beta = -\sqrt{6}A\frac{1}{a\sqrt{a}}\left(\exp\left(\sqrt{\frac{3}{8}}\phi\right) - C\exp\left(-\sqrt{\frac{3}{8}}\phi\right)\right) + p(a),$$

for some function $p = p(a)$. Substituting this form of β into the second equation of (C.12), we easily observe that p must be constant and $p(a) = B \in \mathbb{R}$, that is, β is given by (C.15). The pair α, β is the unique solution of system (C.12). Note that the components α and β of the generator of invariance transformations of ε are related as follows:

$$\beta - B = -4\frac{1}{a}\frac{dh}{d\phi}\alpha,$$

where $h = h(\phi)$ is given by Eq. (C.13). Therefore, X is a generator of invariance transformations of ε, if the infinitesimal generators have the following form:

$$\alpha = A\frac{1}{\sqrt{a}}\left(\exp\left(\sqrt{\frac{3}{8}}\phi\right) + C\exp\left(-\sqrt{\frac{3}{8}}\phi\right)\right), \tag{C.14}$$

$$\beta = -\sqrt{6}A\frac{1}{a\sqrt{a}}\left(\exp\left(\sqrt{\frac{3}{8}}\phi\right) - C\exp\left(-\sqrt{\frac{3}{8}}\phi\right)\right) + B, \tag{C.15}$$

where $A, B \in \mathbb{R}$ and $C > 0$ are real numbers. The forces φ_a, φ_ϕ satisfy the relations

$$\frac{9}{2}\alpha a V + 3\beta a^2\frac{dV}{d\phi} + 6\frac{\partial \alpha}{\partial \phi}a\frac{dV}{d\phi} - \frac{\partial}{\partial a}(\alpha f_a) - \frac{\partial \beta}{\partial a}f_\phi - \beta\frac{\partial f_a}{\partial \phi}$$
$$-\frac{\partial \varphi_a}{\partial \dot{a}}\left(\frac{\partial \alpha}{\partial a}\dot{a} + \frac{\partial \alpha}{\partial \phi}\dot{\phi}\right) - \frac{\partial \varphi_a}{\partial \dot{\phi}}\left(\frac{\partial \beta}{\partial a}\dot{a} + \frac{\partial \beta}{\partial \phi}\dot{\phi}\right) = 0$$

and

$$\frac{3}{2}\alpha a^2\frac{dV}{d\phi}+\beta a^3\frac{d^2V}{d\phi^2}+3\frac{\partial\alpha}{\partial\phi}a^2V-\frac{\partial}{\partial\phi}(\beta f_\phi)-\frac{\partial\alpha}{\partial\phi}f_a-\alpha\frac{\partial f_\phi}{\partial a}$$
$$-\frac{\partial\varphi_\phi}{\partial\dot a}\left(\frac{\partial\alpha}{\partial a}\dot a+\frac{\partial\alpha}{\partial\phi}\dot\phi\right)-\frac{\partial\varphi_\phi}{\partial\dot\phi}\left(\frac{\partial\beta}{\partial a}\dot a+\frac{\partial\beta}{\partial\phi}\dot\phi\right)=0.$$

Regarding the potential, we can further constrain the shape of the infinitesimal generators in order to select those potentials containing symmetries. As a consequence, two cases can be distinguished. We outline them here:

(i) $\alpha=0$ $(A=0)$, $\beta=B\neq 0$, and $V=V_0$. In this case, the equations of motion and the energy condition have a common solution for $V=0$ only, namely the mapping $\gamma:\mathbb{R}\to TM$ given by

$$\gamma(t)=(a(\gamma(t)),\phi(\gamma(t)))=\left(k_1\sqrt[3]{3t+k_2},\pm\sqrt{\frac{2}{3}}\ln|3t+k_2|+k_3\right),\qquad\text{(C.16)}$$

with $k_1,k_2,k_3\in\mathbb{R}$.

(ii) $\alpha\neq 0$ $(A\neq 0)$, $B=0$, and

$$V(\phi)=k\left(\exp\left(\sqrt{\frac{3}{8}}\phi\right)-C\exp\left(-\sqrt{\frac{3}{8}}\phi\right)\right)^2\quad k>0,\qquad\text{(C.17)}$$

which cannot provide any analytic solution for the equations of motion.

References

[1] K. Brading and E. Castellani, "Symmetries in Physics: Philosophical Reflections," Cambridge University Press, Cambridge (2003).

[2] P. G. Fré, "A Conceptual History of Space and Symmetry: From Plato to the Superworld," Springer, Berlin (2018).

[3] H. Stephani, "Differential Equations: Their Solution Using Symmetries," Cambridge University Press, Cambridge (1990), 9780511599941.

[4] O. W. Greenberg, "CPT Violation Implies Violation of Lorentz Invariance," Phys. Rev. Lett. **89** (2002), 231602.

[5] M. E. Peskin and D. V. Schroeder, "An Introduction to Quantum Field Theory," CRC Press, Boca Raton, FL (1995).

[6] S. Weinberg, "The Quantum Theory of Fields. Vol. 1: Foundations," Cambridge University Press, Cambridge (2005).

[7] A. Zee, "Quantum Field Theory in a Nutshell," Princeton University Press, Princeton, NJ (2010).

[8] M. D. Schwartz, "Quantum Field Theory and the Standard Model," Cambridge University Press, Cambridge (2013).

[9] E. Noether, "Invariant Variation Problems," Gott. Nachr. **1918** (1918), 235 [Transp. Theory Statist. Phys. **1** (1971), 186].

[10] D. E. Neuenschwander, "Emmy Noether's Wonderful Theorem," Johns Hopkins University Press, Baltimore (2017).

[11] Y. Kosmann-Schwarzbach and B. E. Schwarzbach, "The Noether Theorems," Springer, New York (2011).

[12] U. Camci and A. Yildirim, "Noether Gauge Symmetry Classes for pp-Wave Spacetimes," Int. J. Geom. Meth. Mod. Phys. **12** (2015), 1550120.

[13] D. Grumiller, M. Irakleidou, I. Lovrekovic, and R. McNees, "Conformal Gravity Holography in Four Dimensions," Phys. Rev. Lett. **112** (2014), 111102.

[14] E. Bessel-Hagen, "Uber die Erhaltungssatze der Electrodynamik," Math. Ann. **84** (1921), 258–276.

[15] D. Krupka, "Introduction to Global Variational Geometry," Atlantis Stud. Var. Geom., vol. 1, Atlantis Press, Amsterdam–Beijing–Paris (2015).

[16] W. Sarlet and F. Cantrijn, "Generalizations of Noether's Theorem in Classical Mechanics," SIAM Review **23**, No. 4 (1981), 467–494.

[17] D. Krupka, "Invariant Variational Structures on Fibered Manifolds," Int. J. Geom. Meth. Mod. Phys. **12** (2015), 1550020.

[18] Z. Urban, F. Bajardi, and S. Capozziello, "The Noether–Bessel-Hagen Symmetry Approach for Dynamical Systems," Int. J. Geom. Meth. Mod. Phys. **17**, No. 14 (2020), 2050215.

[19] A. Trautman, "Invariance of Lagrangian Systems," in L. O'Raifeartaigh (ed.), "General Relativity, Papers in Honour of J. L. Synge," Clarendon Press, Oxford, UK (1972), 85–99.

[20] A. Trautman, "Noether Equations and Conservation Laws," Commun. Math. Phys. **6** (1967), 248–261.

[21] D. Krupka, "A Geometric Theory of Ordinary First Order Variational Problems in Fibered Manifolds. II. Invariance," J. Math. Anal. Appl. **49** (1975), 469–476.

[22] G. Sardanashvily, "Noether's Theorems, Applications in Mechanics and Field Theory," Atlantis Stud. Var. Geom., vol. 3, Atlantis Press, Amsterdam–Beijing–Paris (2016).

[23] P. J. Olver, "Applications of Lie Groups to Differential Equations," GTM 107, Springer-Verlag, New York (1986).

[24] G. W. Bluman and S. Kumei, "Symmetries and Differential Equations," Springer-Verlag, New York (1989).

[25] N. Chien, T. Honein, and G. Herrmann, "Dissipative Systems, Conservation Laws and Symmetries," Int. J. Solids Structures **33**, No. 20–22 (1996), 2959–2968.

[26] T. Honein, N. Chien, and G. Herrmann, "On Conservation Laws for Dissipative Systems," Phys. Lett. A **155** (1991), 223–224.

[27] D. Krupka, "Variational Forces," Lepage Research Institute Library **6** (2018), 1–38.

[28] J. Volná and Z. Urban, "First-Order Variational Sequences in Field Theory," in D. Zenkov (ed.), "The Inverse Problem of the Calculus of Variations, Local and Global Theory," Atlantis Press, Amsterdam–Beijing–Paris (2015), 215–284.

[29] Z. Urban and J. Volná, "On a Global Lagrangian Construction for Ordinary Variational Equations on 2-Manifolds," J. Math. Phys. **60**, No. 9 (2019), 092902.

[30] D. Krupka, Z. Urban, and J. Volná, "Variational Submanifolds of Euclidean Spaces," J. Math. Phys. **59**, No. 3 (2018), 032903.

[31] L. D. Landau and E. M. Lifschits, "The Classical Theory of Fields," Pergamon Press, Oxford (1975).

[32] S. Weinberg, "Gravitation and Cosmology: Principles and Applications of the General Theory of Relativity," John Wiley & Sons Inc, Hoboken, NJ (1972).

[33] R. M. Wald, "General Relativity," University of Chicago Press, Chicago (1984).

[34] C. W. Misner, K. S. Thorne, and J. A. Wheeler, "Gravitation," Princeton University Press, Princeton, NJ (1973).

[35] W.-G. Boskoff and S. Capozziello, "A Mathematical Journey to Relativity", Springer, Berlin (2020).

[36] S. Capozziello and V. Faraoni, "Beyond Einstein Gravity: A Survey of Gravitational Theories for Cosmology and Astrophysics," Fundam. Theor. Phys. **170**, Springer, Berlin (2010).

[37] M. Leclerc, "Canonical and Gravitational Stress-Energy Tensors," Int. J. Mod. Phys. D **15** (2006), 959–990

[38] B. P. Abbott et al. [LIGO Scientific and Virgo], "Observation of Gravitational Waves from a Binary Black Hole Merger," Phys. Rev. Lett. **116**, No. 6 (2016), 061102.

[39] C. M. Will, "The Confrontation between General Relativity and Experiment," Living Rev. Rel. **17** (2014), 4.

[40] K. Akiyama et al. [Event Horizon Telescope], "First M87 Event Horizon Telescope Results. V. Physical Origin of the Asymmetric Ring," Astrophys. J. Lett. **875** No. 1 (2019), L5.

[41] J. A. Peacock, "Cosmological Physics," Cambridge University Press, Cambridge (1999).
[42] E. Di Valentino, L. A. Anchordoqui, O. Akarsu, et al., "Snowmass2021 – Letter of Interest Cosmology Intertwined II: The Hubble Constant Tension," Astropart. Phys. **131** (2021), 102605.
[43] M. Visser, "Essential and Inessential Features of Hawking Radiation," Int. J. Mod. Phys. D **12** (2003), 649.
[44] P. M. Garnavich et al., "Constraints on Cosmological Models from Hubble Space Telescope Observations of High-Z Supernovae," Astrophys. J. **493** (1998), L53.
[45] J. N. Islam, "The Ultimate Fate of the Universe," Cambridge University Press, Cambridge (1983).
[46] M. Bojowald, "What Happened before the Big Bang?", Nat. Phys. **3** (2007), 523.
[47] J. Frieman, M. Turner, and D. Huterer, "Dark Energy and the Accelerating Universe," Ann. Rev. Astron. Astrophys. **46** (2008), 385–432.
[48] J. M. Lattimer and M. Prakash, "Neutron Star Structure and the Equation of State," Astrophys. J. **550** (2001), 426.
[49] J. M. Lattimer and M. Prakash, "Neutron Star Observations: Prognosis for Equation of State Constraints," Phys. Rept. **442** (2007), 109.
[50] A. V. Astashenok, S. Capozziello, and S. D. Odintsov, "Extreme Neutron Stars from Extended Theories of Gravity," JCAP **01** (2015), 001.
[51] J. Beringer et al. [Particle Data Group], "Review of Particle Physics (RPP)," Phys. Rev. D **86** (2012), 010001.
[52] A. Bosma, "21-cm Line Studies of Spiral Galaxies. 2. The Distribution and Kinematics of Neutral Hydrogen in Spiral Galaxies of Various Morphological Types," Astron. J. **86** (1981), 1825.
[53] P. A. R. Ade et al. [Planck], "Planck 2013 Results. I. Overview of Products and Scientific Results," Astron. Astrophys. **571** (2014), A1.
[54] M. Milgrom, "A Modification of the Newtonian Dynamics as a Possible Alternative to the Hidden Mass Hypothesis," Astrophys. J. **270** (1983), 365.
[55] M. Milgrom, "A Modification of the Newtonian Dynamics – Implications for Galaxies," Astrophys. J. **270** (1983), 371.
[56] M. Milgrom, "A Modification of the Newtonian Dynamics: Implications for Galaxy Systems," Astrophys. J. **270** (1983), 384.
[57] K. Bolejko and M. Korzynski, "Inhomogeneous Cosmology and Backreaction: Current Status and Future Prospects," Int. J. Mod. Phys. D **26** (2017), 1730011.
[58] S. Weinberg, "The Cosmological Constant Problem," Rev. Mod. Phys. **61** (1989), 1–23
[59] V. P. Frolov, "Notes on Nonsingular Models of Black Holes," Phys. Rev. D **94**, No. 10 (2016), 104056.
[60] T. Vachaspati, D. Stojkovic, and L. M. Krauss, "Observation of Incipient Black Holes and the Information Loss Problem," Phys. Rev. D **76** (2007), 024005.
[61] S. A. Hayward, "Formation and Evaporation of Regular Black Holes," Phys. Rev. Lett. **96** (2006), 031103.
[62] T. De Lorenzo, C. Pacilio, C. Rovelli, and S. Speziale, "On the Effective Metric of a Planck Star," Gen. Rel. Grav. **47**, No. 4 (2015), 41.
[63] E. Contreras and P. Bargueño, "Scale-Dependent Hayward Black Hole and the Generalized Uncertainty Principle," Mod. Phys. Lett. A **33**, No. 32 (2018), 1850184.
[64] P. O. Mazur and E. Mottola, "Gravitational Vacuum Condensate Stars," Proc. Nat. Acad. of Science **101**, No. 26 (2004), 9545.

[65] G. Chapline, "Dark Energy Stars," in Pisin Chen, Elliott Bloom, Greg Madejski (SLAC), and Vahe Patrosian (Stanford University), eds., Proceedings of the 22nd Texas Symposium on Relativistic Astrophysics (2005), 101–104. Accessed at https://ui.adsabs.harvard.edu/abs/2005tsra.conf..101C [arXiv:astro-ph/0503200].
[66] C. Barceló, S. Liberati, S. Sonego, and M. Visser, "Black Stars, Not Holes," Sci. Am. **301**, No. 4 (2009), 38–45.
[67] A. Perez, "Black Holes in Loop Quantum Gravity," Rept. Prog. Phys. **80**, No. 12 (2017), 126901.
[68] E. Bianchi, "Entropy of Non-Extremal Black Holes from Loop Gravity," [arXiv:1204.5122 [gr-qc]].
[69] N. Bodendorfer, "Black Hole Entropy from Loop Quantum Gravity in Higher Dimensions," Phys. Lett. B **726** (2013), 887–891.
[70] C. Rovelli, "Black Hole Entropy from Loop Quantum Gravity," Phys. Rev. Lett. **77** (1996), 3288–3291.
[71] A. Lasenby, C. Doran, and S. Gull, "Gravity, Gauge Theories and Geometric Algebra," Phil. Trans. Roy. Soc. Lond. A **356** (1998), 487.
[72] F. W. Hehl, P. Von Der Heyde, G. D. Kerlick, and J. M. Nester, "General Relativity with Spin and Torsion: Foundations and Prospects," Rev. Mod. Phys. **48** (1976), 393.
[73] F. Gronwald and F. W. Hehl, "On the Gauge Aspects of Gravity," in P. G. Bergmann, V. de Sabbata, and H.-J. Treder, eds., Quantum Gravity. Proceedings, International School of Cosmology and Gravitation, 14th Course, Erice, Italy, May 11–19, 1995 (World Scientific, Singapore 1996).
[74] E. W. Mielke and P. Baekler, "Topological Gauge Model of Gravity with Torsion," Phys. Lett. A **156** (1991), 399.
[75] M. Gasperini, "Spin Dominated Inflation in the Einstein–Cartan Theory," Phys. Rev. Lett. **56** (1986), 2873.
[76] Y. N. Obukhov, "Poincare Gauge Gravity: Selected Topics," Int. J. Geom. Meth. Mod. Phys. **3** (2006), 95.
[77] J. M. Pawlowski, M. Reichert, C. Wetterich, and M. Yamada, "Higgs Scalar Potential in Asymptotically Safe Quantum Gravity," Phys. Rev. D **99**, No. 8 (2019), 086010.
[78] M. Niedermaier and M. Reuter, "The Asymptotic Safety Scenario in Quantum Gravity," Living Rev. Rel. **9** (2006), 5.
[79] R. Percacci, "Asymptotic Safety," in D. Oriti (ed.), "Approaches to Quantum Gravity: Toward a New Understanding of Space, Time and Matter," Cambridge University Press, Cambridge (2009), 111–128. DOI: https://doi.org/10.1017/CBO9780511575549.009
[80] S. W. Hawking, "Particle Creation by Black Holes," Commun. Math. Phys. **43** (1975), 199–220 [erratum: Commun. Math. Phys. **46** (1976), 206].
[81] P. Horava, "Quantum Gravity at a Lifshitz Point," Phys. Rev. D **79** (2009), 084008.
[82] K. S. Stelle, "Renormalization of Higher Derivative Quantum Gravity," Phys. Rev. D **16** (1977), 953–969.
[83] J. Engle, R. Pereira, and C. Rovelli, "The Loop-Quantum-Gravity Vertex-Amplitude," Phys. Rev. Lett. **99** (2007), 161301.
[84] S. Alexandrov and E. R. Livine, "SU(2) Loop Quantum Gravity Seen from Covariant Theory," Phys. Rev. D **67** (2003), 044009.

[85] J. M. Maldacena, "The Large N Limit of Superconformal Field Theories and Supergravity," Int. J. Theor. Phys. **38** (1999), 1113–1133.

[86] H. P. Nilles, "Supersymmetry, Supergravity and Particle Physics," Phys. Rept. **110** (1984), 1–162.

[87] J. J. Halliwell and S. W. Hawking, "The Origin of Structure in the Universe," Adv. Ser. Astrophys. Cosmol. **3** (1987), 277–291.

[88] F. Bajardi, D. Vernieri, and S. Capozziello, "Bouncing Cosmology in f(Q) Symmetric Teleparallel Gravity," Eur. Phys. J. Plus **135**, No. 11 (2020), 912.

[89] S. Capozziello, M. De Laurentis, and S. D. Odintsov, "Hamiltonian Dynamics and Noether Symmetries in Extended Gravity Cosmology," Eur. Phys. J. C **72** (2012), 2068.

[90] N. Loutrel, N. Yunes, and F. Pretorius "Parametrized Post-Einsteinian Framework for Gravitational Wave Bursts," Phys. Rev. D **90** (2014), 104010.

[91] S. Capozziello and M. De Laurentis, "Extended Theories of Gravity," Phys. Rept. **509** (2011), 167.

[92] A. K. Mishra, M. Rahman, and S. Sarkar, "Black Hole Topology in $f(R)$ Gravity," Class. Quant. Grav. **35**, No. 14 (2018), 145011.

[93] S. Capozziello and M. De Laurentis, "The Dark Matter Problem from f(R) Gravity Viewpoint," Annalen Phys. **524** (2012), 545.

[94] G. J. Olmo, "Palatini Approach to Modified Gravity: f(R) Theories and Beyond," Int. J. Mod. Phys. D **20** (2011), 413.

[95] S. Capozziello, M. De Laurentis, and V. Faraoni, "A Bird's Eye View of f(R)-Gravity," Open Astron. J. **3** (2010), 49.

[96] T. P. Sotiriou and V. Faraoni, "f(R) Theories of Gravity," Rev. Mod. Phys. **82** (2010), 451.

[97] Y. F. Cai, S. Capozziello, M. De Laurentis, and E. N. Saridakis, "f(T) Teleparallel Gravity and Cosmology," Rept. Prog. Phys. **79**, No. 10 (2016), 106901.

[98] M. R. Setare and N. Mohammadipour, "Can $f(T)$ Gravity Theories Mimic ΛCDM Cosmic History," JCAP **1301** (2013), 015.

[99] K. Bamba, S. Capozziello, M. De Laurentis, S. Nojiri, and D. Sáez-Gómez, "No Further Gravitational Wave Modes in $F(T)$ Gravity," Phys. Lett. B **727** (2013), 194–198.

[100] N. Tamanini and C. G. Boehmer, "Good and Bad Tetrads in f(T) Gravity," Phys. Rev. D **86** (2012), 044009.

[101] Y. Zhang, H. Li, Y. Gong, and Z. H. Zhu, "Notes on $f(T)$ Theories," JCAP **1107** (2011), 015.

[102] R. Aldrovandi and J. G. Pereira, "Teleparallel Gravity: An Introduction," Fundam. Theor. Phys. **173** (2013) Springer, Berlin. DOI: http://doi.org/10.1007/978-94-007-5143-9

[103] J. J. Halliwell, "Scalar Fields in Cosmology with an Exponential Potential," Phys. Lett. B **185** (1987), 341.

[104] J. P. Uzan, "Cosmological Scaling Solutions of Nonminimally Coupled Scalar Fields," Phys. Rev. D **59** (1999), 123510.

[105] H. Mohseni Sadjadi, "Notes on Teleparallel Cosmology with Nonminimally Coupled Scalar Field," Phys. Rev. D **87** (2013), 064028.

[106] C. Rubano and P. Scudellaro, "On Some Exponential Potentials for a Cosmological Scalar Field as Quintessence," Gen. Rel. Grav. **34** (2002), 307.

[107] F. Bajardi and S. Capozziello, "$f(\mathcal{G})$ Noether Cosmology," Eur. Phys. J. C **80**, No. 8 (2020), 704.

[108] M. Benetti, S. Santos da Costa, S. Capozziello, J. S. Alcaniz, and M. De Laurentis, "Observational Constraints on Gauss-Bonnet Cosmology," Int. J. Mod. Phys. D **27**, No. 08 (2018), 1850084.

[109] J. L. Blázquez-Salcedo, F. S. Khoo, and J. Kunz, "Quasinormal Modes of Einstein-Gauss-Bonnet-Dilaton Black Holes," Phys. Rev. D **96**, No. 6 (2017), 064008.

[110] J. B. Jiménez, L. Heisenberg, and T. S. Koivisto, "The Geometrical Trinity of Gravity," Universe **5**, No. 7 (2019), 173.

[111] G. M. Tino, L. Cacciapuoti, S. Capozziello, G. Lambiase, and F. Sorrentino, "Precision Gravity Tests and the Einstein Equivalence Principle," Prog. Part. Nucl. Phys. **112** (2020), 103772.

[112] M. Blasone, S. Capozziello, G. Lambiase, and L. Petruzziello, "Equivalence Principle Violation at Finite Temperature in Scalar-Tensor Gravity," Eur. Phys. J. Plus **134**, No. 4 (2019), 169.

[113] H. Shabani and A. H. Ziaie, "Static Vacuum Solutions on Curved Spacetimes with Torsion," Int. J. Mod. Phys. A **33**, No. 16 (2018) 1850095.

[114] S. Capozziello and C. Stornaiolo, "Torsion Tensor and Its Geometric Interpretation," Annales de la Fondation Louis de Broglie **32**, Nos. 2–3 (2007) 196–214.

[115] T. P. Sotiriou, "f(R) Gravity, Torsion and Non-metricity," Class. Quant. Grav. **26** (2009), 152001.

[116] S. Capozziello, "Curvature Quintessence," Int. J. Mod. Phys. D **11** (2002), 483.

[117] S. Capozziello, C. A. Mantica, and L. G. Molinari, "Cosmological Perfect-Fluids in f(R) Gravity," Int. J. Geom. Meth. Mod. Phys. **16**, No. 01 (2018), 1950008.

[118] S. Nojiri and S. D. Odintsov, "Unified Cosmic History in Modified Gravity: From F(R) Theory to Lorentz Non-invariant Models," Phys. Rept. **505** (2011), 59–144 DOI: https://doi.org/10.1016/j.physrep.2011.04.001 [arXiv:1011.0544 [gr-qc]].

[119] M. Benetti, S. Capozziello, and L. L. Graef, "Swampland Conjecture in $f(R)$ Gravity by the Noether Symmetry Approach," Phys. Rev. D **100**, No. 8 (2019), 084013.

[120] S. Capozziello, M. De Laurentis, and O. Luongo, "Connecting Early and Late Universe by $f(R)$ Gravity," Int. J. Mod. Phys. D **24**, No. 04 (2014), 1541002.

[121] J. D. Barrow and A. C. Ottewill, "The Stability of General Relativistic Cosmological Theory," J. Phys. A **16** (1983), 2757.

[122] A. A. Starobinsky, "A New Type of Isotropic Cosmological Models without Singularity," Phys. Lett. B **91** (1980) 99–102.

[123] S. D. Odintsov, V. K. Oikonomou, and L. Sebastiani, "Unification of Constant-Roll Inflation and Dark Energy with Logarithmic R^2-Corrected and Exponential $F(R)$ Gravity," Nucl. Phys. B **923** (2017), 608–632.

[124] I. Quandt and H. J. Schmidt, "The Newtonian Limit of Fourth and Higher Order Gravity," Astron. Nachr. **312** (1991), 97.

[125] L. Modesto, L. Rachwal, and I. L. Shapiro, "Renormalization Group in Super-renormalizable Quantum Gravity," Eur. Phys. J. C **78**, No. 7 (2018), 555.

[126] E. T. Tomboulis, "Renormalization and Unitarity in Higher Derivative and Nonlocal Gravity Theories," Mod. Phys. Lett. A **30**, No. 03n04 (2015), 1540005.

[127] S. Capozziello and F. Bajardi, "Gravitational Waves in Modified Gravity," Int. J. Mod. Phys. D **28**, No. 05 (2019), 1942002.

[128] F. Bajardi, K. F. Dialektopoulos, and S. Capozziello, "Higher Dimensional Static and Spherically Symmetric Solutions in Extended Gauss-Bonnet Gravity," Symmetry **12**, No. 3 (2020), 372.

[129] N. A. Chernikov and E. A. Tagirov, "Quantum Theory of Scalar Fields in de Sitter Spacetime," Ann. Inst. H. Poincare Phys. Theor. A **9** (1968), 109.

[130] C. G. Callan, Jr., S. R. Coleman, and R. Jackiw, "A New Improved Energy – Momentum Tensor," Annals Phys. **59** (1970), 42–73.

[131] A. D. Linde, "A New Inflationary Universe Scenario: A Possible Solution of the Horizon, Flatness, Homogeneity, Isotropy and Primordial Monopole Problems," Adv. Ser. Astrophys. Cosmol. **3** (1987), 149–153.

[132] A. H. Guth, "Inflationary Universe: A Possible Solution to the Horizon and Flatness Problems," Phys. Rev. D **23** (1981), 347–356 [Adv. Ser. Astrophys. Cosmol. **3** (1987) 139].

[133] F. Lucchin, S. Matarrese, and M. D. Pollock, "Inflation with a Nonminimally Coupled Scalar Field," Phys. Lett. B **167** (1986), 163–168.

[134] Y. Shtanov, J. H. Traschen, and R. H. Brandenberger, "Universe Reheating after Inflation," Phys. Rev. D **51** (1995), 5438.

[135] R. Allahverdi, R. Brandenberger, F. Y. Cyr-Racine, and A. Mazumdar, "Reheating in Inflationary Cosmology: Theory and Applications," Ann. Rev. Nucl. Part. Sci. **60** (2010), 27.

[136] P. B. Greene, "Inflationary Reheating and Fermions," AIP Conf. Proc. **478**, No. 1 (1999), 72.

[137] C. de Rham, "Galileons in the Sky," Comptes Rendus Physique **13** (2012), 666–681. DOI: https://doi.org/10.1016/j.crhy.2012.04.006.

[138] E. Babichev and C. Deffayet, "An Introduction to the Vainshtein Mechanism," Class. Quant. Grav. **30** (2013), 184001.

[139] C. Brans and R. H. Dicke, "Mach's Principle and a Relativistic Theory of Gravitation," Phys. Rev. **124** (1961), 925–935.

[140] J. L. Anderson, "Scale Invariance of the Second Kind and the Brans-Dicke Scalar-Tensor Theory," Phys. Rev. D **3** (1971), 1689.

[141] M. P. Dabrowski, T. Denkiewicz, and D. Blaschke, "Puzzles of Isotropic and Anisotropic Conformal Cosmologies," Annalen Phys. **16** (2007), 237.

[142] A. Davidson, "Rippled Cosmological Dark Matter from Damped Oscillating Newton Constant," Class. Quant. Grav. **22**, No. 6 (2005), 1119.

[143] S. Deser, "Scale Invariance and Gravitational Coupling," Annals Phys. **59** (1970), 248.

[144] J. O' Hanlon, "Mach's Principle and a New Gauge Freedom in Brans-Dicke Theory," J. Phys. A **5** (1972), 803.

[145] Y. Fujii and K. Maeda, "The Scalar-Tensor Theory of Gravitation," Cambridge University Press, Cambridge (2003).

[146] G. W. Horndeski, "Second-Order Scalar-Tensor Field Equations in a Four-Dimensional Space," Int. J. Theor. Phys. **10** (1974), 363–384.

[147] C. Deffayet, S. Deser, and G. Esposito-Farese, "Generalized Galileons: All Scalar Models Whose Curved Background Extensions Maintain Second-Order Field Equations and Stress-Tensors," Phys. Rev. D **80** (2009), 064015.

[148] C. Deffayet, G. Esposito-Farese, and A. Vikman, "Covariant Galileon," Phys. Rev. D **79** (2009), 084003.

[149] S. Capozziello, K. F. Dialektopoulos, and S. V. Sushkov, "Classification of the Horndeski Cosmologies via Noether Symmetries," Eur. Phys. J. C **78**, No. 6 (2018), 447.

[150] A. Einstein, "Zur allgemeinen Relativitätstheorie," Sitzungsber. Preuss. Akad. Wiss. Berlin (Math. Phys.) **2** (1915), 778.

[151] A. Einstein, "Die Feldgleichungen der Gravitation," Sitzungsber. Preuss. Akad. Wiss. Berlin (Math. Phys.) **2** (1915), 844.

[152] A. Einstein, "Einheitliche Feldtheorie von Gravitation und Elektrizität." Sitzungsber. Preuss. Akad. Wiss. Berlin (Math. Phys.) **22** (1925), 414.

[153] H. A. Buchdahl, "Representation of the Einstein-Proca Field by an A(4)," J. Phys. A **12** (1979), 1235–1238.

[154] A. Palatini, "Deduzione invariantiva delle equazioni gravitazionali dal principio di Hamilton," Rend. Circ. Mat. Palermo **43** (1919), 203.

[155] G. Allemandi, M. Capone, S. Capozziello, and M. Francaviglia, "Conformal Aspects of Palatini Approach in Extended Theories of Gravity," Gen. Rel. Grav. **38** (2006), 33.

[156] E. Barausse, T. P. Sotiriou, and J. C. Miller, "A No-Go Theorem for Polytropic Spheres in Palatini f(R) Gravity," Class. Quant. Grav. **25** (2008), 062001.

[157] E. Barausse, T. P. Sotiriou, and J. C. Miller, "Curvature Singularities, Tidal Forces and the Viability of Palatini f(R) Gravity," Class. Quant. Grav. **25** (2008), 105008.

[158] E. Barausse, T. P. Sotiriou, and J. C. Miller, "Polytropic Spheres in Palatini f(R) Gravity," EAS Publ. Ser. **30** (2008), 189.

[159] G. J. Olmo, "Hydrogen Atom in Palatini Theories of Gravity," Phys. Rev. D **77** (2008), 084021.

[160] J. L. Synge, "Relativity: The General Theory," North-Holland, Amsterdam (1960).

[161] S. Capozziello, T. Harko, T. S. Koivisto, F. S. N. Lobo, and G. J. Olmo, "Cosmology of Hybrid Metric-Palatini f(X)-Gravity," JCAP **04** (2013), 011.

[162] S. Capozziello, T. Harko, T. S. Koivisto, F. S. N. Lobo, and G. J. Olmo, "Wormholes Supported by Hybrid Metric-Palatini Gravity," Phys. Rev. D **86** (2012), 127504.

[163] B. Danila, T. Harko, F. S. N. Lobo, and M. K. Mak, "Hybrid Metric-Palatini Stars," Phys. Rev. D **95**, No. 4 (2017), 044031.

[164] K. A. Bronnikov, "Spherically Symmetric Black Holes and Wormholes in Hybrid Metric-Palatini Gravity," Grav. Cosmol. **25** (2019), 331–341.

[165] T. Harko and F. S. N. Lobo, "Beyond Einstein's General Relativity: Hybrid metric-Palatini Gravity and Curvature-Matter Couplings," Int. J. Mod. Phys. D **29**, No. 13 (2020), 2030008.

[166] S. Capozziello, T. Harko, F. S. N. Lobo, and G. J. Olmo, "Hybrid Modified Gravity Unifying Local Tests, Galactic Dynamics and Late-Time Cosmic Acceleration," Int. J. Mod. Phys. D **22** (2013), 1342006.

[167] T. Harko, T. S. Koivisto, F. S. N. Lobo, and G. J. Olmo, "Metric-Palatini Gravity Unifying Local Constraints and Late-Time Cosmic Acceleration," Phys. Rev. D **85** (2012), 084016.

[168] S. Capozziello, T. Harko, T. S. Koivisto, F. S. N. Lobo, and G. J. Olmo, "Hybrid Metric-Palatini Gravity," Universe **1**, No. 2 (2015), 199–238.

[169] H. I. Arcos and J. G. Pereira, "Torsion Gravity: A Reappraisal," Int. J. Mod. Phys. D **13** (2004), 2193–2240.

[170] M. Krššák and J. G. Pereira, "Spin Connection and Renormalization of Teleparallel Action," Eur. Phys. J. C **75**, No. 11 (2015), 519.

[171] M. Krššák and E. N. Saridakis, "The Covariant Formulation of f(T) Gravity," Class. Quant. Grav. **33**, No. 11 (2016), 115009.

[172] S. Bahamonde, M. Marciu, and P. Rudra, "Generalised Teleparallel Quintom Dark Energy Non-minimally Coupled with the Scalar Torsion and a Boundary Term," JCAP **1804**, No. 04 (2018), 056.

[173] S. Bahamonde, M. Zubair, and G. Abbas, "Thermodynamics and Cosmological Reconstruction in $f(T, B)$ Gravity," Phys. Dark Univ. **19** (2018), 78.

[174] S. Bahamonde and S. Capozziello, "Noether Symmetry Approach in $f(T, B)$ Teleparallel Cosmology," Eur. Phys. J. C **77**, No. 2 (2017) , 107.

[175] S. Capozziello, M. Capriolo, and L. Caso, "Weak Field Limit and Gravitational Waves in $f(T, B)$ Teleparallel Gravity," Eur. Phys. J. C **80**, No. 2 (2020), 156.

[176] M. R. Setare and N. Mohammadipour, "Cosmological Viability Conditions for $f(T)$ Dark Energy Models," JCAP **1211** (2012), 030.

[177] M. Li, R. X. Miao, and Y. G. Miao, "Degrees of Freedom of $f(T)$ Gravity," JHEP **1107** (2011), 108.

[178] P. Wu and H. W. Yu, "The Dynamical Behavior of $f(T)$ Theory," Phys. Lett. B **692** (2010), 176.

[179] R. Ferraro and F. Fiorini, "Non-trivial Frames for f(T) Theories of Gravity and Beyond," Phys. Lett. B **702** (2011), 75.

[180] S. Capozziello, V. F. Cardone, H. Farajollahi, and A. Ravanpak, "Cosmography in f(T)-Gravity," Phys. Rev. D **84** (2011), 043527.

[181] D. Liu and M. J. Reboucas, "Energy Conditions Bounds on f(T) Gravity," Phys. Rev. D **86** (2012), 083515.

[182] S. Bahamonde, S. Capozziello, M. Faizal, and R. C. Nunes, "Nonlocal Teleparallel Cosmology," Eur. Phys. J. C **77**, No. 9 (2017), 628.

[183] S. Capozziello, M. De Laurentis, and K. F. Dialektopoulos, "Noether Symmetries in Gauss-Bonnet-Teleparallel Cosmology," Eur. Phys. J. C **76**, No. 11 (2016), 629.

[184] F. Bajardi and S. Capozziello, "Noether Symmetries and Quantum Cosmology in Extended Teleparallel Gravity," Int. J. Geom. Meth. Mod. Phys. **18**, No. supp01 (2021), 2140002.

[185] L. Jarv, M. Runkla, M. Saal, and O. Vilson, "Nonmetricity Formulation of General Relativity and Its Scalar-Tensor Extension," Phys. Rev. D **97**, No. 12 (2018), 124025.

[186] M. Rünkla and O. Vilson, "Family of Scalar-Nonmetricity Theories of Gravity," Phys. Rev. D **98**, No. 8 (2018), 084034.

[187] T. Harko, T. S. Koivisto, F. S. N. Lobo, G. J. Olmo, and D. Rubiera-Garcia, "Coupling Matter in Modified Q Gravity," Phys. Rev. D **98**, No. 8 (2018), 084043.

[188] K. F. Dialektopoulos, T. S. Koivisto, and S. Capozziello, "Noether Symmetries in Symmetric Teleparallel Cosmology," Eur. Phys. J. C **79**, No. 7 (2019), 606.

[189] L. Heisenberg, "A Systematic Approach to Generalisations of General Relativity and Their Cosmological Implications," Phys. Rept. **796** (2019), 1.

[190] M. B. Green, J. H. Schwarz, and E. Witten, "Superstring Theory. Vol. 1: Introduction," Cambridge Monographs on Mathematical Physics. Cambridge University Press, Cambridge (1987).

[191] M. B. Green, J. H. Schwarz, and E. Witten, "Superstring Theory. Vol. 2: Loop Amplitudes, Anomalies and Phenomenology," Cambridge Monographs on Mathematical Physics. Cambridge University Press, Cambridge (1987).

[192] J. Polchinski, "String Theory. Vol. 1: An Introduction to the Bosonic String," Cambridge University Press, Cambridge (1998).

[193] J. Polchinski, "String Theory. Vol. 2: Superstring Theory and Beyond," Cambridge University Press, Cambridge (1998).

[194] K. Becker, M. Becker, and J. H. Schwarz, "String Theory and M-Theory: A Modern Introduction," Cambridge University Press, Cambridge (2006).
[195] O. Klein, "Quantum Theory and Five-Dimensional Theory of Relativity. (In German and English)," Z. Phys. **37** (1926), 895–906.
[196] T. Kaluza, "Zum Unitätsproblem der Physik," Sitzungsber. Preuss. Akad. Wiss. Berlin (Math. Phys.) **1921** (1921), 966–972.
[197] T. Han, J. D. Lykken, and R. J. Zhang, "On Kaluza-Klein States from Large Extra Dimensions," Phys. Rev. D **59** (1999), 105006.
[198] C. Rovelli and F. Vidotto, "Covariant Loop Quantum Gravity: An Elementary Introduction to Quantum Gravity and Spinfoam Theory," Cambridge University Press, Cambridge (2014).
[199] A. Ashtekar and P. Singh, "Loop Quantum Cosmology: A Status Report," Class. Quant. Grav. **28** (2011), 213001.
[200] C. Rovelli, "Loop Quantum Gravity," Living Rev. Rel. **1** (1998), 1.
[201] E. Kiritsis and G. Kofinas, "Horava-Lifshitz Cosmology," Nucl. Phys. B **821** (2009), 467.
[202] R. G. Cai, L. M. Cao, and N. Ohta, "Topological Black Holes in Horava-Lifshitz Gravity," Phys. Rev. D **80** (2009), 024003.
[203] T. P. Sotiriou, "Horava-Lifshitz Gravity: A Status Report," J. Phys. Conf. Ser. **283** (2011), 012034.
[204] S. Mukohyama, "Horava-Lifshitz Cosmology: A Review," Class. Quant. Grav. **27** (2010), 223101.
[205] N. Arkani-Hamed, S. Dimopoulos, G. Dvali, and G. Gabadadze, "Non-Local Modification of Gravity and the Cosmological Constant Problem," arXiv e-print (2002) [arXiv: hep-th/0209227].
[206] L. Modesto and S. Tsujikawa, "Non-Local Massive Gravity," Phys. Lett. B **727** (2013), 48.
[207] E. Witten, "Five-Branes and M Theory on an Orbifold," Nucl. Phys. B **463** (1996), 383–397.
[208] M. Dine, "TASI Lectures on M Theory Phenomenology," DOI: https://doi.org/10.1142/9789812799630_0006.
[209] R. L. Arnowitt, S. Deser, and C. W. Misner, "The Dynamics of General Relativity," Gen. Rel. Grav. **40** (2008), 1997.
[210] B. S. DeWitt, "Quantum Theory of Gravity. 1. The Canonical Theory," Phys. Rev. **160** (1967), 1113.
[211] B. S. DeWitt, "Quantum Theory of Gravity. 2. The Manifestly Covariant Theory," Phys. Rev. **162** (1967), 1195.
[212] J. A. Wheeler, "On the Nature of Quantum Geometrodynamics," Annals Phys. **2** (1957), 604–614.
[213] R. Bousso and L. Susskind, "The Multiverse Interpretation of Quantum Mechanics," Phys. Rev. D **85** (2012), 045007.
[214] A. Vilenkin, "The Interpretation of the Wave Function of the Universe," Phys. Rev. D **39** (1989), 1116.
[215] S. W. Hawking, "The Quantum State of the Universe," Adv. Ser. Astrophys. Cosmol. **3** (1987), 236–255.
[216] A. Vilenkin, "Creation of Universes from Nothing," Phys. Lett. **117B** (1982), 25.
[217] A. Vilenkin, "Quantum Creation of Universes," Phys. Rev. D **30** (1984), 509.
[218] A. Vilenkin, "Boundary Conditions in Quantum Cosmology," Phys. Rev. D **33** (1986), 3560.

[219] J. B. Hartle and S. W. Hawking, "Wave Function of the Universe," Phys. Rev. D **28** (1983), 2960 [Adv. Ser. Astrophys. Cosmol. **3** (1987), 174].

[220] S. Capozziello and G. Lambiase, "Selection Rules in Minisuperspace Quantum Cosmology," Gen. Rel. Grav. **32** (2000), 673–696.

[221] J. J. Halliwell, "The Quantum Cosmology of Einstein-Maxwell Theory in Six-Dimensions," Nucl. Phys. B **266** (1986), 228–244.

[222] J. J. Halliwell, "Correlations in the Wave Function of the Universe," Phys. Rev. D **36** (1987), 3626

[223] W. G. Unruh and R. M. Wald, "Time and the Interpretation of Canonical Quantum Gravity," Phys. Rev. D **40** (1989), 2598.

[224] J. S. Schwinger, "Quantized Gravitational Field," Phys. Rev. **130** (1963), 1253.

[225] B. Cvetković and D. Simić, "5D Lovelock Gravity: New Exact Solutions with Torsion," Phys. Rev. D **94**, No. 8 (2016), 084037.

[226] J. Zanelli, "Lecture Notes on Chern-Simons (Super-)Gravities. Second Edition (February 2008)," arXiv e-print (2008), [arXiv: hep-th/0502193].

[227] F. Bajardi, D. Vernieri, and S. Capozziello, "Exact Solutions in Higher-Dimensional Lovelock and AdS $_5$ Chern-Simons Gravity," JCAP **11**, No. 11 (2021), 057.

[228] E. Berti, E. Barausse, V. Cardoso, et al., "Testing General Relativity with Present and Future Astrophysical Observations," Class. Quant. Grav. **32** (2015), 243001.

[229] S. S. Chern and J. Simons, "Characteristic Forms and Geometric Invariants," Annals Math. **99** (1974), 48–69.

[230] O. Aharony, O. Bergman, D. L. Jafferis, and J. Maldacena, "N=6 Superconformal Chern-Simons-Matter Theories, M2-Branes and Their Gravity Duals," JHEP **10** (2008), 091.

[231] E. Witten, "Quantum Field Theory and the Jones Polynomial," Commun. Math. Phys. **121** (1989), 351–399.

[232] A. Achucarro and P. K. Townsend, "A Chern-Simons Action for Three-Dimensional anti-De Sitter Supergravity Theories," Phys. Lett. B **180** (1986), 89.

[233] A. Borowiec, L. Fatibene, M. Ferraris, and M. Francaviglia, "Covariant Lagrangian Formulation of Chern-Simons and BF Theories," Int. J. Geom. Meth. Mod. Phys. **3** (2006), 755–774.

[234] L. Fatibene, M. Francaviglia, and S. Mercadante, "Covariant Formulation of Chern-Simons Theories," Int. J. Geom. Meth. Mod. Phys. **2** (2005), 993–1008.

[235] R. G. Cai and K. S. Soh, "Topological Black Holes in the Dimensionally Continued Gravity," Phys. Rev. D **59** (1999), 044013.

[236] L. Smolin and C. Soo, "The Chern-Simons Invariant as the Natural Time Variable for Classical and Quantum Cosmology," Nucl. Phys. B **449** (1995), 289–316.

[237] S. Giombi, S. Minwalla, S. Prakash, et al., "Chern-Simons Theory with Vector Fermion Matter," Eur. Phys. J. C **72** (2012), 2112.

[238] L. Susskind, "The Quantum Hall Fluid and Noncommutative Chern-Simons Theory," arXiv e-print (2001), [arXiv: hep-th/0101029 [hep-th]].

[239] J. Oliva and S. Ray, "Classification of Six Derivative Lagrangians of Gravity and Static Spherically Symmetric Solutions," Phys. Rev. D **82** (2010), 124030.

[240] G. F. Chapline and N. S. Manton, "Unification of Yang-Mills Theory and Supergravity in Ten-Dimensions," Phys. Lett. B **120** (1983), 105–109.

[241] M. Benna, I. Klebanov, T. Klose, and M. Smedback, "Superconformal Chern-Simons Theories and AdS(4)/CFT(3) Correspondence," JHEP **09** (2008), 072.

[242] A. H. Chamseddine, "Topological Gravity and Supergravity in Various Dimensions," Nucl. Phys. B **346** (1990), 213–234.
[243] S. Capozziello and A. De Felice, "f(R) Cosmology by Noether's Symmetry," JCAP **0808** (2008), 016.
[244] H. R. Fazlollahi, "F(R) Cosmology via Noether Symmetry and Λ-Chaplygin Gas-like Model," Phys. Lett. B **781** (2018), 542.
[245] S. Capozziello and G. Lambiase, "Higher Order Corrections to the Effective Gravitational Action from Noether Symmetry Approach," Gen. Rel. Grav. **32** (2000), 295.
[246] B. Vakili, "Noether Symmetry in f(R) Cosmology," Phys. Lett. B **664** (2008), 16.
[247] S. Capozziello, P. Martin-Moruno, and C. Rubano, "Dark Energy and Dust Matter Phases from an Exact $f(R)$-Cosmology Model," Phys. Lett. B **664** (2008), 12.
[248] G. Allemandi, A. Borowiec, and M. Francaviglia, "Accelerated Cosmological Models in Ricci Squared Gravity," Phys. Rev. D **70** (2004), 103503.
[249] S. Capozziello, R. de Ritis, and A. A. Marino, "Conformal Equivalence and Noether Symmetries in Cosmology," Class. Quant. Grav. **14** (1998), 3259.
[250] S. Capozziello, N. Frusciante, and D. Vernieri, "New Spherically Symmetric Solutions in f(R)-Gravity by Noether Symmetries," Gen. Rel. Grav. **44** (2012), 1881.
[251] M. De Laurentis, "Noether's Stars in $f(\mathcal{R})$ Gravity," Phys. Lett. B **780** (2018), 205.
[252] V. Borka Jovanović, S. Capozziello, P. Jovanović, and D. Borka, "Recovering the Fundamental Plane of Galaxies by $f(R)$ Gravity," Phys. Dark Univ. **14** (2016), 73.
[253] H. Mohseni Sadjadi, "Generalized Noether Symmetry in f(T) Gravity," Phys. Lett. B **718** (2012), 270.
[254] S. Basilakos, S. Capozziello, M. De Laurentis, A. Paliathanasis, and M. Tsamparlis, "Noether Symmetries and Analytical Solutions in f(T)-Cosmology: A Complete Study," Phys. Rev. D **88** (2013), 103526.
[255] K. Atazadeh and F. Darabi, "$f(T)$ Cosmology via Noether Symmetry," Eur. Phys. J. C **72** (2012), 2016.
[256] M. A. Skugoreva and A. V. Toporensky, "On Kasner Solution in Bianchi I $f(T)$ Cosmology," Eur. Phys. J. C **78**, No. 5 (2018), 377.
[257] S. Capozziello, M. De Laurentis, and R. Myrzakulov, "Noether Symmetry Approach for Teleparallel-Curvature Cosmology," Int. J. Geom. Meth. Mod. Phys. **12**, No. 09 (2015), 1550095.
[258] J. Beltrán Jiménez, L. Heisenberg, T. S. Koivisto, and S. Pekar, "Cosmology in $f(Q)$ Geometry," Phys. Rev. D **101**, No. 10 (2020), 103507.
[259] D. Iosifidis and T. Koivisto, "Scale Transformations in Metric-Affine Geometry," Universe **5** (2019), 82.
[260] O. Gron and S. Hervik, "The Weyl Curvature Conjecture," arXiv e-print (2002), [arXiv: gr-qc/0205026].
[261] L. J. Alty, "The Generalized Gauss-Bonnet-Chern Theorem," J. Math. Phys. **36** (1995), 3094.
[262] C. Cherubini, D. Bini, S. Capozziello, and R. Ruffini, "Second Order Scalar Invariants of the Riemann Tensor: Applications to Black Hole Spacetimes," Int. J. Mod. Phys. D **11** (2002), 827.

[263] N. Birrell and P. Davies, "Quantum Fields in Curved Space" (Cambridge Monographs on Mathematical Physics). Cambridge University Press, Cambridge (1982).

[264] A. De Felice and S. Tsujikawa, "Construction of Cosmologically Viable $f(\mathcal{G})$ Dark Energy Models," Phys. Lett. B **675** (2009), 1.

[265] K. Uddin, J. E. Lidsey, and R. Tavakol, "Cosmological Scaling Solutions in Generalised Gauss-Bonnet Gravity Theories," Gen. Rel. Grav. **41** (2009), 2725.

[266] A. De Felice, D. F. Mota, and S. Tsujikawa, "Matter Instabilities in General Gauss-Bonnet Gravity," Phys. Rev. D **81** (2010), 023532.

[267] S. C. Davis, "Solar System Constraints on f(G) Dark Energy," arXiv e-print (2007), [arXiv:0709.4453 [hep-th]].

[268] S. Chakraborty and S. SenGupta, "Spherically Symmetric Brane in a Bulk of $f(R)$ and Gauss-Bonnet Gravity," Class. Quant. Grav. **33**, No. 22 (2016), 225001.

[269] J. de Boer, M. Kulaxizi, and A. Parnachev, "AdS(7)/CFT(6), Gauss-Bonnet Gravity, and Viscosity Bound," JHEP **1003** (2010), 087.

[270] A. V. Astashenok, S. D. Odintsov, and V. K. Oikonomou, "Modified Gauss-Bonnet Gravity with the Lagrange Multiplier Constraint as Mimetic Theory," Class. Quant. Grav. **32**, No. 18 (2015), 185007.

[271] M. De Laurentis and A. J. Lopez-Revelles, "Newtonian, Post Newtonian and Parameterized Post Newtonian Limits of $f(R, G)$ Gravity," Int. J. Geom. Meth. Mod. Phys. **11** (2014), 1450082.

[272] V. A. Rubakov and M. E. Shaposhnikov, "Do We Live Inside a Domain Wall?," Phys. Lett. **125B** (1983), 136.

[273] I. Antoniadis, "A Possible New Dimension at a Few TeV," Phys. Lett. B **246** (1990), 377.

[274] L. Randall and R. Sundrum, "A Large Mass Hierarchy from a Small Extra Dimension," Phys. Rev. Lett. **83** (1999), 3370.

[275] L. Randall and R. Sundrum, "An Alternative to Compactification," Phys. Rev. Lett. **83** (1999) 4690.

[276] N. Arkani-Hamed, S. Dimopoulos, and G. R. Dvali, "The Hierarchy Problem and New Dimensions at a Millimeter," Phys. Lett. B **429** (1998), 263.

[277] D. Lovelock, "The Einstein Tensor and Its Generalizations," J. Math. Phys. **12** (1971), 498.

[278] S. Capozziello, C. Corda, and M. F. De Laurentis, "Massive Gravitational Waves from f(R) Theories of Gravity: Potential Detection with LISA," Phys. Lett. B **669** (2008), 255–259.

[279] A. De Felice and T. Suyama, "Vacuum Structure for Scalar Cosmological Perturbations in Modified Gravity Models," JCAP **06** (2009), 034.

[280] S. Nojiri, S. D. Odintsov, and V. K. Oikonomou, "Ghost-Free Gauss-Bonnet Theories of Gravity," Phys. Rev. D **99**, No. 4 (2019), 044050.

[281] S. Capozziello, M. De Laurentis, and S. D. Odintsov, "Noether Symmetry Approach in Gauss-Bonnet Cosmology," Mod. Phys. Lett. A **29**, No. 30 (2014), 1450164.

[282] U. Camci, "F (R,G) Cosmology through Noether Symmetry Approach," Symmetry **10**, No. 12 (2018), 719.

[283] S. Santos Da Costa, F. V. Roig, J. S. Alcaniz, et al., "Dynamical Analysis on $f(R, \mathcal{G})$ Cosmology," Class. Quant. Grav. **35**, No. 7 (2018), 075013.

[284] S. Capozziello, C. A. Mantica, and L. G. Molinari, "Cosmological Perfect Fluids in Gauss–Bonnet Gravity," Int. J. Geom. Meth. Mod. Phys. **16**, No. 09 (2019), 1950133.

[285] S. Nojiri and S. D. Odintsov, "Modified Gauss-Bonnet Theory as Gravitational Alternative for Dark Energy," Phys. Lett. B **631** (2005) 1.

[286] C. Charmousis and J. F. Dufaux, "General Gauss-Bonnet Brane Cosmology," Class. Quant. Grav. **19** (2002), 4671.

[287] B. Li, J. D. Barrow, and D. F. Mota, "The Cosmology of Modified Gauss-Bonnet Gravity," Phys. Rev. D **76** (2007), 044027.

[288] S. Y. Zhou, E. J. Copeland, and P. M. Saffin, "Cosmological Constraints on $f(G)$ Dark Energy Models," JCAP **0907** (2009), 009.

[289] R. Myrzakulov, D. Saez-Gomez, and A. Tureanu, "On the ΛCDM Universe in $f(G)$ Gravity," Gen. Rel. Grav. **43** (2011), 1671.

[290] M. Gurses, T. Ç. Şişman, and B. Tekin, "Comment on 'Einstein-Gauss-Bonnet Gravity in Four-Dimensional Spacetime'," Phys. Rev. Lett. **125**, No. 14 (2020), 149001.

[291] D. Glavan and C. Lin, "Einstein-Gauss-Bonnet Gravity in Four-Dimensional Spacetime," Phys. Rev. Lett. **124**, No. 8 (2020), 081301.

[292] M. Sharif and K. Nazir, "Study of Static Wormhole Solutions in $F(T, T_{\mathcal{G}})$ Gravity," Annals Phys. **393** (2018), 145.

[293] S. Bahamonde and C. G. Böhmer, "Modified Teleparallel Theories of Gravity: Gauss–Bonnet and Trace Extensions," Eur. Phys. J. C **76**, No. 10 (2016), 578.

[294] M. Zubair and A. Jawad, "Generalized Second Law of Thermodynamics in $f(T, T_G)$ Gravity," Astrophys. Space Sci. **360** (2015), 11.

[295] G. Kofinas, G. Leon, and E. N. Saridakis, "Dynamical Behaviour in $f(T, T_G)$ cosmology," Class. Quant. Grav. **31** (2014), 175011.

[296] G. Kofinas and E. N. Saridakis, "Teleparallel Equivalent of Gauss-Bonnet Gravity and Its Modifications," Phys. Rev. D **90** (2014), 084044.

[297] G. Kofinas and E. N. Saridakis, "Cosmological Applications of $F(T, T_G)$ Gravity," Phys. Rev. D **90** (2014), 084045.

[298] S. Bahamonde, U. Camci, and S. Capozziello, "Noether Symmetries and Boundary Terms in Extended Teleparallel Gravity Cosmology," Class. Quant. Grav. **36**, No. 6 (2019), 065013.

[299] F. C. Adams, K. Freese, and A. H. Guth, "Constraints on the Scalar Field Potential in Inflationary Models," Phys. Rev. D **43** (1991), 965.

[300] S. Cotsakis and A. Tsokaros, "Flat, Radiation Universes with Quadratic Corrections and Asymptotic Analysis," arXiv e-print (2006), [arXiv:gr-qc/0612190].

[301] L. Amendola, A. Battaglia Mayer, S. Capozziello, et al., "Generalized Sixth Order Gravity and Inflation," Class. Quant. Grav. **10** (1993), L43.

[302] D. Wands, "Extended Gravity Theories and the Einstein-Hilbert Action," Class. Quant. Grav. **11** (1994), 269.

[303] S. Kluske and H. J. Schmidt, "Towards a No Hair Theorem for Higher Order Gravity," Astron. Nachr. **317** (1996), 337.

[304] S. Gottlöber, H. J. Schmidt, and A. A. Starobinsky, "Sixth Order Gravity and Conformal Transformations," Class. Quant. Grav. **7** (1990), 893.

[305] A. K. Sanyal, B. Modak, C. Rubano, and E. Piedipalumbo, "Noether Symmetry in the Higher Order Gravity Theory," Gen. Rel. Grav. **37** (2005), 407.

[306] A. L. Berkin and K. i. Maeda, "Effects of R**3 and R Box R Terms on R**2 Inflation," Phys. Lett. B **245** (1990), 348–354.

[307] Y. B. Zeldovich and A. A. Starobinsky, "Particle Production and Vacuum Polarization in an Anisotropic Gravitational Field," Sov. Phys. JETP **34** (1972), 1159 [Zh. Eksp. Teor. Fiz. **61** (1971)].

[308] L. Parker and J. Z. Simon, "Einstein Equation with Quantum Corrections Reduced to Second Order," Phys. Rev. D **47** (1993), 1339–1355.
[309] A. R. Liddle and F. Mellor, "De Sitter Solutions in Higher Order Gravity," Gen. Rel. Grav. **24** (1992), 897–909.
[310] S. Capozziello, G. Lambiase, M. Sakellariadou, and A. Stabile, "Constraining Models of Extended Gravity Using Gravity Probe B and LARES Experiments," Phys. Rev. D **91**, No. 4 (2015), 044012.
[311] F. Bajardi and S. Capozziello, "Equivalence of Nonminimally Coupled Cosmologies by Noether Symmetries," Int. J. Mod. Phys. D **29**, No. 14 (2020), 2030015.
[312] S. Capozziello and R. de Ritis, "Minisuperspace and Wheeler-DeWitt Equation for String Dilaton Cosmology," Int. J. Mod. Phys. D **2** (1993), 373.
[313] S. Capozziello, R. de Ritis, and C. Rubano, "String Dilaton Cosmology with Exponential Potential," Phys. Lett. A **177** (1993), 8–12.
[314] S. Capozziello, R. de Ritis, and P. Scudellaro, "Noether's Symmetries in (n+1)-Dimensional Nonminimally Coupled Cosmologies," Int. J. Mod. Phys. D **2** (1993), 463.
[315] S. Capozziello, S. Nesseris, and L. Perivolaropoulos, "Reconstruction of the Scalar-Tensor Lagrangian from a LCDM Background and Noether Symmetry," JCAP **12** (2007), 009.
[316] S. Capozziello, R. De Ritis, and P. Scudellaro, "Noether's Symmetries in Nonflat Cosmologies," Nuovo Cim. B **109** (1994), 159.
[317] J. A. Belinchón, T. Harko, and M. K. Mak, "Exact Scalar-Tensor Cosmological Models," Int. J. Mod. Phys. D **26**, No. 07 (2017), 1750073.
[318] S. Capozziello and R. de Ritis, "Relation between the Potential and Nonminimal Coupling in Inflationary Cosmology," Phys. Lett. A **177** (1993), 1.
[319] A. Borowiec, S. Capozziello, M. De Laurentis, et al., "Invariant Solutions and Noether Symmetries in Hybrid Gravity," Phys. Rev. D **91**, No. 2 (2015), 023517.
[320] F. Bajardi and S. Capozziello, "Noether Symmetries and Quantum Cosmology in Extended Teleparallel Gravity," Int. J. Geom. Meth. Mod. Phys. **18**, No. supp01 (2021), 2140002.
[321] C. Deffayet, X. Gao, D. A. Steer, and G. Zahariade, "From k-Essence to Generalised Galileons," Phys. Rev. D **84** (2011), 064039.
[322] E. Babichev, C. Charmousis, and A. Lehébel, "Black Holes and Stars in Horndeski Theory," Class. Quant. Grav. **33**, No. 15 (2016), 154002.
[323] E. Babichev, C. Charmousis, and M. Hassaine, "Charged Galileon Black Holes," JCAP **05** (2015), 031.
[324] S. Bhattacharya, K. F. Dialektopoulos, A. E. Romano, and T. N. Tomaras, "Brans-Dicke Theory with $\Lambda > 0$: Black Holes and Large Scale Structures," Phys. Rev. Lett. **115**, No. 18 (2015), 181104.
[325] A. Maselli, H. O. Silva, M. Minamitsuji, and E. Berti, "Neutron Stars in Horndeski Gravity," Phys. Rev. D **93**, No. 12 (2016), 124056.
[326] A. Cisterna, T. Delsate, and M. Rinaldi, "Neutron Stars in General Second Order Scalar-Tensor Theory: The Case of Nonminimal Derivative Coupling," Phys. Rev. D **92**, No. 4 (2015), 044050.
[327] S. Tsujikawa, "The Effective Field Theory of Inflation/Dark Energy and the Horndeski Theory," Lect. Notes Phys. **892** (2015), 97–136 DOI: https://doi.org/10.1007/978-3-319-10070-8_4.

[328] T. Kobayashi, M. Yamaguchi, and J. Yokoyama, "Generalized G-Inflation: Inflation with the Most General Second-Order Field Equations," Prog. Theor. Phys. **126** (2011), 511–529.

[329] A. Barvinsky, "Aspects of Nonlocality in Quantum Field Theory, Quantum Gravity and Cosmology," Mod. Phys. Lett. A **30**, No. 03n04 (2015), 1540003.

[330] L. Modesto, "Super-renormalizable Quantum Gravity," Phys. Rev. D **86** (2012), 044005.

[331] L. Modesto, "Super-renormalizable Gravity," arXiv e-print (2013), [arXiv:1302.6348 [hep-th]].

[332] L. Modesto, "Super-renormalizable Multidimensional Quantum Gravity," Astron. Rev. **8**, No. 2 (2013), 4.

[333] S. Nojiri and S. D. Odintsov, "Modified Non-Local-F(R) Gravity as the Key for the Inflation and Dark Energy," Phys. Lett. B **659** (2008), 821.

[334] E. Elizalde, S. D. Odintsov, E. O. Pozdeeva, and S. Y. Vernov, "De Sitter and Power-Law Solutions in Non-local Gauss-Bonnet Gravity," Int. J. Geom. Meth. Mod. Phys. **15**, No. 11 (2018), 1850188.

[335] F. Briscese, A. Marcianò, L. Modesto, and E. N. Saridakis, "Inflation in (Super-)renormalizable Gravity," Phys. Rev. D **87**, No. 8 (2013), 083507.

[336] S. Deser and R. Woodard, "Nonlocal Cosmology," Phys. Rev. Lett. **99** (2007), 111301.

[337] T. Biswas, E. Gerwick, T. Koivisto, and A. Mazumdar, "Towards Singularity and Ghost Free Theories of Gravity," Phys. Rev. Lett. **108** (2012), 031101.

[338] T. Biswas, A. S. Koshelev, and A. Mazumdar, "Gravitational Theories with Stable (Anti-)de Sitter Backgrounds," Fundam. Theor. Phys. **183** (2016), 97–114.

[339] T. Biswas, A. S. Koshelev, and A. Mazumdar, "Consistent Higher Derivative Gravitational Theories with Stable de Sitter and Anti–de Sitter Backgrounds," Phys. Rev. D **95**, No. 4 (2017), 043533.

[340] T. Biswas, A. Mazumdar, and W. Siegel, "Bouncing Universes in String-Inspired Gravity," JCAP **03** (2006), 009.

[341] L. Buoninfante, A. S. Koshelev, G. Lambiase, and A. Mazumdar, "Classical Properties of Non-local, Ghost- and Singularity-Free Gravity," JCAP **09** (2018), 034.

[342] L. Buoninfante, G. Lambiase, and A. Mazumdar, "Ghost-Free Infinite Derivative Quantum Field Theory," Nucl. Phys. B **944** (2019), 114646.

[343] F. Briscese, L. Modesto, and S. Tsujikawa, "Super-renormalizable or Finite Completion of the Starobinsky Theory," Phys. Rev. D **89**, No. 2 (2014), 024029.

[344] S. Bahamonde, S. Capozziello, and K. F. Dialektopoulos, "Constraining Generalized Non-local Cosmology from Noether Symmetries," Eur. Phys. J. C **77**, No. 11 (2017), 722.

[345] A. Acunzo, F. Bajardi, and S. Capozziello, "Non-Local Curvature Gravity Cosmology via Noether Symmetries," Phys. Lett. B **826** (2022), 136907.

[346] F. Bajardi, S. Capozziello, and D. Vernieri, "Non-Local Curvature and Gauss–Bonnet Cosmologies by Noether Symmetries," Eur. Phys. J. Plus **135**, No. 12 (2020), 942.

[347] S. Capozziello, E. Elizalde, S. Nojiri, and S. D. Odintsov, "Accelerating Cosmologies from Non-local Higher-Derivative Gravity," Phys. Lett. B **671** (2009), 193–198.

[348] P. Channuie and D. Momeni, "Noether Symmetry in a Nonlocal f(T) Gravity," Nucl. Phys. B **935** (2018), 256.

[349] S. Capozziello, G. Marmo, C. Rubano, and P. Scudellaro, "Noether Symmetries in Bianchi Universes," Int. J. Mod. Phys. D **6** (1997), 491–503.
[350] T. Rothman and G. F. R. Ellis, Phys. Lett. B **180** (1986), 19–24.
[351] J. D. Barrow, "Why the Universe Is Not Anisotropic," Phys. Rev. D **51** (1995), 3113–3116.
[352] J. F. Carinena, A. Ibort, G. Marmo, and A. Perelomov, "On the Geometry of Lie Algebras and Poisson Tensors," J. Phys. A **27** (1994), 7425–7450.
[353] J. Grabowski, G. Marmo, and A. M. Perelomov, "Poisson Structures: Towards a Classification," Mod. Phys. Lett. A **8** (1993), 1719–1733.
[354] M. P. Ryan Jr., "Hamiltonian Cosmology: Death and Transfiguration," J. Math. Phys. **15** (1974), 812.
[355] G. F. R. Ellis and M. A. H. McCallum, "A Class of Homogeneous Cosmological Models," Comm. Math. Phys. **12** (1969), 108.
[356] M. Demianski, R. de Ritis, C. Rubano, and P. Scudellaro, "Scalar Fields and Anisotropy in Cosmological Models," Phys. Rev. D **46** (1992), 1391–1398.
[357] S. Benic, D. Blaschke, D. E. Alvarez-Castillo, T. Fischer, and S. Typel, "A New Quark-Hadron Hybrid Equation of State for Astrophysics - I. High-Mass Twin Compact Stars," Astron. Astrophys. **577** (2015), A40.
[358] D. Alvarez-Castillo, A. Ayriyan, S. Benic, et al., "New Class of Hybrid EoS and Bayesian M-R Data Analysis," Eur. Phys. J. A **52**, No. 3 (2016), 69.
[359] S. Capozziello, M. De Laurentis, R. Farinelli, and S. D. Odintsov, "Mass-Radius Relation for Neutron Stars in f(R) Gravity," Phys. Rev. D **93**, No. 2 (2016), 023501.
[360] M. Persic, P. Salucci, and F. Stel, "The Universal Rotation Curve of Spiral Galaxies: 1. The Dark Matter Connection," Mon. Not. Roy. Astron. Soc. **281** (1996) 27.
[361] J. F. Navarro, C. S. Frenk, and S. D. M. White, "The Structure of Cold Dark Matter Halos," Astrophys. J. **462** (1996), 563.
[362] S. Capozziello, V. F. Cardone, and A. Troisi, "Low Surface Brightness Galaxies Rotation Curves in the Low Energy Limit of Rn Gravity: No Need for Dark Matter?," Mon. Not. Roy. Astron. Soc. **375** (2007), 1423.
[363] C. F. Martins and P. Salucci, "Analysis of Rotation Curves in the Framework of R**n Gravity," Mon. Not. Roy. Astron. Soc. **381** (2007), 1103.
[364] V. Vikram, J. Sakstein, C. Davis, and A. Neil, "Astrophysical Tests of Modified Gravity: Stellar and Gaseous Rotation Curves in Dwarf Galaxies," Phys. Rev. D **97**, No. 10 (2018), 104055.
[365] N. R. Napolitano, S. Capozziello, A. J. Romanowsky, M. Capaccioli, and C. Tortora, "Testing Yukawa-like Potentials from f(R)-Gravity in Elliptical Galaxies," Astrophys. J. **748** (2012), 87.
[366] S. Capozziello, A. Stabile, and A. Troisi, "Spherically Symmetric Solutions in f(R)-Gravity via Noether Symmetry Approach," Class. Quant. Grav. **24** (2007), 2153.
[367] M. Farasat Shamir, A. Jhangeer, and A. A. Bhatti, "Conserved Quantities in $f(R)$ Gravity via Noether Symmetry," Chin. Phys. Lett. **29** (2012), 080402.
[368] S. Capozziello, M. De Laurentis, and A. Stabile, "Axially Symmetric Solutions in f(R)-Gravity," Class. Quant. Grav. **27** (2010), 165008.
[369] T. Clifton and J. D. Barrow, "The Power of General Relativity," Phys. Rev. D **72**, No. 10 (2005), 103005. Erratum: [Phys. Rev. D **90**, No. 2 (2014), 029902].
[370] T. Clifton, "Spherically Symmetric Solutions to Fourth-Order Theories of Gravity," Class. Quant. Grav. **23** (2006), 7445.

[371] S. Capozziello, P. Jovanović, V. B. Jovanović, and D. Borka, "Addressing the Missing Matter Problem in Galaxies Through a New Fundamental Gravitational Radius," JCAP **1706**, No. 06 (2017), 044.
[372] T. Bernal, S. Capozziello, J. C. Hidalgo, and S. Mendoza, "Recovering MOND from Extended Metric Theories of Gravity," Eur. Phys. J. C **71** (2011), 1794.
[373] S. P. Drake and P. Szekeres, "Uniqueness of the Newman-Janis Algorithm in Generating the Kerr-Newman metric," Gen. Rel. Grav. **32** (2000), 445–458.
[374] R. Canonico, L. Parisi, and G. V. Vilasi, "Theoretical Models for Astrophysical Objects and the Newman-Janis Algorithm," Proc. Geom. Int. Quant. **11** (2010), 85–96.
[375] S. Capozziello, V. F. Cardone, and A. Troisi, "Gravitational Lensing in Fourth Order Gravity," Phys. Rev. D **73** (2006), 104019.
[376] D. B. Fisher, "Central Star Formation in Pseudobulges and Classical Bulges," Astrophys. J. **642** (2006), L17.
[377] R. C. Nunes, M. E. S. Alves, and J. C. N. de Araujo, "Forecast Constraints on $f(T)$ Gravity with Gravitational Waves from Compact Binary Coalescences," Phys. Rev. D **100**, No. 6 (2019), 064012.
[378] H. Abedi and S. Capozziello, "Gravitational Waves in Modified Teleparallel Theories of Gravity," Eur. Phys. J. C **78**, No. 6 (2018), 474.
[379] K. Bamba, C. Q. Geng, C. C. Lee, and L. W. Luo, "Equation of State for Dark Energy in $f(T)$ Gravity," JCAP **01** (2011), 021.
[380] G. R. Bengochea and R. Ferraro, "Dark Torsion as the Cosmic Speed-up," Phys. Rev. D **79** (2009), 124019.
[381] Y. F. Cai, S. H. Chen, J. B. Dent, S. Dutta, and E. N. Saridakis, "Matter Bounce Cosmology with the f(T) Gravity," Class. Quant. Grav. **28** (2011), 215011.
[382] R. Ferraro and F. Fiorini, "Spherically Symmetric Static Spacetimes in Vacuum f(T) Gravity," Phys. Rev. D **84** (2011), 083518.
[383] A. Paliathanasis, S. Basilakos, E. N. Saridakis, et al., "New Schwarzschild-like Solutions in f(T) Gravity through Noether Symmetries," Phys. Rev. D **89** (2014), 104042.
[384] S. Bahamonde and U. Camci, "Exact Spherically Symmetric Solutions in Modified Teleparallel Gravity," Symmetry **11**, No. 12 (2019), 1462.
[385] A. N. Nurbaki, S. Capozziello, and C. Deliduman, "Spherical and Cylindrical Solutions in $f(T)$ Gravity by Noether Symmetry Approach," Eur. Phys. J. C **80**, No. 2 (2020), 108.
[386] A. K. Ahmed, M. Azreg-Aïnou, S. Bahamonde, S. Capozziello, and M. Jamil, "Astrophysical Flows Near $f(T)$ Gravity Black Holes," Eur. Phys. J. C **76**, No. 5 (2016), 269.
[387] S. Bahamonde, K. Flathmann, and C. Pfeifer, "Photon Sphere and Perihelion Shift in Weak $f(T)$ Gravity," Phys. Rev. D **100**, No. 8 (2019), 084064.
[388] C. G. Boehmer, T. Harko, and F. S. N. Lobo, "Wormhole Geometries in Modified Teleparallel Gravity and the Energy Conditions," Phys. Rev. D **85** (2012), 044033.
[389] N. Deppe, C. D. Leonard, T. Taves, G. Kunstatter, and R. B. Mann, "Critical Collapse in Einstein-Gauss-Bonnet Gravity in Five and Six Dimensions," Phys. Rev. D **86** (2012), 104011.
[390] S. Ray, "Birkhoff's Theorem in Lovelock Gravity for General Base Manifolds," Class. Quant. Grav. **32**, No. 19 (2015), 195022.
[391] R. Zegers, "Birkhoff's Theorem in Lovelock Gravity," J. Math. Phys. **46** (2005), 072502.

[392] G. Dotti, J. Oliva, and R. Troncoso, "Static Solutions with Nontrivial Boundaries for the Einstein-Gauss-Bonnet Theory in Vacuum," Phys. Rev. D **82** (2010), 024002.
[393] C. Bogdanos, "Extensions of Birkhoff's Theorem in 6D Gauss-Bonnet Gravity," AIP Conf. Proc. **1241**, No. 1 (2010), 521.
[394] A. Stabile, "The Most General Fourth Order Theory of Gravity at Low Energy," Phys. Rev. D **82** (2010), 124026.
[395] I. I. Shapiro, "Fourth Test of General Relativity." Phys. Rev. Lett. **13** (1964), 789.
[396] A. F. Zakharov, A. A. Nucita, F. De Paolis, and G. Ingrosso, "Solar System Constraints on R^n Gravity," Phys. Rev. D **74** (2006), 107101.
[397] G. Covone, M. Sereno, and R. de Ritis, "Statistics of Gravitational Lenses in the Clumpy Universe," Mon. Not. Roy. Astron. Soc. **357** (2005), 773.
[398] A. V. Astashenok, S. Capozziello, S. D. Odintsov, and V. K. Oikonomou, "Extended Gravity Description for the GW190814 Supermassive Neutron Star," Phys. Lett. B **811** (2020), 135910.
[399] S. Capozziello, M. De Laurentis, and M. Francaviglia, "Higher-Order Gravity and the Cosmological Background of Gravitational Waves," Astropart. Phys. **29** (2008), 125.
[400] P. Haensel and A. Y. Potekhin, "Analytical Representations of Unified Equations of State of Neutron-Star Matter," Astron. Astrophys. **428** (2004), 191.
[401] R. Abbott et al. [LIGO Scientific and Virgo], "GW190814: Gravitational Waves from the Coalescence of a 23 Solar Mass Black Hole with a 2.6 Solar Mass Compact Object," Astrophys. J. Lett. **896**, No. 2 (2020), L44.
[402] A. D. Dolgov and M. Kawasaki, "Can Modified Gravity Explain Accelerated Cosmic Expansion?," Phys. Lett. B **573** (2003), 1–4.
[403] V. Faraoni, "Matter Instability in Modified Gravity," Phys. Rev. D **74** (2006), 104017.
[404] J. Ehlers and P. Schneider, "Gravitational Lensing," Springer (1992).
[405] T. van Putten, A. L. Watts, C. R. D'Angelo, M. G. Baring, and C. Kouveliotou, "Models of Hydrostatic Magnetar Atmospheres at High Luminosities," Mon. Not. Roy. Astron. Soc. **434** (2013), 1398.
[406] K. F. Dialektopoulos, D. Borka, S. Capozziello, V. Borka Jovanović, and P. Jovanović, "Constraining Nonlocal Gravity by S2 Star Orbits," Phys. Rev. D **99**, No. 4 (2019), 044053.
[407] S. Gillessen, F. Eisenhauer, S. Trippe, et al., "Monitoring Stellar Orbits around the Massive Black Hole in the Galactic Center," Astrophys. J. **692** (2009), 1075.
[408] V. Borka-Jovanović, P. Jovanović, D. Borka, et al., "Constraining Scalar-Tensor Gravity Models by S2 Star Orbits around the Galactic Center," Facta Univ. Ser. Phys. Chem. Tech. **17**, No. 1 (2019), 11–20.
[409] F. Zwicky, "Die Rotverschiebung von extragalaktischen Nebeln" [The red shift of extragalactic nebulae], Helv. Phys. Acta, **6** (1933) 110.
[410] F. Zwicky, "On the Masses of Nebulae and of Clusters of Nebulae," Astroph. Jou. **86** (1937), 217.
[411] A. Finch and J. L. Said, "Galactic Rotation Dynamics in f(T) Gravity," Eur. Phys. J. C **78**, No. 7 (2018), 560.
[412] S. Capozziello, E. De Filippis, and V. Salzano, "Modelling Clusters of Galaxies by f(R)-Gravity," Mon. Not. Roy. Astron. Soc. **394** (2009), 947.
[413] S. Capozziello, M. Faizal, M. Hameeda, et al., "Clustering of Galaxies with f(R) Gravity," Mon. Not. Roy. Astron. Soc. **474** (2018), 2430.

[414] K. G. Begeman, A. H. Broeils, and R. H. Sanders, "Extended Rotation Curves of Spiral Galaxies: Dark Haloes and Modified Dynamics," Mon. Not. Roy. Astron. Soc. **249** (1991), 523.

[415] H. Kojima, H. M. Antia, S. R. Dugad, et al., "Dependence of Cosmic Ray Intensity on Variation of Solar Wind Velocity Measured by the GRAPES-3 Experiment for Space Weather Studies," Phys. Rev. D **91**, No. 12 (2015), 121303.

[416] K. C. Freeman, "On the Disks of Spiral and SO Galaxies," Astrophys. J. **160** (1970), 811.

[417] R. Bender, D. Burstein, and S. M. Faber, "Dynamically Hot Galaxies. 1. Structural Properties," Astrophys. J. **399** (1992), 462.

[418] R. Bender, D. Burstein, and S. M. Faber, "Dynamically Hot Galaxies. 2. Global Stellar Populations," Astrophys. J. **411** (1993), 153.

[419] D. Taranu, J. Dubinski, and H. Yee, "The Fundamental Plane of Galaxy Group Mergers," ASP Conf. Ser. **477** (2013), 105.

[420] M. D'Onofrio et al., "The Hybrid Solution for the Fundamental Plane," Mon. Not. Roy. Astron. Soc. **435** (2013), 45.

[421] S. Capozziello, R. De Ritis, C. Rubano, and M. Demianski, "Cosmological Perturbations in Exact Noether Background Solutions," Phys. Rev. D **52** (1995), 3288.

[422] M. De Laurentis, M. Paolella, and S. Capozziello, "Cosmological Inflation in $F(R,\mathcal{G})$ Gravity," Phys. Rev. D **91**, No. 8 (2015), 083531.

[423] A. Albrecht and P. J. Steinhardt, "Cosmology for Grand Unified Theories with Radiatively Induced Symmetry Breaking," Adv. Ser. Astrophys. Cosmol. **3** (1987), 158–161.

[424] K. Freese, J. A. Frieman, and A. V. Olinto, "Natural Inflation with Pseudo-Nambu-Goldstone Bosons," Phys. Rev. Lett. **65** (1990), 3233–3236.

[425] D. Polarski and A. A. Starobinsky, "Spectra of Perturbations Produced by Double Inflation with an Intermediate Matter Dominated Stage," Nucl. Phys. B **385** (1992), 623–650.

[426] A. D. Linde, "Hybrid Inflation," Phys. Rev. D **49** (1994), 748–754.

[427] S. Capozziello, F. S. N. Lobo, and J. P. Mimoso, "Generalized Energy Conditions in Extended Theories of Gravity," Phys. Rev. D **91**, No. 12 (2015), 124019.

[428] G. F. R. Ellis, M. Bruni, and J. Hwang, "Density Gradient-Vorticity Relation in Perfect Fluid Robertson-Walker Perturbations," Phys. Rev. D **42** (1990), 1035.

[429] J. C. Hwang, "Cosmological Perturbations in Generalized Gravity Theories: Solutions," Phys. Rev. D **42** (1990), 2601.

[430] J. C. Hwang, "Cosmological Perturbations in Generalized Gravity Theories: Formulation," Class. Quant. Grav. **7** (1990), 1613.

[431] I. de Martino, M. De Laurentis, and S. Capozziello, "Tracing the Cosmic History by Gauss-Bonnet Gravity," Phys. Rev. D **102**, No. 6 (2020), 063508.

[432] S. W. Hawking and G. F. R. Ellis, "The Large Scale Structure of Space-Time," Cambridge Monographs on Mathematical Physics, Cambridge University Press, Cambridge (1973).

[433] E. Papantonopoulos and C. Vlachos, "Wormhole Solutions in Modified Brans-Dicke Theory," Phys. Rev. D **101**, No. 6 (2020), 064025.

[434] F. S. N. Lobo and M. A. Oliveira, "Wormhole Geometries in f(R) Modified Theories of Gravity," Phys. Rev. D **80** (2009), 104012.

[435] T. Harko, F. S. N. Lobo, M. K. Mak, and S. V. Sushkov, "Modified-Gravity Wormholes without Exotic Matter," Phys. Rev. D **87**, No. 6 (2013), 067504.

[436] M. Jamil, D. Momeni, and R. Myrzakulov, "Wormholes in a Viable f(T) Gravity," Eur. Phys. J. C **73** (2013), 2267.
[437] N. Furey and A. DeBenedictis, "Wormhole Throats in R**m Gravity," Class. Quant. Grav. **22** (2005), 313.
[438] T. Azizi, "Wormhole Geometries in $f(R,T)$ Gravity," Int. J. Theor. Phys. **52** (2013), 3486.
[439] S. Capozziello, O. Luongo, and L. Mauro, "Traversable Wormholes with Vanishing Sound Speed in $f(R)$ Gravity," Eur. Phys. J. Plus **136**, No. 2 (2021), 167.
[440] N. M. Garcia, T. Harko, F. S. N. Lobo, and J. P. Mimoso, "Energy Conditions in Modified Gauss-Bonnet Gravity," Phys. Rev. D **83** (2011), 104032.
[441] D. Rapetti, S. W. Allen, M. A. Amin, and R. D. Blandford, "A Kinematical Approach to Dark Energy Studies," Mon. Not. Roy. Astron. Soc. **375** (2007), 1510.
[442] J. Santos, J. S. Alcaniz, M. J. Reboucas, and F. C. Carvalho, "Energy Conditions in f(R)-Gravity," Phys. Rev. D **76** (2007), 083513.
[443] J. Santos, M. J. Reboucas, and J. S. Alcaniz, "Energy Conditions Constraints on a Class of f(R) Gravity," Int. J. Mod. Phys. D **19** (2010), 1315–1321.
[444] S. Capozziello, "Torsion Quintessence," Mod. Phys. Lett. A **17** (2002), 1621.
[445] S. Capozziello, C. A. Mantica, and L. G. Molinari, "Geometric Perfect Fluids from Extended Gravity," EPL **137**, No. 1 (2022), 19001.
[446] S. Capozziello and G. Lambiase, "Cosmological Curvature Acceleration," Eur. Phys. J. ST **230**, No. 9 (2021), 2123–2138.
[447] S. Nojiri and S. D. Odintsov, "The Future Evolution and Finite-Time Singularities in F(R)-Gravity Unifying the Inflation and Cosmic Acceleration," Phys. Rev. D **78** (2008), 046006.
[448] K. Bamba and S. D. Odintsov, "Inflation and Late-Time Cosmic Acceleration in Non-minimal Maxwell-$F(R)$ Gravity and the Generation of Large-Scale Magnetic Fields," JCAP **04** (2008), 024.
[449] L. Amendola, D. Polarski, and S. Tsujikawa, "Are f(R) Dark Energy Models Cosmologically Viable?," Phys. Rev. Lett. **98** (2007), 131302.
[450] L. Amendola, R. Gannouji, D. Polarski, and S. Tsujikawa, "Conditions for the Cosmological Viability of f(R) Dark Energy Models," Phys. Rev. D **75** (2007), 083504.
[451] A. W. Brookfield, C. van de Bruck, and L. M. H. Hall, "Viability of f(R) Theories with Additional Powers of Curvature," Phys. Rev. D **74** (2006), 064028.
[452] S. Capozziello, S. Nojiri, S. D. Odintsov, and A. Troisi, "Cosmological Viability of f(R)-Gravity as an Ideal Fluid and Its Compatibility with a Matter Dominated Phase," Phys. Lett. B **639** (2006), 135–143.
[453] S. Capozziello, V. F. Cardone, and A. Troisi, "Reconciling Dark Energy Models with f(R) Theories," Phys. Rev. D **71** (2005), 043503.
[454] M. Jamil, F. M. Mahomed, and D. Momeni, "Noether Symmetry Approach in $f(R)$-Tachyon Model," Phys. Lett. B **702** (2011), 315.
[455] S. Capozziello, R. De Ritis, and P. Scudellaro, "Noether's Symmetries in Quantum Cosmology," Int. J. Mod. Phys. D **3** (1994), 609.
[456] S. Capozziello and R. de Ritis, "Noether's Symmetries and Exact Solutions in Flat Nonminimally Coupled Cosmological Models," Class. Quant. Grav. **11** (1994), 107.
[457] S. Capozziello and M. De Laurentis, "Noether Symmetries in Extended Gravity Quantum Cosmology," Int. J. Geom. Meth. Mod. Phys. **11** (2014), 1460004.

[458] A. Paliathanasis, M. Tsamparlis, S. Basilakos, and S. Capozziello, "Scalar-Tensor Gravity Cosmology: Noether Symmetries and Analytical Solutions," Phys. Rev. D **89**, No. 6 (2014), 063532.

[459] G. Gionti, S.J., "O(d,d) Duality Transformations in F(R) Theories of Gravity," Proceedings, 14th Marcel Grossmann Meeting on Recent Developments in Theoretical and Experimental General Relativity, Astrophysics, and Relativistic Field Theories (MG14) (in 4 Volumes): Rome, Italy, July 12–18, 2015, arXiv:1608.05223 [gr-qc].

[460] S. Capozziello, G. Gionti, S.J., and D. Vernieri, "String Duality Transformations in $f(R)$ Gravity from Noether Symmetry Approach," JCAP **1601** (2016), 015.

[461] J. Scherk and J. H. Schwarz, "Dual Models for Nonhadrons," Nucl. Phys. B **81** (1974), 118.

[462] E. S. Fradkin and A. A. Tseytlin, "Quantum String Theory Effective Action," Nucl. Phys. B **261** (1985), 1. Erratum: [Nucl. Phys. B **269** (1986), 745].

[463] C. G. Callan, Jr., E. J. Martinec, M. J. Perry, and D. Friedan, "Strings in Background Fields," Nucl. Phys. B **262** (1985), 593.

[464] S. Capozziello and R. de Ritis, "Scale Factor Duality and General Transformations for String Cosmology," Int. J. Mod. Phys. D **2** (1993), 367.

[465] F. Benini, S. Benvenuti, and S. Pasquetti, "SUSY Monopole Potentials in 2+1 Dimensions," JHEP **08** (2017), 086.

[466] N. Arkani-Hamed, L. Motl, A. Nicolis, and C. Vafa, "The String Landscape, Black Holes and Gravity as the Weakest Force," JHEP **06** (2007), 060.

[467] H. Ooguri and C. Vafa, "Non-supersymmetric AdS and the Swampland," Adv. Theor. Math. Phys. **21** (2017), 1787–1801.

[468] R. Brandenberger, L. L. Graef, G. Marozzi, and G. P. Vacca, "Backreaction of Super-Hubble Cosmological Perturbations beyond Perturbation Theory," Phys. Rev. D **98**, No. 10 (2018), 103523.

[469] L. Heisenberg, M. Bartelmann, R. Brandenberger and A. Refregier, "Dark Energy in the Swampland," Phys. Rev. D **98**, No. 12 (2018), 123502.

[470] M. C. David Marsh, "The Swampland, Quintessence and the Vacuum Energy," Phys. Lett. B **789** (2019), 639–642.

[471] A. Kehagias and A. Riotto, "A Note on Inflation and the Swampland," Fortsch. Phys. **66**, No. 10 (2018), 1800052.

[472] M. Ibe, M. Yamazaki, and T. T. Yanagida, "Quintessence Axion Revisited in Light of Swampland Conjectures," Class. Quant. Grav. **36**, No. 23 (2019), 235020.

[473] F. Denef, A. Hebecker, and T. Wrase, "De Sitter Swampland Conjecture and the Higgs Potential," Phys. Rev. D **98**, No. 8 (2018), 086004.

[474] R. Blumenhagen, I. Valenzuela, and F. Wolf, "The Swampland Conjecture and F-term Axion Monodromy Inflation," JHEP **07** (2017), 145.

[475] B. Heidenreich, M. Reece, and T. Rudelius, "Emergence of Weak Coupling at Large Distance in Quantum Gravity," Phys. Rev. Lett. **121**, No. 5 (2018), 051601.

[476] D. Andriot, "On the de Sitter Swampland Criterion," Phys. Lett. B **785** (2018), 570–573.

[477] S. K. Garg and C. Krishnan, "Bounds on Slow Roll and the de Sitter Swampland," JHEP **11** (2019), 075.

[478] A. Achúcarro and G. A. Palma, "The String Swampland Constraints Require Multifield Inflation," JCAP **02** (2019), 041.

[479] M. Motaharfar, V. Kamali, and R. O. Ramos, "Warm Inflation as a Way out of the Swampland," Phys. Rev. D **99**, No. 6 (2019), 063513.

[480] V. Kamali, "Reheating after Swampland Conjecture," JHEP **01** (2020), 092.

[481] U. H. Danielsson and T. Van Riet, "What If String Theory Has No de Sitter Vacua?," Int. J. Mod. Phys. D **27**, No. 12 (2018), 1830007.

[482] L. Heisenberg, M. Bartelmann, R. Brandenberger, and A. Refregier, "Horndeski Gravity in the Swampland," Phys. Rev. D **99**, No. 12 (2019), 124020.

[483] S. Brahma and M. W. Hossain, "Dark Energy beyond Quintessence: Constraints from the Swampland," JHEP **06** (2019), 070.

[484] M. Artymowski and I. Ben-Dayan, "f(R) and Brans-Dicke Theories and the Swampland," JCAP **05** (2019), 042.

[485] P. Agrawal, G. Obied, P. J. Steinhardt, and C. Vafa, "On the Cosmological Implications of the String Swampland," Phys. Lett. B **784** (2018), 271.

[486] S. Capozziello, V. F. Cardone, S. Carloni, and A. Troisi, "Curvature Quintessence Matched with Observational Data," Int. J. Mod. Phys. D **12** (2003), 1969.

[487] M. Niedermaier, "The Asymptotic Safety Scenario in Quantum Gravity: An Introduction," Classical and Quantum Gravity **24** (2007), R171–R230.

[488] M. Shaposhnikov and C. Wetterich, "Asymptotic Safety of Gravity and the Higgs Boson Mass," Phys. Lett. B **683** (2010), 196.

[489] M. Asorey, J. L. Lopez, and I. L. Shapiro, "Some Remarks on High Derivative Quantum Gravity," Int. J. Mod. Phys. A **12** (1997), 5711–5734.

[490] C. F. Steinwachs, "Towards a Unitary, Renormalizable and Ultraviolet-Complete Quantum Theory of Gravity," DOI: https://doi.org/10.3389/fphy.2020.00185 [arXiv:2004.07842 [hep-th]].

[491] M. S. Ruf and C. F. Steinwachs, "One-Loop Divergences for $f(R)$ Gravity," Phys. Rev. D **97**, No. 4 (2018), 044049.

[492] C. F. Steinwachs and A. Y. Kamenshchik, "One-Loop Divergences for Gravity Nonminimally Coupled to a Multiplet of Scalar Fields: Calculation in the Jordan Frame. I. The Main Results," Phys. Rev. D **84** (2011), 024026.

[493] M. S. Ruf and C. F. Steinwachs, "Quantum Equivalence of $f(R)$ Gravity and Scalar-Tensor Theories," Phys. Rev. D **97**, No. 4 (2018), 044050.

[494] M. H. Goroff and A. Sagnotti, "The Ultraviolet Behaviour of Einstein Gravity," Nucl. Phys. B **266** (1986) 709. DOI: https://doi.org/10.1016/0550-3213(86)90193-8.

[495] I. Kuntz, "Quantum Corrections to the Gravitational Backreaction," Eur. Phys. J. C **78**, No. 1 (2018), 3.

[496] F. Bajardi, F. Bascone, and S. Capozziello, "Renormalizability of Alternative Theories of Gravity: Differences between Power Counting and Entropy Argument," Universe **7**, No. 5 (2021), 148.

[497] T. Chiba, T. L. Smith, and A. L. Erickcek, "Solar System Constraints to General f(R) Gravity," Phys. Rev. D **75** (2007), 124014.

[498] B. Bertotti, L. Iess and P. Tortora, "A Test of General Relativity Using Radio Links with the Cassini Spacecraft," Nature **425** (2003), 374–376.

[499] S. Capozziello and S. Tsujikawa, "Solar System and Equivalence Principle Constraints on f(R) Gravity by Chameleon Approach," Phys. Rev. D **77** (2008), 107501.

[500] I. Navarro and K. Van Acoleyen, "Consistent Long Distance Modification of Gravity from Inverse Powers of the Curvature," JCAP **03** (2006), 008.

[501] T. Faulkner, M. Tegmark, E. F. Bunn, and Y. Mao, "Constraining f(R) Gravity as a Scalar Tensor Theory," Phys. Rev. D **76** (2007), 063505.

[502] G. Cognola and S. Zerbini, "Homogeneous Cosmologies in Generalized Modified Gravity," Int. J. Theor. Phys. **47** (2008), 3186–3200.

[503] G. Cognola, E. Elizalde, S. Nojiri, et al., "A Class of Viable Modified f(R) Gravities Describing Inflation and the Onset of Accelerated Expansion," Phys. Rev. D **77** (2008), 046009.

[504] V. Faraoni, "de Sitter Space and the Equivalence between f(R) and Scalar-Tensor Gravity," Phys. Rev. D **75** (2007), 067302.

[505] T. P. Sotiriou, "Curvature Scalar Instability in f(R) Gravity," Phys. Lett. B **645** (2007), 389–392.

[506] D. Comelli, "Born-Infeld Type Gravity," Phys. Rev. D **72** (2005), 064018.

[507] Y. S. Song, W. Hu, and I. Sawicki, "The Large Scale Structure of f(R) Gravity," Phys. Rev. D **75** (2007), 044004.

[508] W. Hu and I. Sawicki, "A Parameterized Post-Friedman Framework for Modified Gravity," Phys. Rev. D **76** (2007), 104043.

[509] A. de la Cruz-Dombriz and A. Dobado, "A f(R) Gravity without Cosmological Constant," Phys. Rev. D **74** (2006), 087501.

[510] S. Fay, S. Nesseris, and L. Perivolaropoulos, "Can f(R) Modified Gravity Theories Mimic a LCDM Cosmology?," Phys. Rev. D **76** (2007), 063504.

[511] D. R. Noakes, "The Initial Value Formulation of Higher Derivative Gravity," J. Math. Phys. **24** (1983), 1846–1850.

[512] S. Capozziello and S. Vignolo, "The Cauchy Problem for f(R)-Gravity: An Overview," Int. J. Geom. Meth. Mod. Phys. **9** (2012), 1250006.

[513] S. Capozziello and S. Vignolo, "The Cauchy Problem for Metric-Affine f(R)-Gravity in Presence of Perfect-Fluid Matter," Class. Quant. Grav. **26** (2009), 175013.

[514] S. Capozziello and S. Vignolo, "The Cauchy Problem for Metric-Affine f(R)-Gravity in Presence of a Klein-Gordon Scalar Field," Int. J. Geom. Meth. Mod. Phys. **8** (2011), 167–176.

[515] M. Sharif and I. Nawazish, "Viable Wormhole Solutions and Noether Symmetry in f(R,T) Gravity," Annals Phys. **400** (2019), 37.

[516] M. Sharif and I. Nawazish, "Wormhole Geometry and Noether Symmetry in $f(R)$ Gravity," Annals Phys. **389** (2018), 283.

[517] A. K. Sanyal and R. Mandal, "Cosmological Lorentzian Wormholes via Noether Symmetry Approach," Phys. Astron. Int. J. **6** (2018), 498.

[518] B. Tajahmad, "Exact Solutions of an Anisotropic Universe in a Modified Teleparallel Gravity Model via the Noether and B.N.S. Approaches," arXiv e-print (2016), [arXiv:1610.08707 [gr-qc]].

[519] M. F. Shamir and F. Kanwal, "Noether Symmetry Analysis of Anisotropic Universe in Modified Gravity," Eur. Phys. J. C **77**, No. 5 (2017), 286.

[520] S. Zhang and Y. Q. Dong, "Anisotropic Cosmology with Vector Field in the Noether Approach," Astrophys. Space Sci. **359**, No. 1 (2015), 16.

[521] Y. Kucukakca, "Anisotropic Solution via Noether Symmetry for $f(R)$ Palatini Gravity," Astrophys. Space Sci. **361**, No. 2 (2016), 80.

[522] S. Mondal, S. Dutta, M. Tarafdar, and S. Chakraborty, "A Cosmological Study of Einstein-Skyrme Model in Anisotropic Kantowski-Sachs Spacetime Using Lie and Noether Symmetries," Int. J. Geom. Meth. Mod. Phys. **15**, No. 06 (2018), 1850089.

[523] U. Camci, A. Yildirim, and I. Basaran Oz, "New Exact Solutions of Bianchi I, Bianchi III and Kantowski-Sachs Spacetimes in Scalar-Coupled Gravity Theories via Noether Gauge Symmetries," Astropart. Phys. **76** (2016), 29.

[524] I. B. Oz, Y. Kucukakca, and N. Unal, "Anisotropic Solution in Phantom Cosmology via Noether Symmetry Approach1," Can. J. Phys. **96**, No. 7 (2018), 677.

[525] M. Tsamparlis and A. Paliathanasis, "Lie and Noether Symmetries of Geodesic Equations and Collineations," Gen. Rel. Grav. **42** (2010), 2957.
[526] M. Tsamparlis and A. Paliathanasis, "The Geometric Nature of Lie and Noether Symmetries," Gen. Rel. Grav. **43** (2011), 1861.
[527] K. F. Dialektopoulos and S. Capozziello, "Noether Symmetries as a Geometric Criterion to Select Theories of Gravity," Int. J. Geom. Meth. Mod. Phys. **15**, (2018).
[528] F. D'Ambrosio, "A Noether Theorem for Discrete Covariant Mechanics," arXiv e-print (2019), [arXiv:1902.08997 [gr-qc]].
[529] R. Bluhm and A. Sehic, "Noether Identities in Gravity Theories with Nondynamical Backgrounds and Explicit Spacetime Symmetry Breaking," Phys. Rev. D **94**, No. 10 (2016), 104034.
[530] M. Tsamparlis and A. Paliathanasis, "Symmetries of Differential Equations in Cosmology," Symmetry **10**, No. 7 (2018), 233.
[531] J. H. Park, S. J. Rey, W. Rim, and Y. Sakatani, "O(D, D) Covariant Noether Currents and Global Charges in Double Field Theory," JHEP **1511** (2015), 131.
[532] G. Gecim, Y. Kucukakca, and Y. Sucu, "Noether Gauge Symmetry of Dirac Field in (2 + 1)-Dimensional Gravity," Adv. High Energy Phys. **2015** (2015), 567395.
[533] S. Kamilya, B. Modak, and S. Biswas, "Induced Gravity Theory from Noether Symmetry," Gen. Rel. Grav. **36** (2004), 661.
[534] B. Tajahmad, "Late-Time-Accelerated Expansion Arisen from Gauge Fields in an Anisotropic Background and a Fruitful Trick for Noether's Approach," JHEP **02** (2020), 084.
[535] S. Bahamonde, K. Bamba, and U. Camci, "New Exact Spherically Symmetric Solutions in $f(R, \phi, X)$ Gravity by Noether's Symmetry Approach," JCAP **1902** (2019), 016.
[536] P. Channuie, D. Momeni, and M. A. Ajmi, "Noether Symmetry in the Nash Theory of Gravity," Eur. Phys. J. C **78**, No. 7 (2018), 588.
[537] G. á. Z. Tóth, "Noether Currents for the Teukolsky Master Equation," Class. Quant. Grav. **35**, No. 18 (2018), 185009.
[538] R. Bluhm, "Gravity Theories with Background Fields and Spacetime Symmetry Breaking," Symmetry **9**, No. 10 (2017), 230.
[539] J. M. Pons, "Noether Symmetries for Fields and Branes in Backgrounds with Killing Vectors," Class. Quant. Grav. **35**, No. 15 (2018), 155014.
[540] G. Gecim and Y. Kucukakca, "Scalar-Tensor Teleparallel Gravity with Boundary Term by Noether Symmetries," Int. J. Geom. Meth. Mod. Phys. **15**, No. 09 (2018), 1850151.
[541] M. K. Dutta and B. Modak, "Can Noether Symmetry in Modified f(G) Gravity Always Yield Cosmic Evolution?," Mod. Phys. Lett. A **32**, No. 1 (2017), 1750046.
[542] N. Dimakis, A. Giacomini, S. Jamal, G. Leon, and A. Paliathanasis, "Noether Symmetries and Stability of Ideal Gas Solutions in Galileon Cosmology," Phys. Rev. D **95**, No. 6 (2017), 064031.
[543] B. Tajahmad, "Studying the Intervention of an Unusual Term in $f(T)$ Gravity via the Noether Symmetry Approach," Eur. Phys. J. C **77**, No. 8 (2017), 510.
[544] M. F. Shamir and M. Ahmad, "Noether Symmetry Approach in $f(\mathcal{G}, T)$ Gravity," Eur. Phys. J. C **77**, No. 1 (2017), 55.
[545] E. Massaeli, M. Motaharfar, and H. R. Sepangi, "General Scalar-Tensor Cosmology: Analytical Solutions via Noether Symmetry," Eur. Phys. J. C **77**, No. 2 (2017), 124.

[546] S. Dutta, M. M. Panja, and S. Chakraborty, "A Study of Dynamical Equations for Non-minimally Coupled Scalar Field Using Noether Symmetric Approach," Mod. Phys. Lett. A **31**, No. 19 (2016), 1650116.

[547] S. Dutta, M. M. Panja, and S. Chakraborty, "A Scalar Field Dark Energy Model: Noether Symmetry Approach," Gen. Rel. Grav. **48**, No. 4 (2016), 54.

[548] K. Myrzakulov, P. Tsyba, and R. Myrzakulov, "Noether Symmetry in $F(T)$ Gravity with f-Essence," arXiv e-print (2016), [arXiv:1601.07357 [physics.gen-ph]].

[549] A. Jhangeer, M. Farasat Shamir, T. Naz, and N. Iftikhar, "Classification of Cosmic Scale Factor via Noether Gauge Symmetries," Int. J. Theor. Phys. **54**, No. 7 (2015), 2343.

[550] B. Vakili, "A Late Time Accelerated FRW Model with Scalar and Vector Fields via Noether Symmetry," Phys. Lett. B **738** (2014), 488.

[551] Y. Kucukakca, "Teleparallel Dark Energy Model with a Fermionic Field via Noether Symmetry," Eur. Phys. J. C **74**, No. 10 (2014), 3086.

[552] F. Darabi, K. Atazadeh, and A. Rezaei-Aghdam, "Quantum Tunneling from Generalized $(2+1)$ Dimensional Black Holes Having Noether Symmetry," Eur. Phys. J. C **74**, No. 7 (2014), 2967.

[553] T. Katsuragawa and S. Nojiri, "Noether Current from Surface Term, Virasoro Algebra and Black Hole Entropy in Bigravity," JPS Conf. Proc. **1** (2014), 013021.

[554] A. Aslam, M. Jamil, D. Momeni, et al., "Noether Gauge Symmetry Approach in Quintom Cosmology," Astrophys. Space Sci. **348** (2013), 533.

[555] B. Vakili and F. Khazaie, "Noether Symmetric Classical and Quantum Scalar Field Cosmology," Class. Quant. Grav. **29** (2012), 035015.

[556] I. Hussain, M. Jamil, and F. M. Mahomed, "Noether Gauge Symmetry Approach in f(R) Gravity," Astrophys. Space Sci. **337** (2012), 373.

[557] B. Vakili, "Quantization of the Schwarzschild Black Hole: A Noether Symmetry Approach," Int. J. Theor. Phys. **51** (2012), 133.

[558] A. Bonanno, G. Esposito, C. Rubano, and P. Scudellaro, "Noether Symmetry Approach to Scalar-Field-Dominated Cosmology with Dynamically Evolving G and Lambda," Int. J. Geom. Meth. Mod. Phys. **5** (2008), 329.

Index

Lightning Source UK Ltd.
Milton Keynes UK
UKHW031947151122
412275UK00007B/50